Lecture Notes in Engineering

Edited by C. A. Brebbia and S. A. Orszag

WITHDRAWN FROM
THE LIBRARY

AF746864

54

T. J. Mueller (Editor)

Low Reynolds Number Aerodynamics

Proceedings of the Conference
Notre Dame, Indiana, USA, 5–7 June 1989

Springer-Verlag
Berlin Heidelberg New York
London Paris Tokyo Hong Kong

1990

Series Editors
C. A. Brebbia · S. A. Orszag

Consulting Editors
J. Argyris · K.-J. Bathe · A. S. Cakmak · J. Connor · R. McCrory
C. S. Desai · K.-P. Holz · F. A. Leckie · G. Pinder · A. R. S. Pont
J. H. Seinfeld · P. Silvester · P. Spanos · W. Wunderlich · S. Yip

Editor
Thomas J. Mueller
Dept. of Aerospace and Mechanical Engineering
University of Notre Dame
Notre Dame, IN 46556
USA

ISBN 3-540-51884-3 Springer-Verlag Berlin Heidelberg New York
ISBN 0-387-51884-3 Springer-Verlag New York Berlin Heidelberg

This work is subject to copyright. All rights are reserved, whether the whole or part of the material is concerned, specifically the rights of translation, reprinting, re-use of illustrations, recitation, broadcasting, reproduction on microfilms or in other ways, and storage in data banks. Duplication of this publication or parts thereof is only permitted under the provisions of the German Copyright Law of September 9, 1965, in its version of June 24, 1985, and a copyright fee must always be paid. Violations fall under the prosecution act of the German Copyright Law.

© Springer-Verlag Berlin, Heidelberg 1989
Printed in Germany

The use of registered names, trademarks, etc. in this publication does not imply, even in the absence of a specific statement, that such names are exempt from the relevant protective laws and regulations and therefore free for general use.

Printing: Mercedes-Druck, Berlin
Binding: B. Helm, Berlin
2161/3020-543210 Printed on acid-free paper.

PREFACE

Continuing interest in a wide variety of low Reynolds number applications has focused attention on the design and evaluation of airfoil sections at chord Reynolds numbers below 500,000. These applications include remotely or robotically piloted vehicles at high altitudes as well as ultra-light and human powered vehicles and mini-RPVs at low altitudes. Other examples include small axial-flow fans used to cool electronic equipment in the unpressurized sections of high-altitude aircraft and gas turbine blades.

High Reynolds number airfoil design strategies attempt to control the onset and development of turbulent boundary layers. This is difficult at low Reynolds numbers because of the increased stability of attached laminar boundary layers. Therefore, laminar separation is common even at small angles of attack at low Reynolds numbers. Under these conditions, the development of a turbulent boundary layer usually depends on the formation of a transitional separation bubble.

This volume is the collection of papers presented at the Conference on Low Reynolds Number Aerodynamics held June 4-7, 1989 at the University of Notre Dame. The Conference was sponsored by the Department of Aerospace and Mechanical Engineering and the College of Engineering at Notre Dame. Over fifty active researchers in this field from Europe, Canada, and the United States were present. This Conference followed the 1986 International Conference in London by about three years and the first Notre Dame Conference of 1985 by four years. It is clear from the papers in this volume that a great deal of progress has been made in understanding the occurrence and behavior of laminar separation and transition as well as their overall effect on the performance of airfoils at low chord Reynolds numbers. The ultimate goals of this understanding are improved analytical methods for the design and evaluation of a variety of practical applications. Significant progress has been made in the achievement of these goals.

I would like to thank the participants for their contributions and the staff of Springer-Verlag for putting together this volume.

Thomas J. Mueller
Notre Dame, IN

July 1989

CONTENTS

XFOIL: An Analysis and Design System for Low Reynolds Number Airfoils

Mark Drela
MIT Dept. of Aeronautics and Astronautics, Cambridge, Massachusetts

Abstract

Calculation procedures for viscous/inviscid analysis and mixed-inverse design of subcritical airfoils are presented. An inviscid linear-vorticity panel method with a Karman-Tsien compressiblity correction is developed for direct and mixed-inverse modes. Source distributions superimposed on the airfoil and wake permit modeling of viscous layer influence on the potential flow. A two-equation lagged dissipation integral method is used to represent the viscous layers. Both laminar and turbulent layers are treated, with an e^9-type amplification formulation determinining the transition point. The boundary layer and transition equations are solved simultaneously with the inviscid flowfield by a global Newton method. The procedure is especially suitable for rapid analysis of low Reynolds number airfoil flows with transitional separation bubbles. Surface pressure distributions and entire polars are calculated and compared with experimental data. Design procedure examples are also presented.

1 Introduction

Effective airfoil design procedures require a fast and robust analysis method for on-design and off-design performance evaluation. Of the various airfoil analysis algorithms which have been developed to date, only the interacted viscous/inviscid zonal approaches have been fast and reliable enough for routine airfoil design work. For low Reynolds number airfoils ($Re < 1/2$ million), the demands on the analysis method become especially severe. Not only must the complex physics of transitional separation bubbles be captured, but the solution algorithm must be able to handle the very strong and nonlinear coupling between the viscous, transition, and inviscid formulations at a separation bubble.

Of the various calculation methods currently in use (GBK code [1] , GRUMFOIL code [2]), only the ISES code [3,4,5] can routinely predict low Reynolds number airfoil flowfields. Its fully compatible laminar and turbulent viscous formulations, a reliable transition formulation, and a global Newton solution method represent the necessary ingredients for prediction of such flows.

ISES has been successfully applied to the design of low Reynolds number airfoils for human-powered aircraft [6], analysis of established airfoils [7], and the design of high Reynolds number transonic transport airfoils, even though it can be demanding in terms of computer time. About two minutes are required to calculate an entire 20-point polar on a dedicated supercomputer. For users limited to a microVAX-class (0.1 MFLOP) machine, this may require several hours, which severely hinders the inherently iterative design process. A major goal in the development of the present XFOIL code was to significantly reduce these computational requirements while retaining the ability to predict low Reynolds number flows. The analysis formulation was also embedded in an interactive driver which also allows the designer to exercise an inverse solver and a geometry-manipulation facility. The overall goal is to improve the productivity of the designer.

The present paper will outline the basic inviscid and viscous formulations of XFOIL, and demonstrate its performance on a number of airfoil cases. The mixed-inverse formulation and associated user interface will also be described. Finally, the code's overall design/analysis environment will be discussed.

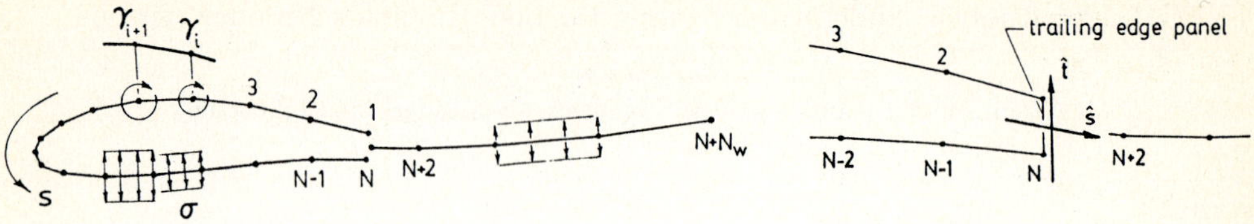

Figure 1: Airfoil and wake paneling with vorticity and source distributions, with trailing edge detail.

2 Inviscid Formulation

Numerous two-dimensional panel methods have been developed in the past [8,9,10], all being more or less successful for inviscid analysis of arbitrary airfoils. The present linear-vorticity streamfunction formulation is designed specifically for compatibility with an inverse mode, and for a natural incorporation of viscous displacement effects.

A general two-dimensional inviscid airfoil flowfield is constructed by the superposition of a freestream flow, a vortex sheet of strength γ on the airfoil surface, and a source sheet of strength σ on the airfoil surface and wake. The streamfunction of this configuration is given by

$$\Psi(x,y) = u_\infty y - v_\infty x + \frac{1}{2\pi}\int \gamma(s)\ln r(s;x,y)\,ds + \frac{1}{2\pi}\int \sigma(s)\,\theta(s;x,y)\,ds \tag{1}$$

where s is the coordinate along the vortex and source sheets, r is the magnitude of the vector between the point at s and the field point x, y , θ is the vector's angle, and $u_\infty = q_\infty \cos\alpha$, $v_\infty = q_\infty \sin\alpha$ are the freestream velocity components.

The airfoil contour and wake trajectory are discretized into flat panels, with N panel nodes on the airfoil, and N_w nodes on the wake as shown in Figure 1. Each airfoil panel has a linear vorticity distribution defined by the node values γ_i $(1 \le i \le N)$. Each airfoil and wake panel also has a constant source strength σ_i $(1 \le i \le N+N_w-1)$ associated with it. These source strengths will be later related to viscous layer quantities.

A panel of uniform source strength σ_{TE} and vortex strength γ_{TE} must be also be placed across the airfoil trailing edge gap if it has a finite thickness. For smooth flow off the trailing edge, the trailing edge panel strengths σ_{TE} , γ_{TE} , must be related to the local airfoil surface vorticity by

$$\sigma_{TE} = \frac{1}{2}(\gamma_1 - \gamma_N)\,|\hat{s}\cdot\hat{t}| \qquad\qquad \gamma_{TE} = \frac{1}{2}(\gamma_1 - \gamma_N)\,|\hat{s}\times\hat{t}| \tag{2}$$

where $\hat{s}$ is the unit vector bisecting the trailing edge angle, and $\hat{t}$ is the unit vector along the trailing edge panel as shown in Figure 1.

For the airfoil with flat panels, equation (1) evaluates to the following expression for the streamfunction at any field point x, y.

$$\begin{aligned}\Psi(x,y) &= u_\infty y - v_\infty x + \frac{1}{4\pi}\sum_{j=1}^{N+N_w-1} \Psi_j^\sigma(x,y)\,2\sigma_j \\ &+ \frac{1}{4\pi}\sum_{j=1}^{N-1} \Psi_j^{\gamma+}(x,y)\,(\gamma_{j+1}+\gamma_j) + \Psi_j^{\gamma-}(x,y)\,(\gamma_{j+1}-\gamma_j) \\ &+ \frac{1}{4\pi}\left(\Psi_N^\sigma(x,y)\,|\hat{s}\cdot\hat{t}| + \Psi_N^{\gamma+}(x,y)\,|\hat{s}\times\hat{t}|\right)(\gamma_1-\gamma_N)\end{aligned} \tag{3}$$

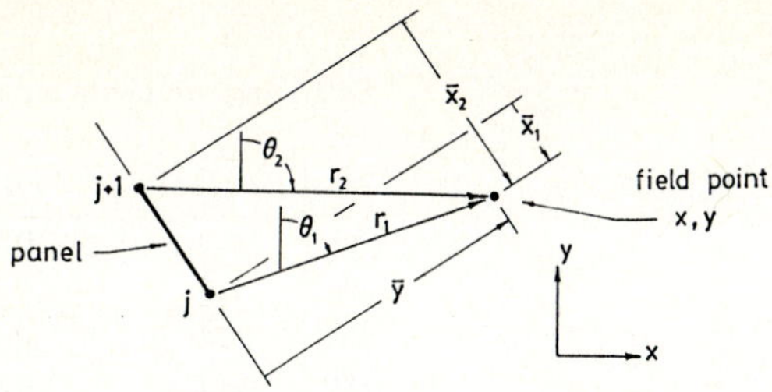

Figure 2: Local panel coordinates.

The unit streamfunctions in equation (3) are readily defined in terms of local panel coordinates $\bar{x}-\bar{y}$ as shown in Figure 2.

$$\Psi_j^{\gamma+}(x,y) = \bar{x}_1 \ln r_1 - \bar{x}_2 \ln r_2 + \bar{x}_2 - \bar{x}_1 + \bar{y}(\theta_1 - \theta_2) \tag{4}$$

$$\Psi_j^{\gamma-}(x,y) = \left[(\bar{x}_1 + \bar{x}_2)\Psi_j^{\gamma+} + r_2^2 \ln r_2 - r_1^2 \ln r_1 + \frac{1}{2}(\bar{x}_1^2 - \bar{x}_2^2)\,\bar{y}\right] \frac{1}{\bar{x}_1 - \bar{x}_2} \tag{5}$$

$$\Psi_j^{\sigma}(x,y) = \bar{x}_2\theta_2 - \bar{x}_1\theta_1 + \bar{y}\,\ln\frac{r_1}{r_2} \tag{6}$$

By requiring that the streamfunction be equal to some constant value Ψ_0 at each node on the airfoil, the following linear system results from the above relations.

$$\sum_{j=1}^{N} a_{ij}\,\gamma_j \;-\; \Psi_0 \;=\; -u_\infty y_i + v_\infty x_i \;-\; \sum_{j=1}^{N+N_w-1} b_{ij}\,\sigma_j \qquad ; \quad 1 \le i \le N \tag{7}$$

The coefficient matrices a_{ij} and b_{ij} are fully determined from the unit streamfunctions (4-6) if all the airfoil panel nodes x_i, y_i and the wake nodes are known. Combining the linear system (7) with a Kutta condition,

$$\gamma_1 + \gamma_N = 0 \tag{8}$$

gives a linear $(N{+}1) \times (N{+}1)$ system for the N node values γ_i and the airfoil surface streamfunction Ψ_0.

A special treatment is required for an airfoil with a sharp trailing edge. In this case the nodes $i = 1$ and $i = N$ coincide, and hence their corresponding equations in (7) are identical. The result is a singular system which cannot be solved for γ_i. To circumvent this problem, the $i = N$ equation in (7) is discarded and replaced by an extrapolation of the mean γ (between top and bottom) to the trailing edge.

$$(\gamma_3 - 2\gamma_2 + \gamma_1) \;-\; (\gamma_{N-2} - 2\gamma_{N-1} + \gamma_N) \;=\; 0 \tag{9}$$

2.1 Inviscid analysis procedure

For an analysis problem where the geometry is known, the linear system formed by the matrix equation (7) and Kutta condition (8) can be readily solved by Gaussian elimination. This gives the solution for the airfoil surface vorticity values as

$$\gamma_i \;=\; \gamma_{0_i}\cos\alpha \;+\; \gamma_{90_i}\sin\alpha \;+\; \sum_{j=1}^{N+N_w-1} b'_{ij}\,\sigma_j \qquad ; \quad 1 \le i \le N \tag{10}$$

where γ_0 and γ_{90} are the vorticity distributions corresponding to a freestream α of 0° and 90°, and $b'_{ij} = -a_{ij}^{-1} b_{ij}$ is the source-influence matrix. By setting $\sigma_i = 0$ in the surface vorticity expression (10) and specifying an angle of attack, an inviscid solution is immediately obtained. For viscous flows the source strengths σ_i are not known a priori, so the equation set (10) must be supplanted with the boundary layer equations to obtain a solvable closed system. This will be dealt with in the viscous analysis section.

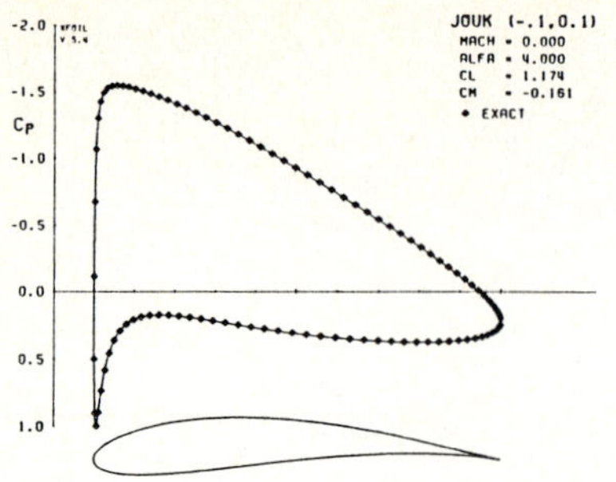

CPU Requirements on MicroVAX II

N	CPU (s)	C_L Error
40	4.0	0.766 %
60	7.5	0.340 %
100	20.0	0.175 %
160	54.0	0.085 %

Figure 3: Joukowsky airfoil test case. $N = 120$ C_P distribution shown.

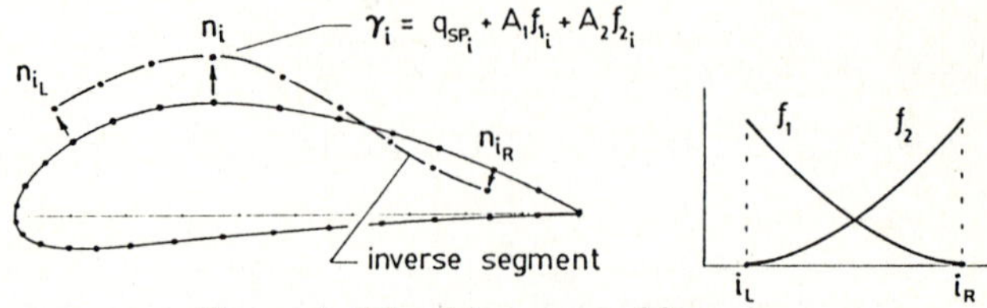

Figure 4: Mixed-inverse problem

Figure 3 shows the comparison between the calculated and exact pressure distributions on a Joukowsky airfoil. Since the flow inside the airfoil is stagnant, the surface velocity is equal to the surface vorticity, and hence the surface pressure coefficient is $C_p = 1 - (\gamma/q_\infty)^2$. Also shown is the accuracy and CPU requirements as functions of the number of panels. The results are typical of most panel methods.

2.2 Inviscid mixed-inverse procedure

A mixed-inverse problem results when the geometry is prescribed over a part of the airfoil surface, and the surface vorticity (or equivalently, speed), is prescribed over the remainder. The local unknown associated with any node i is then either the vorticity γ_i as in the analysis case, or the normal geometric displacement n_i of the node from a seed airfoil geometry as shown in Figure 4. The surface vorticity at node i is specified in the form

$$\gamma_i = q_{SPi} + A_1 f_{1i} + A_2 f_{2i} \tag{11}$$

which introduces two free parameters A_1 , A_2 into the specified vorticity distribution q_{SPi} . The free parameters weight the two specified shape functions f_{1i} , f_{2i} , which are specified to be quadratic over the inverse segment of the airfoil $i_R \leq i \leq i_L$. It is necessary to add the two degrees of freedom to the specified vorticity q_{SPi} in (11) to allow geometric continuity to be enforced at the two joining points between the inverse segment and the fixed part of the airfoil (Figure 4).

$$n_{i_L} = 0 \qquad\qquad n_{i_R} = 0 \tag{12}$$

This is consistent with the Lighthill constraints [10] which do not allow a totally arbitrary speed distribution on the airfoil. The inverse formulation of Kennedy and Marsden [9] does not address this issue, and hence cannot perfectly satisfy the necessary streamfunction constraint (7) at each panel node.

Since the governing streamfunction constraint (7) is nonlinear in the geometry, a Newton-Raphson procedure is used to solve the overall system. Eliminating the source distribution (only the inviscid inverse problem is treated), the equation system (7), the Kutta condition (8), and the geometry-continuity conditions (12) are written as

$$
\begin{aligned}
\sum_{j=1}^{N} a_{ij}\gamma_j \;-\; \Psi_0 \;+\; u_\infty y_i - v_\infty x_i \;\equiv\; R_i \quad &= \; 0 \qquad ; \quad 1 \le i \le N \\
\gamma_1 + \gamma_N \;\equiv\; R_{N+1} \quad &= \; 0 \\
n_{i_L} \;\equiv\; R_{N+2} \quad &= \; 0 \\
n_{i_R} \;\equiv\; R_{N+3} \quad &= \; 0
\end{aligned}
\tag{13}
$$

and are then linearized about the current solution γ_i , x_i , y_i.

$$
\left[\quad J_{ij} \quad \right] \left\{ \begin{array}{c} \delta\gamma_j, \delta n_j \\ \delta\Psi_0 \\ \delta A_1 \\ \delta A_2 \end{array} \right\} = \left\{ -R_i \right\} \qquad ; \quad 1 \le i \le N+3 \tag{14}
$$

The i,j entry of the Jacobian matrix J_{ij} is either $\partial R_i/\partial\gamma_j$ if the speed is unknown at node j, or $\partial R_i/\partial n_j$ if the node position is unknown at node j. To calculate the latter entry, the components n_x , n_y of the direction vector along which a node is to move are first locally specified, typically normal to the current airfoil surface. There follows

$$
\frac{\partial R_i}{\partial n_j} = \frac{\partial R_i}{\partial x_j} n_x + \frac{\partial R_i}{\partial y_j} n_y \tag{15}
$$

and the derivatives with respect to x_j and y_j are obtained in the formal manner from the definition of R_i (13). Ultimately this involves differentiating the unit streamfunction expressions (4 - 6), which is accomplished by a careful application of the chain rule.

The inverse solution proceeds by repeatedly solving the Newton system (14) for the variable changes and updating the current variables.

$$
\Psi_0 \leftarrow \Psi_0 + \delta\Psi_0 \qquad A_1 \leftarrow A_1 + \delta A_1 \qquad A_2 \leftarrow A_2 + \delta A_2 \tag{16}
$$

$$
\gamma_i \leftarrow \gamma_i + \delta\gamma_i \qquad ; \quad i < i_R \ , \ i_L < i \tag{17}
$$

$$
x_i \leftarrow x_i + n_{x_i}\,\delta n_i \qquad ; \quad i_R \le i \le i_L \tag{18}
$$

$$
y_i \leftarrow y_i + n_{y_i}\,\delta n_i \qquad ; \quad i_R \le i \le i_L \tag{19}
$$

$$
\gamma_i \leftarrow q_{SPi} + A_1 f_{1i} + A_2 f_{2i} \qquad ; \quad i_R \le i \le i_L \tag{20}
$$

Typically, two to four Newton iterations are required to converge to machine accuracy, depending on how much the airfoil geometry is to change. Assuming the same number of panels, each iteration requires approximately the same computational effort as that shown in Figure 3 for an inviscid analysis calculation.

The surface speed (or C_P) distribution for inverse calculations is specified as a function of the unwrapped fractional arc length on the as-yet-unknown airfoil. Figure 5 shows the old and specified distributions as viewed by the user, and the resulting calculated airfoil. The seed airfoil is a Wortmann FX63-137 airfoil at an angle of attack of 4°. The input speed distribution can be modified and the airfoil shape recalculated as often as needed.

A useful feature of the mixed-inverse formulation is that absolute control of geometry can be exersized where needed. Nevertheless, for preliminary design it is useful to employ a full-inverse algorithm which permits the speed distribution over the entire airfoil to be specified (subject to the Lighthill constraints). The XFOIL design environment includes such an algorithm, similar to the complex mapping method implemented by Eppler [10], to maximize the designer's flexibility. Prescribed surface speeds are input versus the surface arc length rather than the non-physical circle plane coordinate common to mapping methods. The distribution is input in the format shown in Figure 5, with the speed distribution of a seed airfoil being superimposed. The implementation also allows the trailing edge gap and trailing edge angle to be explicitly controlled.

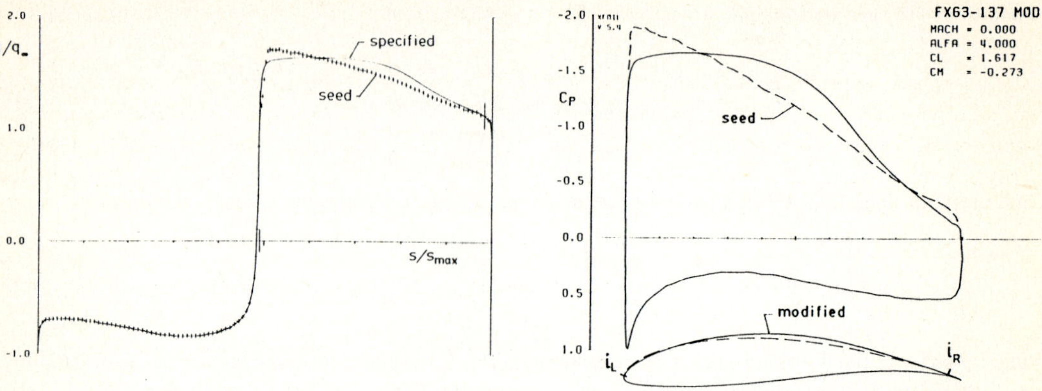

Figure 5: Input q_{SP} vs $s/s_{\max}$ (line) for local-inverse calculation, and calculated airfoil.

3 Viscous Formulation

The present viscous formulation of XFOIL is essentially the same as that of the transonic ISES code [5]. Some changes have been incorporated recently [11] to improve the prediction of "base drag" from blunt trailing edges. Only a summary description of the viscous formulation will be given, with more emphasis placed on the overall viscous/inviscid solution scheme which is unique to the present method.

3.1 Governing Equations and Closure

The viscous formulation employs the following standard compressible integral momentum and kinetic energy shape parameter equations. The streamwise coordinate is ξ.

$$\frac{d\theta}{d\xi} + (2 + H - M_e^2)\frac{\theta}{u_e}\frac{du_e}{d\xi} = \frac{C_f}{2} \tag{21}$$

$$\theta\frac{dH^*}{d\xi} + (2H^{**} + H^*(1 - H))\frac{\theta}{u_e}\frac{du_e}{d\xi} = 2C_D - H^*\frac{C_f}{2} \tag{22}$$

Also, following Green et al [12], a rate equation for the maximum shear stress coefficient C_τ is used to account for deviations of the outer layer dissipation coefficient C_D from the local equilibrium value.

$$\frac{\delta}{C_\tau}\frac{dC_\tau}{d\xi} = 5.6\,(C_{\tau_{EQ}}^{1/2} - C_\tau^{1/2}) + 2\delta\left\{\frac{4}{3\delta^*}\left[\frac{C_f}{2} - \left(\frac{H_k - 1}{6.7H_k}\right)^2\right] - \frac{1}{u_e}\frac{du_e}{d\xi}\right\} \tag{23}$$

The latter equation has been altered slightly from the original formulation described in reference [5], and has resulted in better predictions of lift and drag near stall conditions.

In laminar regions, the shear stress lag equation (23) is replaced by a rate equation which models the growth of the amplitude $\tilde{n}$ of the most-amplified Tollmien-Schlichting wave.

$$\frac{d\tilde{n}}{d\xi} = \frac{d\tilde{n}}{dRe_\theta}(H_k)\,\frac{dRe_\theta}{d\xi}(H_k, \theta) \tag{24}$$

The empirical relation $d\tilde{n}/dRe_\theta\,(H_k)$ is a correlation of spatial growth rates computed from solutions to the Orr-Sommerfeld equation, and $dRe_\theta/d\xi\,(H_k, \theta)$ is obtained from the properties of the Falkner-Skan profile family. Reference [5] gives further details. The transition point is defined by the location where $\tilde{n}$ reaches a user-specified critical value $\tilde{n}_{\text{crit}}$. This parameter in practice is used to represent the background disturbance level and has quite a dramatic effect on low Reynolds number airfoil performance as will be shown in the results section.

The fundamental variables governed by the boundary layer equations are chosen to be θ, δ^*, and C_τ. In laminar regions, the amplification variable $\tilde{n}$ replaces C_τ. In addition, u_e is present as an external unknown which will be related to the global viscous solution via the inviscid outer flow. It is therefore only a convenient intermediate variable and does not constitute an additional unknown. To close the integral boundary layer equations (21), (22), and (23), auxilliary variables are defined in terms of θ, δ^*, C_τ, u_e or their suitable combinations with the following functional dependencies.

$$H_k = H_k(H, M_e) \qquad H^* = H^*(H_k, M_e, Re_\theta) \qquad H^{**} = H^{**}(H_k, M_e) \tag{25}$$

$$U_S = U_S(H^*, H, H_k) \qquad C_{\tau_{EQ}} = C_{\tau_{EQ}}(H^*, H, H_k, U_S) \qquad C_f = C_f(H_k, M_e, Re_\theta) \tag{26}$$

$$C_D = \frac{C_f}{2} U_S + C_\tau (1 - U_S) \tag{27}$$

The actual expressions for the above functional forms are given fully in references [5,11] and will not be repeated here.

The wake is treated as one viscous layer so that only one θ and one δ^* variable is present at each wake station, with the following initial conditions at the trailing edge.

$$\theta_{\text{wake}} = \theta_{\text{upper}} + \theta_{\text{lower}} \qquad \delta^*_{\text{wake}} = \delta^*_{\text{upper}} + \delta^*_{\text{lower}} + h_{TE} \tag{28}$$

This produces a continuous displacement body even with a blunt trailing edge of thickness h_{TE}. The initial wake shear coefficient is taken to be the θ-weighted average of the upper and lower surface values.

$$C_{\tau\,\text{wake}} = \frac{C_{\tau\,\text{upper}}\, \theta_{\text{upper}} + C_{\tau\,\text{lower}}\, \theta_{\text{lower}}}{\theta_{\text{upper}} + \theta_{\text{lower}}} \tag{29}$$

Additional minor corrections are incorporated for the blunt trailing edge case to better account for "base drag" as described in reference [11].

The governing equations (21-24) are discretized using two-point central differences (i.e. the trapezoidal rule). The boundary layer variables θ , δ^* , C_τ or $\tilde{n}$, and u_e are defined to be located at the panel nodes. Each airfoil and wake panel therefore has three coupled nonlinear equations associated with it which are solved by the procedure described below.

3.2 Viscous/Inviscid Coupling

Since the flow inside the airfoil is stagnant, on the airfoil surface u_e is simply equal to the local vorticity γ on the suction side, and $-\gamma$ on the pressure side. No such simple relation exists in the wake, and there it is necessary to relate u_e to the freestream and a sum of all the vorticity and sources on the airfoil,

$$u_{e_i} = \pm \gamma_i \qquad ; \quad 1 \le i \le N \tag{30}$$

$$\begin{aligned} u_{e_i} &= \nabla \Psi \cdot \hat{n} \\ &= u_\infty \hat{n}_x - v_\infty \hat{n}_y + \sum_{j=1}^{N} c^\gamma_{ij} \gamma_j + \sum_{j=1}^{N+N_w-1} c^\sigma_{ij} \sigma_j \qquad ; \quad N+1 \le i \le N+N_w \end{aligned} \tag{31}$$

with $\hat{n}$ being a unit vector locally normal to the wake. Evaluation of the source influence coefficients c^σ_{ij} in equation (31) requires modification of the assumed constant-source distribution on the wake panels, since this results in a logarithmic singularity in the velocity at each node. Using a two-piece linear source distribution over each wake panel defined by using the neighboring panel source strengths eliminates this problem.

The influence of the viscous layer on the potential flow is properly modeled by the wall transpiration concept if the local source strength σ is equal to the local gradient of the mass defect $m \equiv u_e\delta^*$.

$$\sigma_i = \frac{dm}{d\xi} = \pm\frac{m_{i+1} - m_i}{s_{i+1} - s_i} \tag{32}$$

Substituting the general expression (10) for the surface vorticity γ into expressions (30,31), and eliminating the source strength σ in terms of the mass defect m as given by (32), produces the following.

$$u_{e_i} = u_{INV_i} + \sum_{j=1}^{N+N_w-1} d_{ij}\, m_j \qquad ; \qquad 1 \le i \le N+N_w \tag{33}$$

This very general expression gives the potential flow solution about the airfoil for any distribution of mass defect on the airfoil and wake. The "inviscid" edge velocity distribution u_{INV_i} and the mass-influence matrix d_{ij} are uniquely determined by the airfoil/wake geometry and freestream angle of attack, and are determined independently of the viscous solution (the wake node positions are determined by integrating an inviscid streamline trajectory from the trailing edge panel midpoint). It is important to note that d_{ij} embodies the effect of the local m_j near the trailing edge on the global u_{e_i} distribution via its effect on the Kutta condition. This "indirect" effect is quite significant for low Reynolds number flows, especially near stall or if a separation bubble occurs near the trailing edge. Previous expressions similar to (33) derived by Cebeci et al [13] for viscous/inviscid calculations require the repeated recalculation of the potential flow problem to determine this indirect effect. This represents a significant computational effort which is eliminated with the present formulation.

3.3 Newton solution

Since (33) is an explicit expression for u_{e_i} in terms of the boundary layer variables, it closes the discrete boundary layer equations (21-24). This nonlinear system, now rendered elliptic by the global mass influence on u_e, is solved by a full Newton method.

The Newton variables are defined to be $\delta\theta_i$, δm_i , and either $\delta\tilde{n}_i$ or $\delta C_{\tau i}$ depending on whether station i is laminar or turbulent. The Newton system thus has the following form.

$$\Big[\; J_{ij} \;\Big] \left\{ \begin{matrix} \delta\theta_j \\ \delta m_j \\ \delta\tilde{n}_j\, ,\, \delta C_{\tau j} \end{matrix} \right\} = \Big\{ -R_i \Big\} \qquad ; \quad 1 \le i \le N + N_w \tag{34}$$

The specific choice of δm instead of $\delta(\delta^*)$ as one of the Newton variables is made in the interest of efficiency, since in the former case only the δm_j columns in J_{ij} are full. The columns of the other variables $\delta\theta$, δC_τ , and $\delta\tilde{n}$ have entries only near the diagonal. Solving the Newton system (34) therefore requires only about 1/3 as much time as a full system of the same size. A custom solver is used to take full advantage of this feature.

3.4 Viscous Analysis Examples

Figure 6 shows computed polars for the Eppler 387 airfoil at three Reynolds numbers compared with measurements reported in reference [7]. Two calculated curves are shown for each Reynolds number, differing only in the specified disturbance parameter $\tilde{n}_{\text{crit}}$ described earlier. The curves corresponding

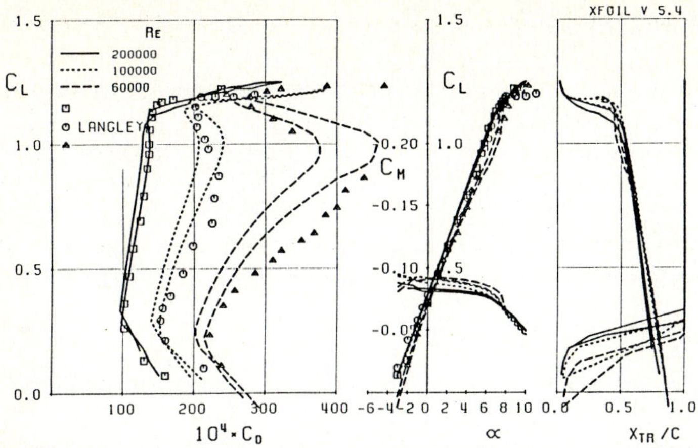

Figure 6: Calculated and measured Eppler 387 polars for $n_{\text{crit}} = 9, 12$ (lower, higher C_D).

to the lower disturbance levels have larger separation bubbles and show a higher drag, except near maximum lift where separation bubble loss has a relatively minor effect. This is consistent with the observed behavior of low Reynolds number airfoil flows. Overall the agreement is quite satisfactory, with accuracy degrading somewhat at the lowest Reynolds number.

Figure 7 shows computed and measured C_P distributions for one operating point and the MicroVAX II CPU time required. The dashed C_P curve is the "inviscid" u_{INV} distribution for that same α. As expected of a panel method, the CPU time is quite sensitive to panel density. For this case, the calculated result did not change significantly even for the smallest number of panels listed, indicating the method is quite fast. However, the number of panels appropriate for any given application depends primarily on the panel density required to resolve separation bubbles. A particularly severe test case in this regard is provided by an FX67-K-170 sailplane airfoil. At a Reynolds number of 2 million, this airfoil has small, but quite strong separation bubbles which must be resolved adequately for accurate drag prediction. Figure 8 shows the fractional error in C_L, C_D, C_m as a function of the number of panels on the entire airfoil. The "noise" in the error is due to panel nodes "moving" across the transition point as the number of panels is increased. For this case, $N = 120$ appears to be necessary to produce reasonably converged values. At lower Reynolds numbers where the bubbles are physically larger, and at higher Reynolds numbers where the bubbles contribute little to profile drag, the panel density requirements are lower. Local panel clustering in the vicinity of an anticipated bubble, which is readily specified with XFOIL's paneling routine, can also reduce panel number requirements and the associated computational effort.

4 Compressibility Correction

In the Karman-Tsien correction (derived in [14]), the compressible speed q and pressure coefficient C_P on an airfoil in compressible flow can be approximately determined from the incompressible flow values q_{inc} , $C_{P\text{inc}}$ by

$$C_P = \frac{C_{P\text{inc}}}{\beta + \lambda(1+\beta)\, C_{P\text{inc}}/2} \qquad\qquad q = \frac{q_{\text{inc}}\,(1-\lambda)}{1 - \lambda\,(q/q_\infty)^2_{\text{inc}}} \tag{35}$$

where $\beta = \sqrt{1 - M_\infty^2}$ and $\lambda = M_\infty^2/(1+\beta)^2$.

In the code implementation, all calculations are performed assuming the airfoil surface vorticity represents the incompressible speed q_{inc}. Relations (35) are used only to define the actual compressible

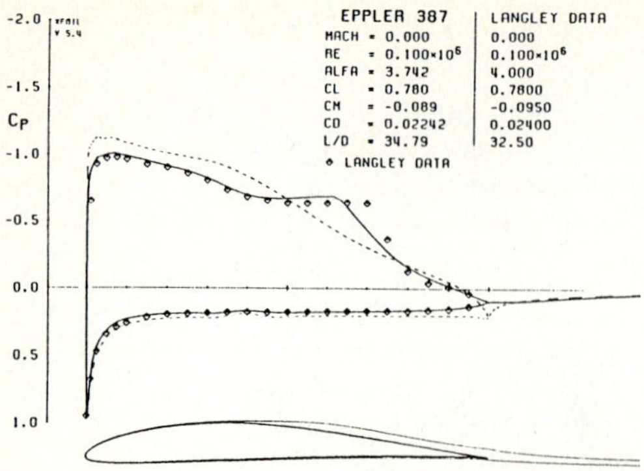

CPU Requirements on MicroVAX II

N	Startup CPU (s)	Newton it. CPU (s)	5 it. total CPU (s)
80	22.0	11.0	77.0
120	58.0	30.0	208.0
160	115.0	62.0	425.0

Figure 7: Calculated and measured C_P distributions and CPU requirements for Eppler 387 airfoil.

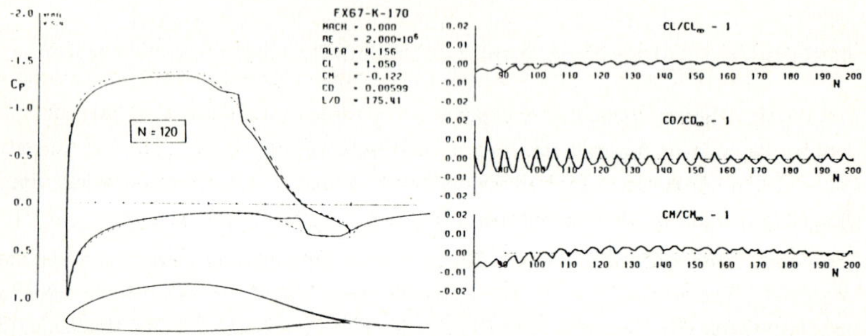

Figure 8: Error in C_L, C_D, C_m as a function of number of panels on an FX67-K-170 airfoil.

C_P for lift and moment calculations, and in conjunction with expression (33) to determine the actual compressible u_e seen by the boundary layer. The boundary layer formulation itself is already valid for compressible flows and requires no corrections.

Figure 9 shows a viscous calculation of a RAE 2822 airfoil compared with the experimental data of reference [15]. Reasonable agreement is obtained, despite the fact that the Karman-Tsien correction formally breaks down in supersonic flow. In practice, the method is reliable right up to sonic conditions, but rapidly degrades as significant supersonic regions appear. Hence, this correction extends the validity of XFOIL into the high-subsonic regime, and greatly extends its range of applicability.

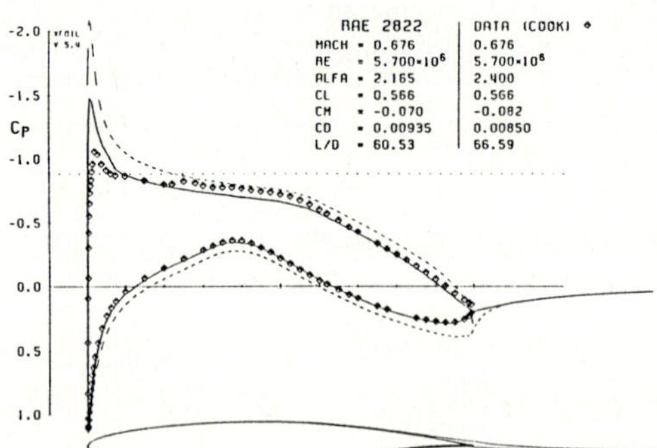

Figure 9: Calculated and experimental results for RAE 2822 airfoil near critical conditions.

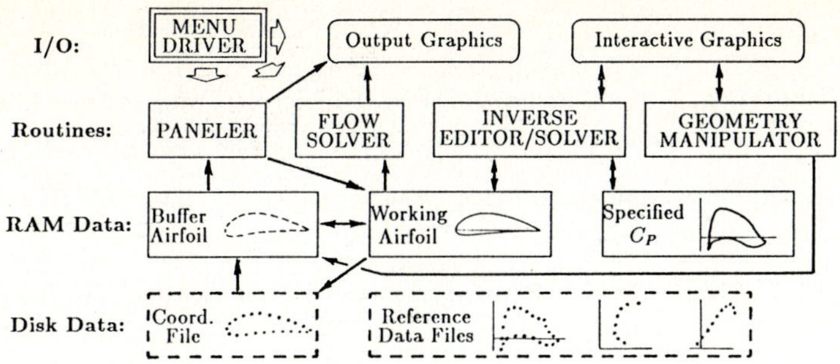

Figure 10: The XFOIL airfoil analysis/design system.

5 The XFOIL Design Environment

The overall XFOIL design/analysis system (Figure 10) consists of a collection of menu-driven routines which perform the viscous or inviscid analysis and inverse calculations. Supporting routines such as panel distribution generation, explicit geometry manipulation, disk I/O, and graphics are also driven from the same menu structure. The "Geometry Manipulator" allows explicit control of camber, thickness, leading edge radius, trailing edge gap, etc. This is very useful if inviolable geometric constraints must be met, and can be very effective for achieving certain aerodynamic specifications as well. For example, if a candidate airfoil's drag bucket is centered about $C_L = 0.5$, increasing its camber by a factor of 1.5 will center the bucket at approximately $C_L = 1.5 \times 0.5 = 0.75$. The Geometry Manipulator performs such a modification with one keyboard entry. Other capabilities include specifying flap deflections, changing camber via a specified loading change, and explicit contour input via a screen cursor.

Not shown in Figure 10 is a facility for calculating polars in batch mode, a convenience and a necessity if only modest computing power is available. The calculated polar points and associated surface pressure and boundary layer parameter distributions are saved in disk files and can be plotted later. The calculated polars shown in Figure 6 were generated in this manner.

6 Conclusions

The XFOIL airfoil design system outlined in this paper has proven to be a powerful and very useful tool for subcritical airfoil design, and is particularly applicable to low Reynolds number airfoils. Various graphics-oriented routines perform assorted analysis, inverse, and geometry-modification functions on a common airfoil representation, and can be easily exercised by the designer from a unified menu stucture with complete flexibility. Since all the important physical mechanisms which affect airfoil performance (instability, separation, bubble losses, etc.) are represented in the overall computational model, any airfoil modifications which are found to affect performance can be relied on to have a physical rather than numerical basis. The system thus allows the designer to quickly try out new design approaches and develop design philosophies for any given type of application.

Acknowledgements

This research was supported by the MIT Carl Richard Souderton Faculty Development Chair and the National Science Foundation Presidential Young Investigator Program.

References

[1] F. Bauer, P. Garabedian, D. Korn, and A. Jameson. Supercritical wing sections I, II, III. In *Lecture Notes in Economics and Mathematical Systems*, Springer-Verlag, New York, 1972, 1975, 1977.

[2] R. E. Melnik, R. R. Chow, and H. R. Mead. *Theory of Viscous Transonic Flow Over Airfoils at High Reynolds Number.* AIAA-77-680, Jun 1977.

[3] M. Drela and M. B. Giles. *ISES: A Two-Dimensional Viscous Aerodynamic Design and Analysis Code.* AIAA-87-0424, Jan 1987.

[4] M. B. Giles and M. Drela. Two-dimensional transonic aerodynamic design method. *AIAA Journal*, 25(9), Sep 1987.

[5] M. Drela and M. B. Giles. Viscous-inviscid analysis of transonic and low Reynolds number airfoils. *AIAA Journal*, 25(10), Oct 1987.

[6] M. Drela. Low-Reynolds number airfoil design for the MIT Daedalus prototype: A case study. *Journal of Aircraft*, 25(8), Aug 1988.

[7] R. J. McGhee, G. S. Jones, and R. Jouty. *Performance Characteristics from Wind-Tunnel Tests of a Low-Reynolds-Number Airfoil.* AIAA-88-6070, Jan 1988.

[8] E. Soinne and S Laine. An inverse boundary element method for single component airfoil design. *Journal of Aircraft*, 22(6):541–543, Jun 1985.

[9] J. L. Kennedy and D. J. Marsden. A potential flow design method for multicomponent airfoil sections. *Journal of Aircraft*, 15(1):47–52, Jan 1978.

[10] R. Eppler and D. M. Somers. *A Computer Program for the Design and Analysis of Low-Speed Airfoils.* NASA TM 80210, Aug 1980.

[11] M. Drela. *An Integral Boundary Layer Formulation for Blunt Trailing Edges.* AIAA-89-2200, Aug 1989.

[12] J. E. Green, D. J. Weeks, and J. W. F. Brooman. *Prediction of Turbulent Boundary Layers and Wakes in Compressible Flow by a Lag-Entrainment Method.* R & M Report 3791, Aeronautical Research Council, HMSO, London, 1977.

[13] T. Cebeci and R. W. Clark. An interactive approach to subsonic flows with separation. In *Second Symposium on Numerical and Physical Aspects of Aerodynamic Flows*, Long Beach, California, Jan 1983.

[14] A. H. Shapiro. *Compressible Fluid Flow I.* Wiley, New York, 1953.

[15] P. H. Cook, M. A. McDonald, and M. C. P. Firmin. Aerofoil RAE 2822 pressure distributions and boundary layer and wake measurements. In *Experimental Data Base for Computer Program Assessment, AR-138*, AGARD, 1979.

PREDICTION OF AERODYNAMIC PERFORMANCE OF AIRFOILS IN LOW REYNOLDS NUMBER FLOWS

Domenico P. Coiro
CIRA - Italian Center for Aerospace Research
Via Maiorise, 81041 Capua (CE) - ITALY

Carlo de Nicola
Gasdynamics Institute, University of Naples
Piazzale Tecchio 80, 80125, Napoli - ITALY

ABSTRACT

In many industrial applications the evaluation of the influence of laminar separation can be fundamental for the prediction of accurate global and local aerodynamic performances; this is the case of wing sections operating at low Reynolds numbers.

In this paper a viscous/inviscid interaction procedure for the solution of incompressible flowfields with large laminar separated regions is presented, based on the approach previously followed by the Authors for the prediction of the performances of airfoils up to stall conditions.

The extension of the interaction procedure to take into account the existence of laminar separation bubbles is described, with an eye toward the question of the numerical determination of the transition.

The application of the computational procedure for two airfoils widely investigated experimentally at low Reynolds number is then discussed; numerical results show the reliability of the method in these conditions. A critical analysis of the solutions indicates the guidelines for the future improvements of the method.

INTRODUCTION

An efficient viscous/inviscid interaction method should be able to predict accurate aerodynamic characteristics of an airfoil either when large regions of turbulent separated flow exist, as for high lift condition case, or when extensive bubbles of laminar separated flow are present. This second kind of calculations is necesssary for the prediction of the performances of small chord airfoils, i.e., a turbine blade profile, or wing used for long endurance airplanes or gliders, /1/; in all these cases the chord Reynolds number may fall below some limit, and thus laminar separated flows occur.

Different methods of solution can be used to treat such flows. Most commonly used are:

- semi-empirical methods;
- viscous/inviscid interaction methods;
- the solution of the Navier-Stokes equations.

Semi-empirical methods have been extensively used in the past. Their advantage is the small need of computational effort, but essentially they cannot take into account the influence of the local separation on the remaining flowfield.

Navier-Stokes solutions introduce high level of computational difficulties, and in any case they need an accurate modelling of the separated region, where a very complex transitional regime, with subsequent turbulent reattachement, is present. To date, Navier-Stokes solution is almost unusable for industrial design purposes.

Indeed, viscous/inviscid interaction techniques can predict in accurate way, and with relatively small computational costs, the extent and burst of separation bubbles. The major difficulty of this approach is the prediction of the transition point in laminar separated flows, which can greatly influence the prediction of the characteristics of the bubble flows, and thus the global aerodynamic coeffícents.

MULTIM code based on the viscous/inviscid interaction approach was developed by the Authors for the prediction of airfoil aerodynamic performances. It is a multi-methods computer code, since it can use many different methods of solution for both inviscid and viscous fields. Panel methods were employed for the potential solution, and integral boundary layer equations, written in direct and inverse form, were used for the viscous part.

Such a code has shown a good capability in predicting airfoil flows up to stall conditions, but only for Reynolds number above 1 million, since no laminar separation region analysis was present. This limitation is reasonable, because at low Reynolds number conditions the boundary layer approximation is less accurate, and the interaction between external flow and viscous region becomes stronger due to boundary layer thickening.

The aim of this work is the extension of the range of applicability of the MULTIM code to low Reynolds number flows by including the presence of large separated bubbles into the model.

THE METHOD OF SOLUTION

The interaction procedure of MULTIM code is based on the iterative coupling of potential and boundary layer integral equations using as matching parameter the transpiration, boundary layer induced, normal velocity,

$$V_n = d(Ue\delta^*)/ds$$

as reported in /2/.

Initially, the method was developed assuming that only turbulent separated regions could exist in the flowfield: the development of a laminar separation could not be analyzed in detail because only direct solution was provided for the laminar boundary layer; when a laminar separation was detected, the characteristics of the bubble could be evaluated using a typical semi-empirical approach (i.e. Horton's method, /3/) but the transition to turbulent flow was assumed at the point where the laminar separation occurred. This model provided satisfactory prediction of airfoil flows for high values of the Reynolds number up to stall condition, and multi-element geometries could be analized without any difficulty /4/.

To circumvent the difficulties occuring when MULTIM code was used to predict the aerodynamic characteristics of airfoils at low Reynolds numbers, it was sufficient to provide the solution of laminar boundary layer equations in both direct and inverse mode; the (integral) momentum and kynetic energy equations were used, while the laminar closure relationships were obtained using Falkner-Skan one parameter velocity profiles family, /5/. The final system to solve is

$$\frac{dH^*}{ds} = f_1 * [2*C_d - (H^*+\beta) * \frac{C_f}{2} + \frac{\beta}{H} * \frac{d\delta^*}{ds}]$$

$$\frac{dUe}{ds} = \frac{Ue}{\theta\,(2+H)} [\frac{C_f}{2} - (\frac{1}{H} * \frac{d\delta^*}{ds} - \frac{\delta^*}{H^2\alpha} * \frac{dH^*}{ds}))]$$

where :

$$f1 = \frac{1.}{\theta + \frac{\beta\delta^*}{H^2\alpha}} \qquad \beta = \frac{H^*(1-H)}{2+H} \qquad \alpha = \frac{dH^*}{dH}$$

and the closure correlations are in the following form:

$$H^*=H^*(H,R_{e\theta}) \qquad C_f=C_f(H,R_{e\theta}) \qquad C_d=C_d(H,R_{e\theta})$$

This approach was also followed to solve the turbulent boundary layer; auxiliary equations and closure correlations characterize the method of solution. We coded Green method, /6/, along with the extentions proposed by East et al., /7/, and also Whitfield method,

/8/, based on the analytical velocity profiles originally proposed by Whitfield, /9/. In any case we did not find any appreciable difference between those methods, due to the fact that the turbulent separated area is normally small compared to the laminar one for the conditions we are dealing with.

The main features of the resulting method are now described.

For any combination of camber and thickness, very efficient low-order panel methods, based on source/vorticity distributions, provide once for all the potential flow velocity matrices.

At each iteration cycle, boundary layer computation on upper and lower streamlines is then performed in direct mode ahead of the separation points; the switch from direct to inverse mode of solution occurs when some boundary layer parameters (e.g., the shape factor H for turbulent flow, or the Thwaites parameter λ in case of laminar boundary layer) assume a critical value indicating the incipient onset of separation.

The Runge-Kutta 4-th order with (variable) stepsize control for the numerical accuracy, /10/, is used for the solution of the boundary layer equations. When inverse computation is required, the displacement thickness distribution is assigned by updating the distribution of the previous iteration, following Carter, /11/:

$$\delta^{*n+1} = \omega * [\ \delta^{*n} * (\ \frac{U_{eBL}}{U_{eIN}}\ )\ ] + (\ 1-\omega\ ) * \delta^{*n}$$

When transition is found, turbulent correlations are used instead of the laminar ones, for both direct and inverse modes of solution.

The interaction is then evaluated by imposing Neumann conditions on the airfoil surface for the onset transpiration flow.

The procedure is then iterated up to the desired accuracy; we used a convergence criterium based on local velocity control, more severe than one based on global coefficients control.

Using this approach, it has shown to be more convenient to solve the boundary layer in direct form as far as possible, avoiding the use of inverse form in regions where the flow is stably attached; the converged solutions could be not independent from the extent of the inverse regions. This fact is mainly due to the stability of the iteration process, which could be reduced using inverse mode for well attached flows.

TRANSITION PREDICTION

The prediction of the transition points seems to be a crucial

aspect for the correct analysis of laminar bubble flow.
While for attached flow no big difference was found between Michel, /12/, and e^9 methods, in laminar separated flows none of them could predict with reasonable accuracy the onset of the transition; furthermore, the application of the method of Crimi and Reeves, /13/, in many cases furnished unsatisfactory results.

So, in a preliminary stage, the transition point has been assigned from experimental results, and keeped fixed during the iterative calculation; in such a way the validation of the method and of the code was possible.

Successively, an analysis has been made to check the trend of the exponent N of a e^N law, from the onset of the laminar separation to determine, if possible, a plausible value for N at transition.

As will be shown in the next paragraph, numerical results indicate that the value of N at transition cannot be uniquely derived, and some uncertainity is present; the ambiguous conclusions reported in /14/ have been confirmed. In any case, we found that a value of N around 15 seems to give, in average, the better agreement with experiments.

NUMERICAL RESULTS

To run the code, 95 points were distributed around the airfoils, with a local refinement in the bubble region. On the CDC 930/13 machine, the code run in about 2 minute of CPU time for a 40 iterations case; about 15 seconds was the CPU time on the ETA 10 computer. Tipically, under-relaxation was used, with a relaxation factor equal, in average, to .6, falling to .2 for strong interaction cases.

EPPLER 387 AIRFOIL

Recently this airfoil has been extensively examined in detail and in different wind tunnels, /15/. We ran the code for Reynolds number Re equal to 100,000, 200,000 and 300,000.

Calculated and experimental pressure coefficient distributions for this airfoil at angles of attack of 0 ,2 ,4 deg. and for Re=300,000 are presented in Fig. 1, 2 and 3. It can be seen that the pressure coefficient distribution is almost everywhere in excellent agreement with experimental values; only in the suction peak region there is a small overstimation of the pressure, which leads to a global lift coefficient slightly greather than that evaluated from the experiments. The drag coefficient was evaluated using Squire-Young formula, and is very close to that indicated in /15/. In Fig. 4 the lift curve is presented; in this case it is also evident that the lift curve slope is very well predicted, with some discrepancy at high angles of attack. The polar is presented in Fig. 5, where it is clearly indicated that the overall agreement is acceptable with an exception for the high lift condition, where the lift and drag coefficients are understimated.

Fig. 6 shows the pressure coefficient distribution at Re=200,000; the agreement is also very good. In Fig. 7, which shows the polar for these conditions, it can be seen that some problems start to appear at intermediate angles of attack due to the fact that the interaction between viscous and inviscid flow becomes bigger than previous cases. These considerations are also valid for Fig. 8, where lift curve is presented.

From Fig. 9 it is very clear the inaccuracy of the numerical results at Re=100,000 due to the big extension of the separation bubble. This is mainly due to the correlation formulas for separated laminar and turbulent flows which cannot bear the growth of boundary layer parameters beyond physical values. It seems very important to improve the separated area modelling to have better prediction of bubble characteristics in these conditions.

WORTMANN FX 63-137 AIRFOIL

Boundary layer experimental analysis were performed for this airfoil at Notre Dame facilities, /16/. In Fig. 10 comparison between numerical and experimental values for the displacement thickness distribution is presented for Re=100,000 and at an angle of attack of 7 deg.; it is clear that the agreement is good up to transition location which is predicted slightly in advance with a value of N equal to 15. After that point some differences appear in predicting the growing of boundary layer, probably due to the incorrect modelling of the separated area when a strong interaction is present.

Fig. 11 shows a typical convergency history for the lift coefficient during the iterative process.

CONCLUSIONS

A critical analysis of the computational results obtained applying the version of MULTIM code extended to the prediction of flows with laminar separated flow indicates the following main conclusions.

By using the viscous/inviscid interaction model it is possible to give a correct representation of low Reynolds number flows. Polars of airfoils can be predicted satisfactorily for Reynolds number not lower than 200,000.

To obtain an higher degree of agreement with the experimental measurements of bubble flows, improvements must be obtained for the prediction of transition point in separated flow. Furthermore, the analysis of the results suggests the need for different correlation formulas when large bubble are present in the flowfield.

REFERENCES

1. Maughmer,M. and Somers,D.M.,"An Airfoil Designed for a High-Altitude, Long-Endurance, Remotely-Piloted Vehicle", AIAA Paper 87-2554CP.
2. de Nicola,C.,Coiro,D. and Losito, V.,"An Efficient Multi-Methods Computer Code for the Prediction of the Inviscid and Viscous Flow over Multi-Component Airfoils", AIAA Paper 88-3562.
3. Horton, H.P., "A Semi-empirical Theory for the Growth and Bursting of Laminar Separation Bubbles", A.R.C. C.P. 1073, 1967.
4. de Nicola, C.,Coiro,D. and Losito,V.,"Prediction of Aerodynamic Characteristics of High Lift Multi-Element Airfoils", Proc. BEM10 Conference, Swansea, U.K., 1988.
5. Drela,M. and Giles, M. B., "Viscous-Inviscid Analysis of Transonic and Low Reynolds Number Airfoils," AIAA J. 25,N. 10, October 1987.
6. Green, J.E.,Weeks,D.J. and Broman,J.W.F.,"Prediction of Turbulent Boundary layers and Wakes in Compressible Flow by a Lag-Entrainment Method", R.A.E. TR 72231, 1972.
7. East, L.F., Smith,P.D. and Merryman, P.J., "Prediction of the Development of Separated Turbulent Boundary Layers by the Lag-Entrainment Method", R.A.E. TR 77046, 1972.
8. Whitfield, D.L., Swafford,T.W. and Jacocks,J.L., "Calculation of Turbulent Boundary Layers with Separation and Viscous-Inviscid Interaction", AIAA J., Vol. 19, no. 10, 1981.
9. Whitfield, D.L., "Analytical Description of the Complete Turbulent Boundary Layer Velocity Profile", AIAA Paper 78-1158.
10. Press,W.H.,Flannery,B.P.,Teukolsky,S.A.,Vetterling,W.T., "**Numerical Recipes**", Cambridge Press, 1986.
11. Carter,J.E., "A New Boundary Layer Inviscid Interaction Technique for Separated Flows", AIAA Paper 79-1450.
12. Michel,R., "Etude de la Transition sur les Profiles d' Aile", ONERA

Report 1/1578a, 1951.

13. Crimi,P. and Rees,B.L.,"Analysis of Leading-Edge Separation Bubbles on Airfoils", AIAA J., Vol. 14, no. 4, 1976.

14. Evangelista, R., and Vemuru, C.S., "Evaluation of an Analysis Method for Low-Speed Airfoils by Comparison with Wind Tunnel Results", AIAA Paper 89-0266.

15. Mc Ghee,R.J.,Walker,B.S. and Millard,B.S.,"Experimental Results for the Eppler 387 Airfoil at Low reynolds Number in the Langley Low-Turbulence Pressure Tunnel", NASA TM 4062, 1988.

16. Brendel,M. and Mueller,T.J.,"Boundary Layer Measurements on an Airfoil at Low Reynolds Number", J. Aircraft, Vol. 25 no. 7, 1988.

NOMENCLATURE

C	airfoil chord
C_f	friction coefficient
C_l	lift coefficient
C_p	pressure coefficient
C_d	drag coefficient
C_d	dissipation coefficient
$H=\delta^*/\theta$	shape factor
$H^*=\theta^*/\theta$	second shape factor
n	iteration counter
N	linear theory stability amplification factor
Ue	velocity at the edge of the boundary layer
U_{eBL}	computed esternal velocity from inverse calculation
U_{eIN}	computed esternal velocity from interaction calculation
Re	chord Reynolds number
$R_{e\theta}$	Reynolds number based on momentum thickness
s	streamwise coordinate along streamlines
x	chordwise coordinate
V_n	transpiration velocity
δ^*	displacement thickness
λ	Thwaithes pressure gradient parameter $[\theta^2\ d(Ue)/ds]/\upsilon$
θ	momentum thickness
θ^*	energy thickness
υ	kynematic viscosity
ω	relaxation factor

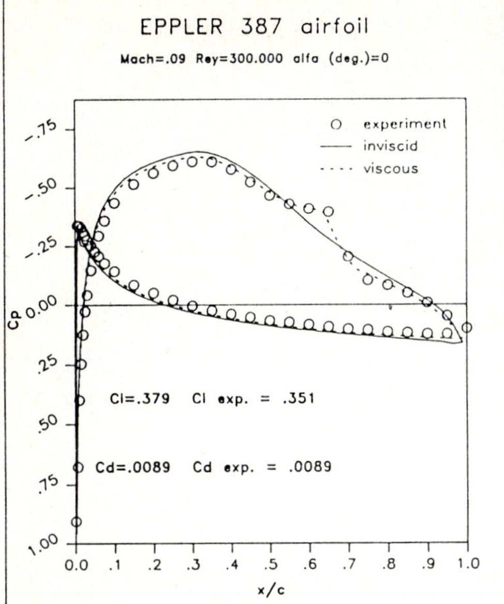

FIG. 1

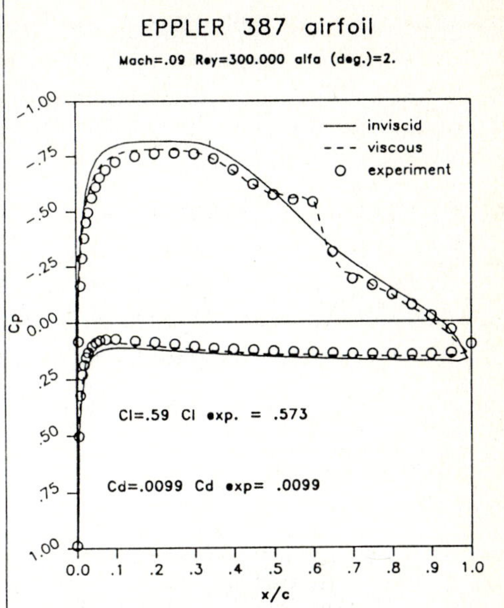

FIG. 2

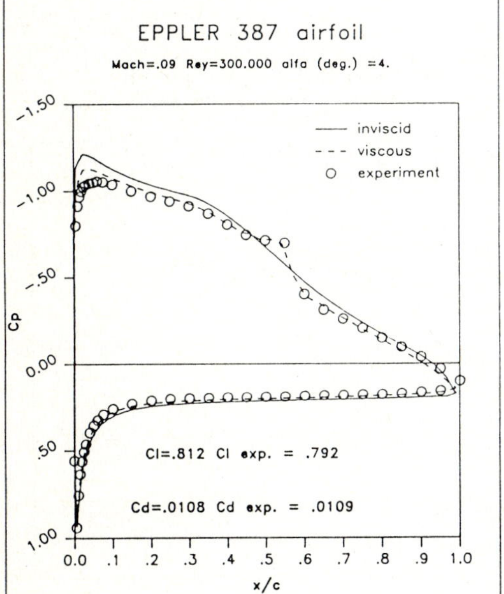

FIG. 3

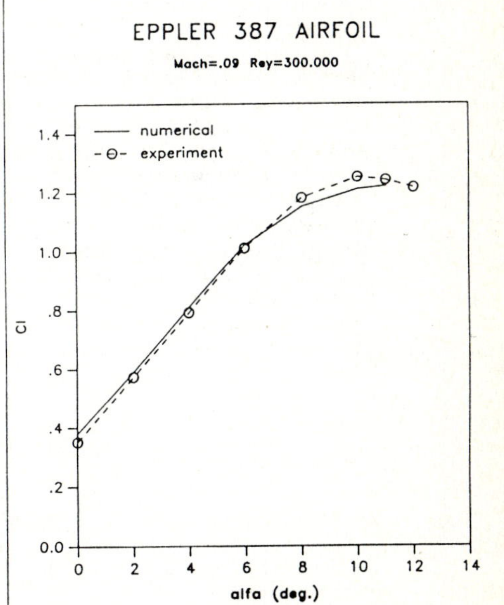

FIG. 4

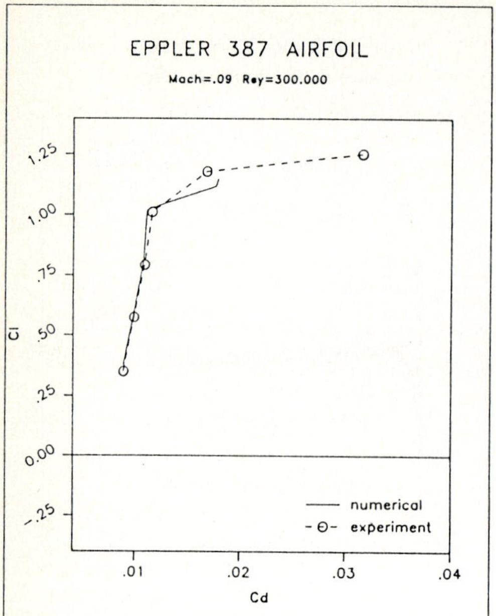

FIG. 5

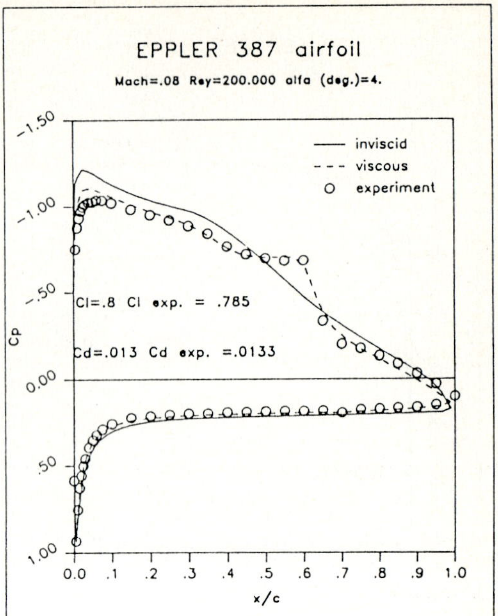

FIG. 6

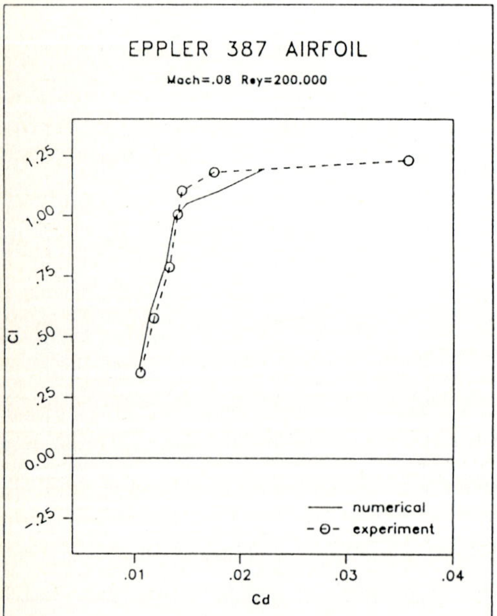

FIG. 7

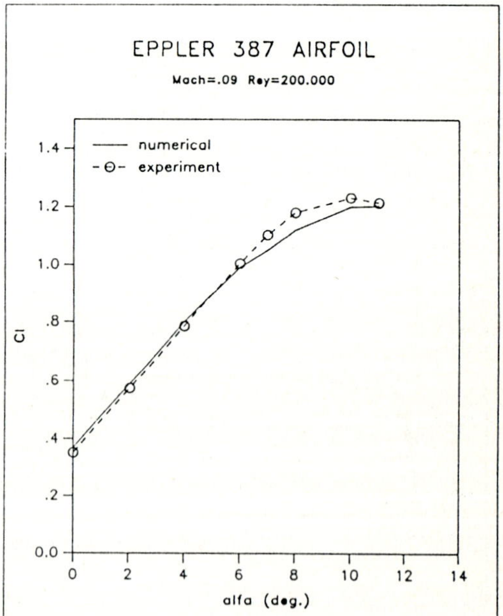

FIG. 8

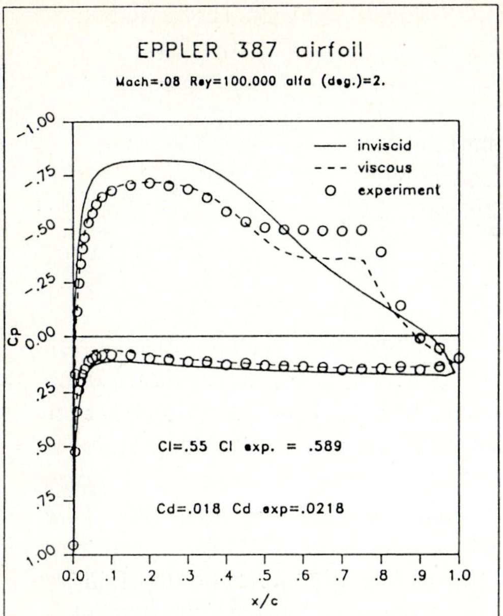

FIG. 9

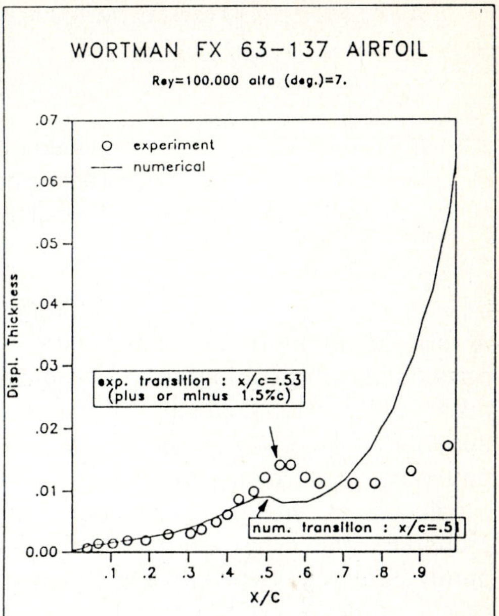

FIG. 10

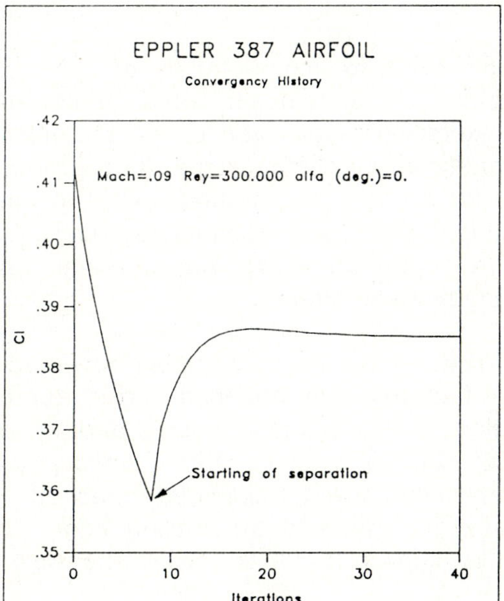

FIG. 11

A FAST METHOD FOR COMPUTATION OF AIRFOIL CHARACTERISTICS

by

Arild Bertelrud
High Technology Corporation
Hampton, Virginia

SUMMARY

A simple method for computation of the flow on airfoils has been developed to help in the development of suitable profiles for turbulence manipulation[1]. Typically these will have chord Reynolds numbers of 300,000 or less, be located in a high subsonic or transonic Mach number boundary layer close to a wall. The analysis does not require a good prediction of maximum lift, but instead must be able to describe flow properties and drag adequately within a CL-range. Since a large number of parameters are involved, and since the turbulence manipulation to a large extent depends on the dynamics of the flow on and behind the profiles,it was necessary to establish a code that was direct, fast, and allowed computation of a typical characteristics in a matter of minutes on a minicomputer or a personal computer.

APPROACH

A variety of methods exist for computation of the flow properties of airfoils at low Reynolds number. Recent years Euler or Navier-Stokes codes requires long execution times even on super-computers[2,3]. In many cases it may be beneficial to compute the flow properties with a fast code that can be run on a personal computer, either to obtain a first guess at the complete solution or to give acceptable results in a certain flow conditions , i.e. when separated areas are so small, that the pressure distribution is only moderately altered.

In inverse codes or codes capable of handling reversed velocity profiles (like in bubbles) the fact that the boundary layer approximation breaks down before separation is taken into account, and in various ways the singularity is avoided[4]. In the present code boundary layer computations are performed wherever the flow is assumed attached, separated areas are handled empirically and close to the trailing edge where the inviscid pressure distribution is known to be in error, the flow development is simply extrapolated.

The method was originally written in FORTRAN for an HP-1000; the current version is adapted to the MacIntosh II environment. It is based on the following observations and assumptions:

- In many cases it is important to have information available immediately to scope out parameter ranges, determine trends or check assumptions or results obtained with a more elaborate code.
- Ease of use is very important - geometry definition and grid generation often prohibits efficient analysis of a variety of geometries.
- The inviscid pressure distribution is determined quite accurately using few panels - this allows a speed up in code execution. However, the boundary layer computations require a much larger number of points on the surface and quite often some sort of smoothing, as pressure gradients are essential to the determination of transition and separation.
- Preferrably all input should be physical quantities, e.g. transition should be influenced by the physically measurable quantities like turbulence intensity and length scale (as well as isotropy and uniformity).
- Most airfoil design starts from the user's previous knowledge concerning profile development. It is common to take an existing geometry, known to behave well under similar conditions, and try to improve upon it.

METHOD

The program consists of two basic modules , as well as a user interface, as is shown in Figure 1. The solution is iterated within and between the two, depending on the flow conditions and the specifications defined by the user. The user interface module only exists on the Mac version, and allows the user to define the geometry and flow conditions interactively. It handles the graphic presentation of the results and is used to define the mode of operation, i.e. if iterations are required. This module also contains the possibility of comparing with experimental (or "known") drag polar data.

INVISCID MODULE

The geometry is defined through a designation of NACA profile name, using a profile library or adding coordinates. If the angle of attack and Mach number are specified the panel method yields the inviscid pressure distribution. The panel method is adapted from Ref. 5 , and uses vortex panels of linearly varying strength. The compressibility is taken into account using the Prandtl-Glauert transformation. As a standard 35 points on the airfoil are used to compute the pressure distribution - for most configurations this yields the same pressure distribution as 120 points (Ref.5), albeit at fewer control points.

In general, i.e. at higher Reynolds numbers, the inviscid flow is computed only once, as this yields a good approximation even if small, local separations exist on the airfoil. However, the experimental data available in the user interface, may alert the user that discrepancies from known , global data are large, suggesting a check into the causes.

VISCOUS MODULE

For a given Reynolds number, turbulence intensity and length scale, the viscous flow is then computed. The stagnation point is determined from the vorticity distribution, and upper and lower sides are computed separately. As mentioned, the boundary layer calculations require more closely spaced points than feasible to compute with the panel method, so for the laminar boundary layer calculations , the number of points is increased through interpolation.The laminar boundary layer is computed using Thwaites' integral method[6]. Transition is predicted using Michel's[7] or Wazzan's[8] transition criteria - the former depending on the development of the momentum thickness, whereas the latter incvludes the pressure gradient effect. For application of LEBUs, the turbulence intensity should be considered - neither of the two methods above do this. So for comparison the e^n-transition criterion has been included, as simplified by Drela and Giles[3]. In this criterion the transition is considered to occur where an initially small disturbance has grown e^n times. Mack's[9] correlation between the turbulence intensity and the exponent n is used in the code.

A laminar separation is followed by a bubble prediction based on criteria for different parts of the bubble by Horton[10], Roberts[11], and van Ingen[12]. One of the crucial parts of the prediction is the estimate of the laminar portion of the bubble. While many codes incorporate an extensive computation in this region with separated flow profiles etc., the present approach is to make an overall estimate of the length based on empirical criteria that can take into account both the properties of the incoming laminar boundary layer and the influence of free stream turbulence on the stability of the free shear layer in the bubble.

Once transition is assumed to have occured in the bubble, the code steps downstream in small steps, checking to see if the reattachment criterion by Horton has been met. If the 0.97 % chord point is passed without an indication of reattachment, the bubble is assumed to have burst. Depending on the purpose of the computations, the computations may either stop or computation of a viscous/inviscid interaction may be invoked.

If flow reattachment is predicted (or if natural transition is assumed to have occurred,) Bradshaw/Mizner/Unsworth's turbulent finite difference code[13] is used for prediction of the turbulent boundary layer development. This code allows for heat transfer, surface roughness, suction and blowing, and is therefore suitable for computation of the flow on airfoils with boundary layer manipulation. It uses a pressure distribution described in equidistant points along the surface , and performs its own smoothing. It does not make any special provision for low Reynolds number flows, however.

As the current emphasis is placed on two-dimensional flows, the MacIntosh version has replaced Bradshaw's finite difference code by a turbulent integral code according to Walz[14]. Also , the Ludwieg-Tillmann formula commonly used to relate the local skin friction to momentum thickness Reynolds number and the shape parameter, has been replaced by a formula according to White[6], better suited both at low and high Reynolds numbers.

Taking the predicted flow properties at 90-95 % chord from both sides of the airfoil[15], the drag coefficient is predicted using Squire-Young's[16] method as developed by Lock[17] for compressible flow. If the turbulent code indicates a turbulent separation before 95 % of the chord, the flow field can be computed analogous to the laminar separation cases.

VISCOUS/INVISCID INTERACTION

In many cases the viscous/inviscid ineraction may have a large influence on both pressure distribution and the boundary layer development, and to get a proper prediction, it is necessary to iterate between the modules. Cebeci[18] has recently argued that the reattachment computations for a bubble may be in error due to numerical problems, and it seems that a large number of iterations may actually distort the solution.

A distinction is made between two different cases:

- A closed separation bubble. See Figure 2 from Ref. 19. Experimentally two characteristics are evident: There is quite often an almost constant pressure under the laminar part of the bubble; however appreciable pressure gradients, of both signs, may occur for various reasons. In the reattachment region there is an "undershoot" of pressure where the reattaching layer adjusts to the airfoil surface and the start of a new attached turbulent boundary layer is happening. Most computational codes fails to model these effects, and the current version of the present code does not address them either.
- Open separation. This generally means a recirculation zone extending back into the wake, but for modelling purposes the static pressure imprint the zone causes on the airfoil surface is that of a constant pressure level.

The viscous influence may be simulated either by retaining the body shape and imposing a normal velocity at each point , or through modification of the body based on the displacement thickness in the boundary layer or on the shape of the separating streamline. Here the latter option was chosen, and the first modification to the airfoil shape is obtained through adding the displacement thickness to the airfoil geometry up to the laminar separation point. If it reattaches, a turbulent boundary layer is computed and in the viscid/invidscid interaction the corresponding displacement thickness is added. If a turbulent separation occurs, the flow is considered non-reattaching. From a laminar or turbulent separation point, the wake shape is found by iteration. The procedure follows Dvorak et.al.[20,21]:

• Two main assumptions are made, allowing a simplification of the procedure:

1. Although a separation may be massive and change the pressure level appreciably, the location of the separation generally does not move very far. Thus a viscous computation is generally performed using the inviscid pressure distribution, and another may be performed after the inviscid iterations to reassess the viscous drag.

2. The pressure level in the separated region is determined from the second potential flow calculation. Thus it includes the viscous displacement effects in the attached flow regions but lets the further iterations in the separated region iterate towards a fixed point.

•If the separation is an open one, the new shape is defined through addition of 6 points, three on each side, extending back to x/c=1.3. This is based on an assumption that the actual wake region is short - the wake closing point located only 30 % behind the airfoil is a realistic one.

•The initial slope is taken to be the local airfoil slope, and parabolic shapes are used to estimate the two streamlines on the upper and lower side. The target pressure distribution is constant from the separation point on, and after the first iteration an improved next shape is based on the sensitivities between shape and pressure distribution obtained from the inviscid cycle and the first viscous shape.

COMPARISONS WITH EXPERIMENT

Verification of the code predictions must be done through comparison with well documented two-dimensional airfoil experiments. Although airfoil data is abundant, only few investigations give all required information for a comprehensive comparison.

One of the airfoils chosen for the current comparison, is the Eppler 387 airfoil, where experimental as well as computed data is available[22,23] to compare both overall characteristics (i.e. lift & drag coefficients) as well as local flow conditions such as separations etc. Recently comparisons were made between the experimental results and computations[24].
Figure 3 shows the overall comparison of lift and drag coefficients for Re=300,000. The current code was run without iterations, which would make it overpredict the lift coefficient , and yielded results roughly comparable to Eppler's code[25] for cases when the bubble is smaller than 10-20 % of the chord. The explanation for this is to be found in Figure 4 where the development of transition regions and separated regions have been plotted.

In general, the prediction of laminar separation is seen to be quite good over the entire angle of attack range. Also the predicted bubble length was quite well predicted. The computations were performed with Tu=0.1 % , which may be seen as an average for the experiment. The length scale was unknown , and was input as Ls/C = 0.1 - the effect of this assumption is illustrated in Figures 5 and 6, where the length of the laminar region , scaled by the momentum thickness at separation has been plotted against Re.theta.sep. As can be seen from the experimental points and the curves, there is a large scatter. Mueller et.al.[26,27] has correlated the laminar length with the height at transition. In the present case Mueller's data has been replotted together with other available data , and for the purpose of using in a computational code the scatter may seem too large. However, if the equations for estimate of a Turbulence Factor (Roberts[11]) are used, the turbulence intensity and the length scale yields an appreciable variations, as can be seen in Figure 6. In the current code the basic laminar length Reynolds number is chosen to be 25 000 corresponding to the reference point of Tu=1% , Ls/C=1 . For the E387 computations ,use of Tu=0.1 % and choice of Ls/C=0.1 means a Reynolds number of more than 40000. The curve is analogous to Mack's results[9] for attached flow laminar instability,.
As the Reynolds number of the E387 is decreased , the relative size of the bubble increases; until the viscous flow starts to modify the pressure peak and the trailing edge pressure level appreciably, the bubble start and the Reynolds number length of the bubble does not change much. Figure 7 shows an interactive computation where the viscous effects are large, Re=100,000 . The effective airfoil shape has changed appreciably due to the displacement thickness addition, and the lift is seen to be decrease substantially. It should be noted that the computations are performed without fudge factors of any kind, except for the arbitrary choice of the length scale being Ls/C=0.1 used throughout the study. In this case as few as 25 points were used to describe the airfoil geometry.

While the previous airfoil has a moderately large separation bubble on the upper side for most moderate angles of attack , the FX 63-137 airfoil has a separation on both sides at low angles of attack, and these developed into open separations as the Reynolds number is decreased. Brendel & Mueller[26] investigated the flow around this airfoil at Re=100-200,000. It may be of interest to see that in the computations performed with the present code the location of the separation point is quite accurately predicted without any iterations at all, see Figure 8. The indicated lift coefficient is grossly overpredicted, but for information where to put a transition trip, where to put instrumentation etc. the information is quite adequate. It may also be of interest to note, that the agreement between the Re at separation as measured and according to the computations agree well.

The NACA0009 was used to obtain a comparison of estimated drag levels for comparison with experiment - thus the data mostly concerns flow conditions above Reynolds numbers of 300,000. As can be seen in Figure 10, the agreement and the trend with Reynolds number is quite good. In this case we chose experimental data obtained with high turbulence intensity, to approach the condition a LEBU profile would operate under.

Good agreement with experiment was obtained in the Reynolds number region of main interest for LEBU development. Recent experiments at high Reynolds numbers (Anders[28]) has led to the conclusion that use of ordinary profiles do not yield neither appreciable local nor substantial net drag reduction at high Re_θ. This has been attributed to the difference in scales as the Reynolds number is increased. Another hypothesis (Bandyopadhyay[29]) is that it is the von Karman vortex shedding that is missing or altered for the high Reynolds profiles tested. For cylinder flows the existence of the vortices and their behavior as a function of Reynolds number has been well established[30], and the importance of transition on the cylinders has been documented. For airfoils their existence has been shown[31], but it is currently unclear how they are related to the flow on the airfoil.

Computations were performed with the present code for the Reynolds numbers and the turbulence intensities and length scales appropriate for tests performed in the various facilities. At low Reynolds numbers experiments have given widely varying results. The extreme sensitivity to the trailing edge shape observed in some tests were confirmed in the computations. The key to the downstream behaviour, and possibly the drag reduction feature, appears to be the existence of a separated region where disturbances at the neutral stability frequency may be amplified and shed into the wake where they are developing through subharmonic evolution. Although the results differed substantially between the different profiles and lift coefficients, there was a clear trend that the transition and separation/reattachement characteristics were different at high Reynolds numbers. As the Reynolds number was increased the bubbles got shorter and over most of the Reynolds number region investigated by Anders, the airfoils would have natural transition. On-going tests in the NASA 7x11 inch low speed tunnel confirms the drag and wake characteristics predicted (Liandrat[32]) at low Reynolds number .

CONCLUSIONS AND FUTURE WORK

A fast and reasonably reliable method for computation of flow around airfoils at moderately low Reynolds numbers has been described. It offers an ease of interpretation that makes it attractive compared to more complex, comprehensive methods.

The key to a good prediction of drag and flow characteristics, is a good estimate of transition and laminar bubble length. An acceptable average of laminar layer lengths appear to be 160 times the momentum thickness at separation - information concerning turbulence intensity and length scale can then be used to modify the value used in the computations.
Ongoing work covers inclusion of a method to improve computation of the flow after open laminar or turbulent separation. It seems that one essential point to improve is a better description of the laminar length of the separation bubble. It is clear that both turbulence intensity and length scale must be included and empirical correlations made. The code is also being modified to handle infinite swept condition; the laminar and turbulent codes are in essence three-dimensional, while transition- and bubble criteria have to be defined.

REFERENCES

1. Bushnell, D.M.: " Turbulent drag reduction for external flows." AIAA Paper 83-0227 (1983)

2. Kothari, A.P. & Anderson,Jr., J.D. "Flows over low Reynolds number airfoils - Compressible Navier-Stokes Numerical Solutions." AIAA Paper 85-0107, AIAA 23rd Aerospace Sciences Meeting, Reno, NV, January 14-17, 1985

3. Drela,M. & Giles, M. B.: " Viscous-inviscid analysis of transonic and low Reynolds number airfoils." AIAA Journal , Vol. 25, No. 10, pp. 1347-1355 (1987)

4. Coton, F.N. & Galbraith, Mc D.: " A direct aerofoil performance code incorporating laminar separation bubble effects." ICAS Paper 88-4.2.4 , Jerusalem, Israel, September 1988.

5. Kuethe,A.M. & Chow, C.-Y.: ' Foundations of Aerodynamics - Bases of Aerodynamic Design.' John Wiley & Sons, New York, 1986.

6. White, F.M.: ' Viscous Fluid Flow.' McGraw Hill, 1974.

7. Michel, R.: "Etude de la transition sur le profile d'aile; établissement d'un critere de determination de point de la transition et calcul de la trainee de profile incompressible." ONERA Rep.1/157A (1951)

8. Wazzan,A.R. , Gazley,Jr.,C. and Smith,A.M.O.: " H-R_x method for predicting transition." AIAA Journal, Vol.19, No 6, pp. 810-812 (1981)

9. Mack, L.M. " Transition and laminar instability." JPL Report 77-15 (1977)

10. Horton, H.P.:" A semi-empirical theory for the growth and bursting of laminar separation bubbles." ARC CP-1073 (1969)

11. Roberts, W.B.: " Calculation of laminar separation bubbles." Journal of Aircraft , Vol. 18 , No. 1 , 1980.

12. van Ingen,J.L.: " Theoretical and experimental investigations of incompressible laminar boundary layers with and without suction." Technological University Delft Report VTH-124 (1965)

13. Bradshaw,P. , Mizner,G.A. & Unsworth,K.:" Calculation of compressible turbulent boundary layers with heat transfer on straight-tapered swept wings." Imperial College IC Aero Report 75-04 , 1974. See also: AIAA J. Vol. 14, 1976, p.399.

14. Walz, A.: "Boundary layers of flow and temperature." MIT Press, Cambridge, (1969)

15. Smith,A.M.O. & Cebeci,T.: ' Remarks on methods for predicting viscous drag.' In: AGARD CP-124 , 'Aerodynamic Drag.' , Izmir, Turkey, April 1973.

16. Squire, H.B. & Young,A.D. ' The calculation of profile drag of aerofoils.' ARC R&M 1838 (1937)

17. Lock, R.C.: ' The prediction of the drag of aerofoils and wings at high subsonic speeds.' Aeronautical Journal, June/July 1986, pp. 207-226.

18. Cebeci, T.: " Numerical instabilities in the calculation of laminar separation bubbles and their implications." AIAA Journal Vol. 27, No. 5 , pp.656-658 (1989)

19. Weibust,E. , Bertelrud, A. and Ridder, S.O.: " Experimental and theoretical analysis of laminar separation bubbles." AIAA Paper 84-2201 (1984) See also Journal of Aircraft.

20. Dvorak,F.A. & Maskew, B.: ' The modeling of airfoil separation in subsonic and transonic flow.' AFFDL-TR-80-3088 (1980)

21. Dvorak,F.A. & Choi,D.H.: ' Separation model for two-dimensional airfoils in transonic flow.' AIAA Journal , Vol.22 , No 8, pp. 1064-1070 (1984)

22. McGhee, R.J. , Walker,B.S. & Millard,B.F.: ' Experimental results for the Eppler 387 airfoil at low Reynolds numbers in the Langley Low-Turbulence pressure tunnel.' NASA TM 4062 (October 1988)

23. Miley, S.J. " A catalog of low Reynolds number data for wind tunnel turbine applications." Department of Aerospace Engineering, Texas A & M University, College Station, Texas, RFD-3387, February 1982.

24. Evangelista, R. and Vemuru, C.S. : " Evaluation of an Analysis Method for Low-Speed Airfoils by Comparison with Wind Tunnel Results." AIAA Paper 89-0266 (1989)

25. Eppler,R. and Somers, D.M.: " A computer program for the design and analysis of low-speed airfoils." NASA TM-80210 (1980)

26. Brendel, M. and Mueller,T.J.: " Boundary-layer measurements on an airfoil at low Reynolds numbers." Journal of Aircraft , Vol. 25, No. 7, July 1988, pp612-617.

27. O'Meara, M.M. and Mueller,T.J. : " Experimental determination of laminar separation bubbles on the airfoil at low Reynolds numbers." AIAA Journal, Vol. 25, August 1987.

28. Anders,J.B. "LEBU drag reduction in high Reynolds number boundary layers." AIAA Paper 89-1011 , AIAA 2nd Shear Flow Conference, Tempe, AZ,March 1989

29. Bandyophadyay, P. : Private communication.

30. Schlichting, H.:" Boundary-Layer Tehory " MacGraw-Hill, New York,(1985)

31. Heinemann,H.J. , Lawaczeck, O. and BütefischK.A. : " V. Karman Vortices and Their Frequency Determination in the Wakes of Profiles in the Sub- and Transonic Regimes."

32. Liandrat, Marie-Pierre : Private communication.

Figure 1 Flow chart of computational method.

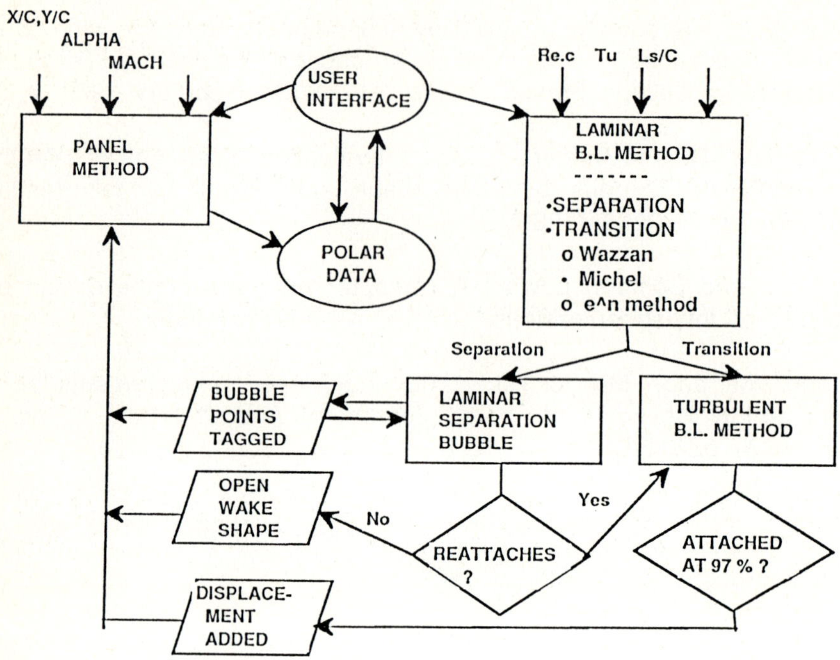

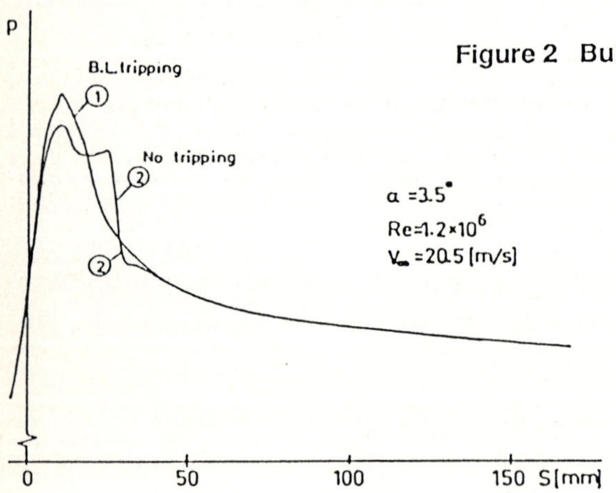

Figure 2 Bubble effect on pressure distribution.

From Ref. 19.

Figure 3 E387 Re=300,000
Drag polar - experiment & computations

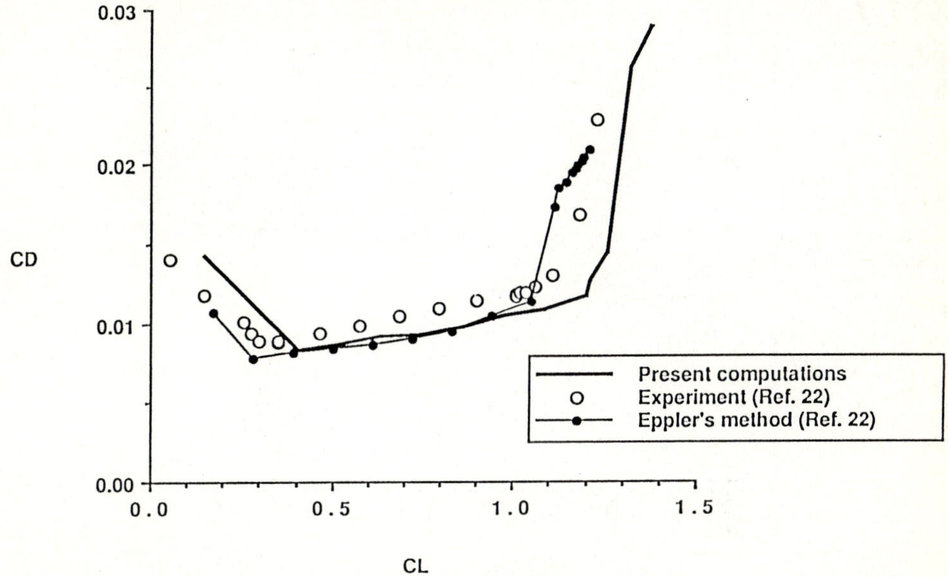

Figure 4 Comparison of experimental and computational flow characteristics. E387 Re=300,000

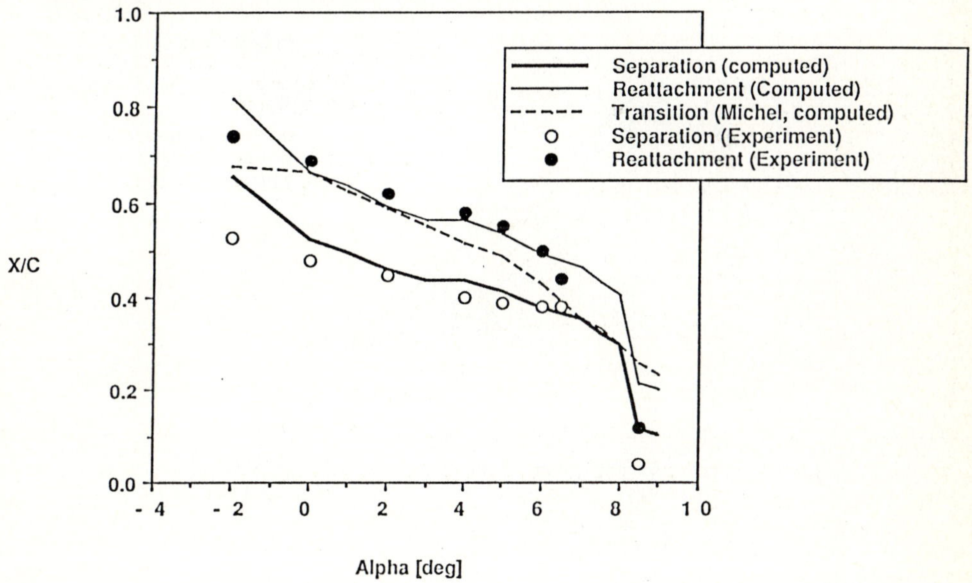

Figure 5 Laminar bubble length versus Re.theta at separation.

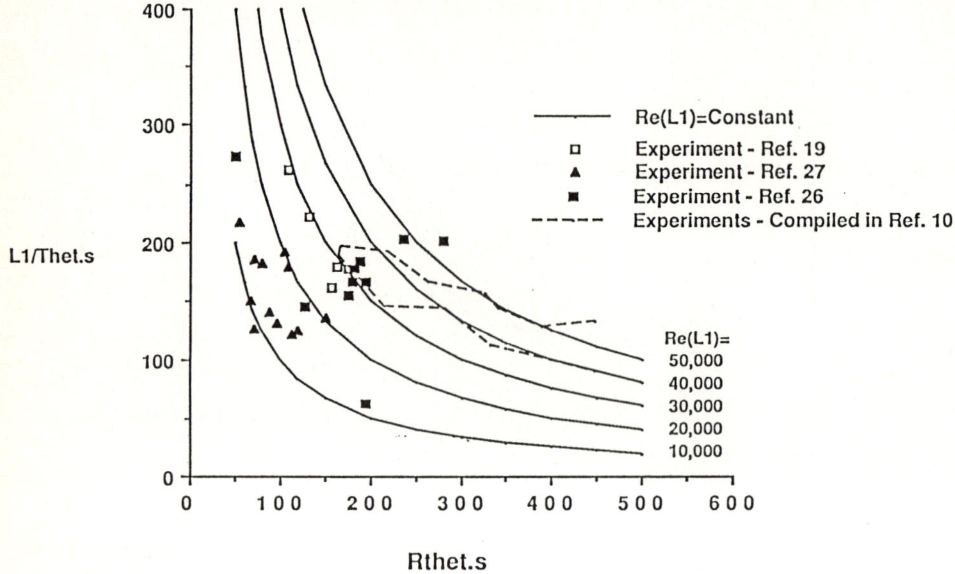

Figure 6 Relative laminar length

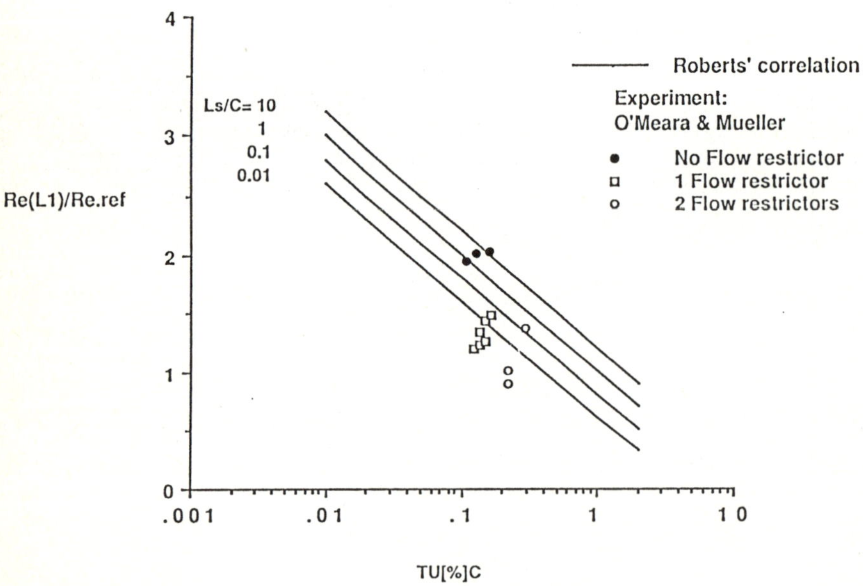

Figure 7 E387, Alpha=2deg ,Re=100,000

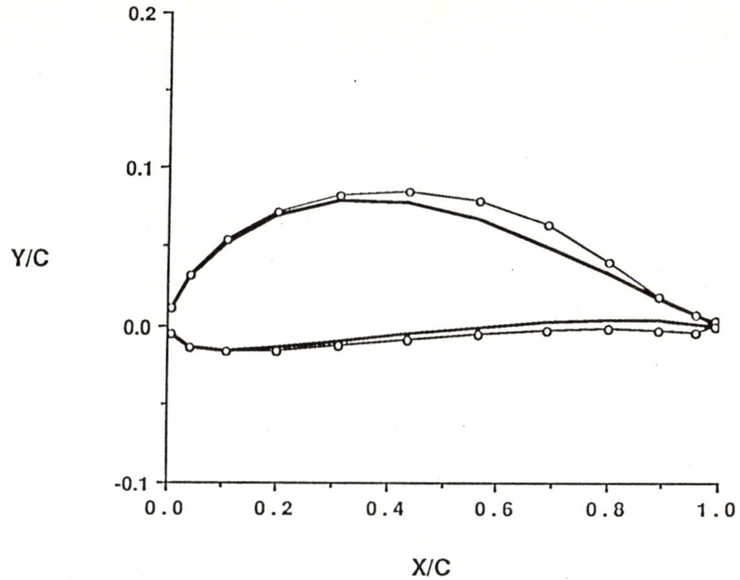

Figure 8 Separation location
Wortmann FX 63-137 Re=100,000

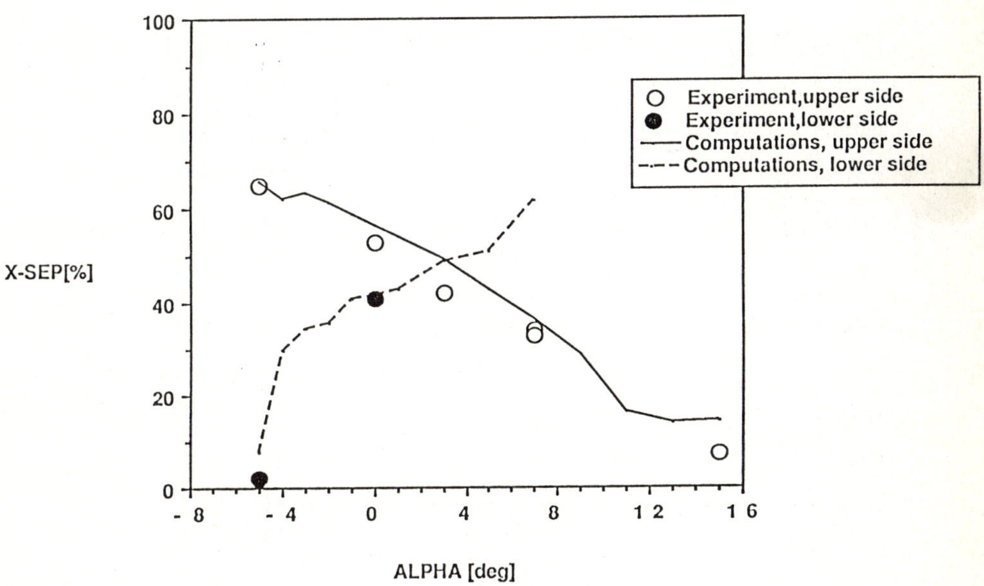

Figure 9 Momentum thickness Reynolds number at separation .
Wortmann FX 63-137 Re=100,000 Experiment: Brendel & Mueller

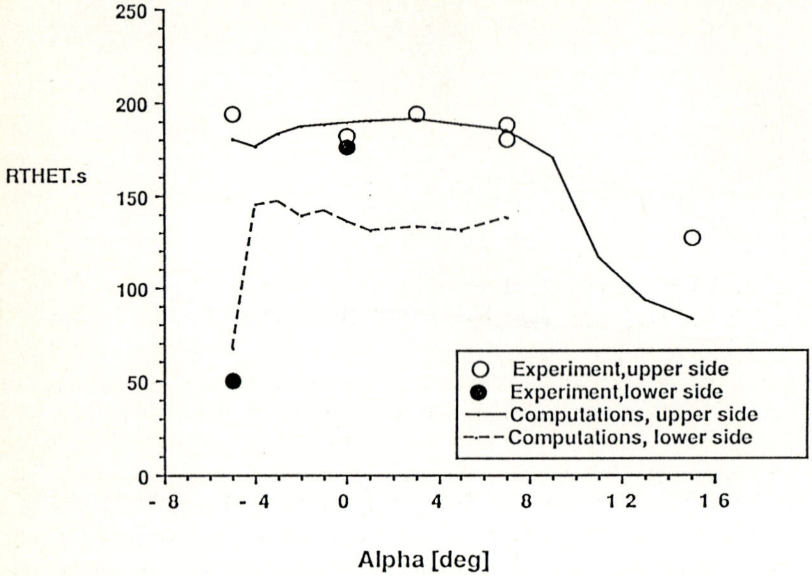

Figure 10 Comparison computations and experiment
NACA0009 , Alpha=0 deg

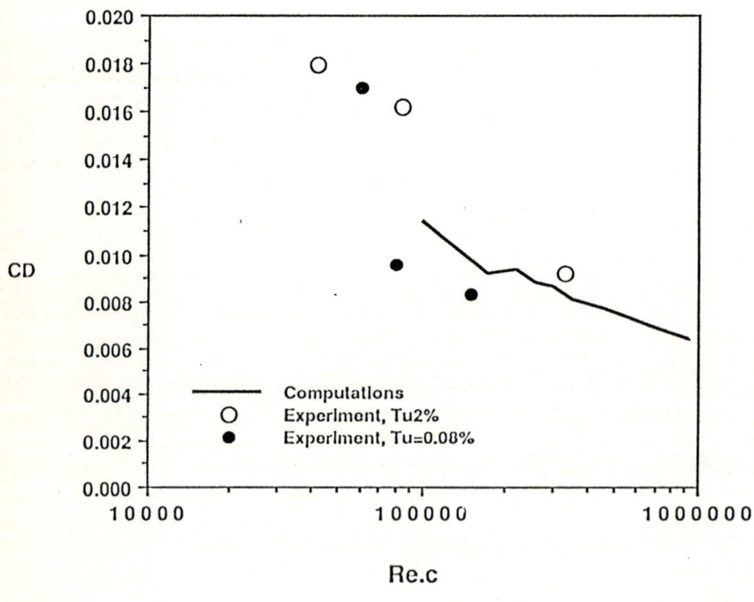

Low Reynolds Number Airfoil Design and Wind Tunnel Testing at Princeton University

J. F. Donovan† & M. S. Selig‡

Princeton University
Gas Dynamics Laboratory
Princeton, NJ 08544

Abstract

This paper focuses on the development of efficient low Reynolds number airfoils. Both experimental and computational techniques were used. The experimental facility and measurement technique are discussed in detail, and turbulence measurements in the tunnel freestream are presented. Lift and Drag data were taken at chord Reynolds numbers between 0.6×10^5 and 3.0×10^5. Comparisons of data obtained in the Princeton facility with that in several others are presented and show good agreement. Based on the results of over 40 airfoils tested during the first phase of this program (including the DAE51, FX63-137, E205, E374, E214, E387, Miley, NACA 0009, S3021, S2091, S4233), several new airfoils were designed using the Eppler and Somers code and screened using the Drela and Giles ISES code. Seventeen of the most promising designs were actually wind tunnel tested. The design philosophy is discussed and verified experimentally. Several of the new airfoils show significant performance improvements over previous airfoils. Boundary layer trips were also investigated as a means of reducing drag. Several types of trips were compared (zig-zag trips, bump tape, blowing, and two-dimensional trips), and the simple two-dimensional trip was found to yield the greatest improvement. The effects of model inaccuracies are also discussed, as well as the importance of a thin trailing edge in achieving low drag.

1. Introduction

The distinguishing characteristic of an airfoil operated at low chord Reynolds numbers ($Re < 5.0 \times 10^5$) is the formation of an extensive laminar separation bubble on either the upper or lower surface or both. This bubble can significantly increase the drag. As the laminar boundary layer negotiates the adverse pressure gradient of recovery it separates, thereby defining the beginning of the bubble. Amplified by the adverse pressure gradient, instabilities in the resulting free shear layer cause it to become turbulent. In most cases, the enhanced momentum transfer provided by the turbulent free shear layer allows it to reattach. Laminar separation bubbles not only increase airfoil drag but also cause hysteresis in both lift and drag with angle of attack.

In the past, developing methods of dealing with laminar separation bubbles to reduce drag has received little attention. Recently, however, a growing need for efficient

---

† Research Scientist, McDonnell Douglas Research Laboratories, P. O. Box 516, St Louis, MO 63166

‡ Graduate Student, The Pennsylvania State University, 233 Hammond Bldg., University Park, PA 16802

low Reynolds number airfoils has prompted interest in this area. Much of the experimental and theoretical effort has been concentrated on a fundamental understanding of laminar separation bubbles[1,2] in the hopes of predicting their behavior. This type of investigation has not yet led to the accurate prediction of low Reynolds number airfoil performance, and thus, the design problem still remains. However, efficient designs can be developed based on a limited knowledge of laminar separation bubble behavior.

Performance may be improved by reducing the size of the laminar separation bubble through the use of (1) transition ramps[3] (or more appropriately called a "bubble ramp") or (2) boundary layer trips[1]. Presently it is unclear if the optimum airfoil for a given task would make use of either or both of these techniques. Moreover, little data exists on the effects of boundary layer trips on low Reynolds number airfoil performance.

The approach of this investigation was to evaluate existing airfoil designs as well as the effects of trips on their performance, and then, based upon this limited information, to design more efficient airfoils. These new airfoils were developed through the use of the Eppler and Somers code[4] and the Drela and Giles ISES[5,6] code. In all, 62 airfoil sections were tested (17 of which were new designs) under various conditions resulting in 130 polars. Measurements were generally taken at Reynolds numbers of 0.6×10^5, 1.0×10^5, 1.5×10^5, 2.0×10^5 and 3.0×10^5.

An important concern when taking measurements at low Reynolds numbers is the high sensitivity of the flow to freestream disturbances. A high freestream turbulence level can cause transition to occur sooner in the free shear layer and, as a result, reduce drag. In this investigation, the freestream turbulence properties have been well documented.

This paper is broken into two major sections. First, the experimental facility and measurement technique are discussed. Second, some important highlights of these experiments are presented, and finally, several conclusions are drawn. Further details and a complete compilation of this data (including the over 130 polars as well as the tabulated data) may be found in Reference 7.

2. Experimental Facility and Measurement Technique

The tests were performed in the Princeton University 3 ft by 4 ft Smoke Tunnel. A sketch of the tunnel is shown in Figure 1. It consists of an inlet and stilling chamber 9 ft high by 12 ft wide containing screens and flow straighteners. The flow straighteners are 3 in square and 12 in long. This section is followed by a 9:1 contraction leading to the test section 4 ft wide by 3 ft high. Downstream of the test section the flow is turned by 90° and exits through a 50 HP fan. The tunnel speed in the test section was variable from 5 ft/s to 70 ft/s.

2.1 Flow Quality

Constant-temperature hot-wire anemometry[8] (using Dantec model 55M01) was used to determine turbulence levels in the freestream. At all conditions, the wire was operated at an overheat of 0.8. Frequency response was optimized using the standard square-wave test in which a square wave voltage is injected at the Wheatstone bridge to simulate an impulse in velocity. The −3 dB point of the response curve was at 33 kHz for chord Reynolds numbers of 1.0×10^5, 2.0×10^5, and 3.0×10^5, and at 25 kHz for a

Reynolds number of 0.6 $\times 10^5$. These frequencies are well above the energy containing frequencies as will be shown shortly.

A common problem when measuring turbulence levels in low-speed facilities is determining the lowest frequency of interest. Usually, the anemometer signal is high-pass filtered, leaving only the fluctuating component. This technique removes possibly important contributions of low frequency turbulent fluctuations to the RMS. In this work, however, no high-pass filter was used. Instead, the DC component (the mean) of the anemometer signal was subtracted off ("bucked off") using an OP-amp summer of an analog computer. The remaining signal was then amplified to fill the range of the A/D converter and sampled at frequencies from 10 Hz to 10 kHz. By sampling over a range of frequencies, high resolution of the spectra was obtained. In each case the low-pass frequency of the filter was set to a frequency somewhat less than the Nyquist sampling frequency to eliminate alaising errors.

An example spectrum is shown in Figure 2 for a Re of 1.0 $\times 10^5$. Power spectral density multiplied by frequency is plotted against the logarithm of frequency. In this way, the area under the curve is directly proportional to $(u'_{rms})^2$. The sampling frequency was 100 Hz, and the hot wire was located along the tunnel horizontal centerline and 3 in below the vertical centerline. In this case the turbulence level, u'_{rms}/U_∞, is 0.36%, including all electrical noise that was present. However, the majority of this energy is contained below 1 Hz. If the signal was high-passed above 1 Hz, this contribution to the turbulence would be lost. Perhaps, frequencies this low could be considered to have quasi-steady effects. The computed turbulence level, excluding turbulence below 1 Hz, drops to 0.034%. It is currently unclear which number is significant (0.36% or 0.034%), and thus, both numbers are presented.

Spectra at other Reynolds numbers are similar to that in Figure 2 (see Reference 7 for the complete set of spectra). The unfiltered turbulence levels at various Reynolds numbers are: 3.0 $\times 10^5$, 0.17%; 2.0 $\times 10^5$, 0.188%; 1.0 $\times 10^5$, 0.358%; and 0.6 $\times 10^5$, 0.563%. If the contribution to these levels from frequencies below 1 Hz is excluded, the turbulence levels then become: 3.0 $\times 10^5$, 0.0077%; 2.0 $\times 10^5$, 0.0174%; 1.0 $\times 10^5$, 0.064%; and 0.6 $\times 10^5$, 0.050%.

2.2 Wind Tunnel Models

In selecting the model size to obtain the desired Reynolds number, several tradeoffs were considered. To achieve a given test Reynolds number, the measured forces increase with decreasing chord. While large forces are desirable, models with small chords are difficult to build accurately. For this work, a model shop was not used; rather, experienced model sailplane enthusiasts were solicited to build the models. Consequently, construction tolerances were on the order of that found on model sailplanes. A 12 in chord was selected as a compromise between the two competing effects. The model span was 33 3/8 in. Construction techniques ranged from an all-balsa construction with ribs and spars, to fiberglass-covered foam.

As a check for model accuracy and for later computations, every model was profiled using a digitizer to obtain the actual airfoil shape. A comparison was then made with the desired airfoil shape to determine the accuracy of the model. A plot of the error normal to the chord line for the E205 is shown in Figure 3. The error is only determined for points that were digitized and the best fit between the airfoils was determined using a least squares method developed by Fraser[9]. The root mean square (RMS) of this error ranged from 0.004 in. (0.03%) to 0.035 in. (0.29%) for the 62 models tested, with a median of 0.01 in. (0.08%). For a more complete discussion concerning the accuracy of the models and the actual model coordinates, see Reference 7.

2.3 Measurement Technique and Instrumentation

Lift was measured directly using a mechanical force balance; whereas, drag was found indirectly using the momentum method[10]. Rather than computing the drag based on just one vertical survey, the wake was surveyed and drag was computed at four spanwise locations and then averaged. Measurement of lift will be discussed first.

A sketch of the apparatus to measure lift is shown in Figure 4. The airfoil model was mounted horizontally in the tunnel between two 0.375 in thick Plexiglas end plates (not shown for clarity) to isolate the model ends from the tunnel side-wall boundary layers and the support hardware shown. One side pivoted and the other was free to move vertically on a precison-ground shaft. Two linear ball bearings spaced 8 in apart provided essentially frictionless movement for a carriage (see Figure 4) which held the airfoil and angle of attack control hardware. Spherical bearings were used to minimize moments transmitted to each linear bearing. A force transducer coupled to the carriage through a pushrod sensed the lift (actually 1/2 of the lift was transmitted to the transducer).

The lift force balance consists of a torque motor, lever arm, angular position sensor (angular transformer), and associated control electronics. This device operates by sensing the angular position and feeding an error signal back to the torque motor to keep the angular position fixed. The motor current is linearly related to the torque and thus the device provided excellent linearity. Nine-point calibrations of the force balance were performed frequently to minimize the effects of drift. The overall system had an accuracy of $\pm 0.25\%$ of full scale or ± 0.002 lb_f, whichever is larger. The term full scale refers to the maximum force experienced over a given run at constant Re.

The drag was measured using the momentum method rather than a mechanical force balance which is both difficult and expensive. In addition, if drag is obtained by mechanical means it includes three-dimensional effects due to the side walls. These effects can be reduced by using a three-piece model with only the central panel connected to the force balance. In this case, the angle of attack of the two tips must be kept equal to that of the central portion and the gaps must be minimized. Althaus[11] investigated the effect of a gap on the drag and found that with a 0.5 mm (0.3%) gap and 250 mm (1.56 chord) span, the drag was increased 12% at an α of 9°.

To compute drag using the momentum method, a pitot probe was surveyed through the wake 1.25 chord lengths downstream of the trailing edge to find the velocity deficit. Based on the application of the two-dimensional momentum and continuity equations to a control volume about the airfoil[10], the drag force can be found as:

$$D = b\rho \int_{-\infty}^{\infty} u(U_\infty - u)dy \ , \tag{1}$$

where the integral is performed perpendicular to the freestream, downstream of the airfoil. The freestream velocity is U_∞, y is in the direction normal to the freestream, u is the x-component of velocity at the downstream location, and b is the airfoil span. A typical survey through the wake took 2 minutes, which effectively yielded a time-averaged drag value for each spanwise station. This method of determining the drag is valid only if the wake survey is made at a region where the static pressure is equal to that in the freestream. Wake static pressures at the survey location were found to be very nearly constant.

For pitot probe misalignments of less than 10° the measured total pressure is essentially independent of flow angle. The drag calculation requires only the streamwise component of the velocity and thus transverse velocity components at the survey location can decrease the measured drag. Drag values were found to remain constant as the survey location was moved farther downstream, indicating that the chosen location was sufficiently far from the trailing edge so that transverse velocity components were negligible.

Drag was actually calculated by measuring the difference between the total pressure upstream of the airfoil and that in the wake. Equation (1) may be rewritten to give:

$$D = 2b \int_{-\infty}^{\infty} \sqrt{P_{d\infty} - \Delta P_0}\left(\sqrt{P_{d\infty}} - \sqrt{P_{d\infty} - \Delta P_0}\right) dy \ , \qquad (2)$$

where $P_{d\infty}$ is the freestream dynamic pressure and ΔP_0 is the difference between the total pressure in the freestream and the total pressure in the wake. This pressure difference is small and difficult to measure, requiring a sensitive transducer. A unit made by MKS was used for this purpose with a full scale of 1 mm of Hg and an accuracy of 0.15% of reading. It was factory calibrated against a standard traceable to the National Bureau of Standards.

The spanwise non-uniformity in the wake is well known[11,1]. The drag variation can be on the order of 50% or more at the lower Reynolds numbers, and thus several stations were averaged to provide a more realistic measure of the airfoil performance. As mentioned previously, four spanwise stations were used, spaced uniformly over the central 1 ft of the airfoil.

Wind tunnel corrections[10] were applied to values of C_l and C_d and were approximately 4% and 2% respectively. Error estimates indicate that the accuracy of the resulting C_l value is $\pm 1\%$ and that of the C_d value is $\pm 2\%$.

The angle of attack of the airfoil was controlled using a gear motor with a worm drive and a sector gear and sensed using an angular transformer, shown in Figure 4, similar to that used in the force balance. The accuracy in determining α was $\pm 0.25°$.

All transducer voltages were recorded using a Scientific Solutions A/D interfaced to an IBM PC. The PC controlled the wake pitot probe and airfoil angle of attack. After manually setting the tunnel speed to achieve the desired Reynolds number, the data taking was completely automated and proceeded as follows. The first angle of attack was set, and the location of the wake was found. Next, the four wake surveys were performed, and the angle of attack was increased. Usually, a polar at a given Reynolds number consisted of between 15 and 20 angles of attack from $-3°$ to $15°$. This process continued until stall and took typically 1.7 hours. The drag was only measured for increasing angles of attack, in which case hysteresis was not examined. This was done for two reasons. First, the amount of run time would have been doubled to 3.4 hours on average. Second, hysteresis is a sign of gross laminar separation — a high drag condition. Interest in this investigation was on examining the characteristics of low drag airfoils in application to RC sailplanes. Hence, high drag conditions were of little interest.

In addition to taking lift and drag data simultaneously, which was relatively slow, in many cases a second run was made in which just lift was measured, allowing the angle of attack to be incremented relatively rapidly. In this mode of operation, α was increased up to a pre-set value and then decreased. Hysteresis loops present in the lift behavior were then observed[7]. Approximately 40 angles of attack were used and this process usually required 5 minutes — much less than the 3.4 hours required to obtain a complete drag polar.

2.4 Comparison with Other Facilities

Measurements in other facilities provide a basis of comparison for the lift and drag obtained in this study. Figures 5 *(a–c)* show drag polars obtained in the Princeton tunnel and those in the Delft University tunnel[12] and in the Model Wind Tunnel at the University of Stuttgart[13] for the E205 at Reynolds numbers of 0.6×10^5, 1.0×10^5, and 2.0×10^5. At 2.0×10^5 all three facilities agree to within 10% over the central region of the lift range. The agreement between Delft and Princeton data at 1.0×10^5 is also favorable. However, at 0.6×10^5, the agreement becomes worse. The discrepancies present are primarily due to differences in (1) flow quality, (2) accuracy of measurements, and (3) methods of measurement. In light of the these important differences, the overall agreement is reasonably good between the three facilities. The remainder of this paper will discuss the results.

3. Discussion of Results

There are several characteristics of an airfoil which can be changed to affect its performance. For example, the contour of the upper and lower surfaces can be altered, boundary layer trips can be added, the trailing edge may be thickened etc. Boundary layer trips are used to reduce the extent of the laminar separation bubble by causing the boundary layer to become turbulent earlier than in the untripped case. However, there is a tradeoff between the drag reduction due to the smaller bubble and the drag increase caused by the presence of the trip as well as the longer region of turbulent flow. Airfoil shape and boundary layer trips as a means of improving performance have been considered and the following sections discuss some of the more important results of this study. For a more complete discussion of all of the results, see Reference 7. First, the airfoil shape is discussed, followed by an illustration of the effects of trips, model accuracy, and finally, the effects of trailing edge thickness.

3.1 Airfoil Design

Three main tools were used to design new airfoils: the Eppler and Somers design code[4], the ISES code written by Drela and Giles[5,6], and the wind tunnel described above. The Eppler and Somers code formulates the design problem in such a way which allows quick and easy manipulation of the airfoil shape. With a minimum number of parameters, almost any desired velocity distribution can be obtained. However, this code does not accurately predict the performance of airfoils in the Reynolds number range considered here. The Eppler and Somers code was used mainly to obtain the inviscid velocity distributions and to give an estimate of the transition point behavior.

The ISES code solves the two-dimensional Euler equations coupled with a momentum integral boundary layer formulation using a global Newton method. Over the Reynolds number range considered in this investigation it predicts airfoil performance more accurately than the Eppler and Somers code. In particular, the agreement with the experiment at Reynolds numbers of 2.0×10^5 and greater is surprisingly good. However, the agreement depends heavily on the choice of the n value used in the e^n transition criterion. While the ISES code provides a relatively good estimate of the performance, wind tunnel results are the ultimate test of an airfoil.

The design approach was to generate an airfoil with the desired inviscid velocity distribution using the Eppler and Somers code and then predict the performance at a Reynolds number of 2.0×10^5 using the ISES code. If the performance was poor, the new airfoil was redesigned and the process repeated. Upon reaching a suitable design through this iteration process, a wind tunnel model was subsequently built and tested. Based upon the wind tunnel results, the new airfoils were further refined and the process repeated.

Before discussing airfoil design, it should be pointed out that for a given chord, wing loading and atmospheric conditions, the relation between C_l and chord Reynolds number is

$$Re \propto \frac{1}{\sqrt{C_l}} .$$

This relation emphasizes the fact that the drag should be minimized for a value of C_l at a given Reynolds number, and this Reynolds number depends on the value of C_l. Thus, the optimum airfoil design is clearly dependent upon the configuration and desired tasks of the aircraft for which it is designed. The designs discussed below were based upon RC sailplane configurations; however, the general conclusions apply to any type of low-Reynolds number aircraft.

A popular RC soaring cross-country airfoil is the Eppler 374. It is commonly used on aircraft intended for high speeds, with little importance placed on the performance at low speeds. The experimentally determined drag polars for this airfoil are shown in Figure 6. This airfoil works well at high speeds because of the small values of the drag coefficient at the higher Reynolds numbers throughout a range of low C_l values. At lower Reynolds numbers, the drag increases dramatically as C_l moves from 0.0 to 0.5, and then decreases again from 0.5 to 0.8. This behavior indicates the formation of a large laminar separation bubble on the upper surface.

The inviscid velocity distribution about the E374 (as predicted by the Eppler and Somers code) for a angle of attack of 5° (with respect to the zero-lift angle of attack) is shown in Figure 7. A "kink" in the upper surface velocity distribution beginning at 40% separates it into two distinct region. Over the first 40% of the upper surface, the velocity changes little, and the majority of recovery takes place over the latter 50% with a relatively strong adverse pressure gradient. At low Reynolds numbers, this strong adverse pressure gradient results in a large laminar separation bubble. To reduce the drag, the strength of the adverse pressure gradient should be reduced. If the same pressure differential is to be recovered, then the recovery region must start farther upstream, as shown by the dashed line in Figure 7. This longer region of smaller adverse pressure gradient is termed a "bubble ramp." Before this point is discussed further, it is important to observe the behavior of the transition point on the upper surface with increasing C_l.

As a result of the kink in the velocity distribution at 40% chord, the transition point moves rapidly forward with C_l as shown in Figure 8. Of course, transition does not occur at a point but rather over some finite distance. In this case the point refers to the location at which transition was predicted to occur by the Eppler and Somers code using a method based on the shape factor. Knowledge of the shape of the transition point curve is helpful when designing with the Eppler and Somers code because it is similar to the distribution of the design parameter, α^*. (The airfoil design is specified with α^* and ν^4, where α^* is the angle of attack at which the velocity is constant on the surface of the airfoil at a given value of ν, where ν is related the distance along the surface of the airfoil.)

The kink in the velocity distribution was removed by deriving a smooth α^* distribution to define a new airfoil, the SD6060 (see Table 1). The resulting transition point behavior and velocity distribution are shown by the dashed lines in Figures 8 and 7, respectively. Removing the kink shifted the transition point farther forward for C_l greater than 0.5. In this case, separation will occur earlier because of the steeper initial gradient, but, with the transition point farther forward, the separation bubble will be shorter and the drag will be lower. A comparison between the experimentally determined drag polars for the E374 and SD6060 is shown in Figure 9. There has been a reduction in drag throughout the central portion of the polars for all Reynolds numbers, thus, the bubble ramp has reduced the length of the separation bubble. Some of this reduction in drag is due to a thinning of the airfoil; however, the E374 is 10.9% thick and the SD6060 is 10.4% thick so this effect is small. In addition, the increase in drag as C_l approaches 1.0 is more gradual in the case of the SD6060, which is consistent with the more smooth forward movement of the transition point.

A further example illustrating the effectiveness of a bubble ramp in the upper surface velocity distributions can be seen by comparing the E205 and the S3021[14]. The E205 is usually used as a "multi-task" airfoil because of its relatively good performance at both high and low lift. This airfoil has an upper-surface velocity distribution which is similar to the E374 in that it also contains a kink. The velocity distribution of the S3021 is essentially the same as that of the E205 except the kink has been replaced with a bubble ramp as in the SD6060. Figure 10 shows a comparison between the drag polars of the E205 and S3021 at several Reynolds numbers. The differences are similar to those noted between the E374 and SD6060. At all Reynolds numbers, the drag of the S3021 is lower than that of the E205 in the central region of the polars. However, at the highest Reynolds number, 3.0×10^5, the E205 has lower drag than the S3021 for $C_l = 0.9$. As discussed earlier, as the speed increases, the lift coefficient decreases so that, for typical low Reynolds number configurations, at a Reynolds number of 3.0×10^5, the lift coefficient would be considerably less than 0.9. Thus, for low Reynolds number aircraft, the S3021 will perform better than the E205. Further examples of performance improvements by the use of bubble ramps are given in Reference 7. These examples illustrate improvements, but the optimum bubble ramp shape and location remains to be determined.

3.2 Boundary Layer Trips

In the last section, altering the shape of the upper surface to reduce the size of the laminar separation bubble was discussed. This section deals with an alternate method, the use of boundary layer turbulators or trips. Throughout this investigation, several different types of trips were used on a variety of airfoils at different locations. Only a few of them will be discussed here.

Two-dimensional trips were made from automobile pin-striping tape placed along the span at a constant chord location. A trip height of 0.17% chord was found to be close to an optimum in terms of its ability to promote transition at low Reynolds numbers and not cause excessive drag at high Reynolds numbers. The typical effect of such a trip was to decrease the drag at Reynolds numbers below about 1.5×10^5 and increase the drag at higher Reynolds numbers. Of course these trends are for the particular class of airfoils tested. Most of the airfoils were on the order of 10% thick and had approximately 2.5% camber. An example of the use of trips is shown in Figure 11 which depicts drag polars of the E374 with and without a boundary layer trip at two Reynolds numbers. In this case, the trip was placed at 20% chord and is 1.0% wide. Note that the drag at 1.5×10^5 is decreased by the presence of the trip while the drag at 3.0×10^5 is increased. Similar behavior was seen in the case of many other airfoils[7]. It was also found that using trips on airfoils which already had low drag at low Reynolds numbers yielded a smaller benefit. For example, boundary layer trips improved the performance of the E374 by a greater amount than for the SD6060.

Zig-zag tape, as used on full-scale sailplanes, was also investigated as a means of tripping the boundary layer. Based upon data on the S4061[7,14], there was no advantage to using the zig-zag tape. While it may be more effective at causing transition[15], the benefit of shortening the laminar separation bubble was balanced by a higher trip drag.

Hemispheres (commonly called upper-surface bumps) with a diameter of 0.15 in (10 viscous units) and a spacing of 0.45 in (100 viscous units at $Re = 2.0 \times 10^5$) were also used to trip the boundary layer. These dimensions correspond to the width and spacing of low-speed streaks observed in transitional and turbulent boundary layers and thus are reasonable choices for the generation of artificial low-speed streaks. Further discussions of this subject can be found in many references, including 16 and 17. The viscous units were calculated based upon the local boundary layer properties at the trip location. Results from the Miley airfoil[18] showed that hemispheres (bumps) were somewhat less effective that the simple two-dimensional trip[7]. Other airfoils exhibited similar trends[7].

In addition to surface protuberances, upper-surface blowing was also used to trip the boundary layer. A hollow model of the HQ2/9 was tested with holes drilled normal to the surface across the span at 50% chord and spaced at approximately 100 viscous units (0.45 in at $Re = 2.0 \times 10^5$). (The designation "HQ"is for H. Quabeck — not to be confused with airfoils designed by K. H. Horstmann and A. Quast.) A ram inlet pressurized the airfoil to feed the blowing holes. Comparisons with a two-dimensional trip at the same location indicated that the blowing was less effective in reducing drag, not considering the drag of the ram inlet.

It is clear from this work that an airfoil which performs poorly at low Reynolds numbers can be improved through the use of trips. This will often degrade the performance at higher Reynolds numbers however. The choice of trip appears to be the simple two-dimensional strip with a height of 0.17% for the airfoils considered here (approximately 10% thick with 2.5% camber). In this study the location for the trip was chosen based upon both intuition and reason. Using the ISES code, an estimate of the separation point at a given angle of attack could be found. The distance upstream of this point necessary for a trip to prevent separation was generally chosen, rather arbitrarily, as 5–10% of the chord. This method appeared to work reasonably well. However, this distance is certainly a function of Reynolds number, trip type and size. The current inability to predict the transitional flows near a trip and near separation indicate that trip location may be best guided by experiment. More importantly, the question: "Should the optimum airfoil for a given task make use of trips?", has not yet been answered.

3.3 Effects of Contour Inaccuracies

Often when constructing an airfoil, an estimate of the accuracy required to achieve the expected performance is necessary. In the current study, two models of a given airfoil were tested in several cases, allowing the effect of accuracy and surface finish to be examined.

Two models of the E205, E214, E374, E387, S3021, S4061, and SD7032 were tested. For a given airfoil, differences between the models were compared to differences in the resulting performance. None of the models had surface discontinuities so that all errors were distributed over some chordwise distance. The correlation between accuracy and performance is difficult to make; however, the following observations were made. First, if the error in surface contour is greater than 0.2% of chord, the performance will be affected. Differences in the trailing edge included angle seemed to have little effect in comparison to other errors. The two E205 models performed the most alike and had upper surfaces which were quite similar from 10% to 60% of chord. Further conclusions are difficult to draw, but it may be assumed that errors in the region where a separation bubble forms will have the largest effect on performance. Additional details can be found in Reference 7.

It is often believed that the accuracy near the leading edge is quite important[13]. Modeling clay was applied to the first 15% of the upper surface of the E374 in an irregular pattern to investigate this claim. The clay was roughly 0.2% chord thick and extended over the entire span of the model. Although the clay was wavy, all edges were carefully blended to the surface. In comparison with the clean E374 data, there was almost no difference at Reynolds numbers of 1.5×10^5 and above. At $Re = 1.0 \times 10^5$ there was a slight decrease in the drag peak near $C_l = 0.5$, which was probably due to an enhancement of transition causing the bubble to become shorter.

A common error when constructing airfoils is a thick trailing edge. The adverse effects of such an error were investigated on the DAE51, the airfoil used on the propeller of the Daedalus human powered aircraft[19] This airfoil was first tested as originally constructed, and then the trailing edge was thickened by 0.8%. The thickness was fared into the original contour over the latter 14% of the upper surface. A comparison between the drag polars of the DAE51 with and without the modification are shown in Figure 12 for three Reynolds numbers. The drag is clearly increased for the higher Reynolds numbers, but at a *Re* of 1.0 $\times 10^5$there is little difference. At this Reynolds number, the flow is probably separated at the trailing edge so the thickness has little effect. This increased drag is in agreement with Althaus[20] who found similar results for thickened trailing edges. Therefore, under most conditions, a thin trailing edge is desirable.

4. Conclusions

Several airfoils were tested in efforts to design new and improved airfoils. Two methods to reduce the drag of an airfoil operating at low Reynolds numbers were examined. First, a long region of roughly constant adverse pressure gradient on the upper surface (termed a bubble ramp) achieved a lower drag than the more conventional laminar flow-type velocity distribution in which the pressure remains approximately constant initially, and then more rapidly recovers. Second, trips were used to reduce drag by shortening the laminar separation bubble. Several different methods of tripping the boundary layer were investigated. A simple two-dimensional trip performed as well or better than zig-zag tape used on full-size sailplanes, hemispheres (bumps) attached to the surface, and blowing normal to the surface. Boundary layer trips were less effective at improving airfoils which normally had low drag. The question still remains whether the optimum airfoil for a given task requires boundary layer trips or simply an efficient shape, possibly similar to the bubble ramp.

During the course of this work, the effects of geometrical errors were also observed. As a general guideline, accuracy should be held to within 0.2% of chord, with particular attention paid to the region from 20% to 60% on the upper surface where laminar separation bubbles will form. Accuracy at the leading edge is not as critical. Finally, trailing edges should be thin for low drag.

Acknowledgments

This was a privately funded program that could not have been performed without the encouragement, support, and guidance of several individuals. The authors gratefully acknowledge the support and expertise of David Fraser, President, Fraser Volpe Corp., without whom the scale of this work would have been an order of magnitude smaller. Special thanks go to Prof. A. J. Smits at the Princeton University Gas Dynamics Laboratory. Also Prof. Lam and Prof. Curtiss at Princeton are thanked for their support. Thanks are expressed to Prof. M. D. Maughmer at Penn State for his helpful discussions during the course of the work. All of the models tested during this program were made by RC sailplane enthusiasts, and their combined efforts are greatly appreciated. Finally, MKS and Scientific Solutions, Inc. are gratefully acknowledged for their donations of instrumentation.

References

[1] van Ingen, J. L. and Boermans, L. M. M., "Aerodynamics at Low Reynolds Numbers: A Review of Theoretical and Experimental Research at Delft University of Technology," Aerodynamics at Low Reynolds Numbers $10^4 < Re < 10^6$ International Conference, London, October 1986.

[2] Dobbinga, E., van Ingen, J. L., and Kooi, J. W., "Some Research on Two-Dimensional Laminar Separation Bubbles," AGARD CP-102, Paper 2, Lisbon, 1972.

[3] Eppler, R. and Somers, D. M., "Airfoil Design for Reynolds Numbers Between 50,000 and 500,000," Proceedings of the Conference on Low Reynolds Number Airfoil Aerodynamics, Notre Dame, Indiana, June 1986.

[4] Eppler, R. and Somers, D. M., "A Computer Program for the Design and Analysis of Low-Speed Airfoils, Including Transition," NASA TM 80210, August 1980.

[5] Drela, M. and Giles, M. B., "Viscous-Inviscid Analysis of Transonic and Low-Reynolds Number Airfoils," AIAA J., Vol. 25, No. 9, September 1987.

[6] Drela, M. and Giles, M. B., "ISES: A Two-Dimensional Viscous Aerodynamic Design and Analysis Code," AIAA Paper 87-0424, January 1987.

[7] Selig, M. S., Donovan, J. F., and Fraser, D., Airfoils at Low Speeds, Soartech #8, published by H. A. Stokely, 1504 North Horseshoe Circle, Virginia Beach, VA, 23451, U.S.A., July 1989.

[8] Perry, A. E., Hot-Wire Anemometry, Oxford University Press, 1982.

[9] Fraser, D.B., private communications, 1989.

[10] Rae, H. and Pope, A., Low-Speed Wind Tunnel Testing, John Wiley & Sons, second ed., 1984.

[11] Althaus, D., "Recent Wind Tunnel Experiments at Low Reynolds Numbers," Aerodynamics at Low Reynolds Numbers $10^4 < Re < 10^6$ International Conference, London, October 1986.

[12] Volkers, D. F., "Preliminary Results of Wind Tunnel Measurements on Some Airfoil Sections at Reynolds Numbers between 0.6×10^5 and 5.0×10^5," Memo M-276, Delft University of Technology, The Netherlands, 1977.

[13] Althaus, D., Profilpolaren für den Modellflug, Necker-Verlag, Villengen-Schennigen, University of Stuttgart, FRG, 1980.

[14] Selig, M. S., The Design of Airfoils at Low Reynolds Numbers, Soartech #3, published by H. A. Stokely, 1504 North Horseshoe Circle, Virginia Beach, VA, 23451, U.S.A., July 1984. Also AIAA-85-0074.

[15] Hegarty, J. L. and Hama, F. R., "Further Investigation on the Trianglar-Pattern Stimulator," University of Maryland, TN-107, AFOSR TN-57-616, 1957.

[16] Hama, F. R. "An Efficient Tripping Device," Journal of the Aeronautical Sciences, (Readers' Forum), Vol. 24, No. 3, pp 236-237.

[17] Schlichting, H., Boundary-Layer Theory, McGraw-Hill, 1979.

[18] Miley, S. J., "On the Design of Airfoils for Low Reynolds Numbers," AIAA Paper 74-1017, September 1974.

[19] Drela, M., "Low-Reynolds-Number Airfoil Design for the M.I.T. Daedalus Prototype: A Case Study," AIAA J., Vol. 25, No. 8, August 1988.

[20] Althaus, D., "Effects on the Polar Due to Changes or Disturbances to the Contour of the Wing Profile," Technical Soaring, Vol. 10, No. 1, January 1986.

Table 1. - SD6060 Airfoil Coordinates

1	1.00000	0.00000	17	0.43386	0.06866	33	0.00495	—.00647	49	0.62223	—.02527
2	0.99661	0.00023	18	0.38566	0.07003	34	0.01525	—.01148	50	0.67254	—.02231
3	0.98660	0.00108	19	0.33862	0.07020	35	0.03068	—.01612	51	0.72116	—.01906
4	0.97033	0.00283	20	0.29316	0.06922	36	0.05114	—.02025	52	0.76761	—.01568
5	0.94829	0.00559	21	0.24976	0.06715	37	0.07648	—.02381	53	0.81133	—.01236
6	0.92100	0.00941	22	0.20883	0.06402	38	0.10645	—.02678	54	0.85176	—.00922
7	0.88905	0.01419	23	0.17076	0.05988	39	0.14078	—.02919	55	0.88838	—.00638
8	0.85301	0.01977	24	0.13589	0.05480	40	0.17909	—.03105	56	0.92070	—.00399
9	0.81346	0.02595	25	0.10456	0.04887	41	0.22096	—.03238	57	0.94818	—.00214
10	0.77096	0.03248	26	0.07700	0.04218	42	0.26592	—.03321	58	0.97032	—.00090
11	0.72602	0.03912	27	0.05344	0.03486	43	0.31347	—.03354	59	0.98661	—.00024
12	0.67917	0.04563	28	0.03399	0.02710	44	0.36306	—.03338	60	0.99662	—.00002
13	0.63091	0.05177	29	0.01879	0.01913	45	0.41413	—.03273	61	1.00001	0.00000
14	0.58174	0.05738	30	0.00790	0.01132	46	0.46614	—.03159			
15	0.53222	0.06225	31	0.00148	0.00411	47	0.51852	—.02995			
16	0.48283	0.06606	32	0.00025	—.00159	48	0.57073	—.02784			

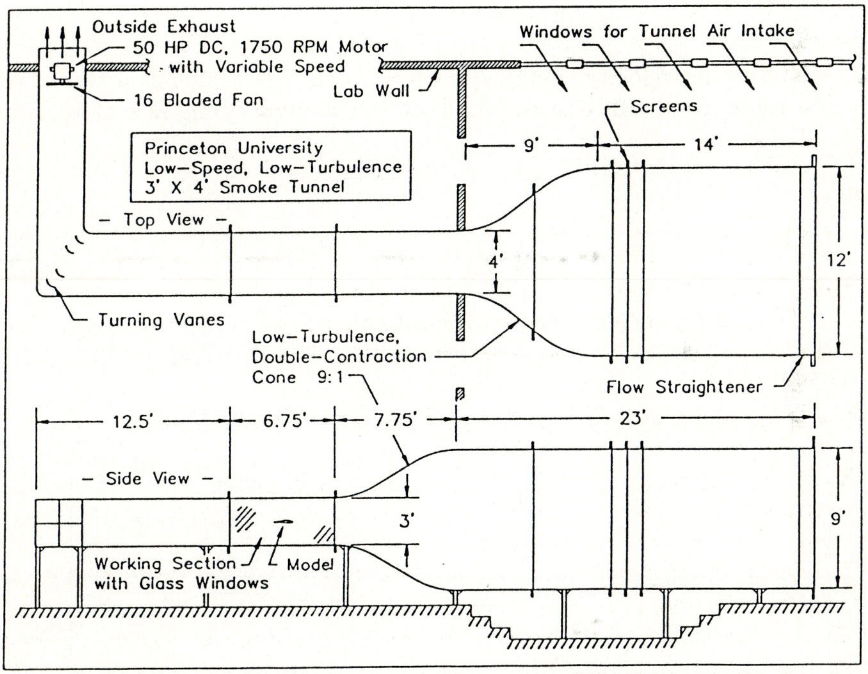

Figure 1. Diagram of the Princeton University low speed wind tunnel (not to scale).

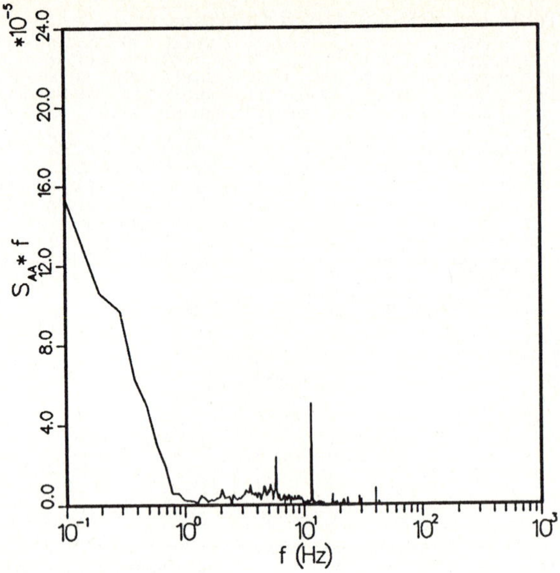

Figure 2. Fluctuating velocity energy spectra from the freestream at a tunnel velocity corresponding to $Re = 1.0 \times 10^5$ with a sampling rate of 100Hz.

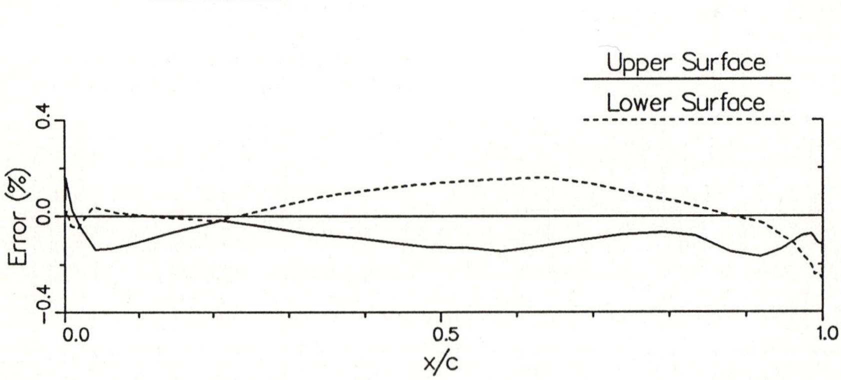

Figure 3. Comparison of the digitized E205 model surface to the desired shape, with the error plotted below as a percentage of the chord.

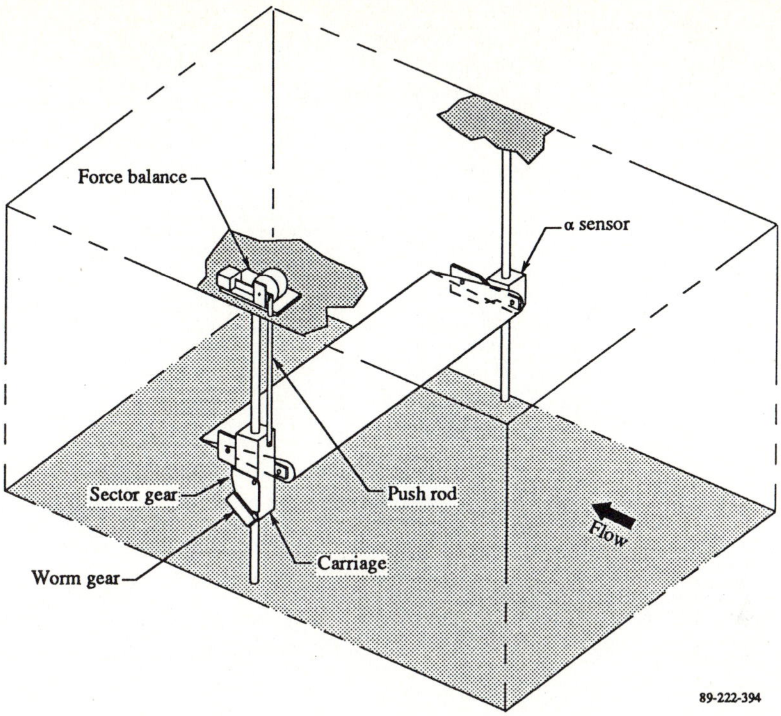

Figure 4. Test rig indicating model orientation and lift measurement method. (Plexiglass end plates are not shown for clarity.)

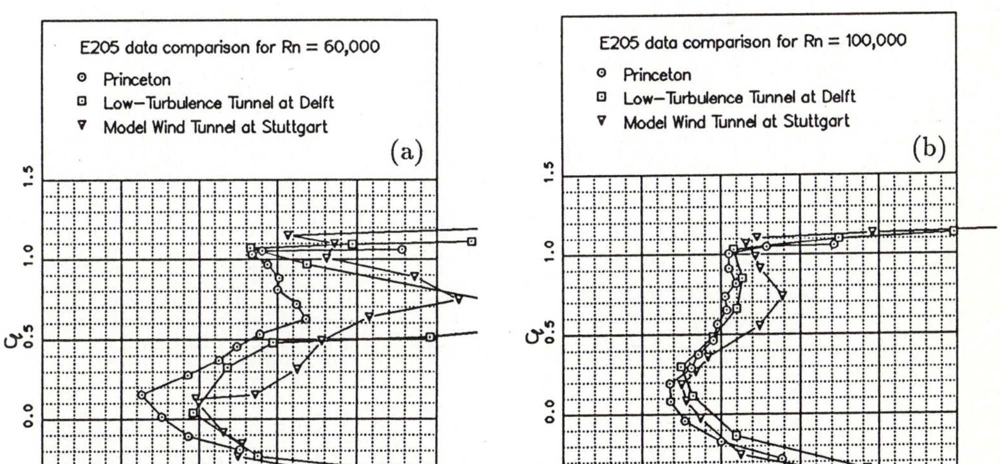

Figure 5. Comparison of the E205 data with that of Delft[12] and Stuttgart[13] at Reynolds numbers of (a) 0.6 $\times 10^5$, (b) 1.0 $\times 10^5$, and (c) 2.0 $\times 10^5$.

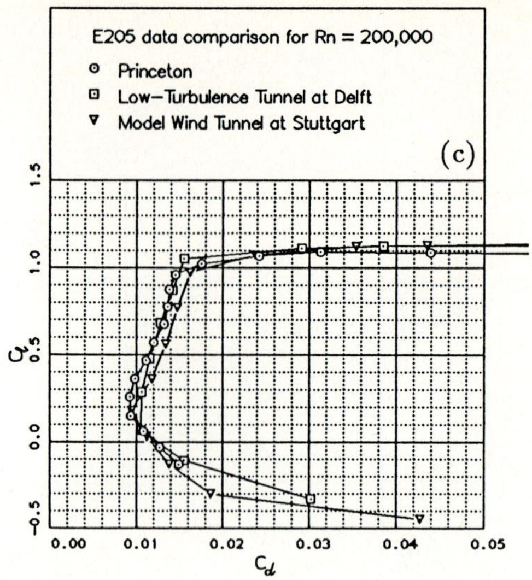

Figure 5. Continued.

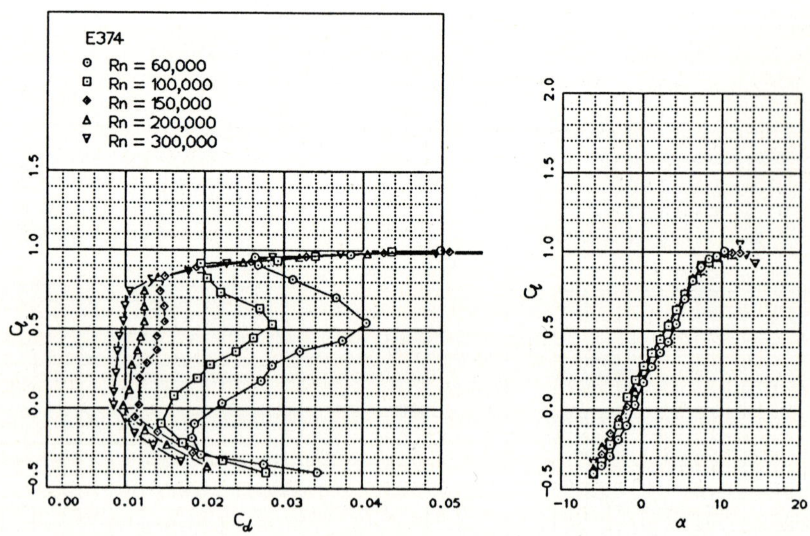

Figure 6. Drag polars for the E374 at several Reynolds numbers.

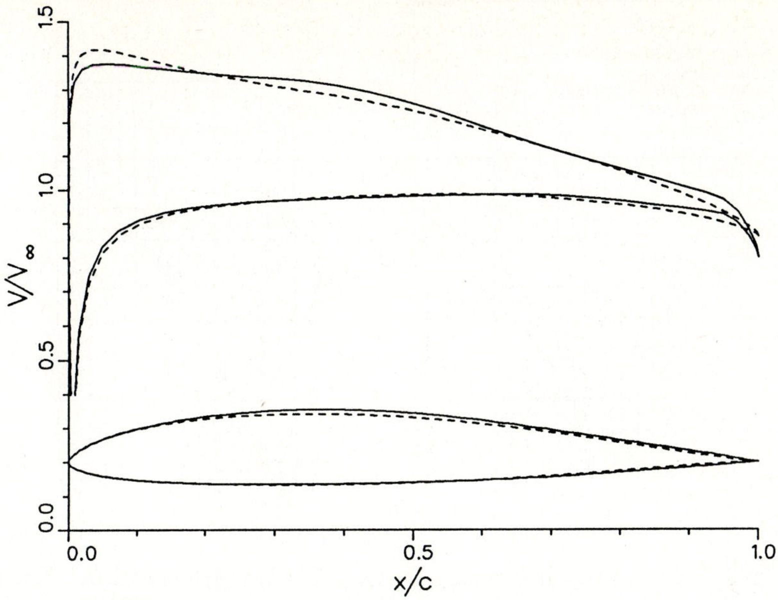

Figure 7. Inviscid velocity distributions for the E374 (solid) and SD6060 (dashed) at $\alpha_{w.r.t.0L} = 5°$ ($C_l = 0.55$).

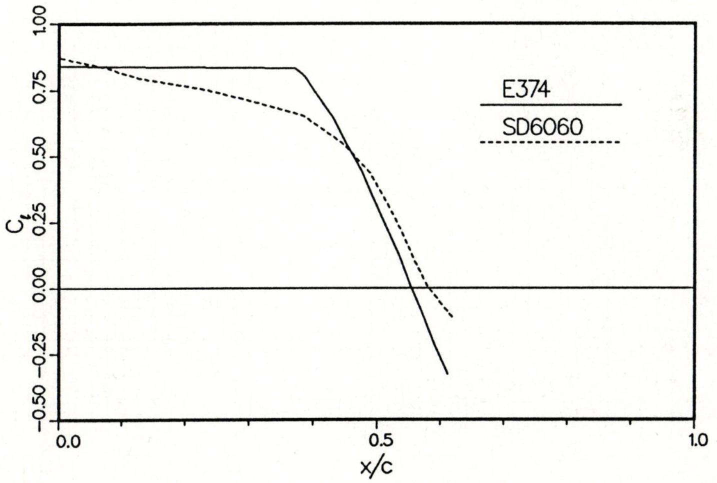

Figure 8. Upper surface transition point location predicted by the Eppler code on the E374 and SD6060 as a function of lift coefficient.

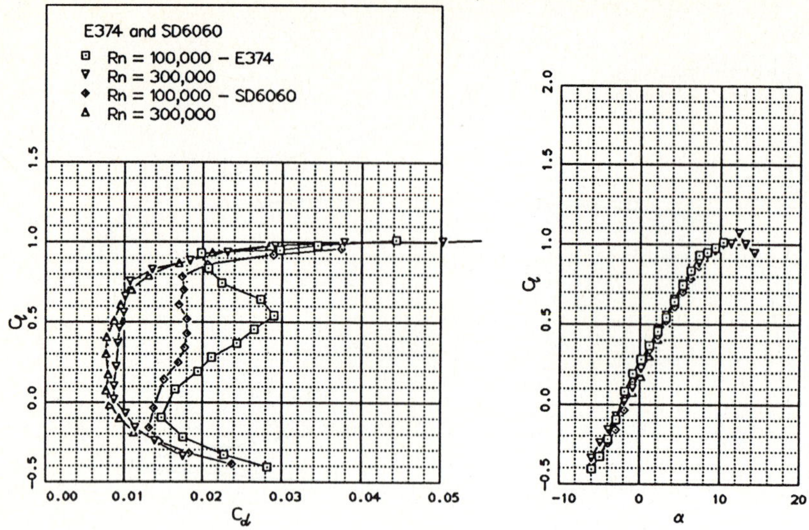

Figure 9. Comparison of the drag polars for the E374 and SD6060 at two Reynolds numbers.

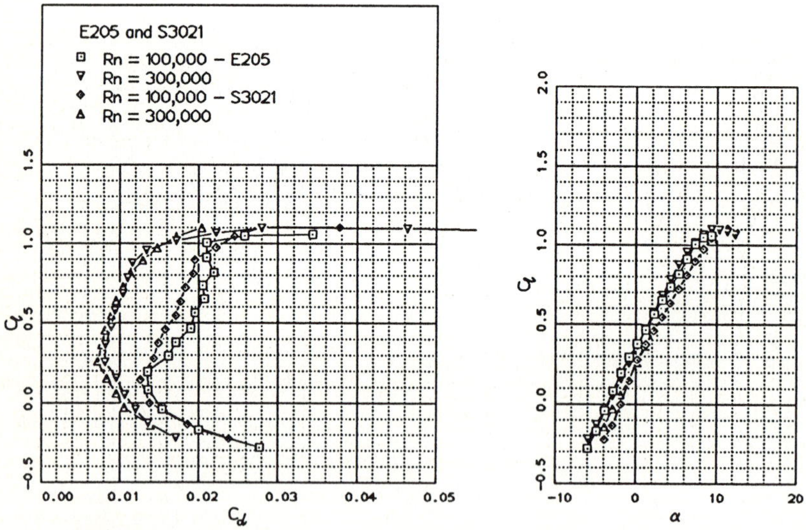

Figure 10. Comparison of the drag polars for the E205 and S3021 at two Reynolds numbers.

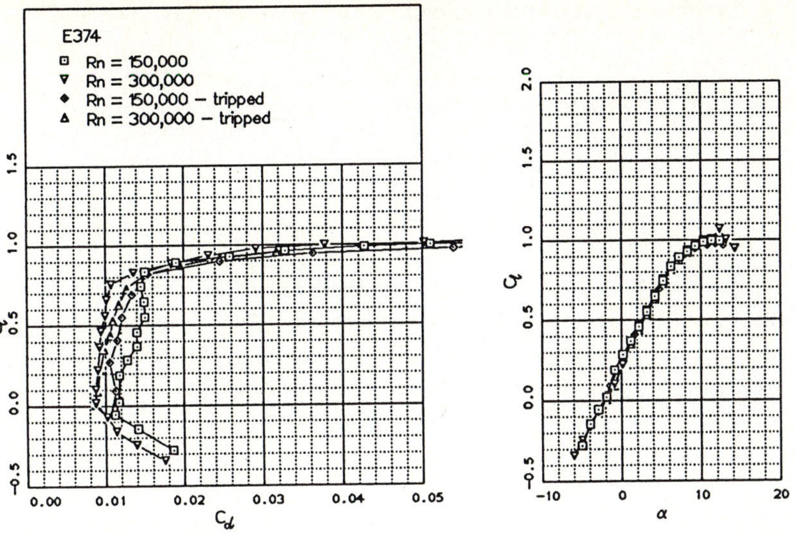

Figure 11. Comparison of the drag polars for the tripped and untripped E374. (simple two-dimensional trip strip: 0.17% high and 1.% wide)

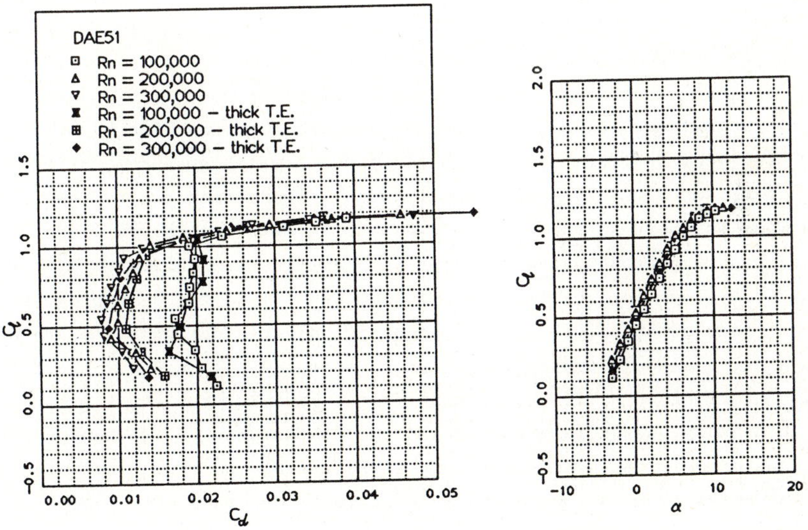

Figure 12. Drag polars of the DAE51 with and without a thickened trailing edge.

STUDY OF LOW-REYNOLDS NUMBER SEPARATED FLOW PAST THE WORTMANN FX 63-137 AIRFOIL

K.N. Ghia, G. Osswald and U. Ghia*
Department of Aerospace Engineering and Engineering Mechanics
*Department of Mechanical and Industrial Engineering
University of Cincinnati, Cincinnati, Ohio

ABSTRACT

Flow past a Wortmann FX 63-137 airfoil is analyzed using the unsteady Navier-Stokes (NS) analysis developed earlier by the authors. The analysis is formulated using conservation form of the governing NS equations in terms of stream function, ψ, and vorticity, ω, in generalized coordinates. A Schwarz-Christoffel mapping technique is developed to provide surface-oriented coordinates for the Wortmann FX 63-137 airfoil. Suitable 1-D clustering transformations are used, not only to obtain a bounded computational domain, but, also to resolve the dominant length scales of the problem. Direct numerical simulation (DNS) methodology is used to obtain flow results at Re = 1,000 with α_f varying from -5° to 10°; simulations are also performed at α_f = 0° with Re = 10,000 and 100,000 without any turbulence model.

1. INTRODUCTION

Significant interest presently exists in the design and performance of airfoils operating at low Reynolds (Re) number. This interest has stemmed from the desire to obtain better performance for civilian as well as military systems requiring efficient airfoil sections in the chord-based Re range of about 100,000 to 1,000,000. These systems include jet-engine compressor and turbine blades, remotely-piloted-vehicles (RPV's) at high altitudes, sail planes, ultra-light human-powered aircraft, and mini-RPV's at low altitudes. The design procedures for airfoil sections at chord-Re above 500,000 appear to be reasonably well developed. Eppler (1985) has suggested that some major difficulties are encountered, however, in single-airfoil flows for chord-Re below ≃ 500,000. To date, some progress has been made, but considerable work remains ahead if reliable design tools are desired. The problem areas include boundary layers with separated regions near the leading edge (LE) and trailing edge (TE), flow transition which is very sensitive to parameters such as Re, pressure gradient and the disturbance environment, hysteresis in the lift and drag forces, etc. For some airfoil-flow configurations, laminar separation may occur without subsequent reattachment, due to a strong adverse pressure gradient after the suction peak. In this case, the airfoil performance will be severely degraded. Eppler (1985) has pointed out that the design of airfoils for Re between 50,000 to 500,000 is governed by the problem of laminar separation which has a predominant effect on the entire airfoil flow field. The characteristics of the laminar separation bubble have been discussed lucidly by Horton (1968) and Mueller and his coworkers (see, e.g., Mueller (1987) and O'Meara and Mueller (1987)). For Re < 500,000, it is much easier to maintain a laminar boundary layer than to achieve a far enough forward transition. It is well recognized that transition is not concentrated at a point but occurs, rather, over some distance. The overall transition process involves amplification of certain Tollmien-Schlichting waves, followed by secondary instabilities, then additional nonlinear processes, and eventually the transition to a completely irregular turbulent motion begins in layers with very high shear velocities as in free-shear layers. It should be noted that the flow in such free-shear layers is much more unstable than the flow in boundary layers.

During the 1980's, the present investigators have extensively studied the flow past two-dimensional airfoils. Their studies were motivated primarily by the need to provide increased lift in the near- and post-stall regimes. K. Ghia, Osswald and U. Ghia (1985) analyzed the structure of massively separated flow past a Joukowski airfoil at large incidence, using a C-grid topology generated using clustered-conformal mapping techniques, and direct simulation methodology. In this approach, attempts were made to resolve as many of the dominant scales as possible without using a turbulence model. The upper limit on Re in the computations was based directly on the availability of supercomputer resources and, in particular, the high-speed memory. Results were obtained for chord-based Re up to 10^4 with the angle of attack α_f being varied up to 15°. Subsequently, Osswald, K. Ghia and U. Ghia (1985) significantly improved their analysis by employing the method of analytical continuation to treat the metric discontinuity at the branch cut, permitting an asymmetric C-grid for which the region of high grid-resolution more closely followed the wake centerline. Further, K. Ghia et al. (1986) studied the massively separated flow that occurs for α_f up to 30° and observed period-doubled limit-cycle solutions. At the still higher incidence of $\alpha_f = 53°$, the asymptotic solution showed bifurcation, with the presence of an incommensurate frequency; see, also, Ref. 6. Osswald et al. (1986) successfully examined buffetting stall for a cambered Joukowski airfoil. The authors have also examined the Schwarz-Christoffel mapping technique of Davis for arbitrary airfoils, applied it to the NACA 0012 airfoil in Ref. 5 and, in Ref. 12, provided the effect of trailing-edge geometry on the overall flow results for NACA 0012 airfoil.

The work reported here is part of an on-going study in which DNS methodology is developed for steady as well as unsteady flows past isolated and multi-element airfoils to characterize steady and unsteady separated flow processes and to predict incipient transition for low-Re flows. The specific objective for this study is to simulate steady as well as unsteady flows past the Wortmann FX 63-137 airfoil using DNS methodology for Re up to 100,000.

2. APPROACH AND NUMERICAL SIMULATION

The conservation form of the unsteady NS equations, in terms of stream function and vorticity in generalized curvilinear coordinates, is used. The boundary condition at the far-field boundary is taken to be the corresponding inviscid flow and is implemented at true infinity. On the airfoil surface, the no-slip condition is enforced, with continuity of the flow variables being maintained at the branch cut through the method of analytical continuation (Ref. 10). The initial condition also corresponds to the inviscid flow solution; hence, the problem formulated is that of an impulsively started airfoil. The Reynolds number used in this study is defined as $Re = U_{\infty}c/\nu$ where c is the airfoil chord. A clustered conformal grid with C-grid topology is generated; the clustering is controlled by appropriate 1-D stretching transformations. An attempt is made to resolve the dominant scales of this flow problem. Typical clustered conformal grids used for the Wortman FX-137 airfoil consist of (301,61) and (441,61) points, the smaller grid being used for flows with Re = 1,000, while the larger grid is used for Re = 10,000 and above. An alternating-direction implicit-block Gaussian elimination (ADI-BGE) method is used to obtain the flow solution. This is a fully implicit time-marching method with second-order spatial and first-order temporal accuracy. The DNS method is applied to the Wortmann FX 63-137 airfoil, which has been generated as suggested by Mueller (1989) from the corresponding 6.3% cambered, 13.7% thick Joukowski airfoil.

3. RESULTS AND DISCUSSION

A total of six flow configurations are investigated. The flow possesses two critical parameters, namely, Re and α. Four configurations are simulated at Re = 1000, with α_f varying from -5°, 0°, 7° to 10°. The two remaining configurations simulated correspond to $\alpha_f = 0°$, with Re varying to 10,000 and 100,000. In the past, the authors have validated their unsteady NS analysis by comparing the predicted results with existing experimental data sets and the agreements have been good for flow past a circular cylinder and an elliptic cylinder at angle of attack α_f.

RESULTS FOR Re = 1000

Steady-State Solutions: Simulation results for flow past the Wortman FX 63-137 airfoil at Re = 1000 and $\alpha_f = -5°$ are depicted in Fig. 2. At a very early time, the flow approaches steady state; however, the calculations were continued further, so as to ascertain that no numerical difficulties are encountered later. Results are plotted here for t=30. The plots are generated using an IRIS 4D/70 GT superworkstation with color graphics. This permits color-coding the counter-clockwise and clockwise spinning fluid. Unfortunately, the final published figures will be in black and white and, as such, the reader will not be able to visualize the additional clarity possible in a color plot. Figure 2a shows the ψ-contours which indicate that the flow has totally detached from the pressure surface at about 0.27c. The ω-contours in Fig. 2b show counter clockwise rotating fluid in the boundary layer on the pressure surface, with $\omega_{max} = 242.8$ in the shear layer. The velocity vectors in Fig. 2c quantify the reversed flow region with a rather small backflow velocity; the figure also shows the development of the near-wake flow. For $\alpha_f = 0°$, the results are similar to those for $\alpha_f = -5°$; the steady-state results obtained at t = 30 are shown in Fig. 3. Separation now occurs as a closed bubble on the suction surface over the last 25% of the chord, as seen in Fig. 3a. Also, $\omega_{max} = 198$ in the shear layer and the velocity vectors show that the reversed flow velocity is very small.

Limit-Cycle Solutions: At $\alpha_f = 7°$, the flow field asymptotes to a limit-cycle solution. Due to limitation of space, only the instantaneous contours of ψ and ω are plotted at t = 32.5 in Fig. 4a-b. To determine the complete state of this flow field, a number of such contours need to be shown at critical event-points over a given cycle. Figure 4a shows a closed separation bubble over approximately 51% of the chord. The complete evolution and interactions of the vortices from the TE and LE can be explained from the detailed simulation results. Solar animation is available on the IRIS workstation. At t = 32.5, there is some clockwise spinning fluid close to the surface along the entire length of the separation bubble; also, ω_{max}=-179.5 in the shear layer on the pressure surface. The phase-space portrait constructed using the embedding technique is shown in Fig. 4f and shows a simple 1-D attractor. The power spectrum density of C_L in Fig. 4e shows the fundamental shedding frequency of 0.968, with 3 higher harmonics participating. The final set of results for Re = 1000 is obtained with $\alpha_f = 10°$ and only limited results are presented here. The instantaneous contours of ψ and ω, as well as the velocity vectors, at t = 50 are shown in Figs. 5a, c and e. The flow separates on the suction surface and the separation extends over approximately 62% of the chord from a point aft of the shoulder to the TE. The shear layer is more intense on the pressure side with $\omega_{max} = -228.45$. The velocity vectors now show that the magnitude of the reverse flow velocity is nearly 25% of the free-stream velocity. Contours of time-averaged ψ and ω are depicted in Figs. 5b and d, whereas the velocity vectors are shown in Fig. 5f. The development of the shear layer and its dissipation are vividly seen in Fig. 5d. The Reynolds stresses $\overline{u'u'}$, $\overline{u'v'}$ and $\overline{v'v'}$ are given in Fig. 5 g,h,i, with Fig. 5i establishing the "foot print" of the body geometry in the wake. The time-history of the lift coefficient is plotted in Fig. 5j and shows that the flow asymptotes to a limit-cycle solution. The attractor geometry is presented in

Fig. 5k, using the three independent observables C_L, C_D and C_M and corresponds to a simple 1-D attractor. The attractor is further characterized by presenting the power spectrum density of C_L, showing the fundamental single shedding frequency of 0.8988, with 4 higher harmonics participating.

Results for α_f=0

To test the analysis at higher Re, simulations are performed for two configurations at $\alpha_f = 0$ with Re=10,000 and 100,000. For these cases, the finer grid with (441,61) points is used. Even with proper clustering transformations, this grid is only marginal for the latter case with Re=100,000. Hence, only a demonstration run is carried out. For the case with Re=10,000, the instantaneous stream-function and vorticity contours are given in Figs. 6a and c, with the velocity vectors being depicted in Fig. 6e. The flow separates and a separation bubble extends over 42% of the chord near the TE, thereby implying that, at fixed α_f, Re increases the extent of the bubble, when compared with Fig. 2d. There is intense interaction between the vortices formed from the TE and LE, as shown in Fig. 6c. The corresponding time-averaged contours of ψ and ω are depicted in Figs. 6b and d, whereas the velocity vectors are shown in Fig. 6f. Figure 6h shows the Reynolds stresses $\overline{v'v'}$, obtained from first principles, whereas the time-history of the lift coefficient showing the limit-cycle nature of the solution is presented in Fig. 6g. The resulting 1-D attractor geometry is not presented here, due to space limitation. It is also apparent that the scales are not fully resolved. This is more so in the case of results for Re=100,000 plotted in Fig. 7a-d. These results are given here to show that the DNS method is capable of simulating flows at higher Re. The authors are fully aware of the fact that the grid clustering used here is not optimal; additional work is required before any detailed results for these configurations will be obtained and verified.

4. CONCLUSION

Flow past the Wortmann FX 63-137 airfoil is analyzed using an unsteady NS analysis and a DNS method. The analysis has provided detailed flow results for flow configurations at Re = 1000 with $\alpha_f = -5°, 0°, 7°$ and 10°. On the other hand, simulations for flow configurations at $\alpha_f = 0°$ with Re = 10,000 and 100,000 were carried out, for demonstration of the capability of the DNS method. This is part of an on-going work and, as such, optimal clustering will be sought for Re $\approx$ 100,000. Thereafter, thorough comparison of the present results is planned with those of Mueller and his associates, as well as with other existing data. After completing this phase, detailed characterization of steady and unsteady separation phenomena for Re $\leq$ 500,000 will be performed.

ACKNOWLEDGEMENT

This work was supported, in part, by AFOSR Grant No. 87-0074, and supercomputer time was provided by the Ohio Supercomputer Center. The authors are thankful to Keith Blodgett, Thomas Rohling and Tim Kuhn for their help in generating the numerical results and in preparation of the figures.

REFERENCES

1. Eppler, R. and Somers, D.M. (1985), "Airfoil Design for Reynolds Numbers Between 50,000 and 500,000," Proceedings of the Conference on Low Reynolds Number Airfoil Aerodynamics, Edited by Thomas J. Mueller, UNDAS-CP-77B123.

2. Horton, H.P., (1968), "Laminar Separation Bubbles in Two and Three Dimensional Incompressible Flow," Ph.D. Thesis, University of London, London.

3. Ghia, K.N., Osswald, G.A. and Ghia, U., (1985), "Analysis of Two-Dimensional Incompressible Flow Past Airfoils Using Unsteady Navier-Stokes Equations," Chapter in Numerical and Physical Aspects of Aerodynamic Flows: Vol. III, Editor: T. Cebeci, Springer-Velag, New York, pp. 318-340.

4. Ghia, K.N., Osswald, G.A. and Ghia, U., (1986), "Simulation of Self-Induced Unsteady Motion in the Near Wake of a Joukowski Airfoil," Lecture Notes in Engineering, Editors: K. Kuwahara et al., Springer-Verlag, New York, Vol. 24, pp. 118-132.

5. Ghia, K.N., Osswald, G.A. and ghia, U., (1989), "Analysis of High-Incidence Separated Flow Past Airfoils," Fourth Symposium on Numerical and Physical Aspects of Aerodynamic Flows, California State University, Long Beach, California.

6. Ghia, K.N., Osswald, G.A. and Ghia, U., (1989), "Chaotic Behavior in High-Incidence Low-Reynolds Number Flows," in Chaotic Dynamics in Fluid Mechanics, FED Vol. 70, ASME Publication.

7. Mueller, T.J., (1987), "The Visualization of Low Speed Separated and Wake Flows," AIAA Paper 87-2422.

8. Mueller, T.J., (1989), Private Communication.

9. O'Meara, M.M. and Mueller, T.J., (1987), "Laminar Separation Bubble Characteristics of an Airfoil at Low Reynolds Numbers," AIAA Journal, Vol. 25, No. 8, pp. 1033-1041.

10. Osswald, G.A., Ghia, K.N. and Ghia, U., (1985), "An Implicit Time-Marching Method for Studying Unsteady Flow with Massive Separation," AIAA CP-834, pp. 25-37.

11 Osswald, G.A., Ghia, K.N. and Ghia, U., (1986), "Simulation of Buffeting Stall for a Cambered Joukowski Airfoil Using Fully Implicit Methods," Lecture Notes in Physics, Editors: F.G. Zhuang and Y.L. Zhu, Springer-Verlag, Vol. 264, 1986, pp. 516-522.

12. Osswald, G.A., Ghia, K.N. and Ghia, U., (1989), "Analysis of Potential and Viscous Flows Past General Two-Dimensional Bodies with Arbitrary Trailing-Edge Geometries," AIAA CP 895, pp. 668-677.

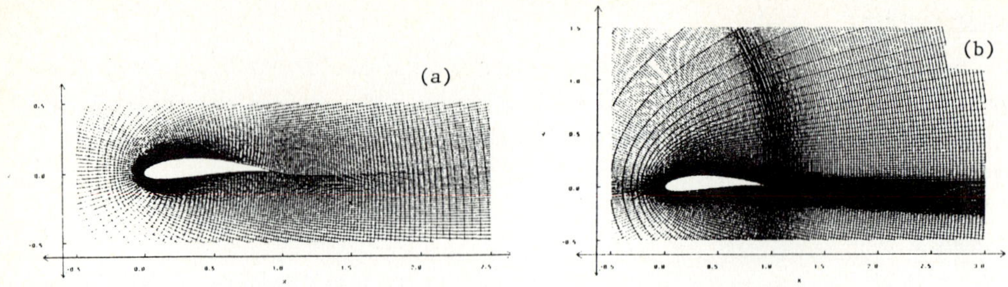

Fig. 1. Typical grid Distributions for the Wortmann FX 63-137 Airfoil.
(a) Re = 1,000, $\alpha_f = 10°$ with (301,61) Points
(b) Re = 10,000, $\alpha_f = 0°$ with (441,61) Points.

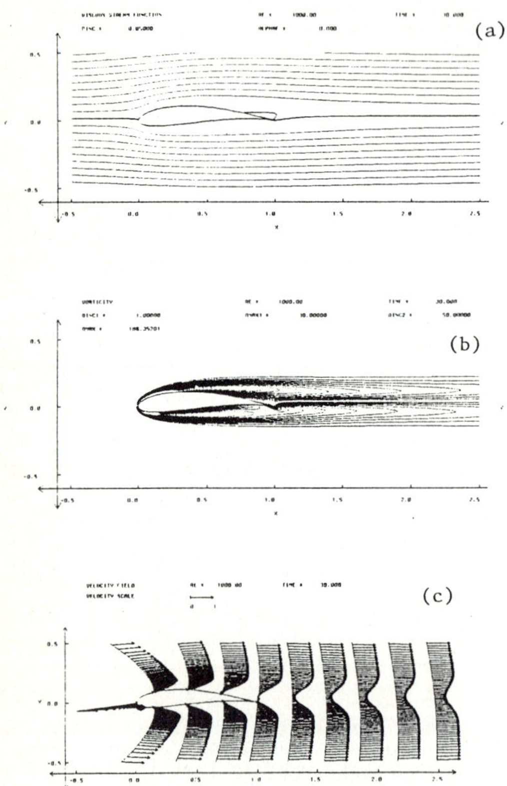

Fig. 2. Steady Flow Past the Wortmann FX 63-137 Airfoil at Re = 1,000, $\alpha_f = -5°$, t = 30.
(a) Stream-Function Contours;
(b) Vorticity Contours;
(c) Velocity Vectors.

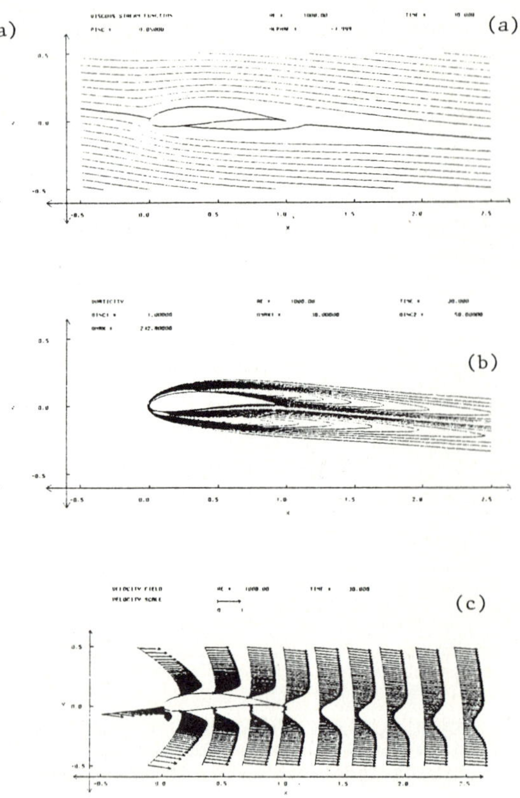

Fig. 3. Steady Flow Past the Wortmann FX 63-137 Airfoil at Re = 1,000, $\alpha_f = 0°$, t = 20.
(a) Stream-Function Contours;
(b) Vorticity Contours;
(c) Velocity Vectors.

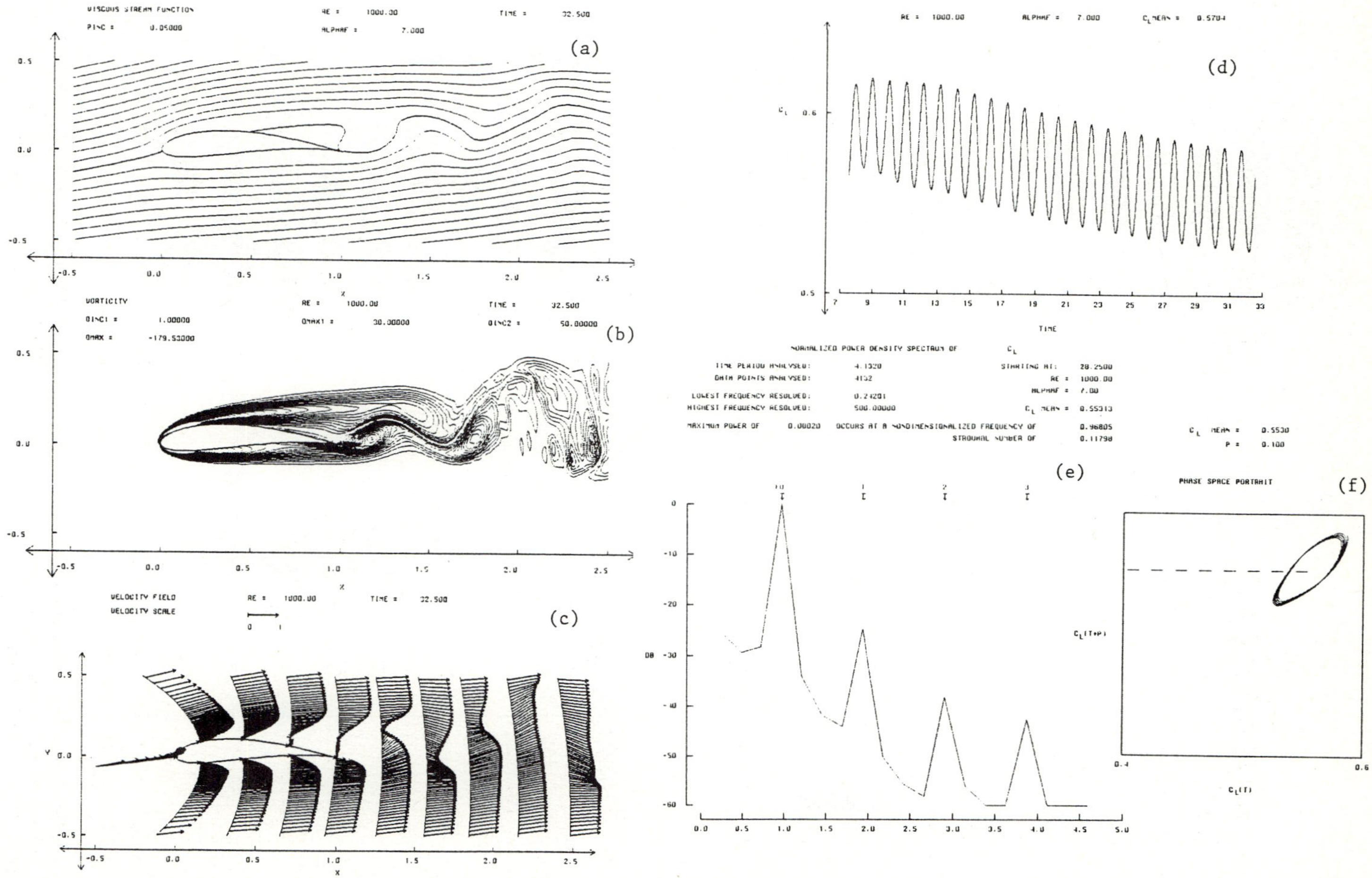

Fig. 4. Limit-Cycle Solution for Flow Past the Wortmann FX 63-137 Airfoil at Re = 1000, $\alpha_f = 7°$. (a) Instantaneous Stream-Function Contours; (b) Instantaneous Vorticity Contours; (c) Velocity Vectors; (d) Time-History of C_L; (e) Power Spectrum of C_L; (f) Phase-Space Portrait.

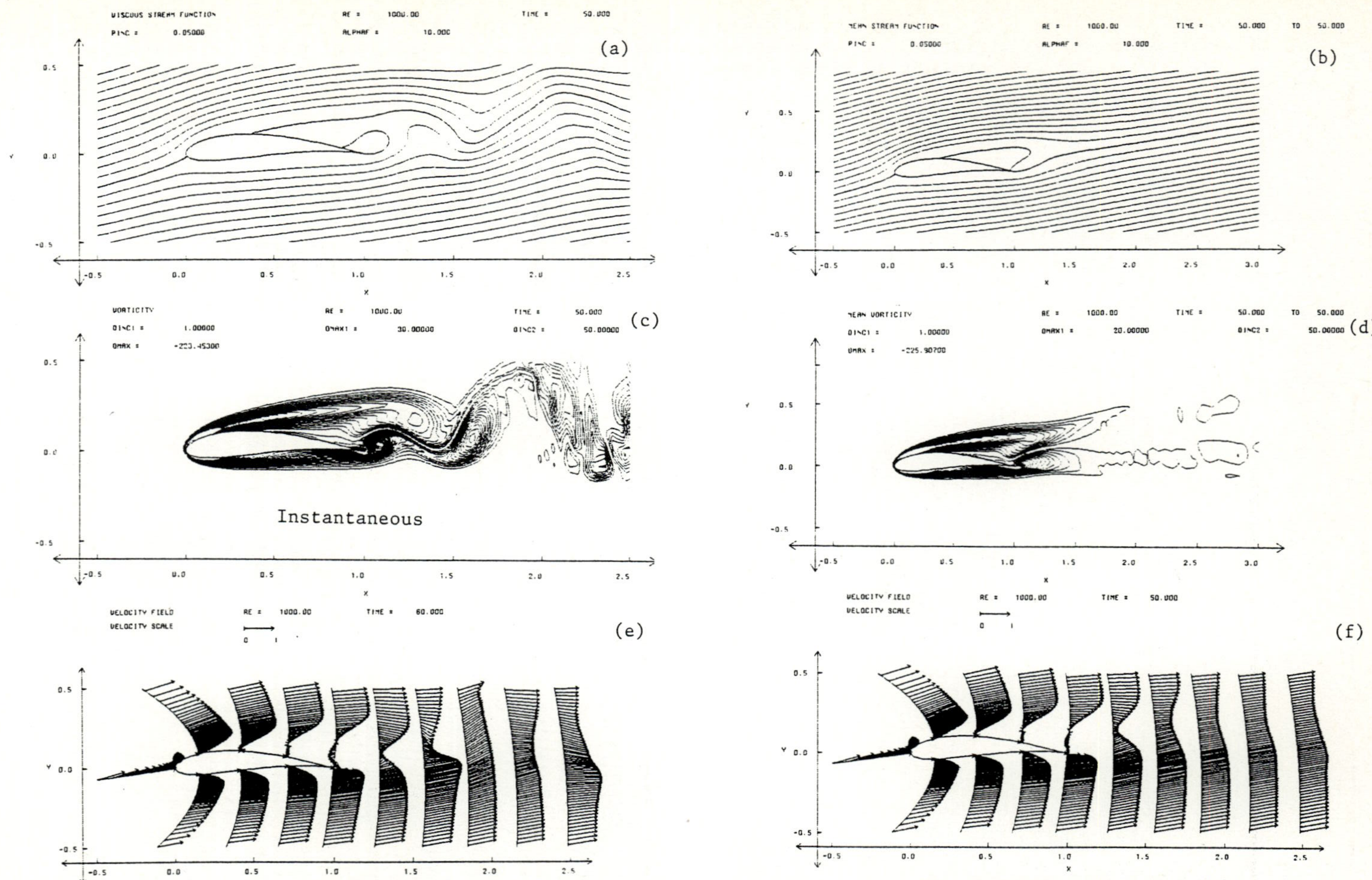

Time-Averaged; $30 \leq t \leq 50$

Fig. 5. Limit-Cycle Solution for Flow Past the Wortmann FX 63-137 Airfoil at Re = 1000, $\alpha_f = 10°$. (a-b) Stream-Function Contours; (c-d) Vorticity Contours; (e-f) Velocity Vectors.

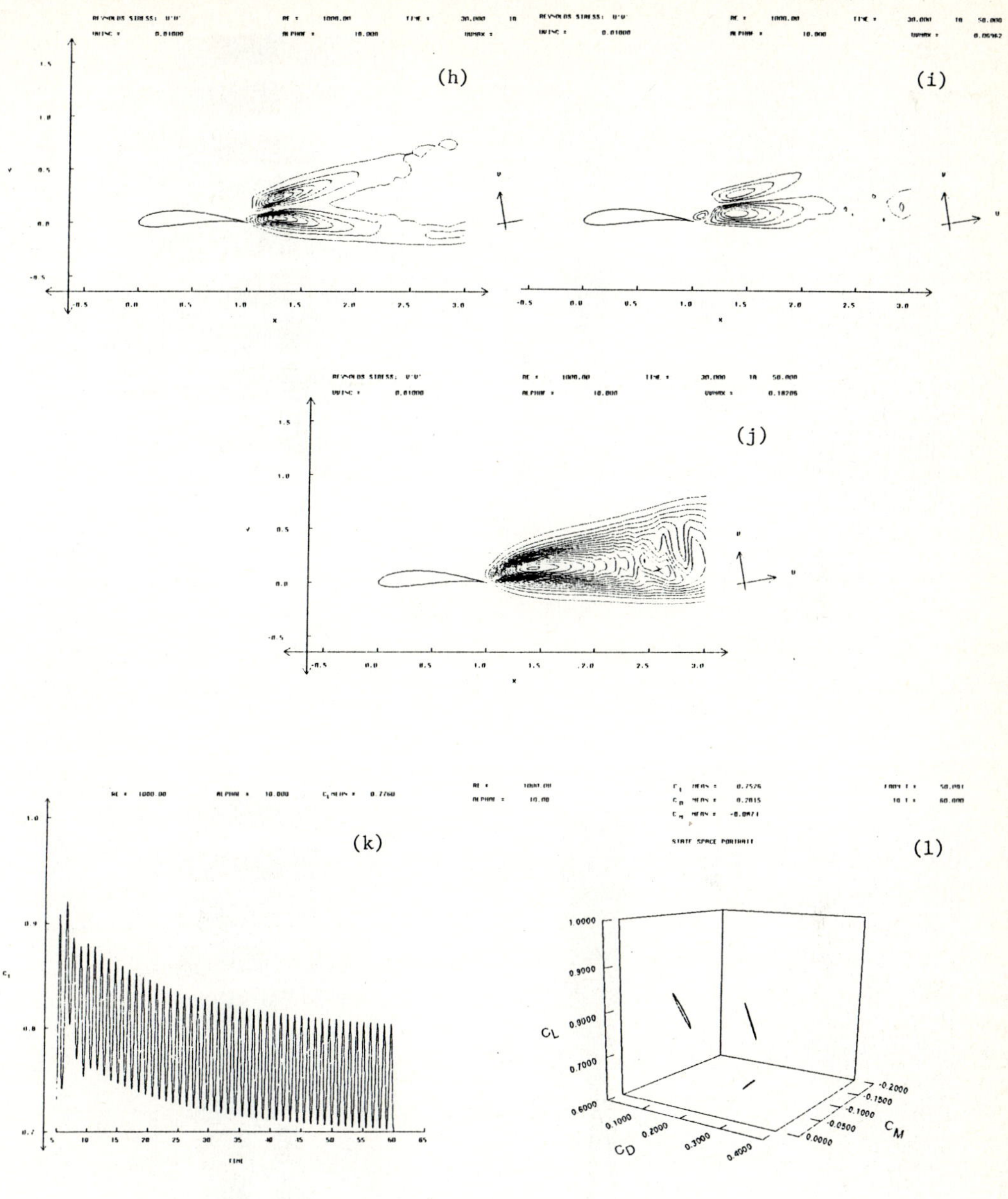

Fig. 5 (Cont'd.). Limit-Cycle Solution for Flow Past the Wortmann FX 63-137 Airfoil at Re = 1000, $\alpha_f = 10°$.

(h,i,j) Reynolds Stresses $\overline{u'u'}$, $\overline{u'v'}$, $\overline{v'v'}$;
(k) Time-History of C_L;
(l) Attractor.

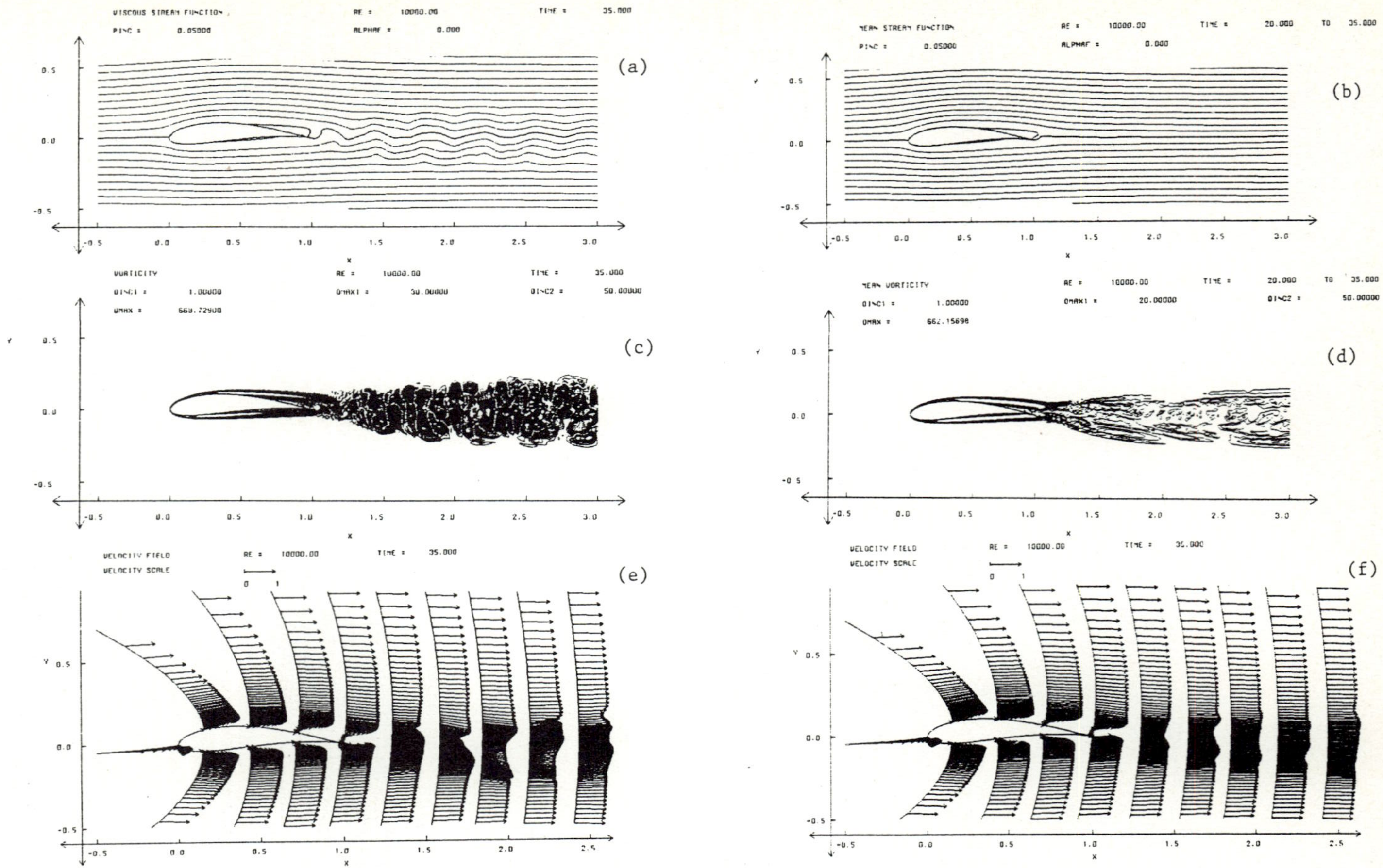

Instantaneous

Time-Averaged; $20 \leq t \leq 35$

Fig. 6. Limit-Cycle Solution for Flow Past the Wortmann FX 63-137 Airfoil at Re = 10,000, $\alpha_f = 0$. (a-b) Stream-Function Contours; (c-d) Vorticity Contours; (e-f) Velocity Vectors.

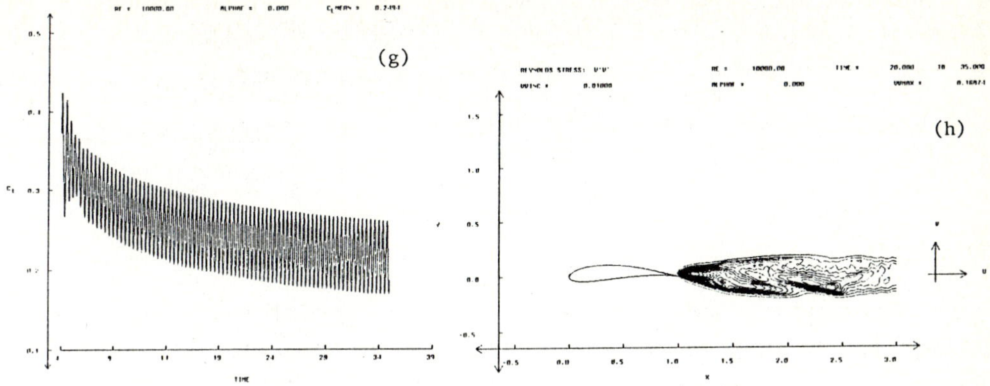

Fig. 6 (Cont'd.). Limit-Cycle Solution for Flow Past the Wortmann FX 63-137 Airfoil at Re = 10,000, $\alpha_f = 0$.
(h) Reynolds Stress $\overline{vv'}$; (g) Time-History of C_L.

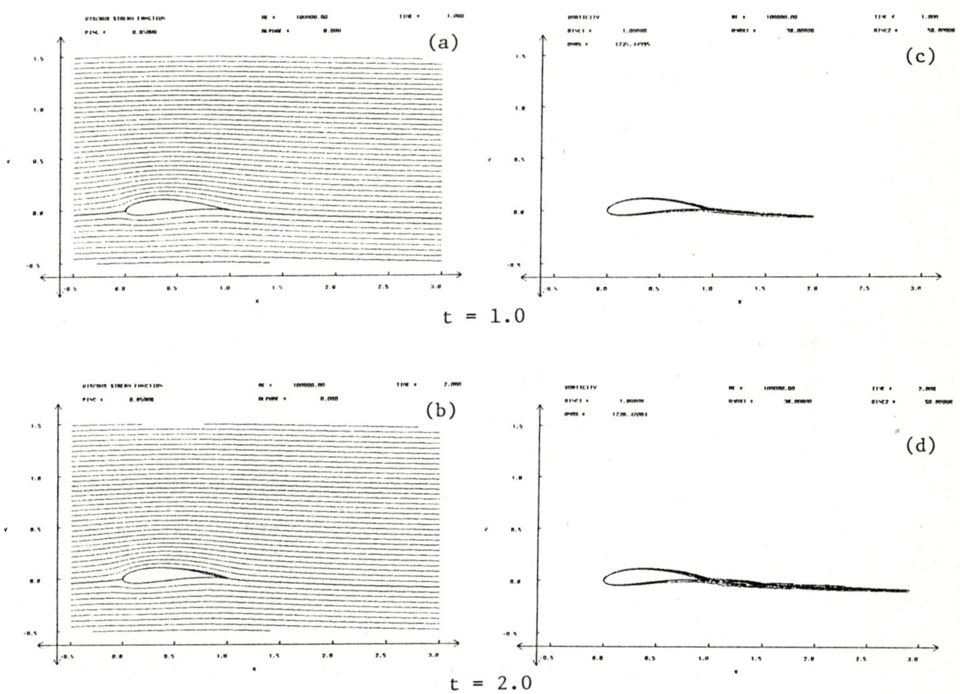

Fig. 7. Transient Flow Past the Wortmann FX 63-137 Airfoil at Re = 100,000, $\alpha_f=0°$.
(a-b) Stream-Function Contours; (c-d) Vorticity Contours.

AN INTERACTIVE BOUNDARY-LAYER STABILITY-TRANSITION APPROACH FOR LOW REYNOLDS-NUMBER AIRFOILS

Tuncer Cebeci* and Margaret McIlvaine**
Aerospace Engineering Department,
California State University, Long Beach
1250 Bellflower Blvd., Long Beach, CA 90840

Abstract

An interactive boundary-layer stability-transition approach for predicting the performance characteristics of airfoils at low Reynolds numbers is described and evaluated in terms of experimental data. The interactive boundary-layer procedure makes use of a panel method and an inverse boundary-layer scheme in which the Reynolds stress term is modeled by an algebraic eddy-viscosity formulation and the governing equations are solved in inverse mode by a two-point finite-difference method. The procedure also makes use of the e^n-method to determine the location of transition. The turbulence model employs an extended version of the intermittency expression in the Cebeci and Smith eddy-viscosity model and this plays an important role in the calculation of low Reynolds numbers flows. The solution procedure also considers the calculation of flow in the wake which becomes increasingly more important with increased flow separation.

1. Introduction

In recent years there has been increased emphasis on low Reynolds number aerodynamics in both civil and military applications such as remotely piloted vehicles, propeller and wind turbine aerodynamics, aircraft with high aspect ratio wings and ultra-light human-powered vehicles [1,2]. The development of theoretical methods for predicting the performance of the airfoils operating at these Reynolds numbers, which are usually below 0.5×10^6, differ from those developed for high Reynolds numbers. The differences are due mainly to the occurrence of separation bubbles whose extent is rather large (10 to 20% of the chord). Contrary to high Reynolds number flows, where the transition location occurs either before or at the flow separation location,

---

*Professor and Chairman
**Graduate Student

the onset of transition in these flows takes place within the separation bubble. The behavior of the transitional region, which follows the onset of transition, also differs from those at high Reynolds numbers, not only because its extent is rather small but also because it is not separation induced. Thus, a computational method developed for high Reynolds numbers must include these additiional ingredients for low Reynolds number flows.

The theoretical methods for predicting the performance of airfoils are based either on the solution of the Navier-Stokes equations or on a combination of inviscid and boundary-layer equations. In both approaches, it is important that conservation equations are solved accurately in full or reduced forms and that they include an accurate method for computing the location of transition and modeling the turbulent flow including the transitional region. Recent studies, conducted at high Reynolds numbers [3] show that as long as there is no, or very little, flow separation on the airfoils, it is sufficient to limit the calculations only on the airfoil. With increasing flow separation, the role of wake flow increases and with near- and post-stall regions, its effect is important, as shown in Fig. 1 for the NACA 0012 airfoil operating at a chord Reynolds number of 3×10^6. As can be seen from Fig. 1a, up to an angle of 10°, the variation of the trailing edge displacement thickness is independent of the wake. With flow separation taking place at higher angles of attack (Fig. 1b), the trailing-edge displacement thickness computed without the wake

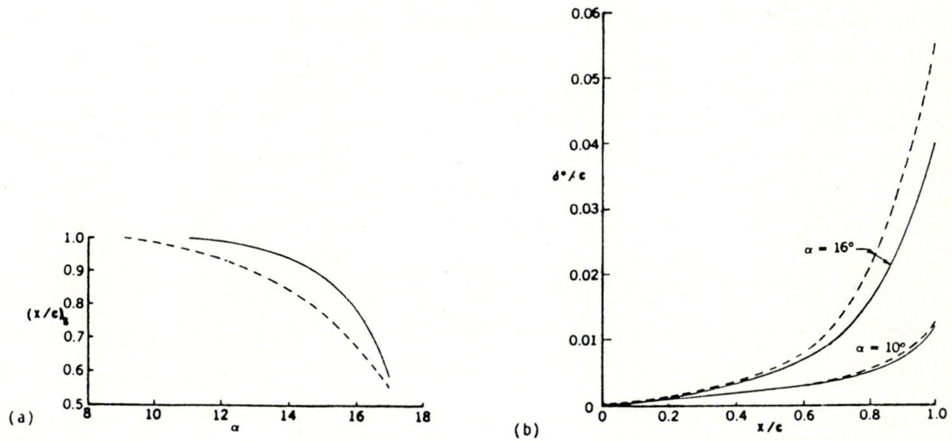

Figure 1. Effect of wake (solid lines) on (a) flow separation and (b) displacement thickness on the NACA 0012 airfoil for $R_c = 3 \times 10^6$. Dashed lines denote calculations without wake.

effect differs from that with the wake effect. The results in Fig. 1 also indicate that inclusion of the wake effect can actually make the calculations at higher angles easier since its effect is to reduce the flow separation on the airfoil.

The present paper addresses the two essential ingredients that need to be included in any theoretical method for predicting the performance of airfoils at low Reynolds numbers regardless of the approach used to solve the conservation equations. The proposed ingredients are included in the inviscid-viscous interaction method reported in [4], which is briefly described in the following section with special emphasis on the model used to represent the transitional flows which are separation induced. In Section 3, we describe the method for predicting the onset of transition which is an extension of the e^n-method proposed by Smith and Gamberoni [5] and Van Ingen [6] from attached flows to separated flows. Section 4 presents the application of this method to an Eppler 387 airfoil measured in the Langley low-turbulence pressure tunnel for Reynolds numbers ranging from 60,000 to 460,000 [7]. The paper ends with a summary of the more important findings.

2. Inviscid/Viscous Interactive Method

The interactive method is described for high and low Reynolds number flows in [3] and [4] respectively, and makes use of an inverse boundary-layer method coupled to a panel method with the interactive formula suggested by Veldman [8]. The boundary-layer equations are solved subject to the usual boundary conditions with the external velocity $u_e(x)$ expressed as the sum of inviscid velocity $u_e^o(x)$ and a perturbation velocity $\delta u_e(x)$ computed from the Hilbert integral

$$\delta u_e(x) = \frac{1}{\pi} \int_{x_a}^{x_b} \frac{d}{d\sigma} (u_e \delta^*) \frac{d\sigma}{x - \sigma} \tag{1}$$

with the interaction region confined to (x_a, x_b). More specifically, the boundary-layer equations are expressed in terms of an eddy viscosity, ε_m, so that the continuity and momentum equations,

$$\frac{\partial u}{\partial x} + \frac{\partial v}{\partial y} = 0 \tag{2}$$

$$u \frac{\partial u}{\partial x} + v \frac{\partial u}{\partial y} = u_e \frac{du_e}{dx} + \frac{\partial}{\partial y} [(\nu + \varepsilon_m) \frac{\partial u}{\partial y}] \tag{3}$$

are solved subject to boundary conditions,

$$y = 0, \quad u = v = 0; \quad y \to \infty, \quad u \to u_e(x) \tag{4}$$

on the airfoil and with y = 0 denoting the dividing streamline that separates the upper and lower parts of the inviscid flow in the wake, subject to the following conditions

$$y \to \pm\infty, \quad u \to u_e(x); \quad y = 0, \quad v = 0 \tag{5}$$

with $u_e(x)$ given by $u_e = u_e^o + \delta u_e(x)$.

The eddy-viscosity formulation of Cebeci and Smith [9] is used with special emphasis on the transitional region

$$\varepsilon_m = \begin{cases} \varepsilon_{m_i} = L^2 \left|\dfrac{\partial u}{\partial y}\right| \gamma_{tr} & (6a) \\ \varepsilon_{m_o} = 0.0168\, u_e \delta^* \gamma_{tr} & (6b) \end{cases}$$

where L and γ_{tr} are given by

$$L = 0.4y[1 - \exp(-y/A)], \quad A = 26\nu u_\tau^{-1}, \quad u_\tau = (\nu \frac{\partial u}{\partial y})^{1/2}_{max} \tag{7a}$$

$$\gamma_{tr} = 1 - \exp[-G(x - x_{tr}) \int_{x_{tr}}^{x} \frac{dx}{u_e}] \tag{7b}$$

Here γ_{tr} corresponds to the expression suggested by Chen and Thyson [10] with x_{tr} denoting the location of the beginning of transition and G a parameter defined by

$$G = (\frac{3}{C^2}) \frac{u_e^3}{\nu^2} R_{x_{tr}}^{-1.34} \tag{8}$$

where the transition Reynolds number $R_{x_{tr}} = (u_e x/\nu)_{tr}$ and C is constant with a recommended value of 60. The expression given by Eq. (7b) was obtained from data based on attached flows and is less likely to be applicable to flows with separation. For this reason, a correlation formula, shown in Fig. 2, was devised by Cebeci [4] to represent C^2 of Eq. (8) in terms of $R_{x_{tr}}$ for the experimental data obtained for airfoils NACA 66_3-0.18, ONERA D, NACA 65-213 and LNV109A. The data encompass a Reynolds number range from $R_c = 2.4 \times 10^5$ to $R_c = 2 \times 10^6$, and fall conveniently on a straight line on a semi-log scale represented by the equation

$$C^2 = 213\,[\log R_{x_{tr}} - 4.7323] \tag{9}$$

For the wake flow calculations following the study of Chang et al. [11], the eddy-viscosity formulation for wall boundary-layer formulas

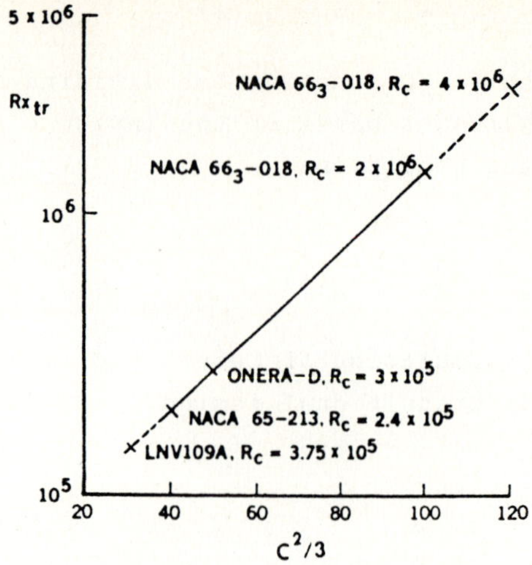

Figure 2. Variation of $C^2/3$ with transition Reynolds number.

was modified and expressed in the form

$$\varepsilon_m = (\varepsilon_m)_w + [(\varepsilon_m)_{te} - (\varepsilon_m)_w] \exp[\frac{x - x_{te}}{20\delta_{te}}] \tag{10}$$

where $(\varepsilon_m)_w$ denotes the eddy viscosity for the far wake given by the maximum of $(\varepsilon_m)_w^{\ell}$ and $(\varepsilon_m)_w^u$ defined by

$$(\varepsilon_m)_w^{\ell} = 0.064 \int_{-\infty}^{y} (u_e - u)dy, \quad (\varepsilon_m)_w^u = 0.064 \int_{y_{min}}^{\infty} (u_e - u)dy \tag{11}$$

with y_{min} as the location where $u = u_{min}$.

3. Transition Method

The e^n-method is used in the present calculations to represent the onset of transition. In this method, the laminar boundary-layer equations are solved first for a given pressure distribution in the inverse model, and the velocity profiles are calculated as a function of surface distance along the body. Next, the dimensionless form of the Orr-Sommerfeld equation

$$\phi^{iv} - 2\alpha^2\phi'' + \alpha^4\phi - iR(\alpha u - \omega)(\phi'' - \alpha^2\phi) + iR\alpha u''\phi = 0 \tag{12}$$

and its boundary conditions, with D denoting d/dy, subscript e the "edge" conditions and ξ_2 defined by

$$\xi_2^2 = \alpha^2 + iR(\alpha u_e - \omega)$$

$$y = 0, \quad \phi = \phi' = 0 \tag{13a}$$

$$y = \delta, \quad (D^2 - \alpha^2)\phi + (\alpha + \xi_2)(D + \alpha)\phi = 0, \quad (D + \xi_2)(D^2 - \alpha^2)\phi = 0 \tag{13b}$$

are solved [12] and the stability properties of the velocity profiles u and u" are examined. The onset of transition is computed by evaluating the integral

$$n = \int_{x_o}^{x} -\alpha_i dx \tag{14}$$

for a set of specified dimensional frequencies ω^* which is related to the dimensionless ω in Eq. (12) by $\omega = \omega^* \ell / u_o$ with ℓ and u_o denoting the length and velocity scales in the Orr-Sommerfeld equation. Here α_i represents the amplification rates determined from Eq. (12) using velocity profiles and their second derivatives obtained from the inviscid-viscous interaction procedure of Section 2, the solution of Eq. (12) begins at a Reynolds number greater than the critical value, R_{cr}, on the lower branch of the neutral stability curve. This provides the desired frequency which, at the subsequent Reynolds numbers, allows the solution of the eigenvalue problem in terms of α and to an amplification curve which rises from 0 to some value either less or greater than n. The process is repeated to obtain similar amplification curves for different values of ω^*. As discussed by Cebeci and Egan [13] for flows with separation, and as in three-dimensional flows, the envelope procedure used to calculate the critical frequency that leads to the most amplified disturbance from which transition is computed by assuming a value of n is not applicable. It is necessary to search for this frequency in the calculations. Since with separation the velocity profiles change significantly from those of attached flows, the amplification rates become very sensitive to the dimensional frequencies computed on the neutral stability curve, and it is necessary to take relatively small Δx increments on the airfoil and implement the eigenvalue procedure used in the numerical method to solve the Orr-Sommerfeld equation. This is done, as in [4], by the use of a continuation method in which the velocity profiles u and u" under consideration at a Reynolds number of R are defined by

$$u = u_{ref} + m(u - u_{ref}), \quad u'' = u''_{ref} + m(u'' - u''_{ref}) \tag{15}$$

Here u_{ref} and u''_{ref} denote reference profiles at a Reynolds number of R_{ref} and have eigenvalues α_o and ω_o. The parameter m is a sequence of specified numbers ranging from 0 to 1. It follows from Eq. (15) that for m = 0, the profiles u and u" correspond to the reference profiles and for m = 1, to those with eigenvalues of α and ω which are unknown.

4. Solution Procedure

The calculations were begun by computing the pressure distribution on the airfoil and in the wake without viscous effects. Next, the displacement thickness resulting from this pressure distribution was obtained from the solution of the boundary-layer equations with the calculations starting at the forward stagnation point, proceeding first with laminar flow only. With velocity profiles now known, the stability calculations were then initiated at the first x-station where the Reynolds number based on the displacement thickness, $R_{\delta *}$ ($\equiv u_e \delta^* / \nu$), exceeded its critical value established for similar boundary layers [4]. Several dimensional frequencies at different x-locations were then computed on the lower branch of the neutral stability curve in order to determine the amplification rates, α_i so that the location of transition could be computed from Eq. (14) for an assumed value of n which was taken to be 10 in our calculations. The boundary-layer calculations were performed until the transition location after which the turbulent flow calculations on the airfoil and in the wake took over. The displacement thickness distribution resulting from these calculations was used to determine a blowing velocity distribution v_n by differentiating the product of external velocity and displacement thickness with respect to the surface distance s, that is,

$$v_n = \frac{d}{ds} (u_e \delta^*) \tag{16}$$

so that the inviscid flow equations were solved again subject to a new boundary condition. This process was repeated on an iterative basis until the solutions of both boundary-layer and inviscid-flow equations converged. The blowing velocity was also determined on the inviscid point distribution in the wake and required a further iteration to ensure that the divided streamline was correctly located. Special care was required in the near wake to ensure that step lengths were sufficiently small so as to avoid convergence problems.

5. Results and Discussion

In the study conducted in [4], the accuracy of the method described in the previous three sections were evaluated for several flow configurations investigated experimentally for five airfoils with chord Reynolds numbers ranging from 3×10^5 to 8×10^6. Except for one airfoil, the range of angle of attack for these airfoils was limited to small angles of attack. The study showed that with a combination of e^n-method and modified transitinal model in the Cebeci-Smith algebraic eddy-viscosity formulation, the lift and drag coefficients of airfoils can be calculated accurately.

In the present study, we consider the experimental data of [7], which contains measurements for the Eppler airfoil in the Langley Low-Turbulence Pressure Tunnel (LTPT). The tests were conducted over a Mach number range from 0.03 to 0.13 and a chord Reynolds number range from 60×10^3 to 460×10^3. Lift and pitching-moment data were obtained from airfoil surface pressure measurements and drag data were obtained from wake surveys. Oil flow visualization was used to determine laminar-separation and turbulentreattachment locations. Comparison of these results with data obtained with the Eppler 387 airfoil in two other facilities, as well as predictions from the Eppler airfoil code, were included in [7].

The calculations reported here are for a chord Reynolds number of 200,000 and for a range of angles of attack from 0 to 6 degrees. At the writing of this paper, additional calculations are being performed for lower and higher Reynolds numbers and angles of attack, including post-stall, and will be reported separately.

The calculation method of the previous three sections requires the coordinates of the airfoil, the chord Reynolds number and a value of n for the e^n-method which, in a way, represents the effect of freestream turbulence on transition and is given by

$$n = -8.43 - 2.4 \ln T \tag{17}$$

where

$$T = \sqrt{\overline{u'^2}}/u_e \tag{18}$$

as proposed by Mack [14]. According to [7], the measured turbulence level was 0.06 percent for a total pressure p_t = 15 psi and for the

Reynolds numbers from 100 x 10^3 which, according to Eq. (17), correspond to a value of $n = 9.4$.

Figure 3 allows a comparison between measured and calculated distributions of pressure coefficients for angles of attack of 0, 2, 4, 5 and 6 degrees. With two small exceptions, the calculated results agree with measurements within experimental accuracy. In the immediate vicinity of the leading edge where the gradient of the pressure coefficient changes sign rapidly, the calculations are slightly in error. Also, the calculated results do not capture perfectly the constant values of pressure coefficient associated with the separation bubbles.

Further details of the results shown in Fig. 3 are presented in Tables 1 and 2. The calculated values of the chordwise location of laminar separation (LS), turbulent reattachment (TR), and the onset of transition are given in Table 1. The experimental results of this table are subject to some uncertainty because of difficulties associated with the surface flow visualization technique. With this proviso, comparison between measured and calculated values must be considered outstanding. It should be noted that the transition location obtained from the e^n-method occurs within the separation bubble in all cases and, in accord with experimental observation, leads to reattachment some distance downstream.

The lift and drag coefficients obtained from the present method are listed in Table 2 together with the experimental results obtained in the Stuttgart and Langley wind tunnels. Results obtained from the well-established calculation method of Eppler, which is based on the solution of integral equations, are also shown. In general, our calculated lift coefficients are higher than those measured and the drag coefficients lie between the two sets of measurements. Discrepancies in the experimental values of lift coefficients are around 10% with the calculated results less than 5% higher than the Langley experiments. This small discrepancy must correspond to integration of the differences shown in Fig. 3. It is interesting that the results of the Eppler code, which contains different and more limited assumptions than the present method, provides similar results.

6. Concluding Remarks

The combination of the interactive boundary layer and e^n-methods has been shown to allow the prediction of the lift and drag coefficients of

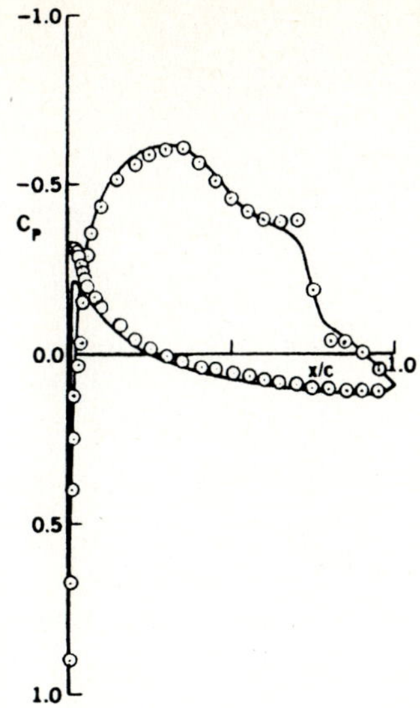

(a) α = 0°

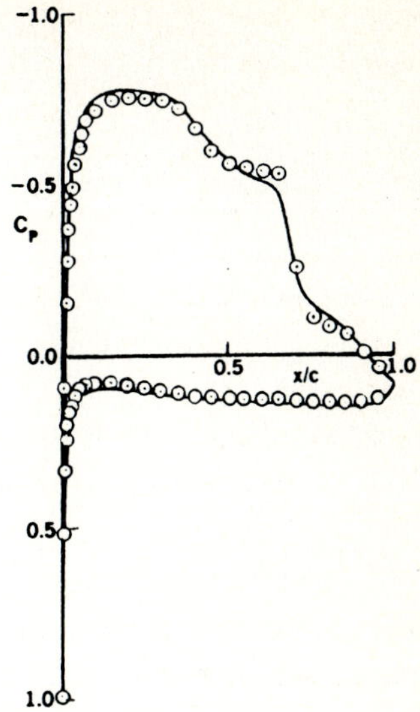

(b) α = 2°

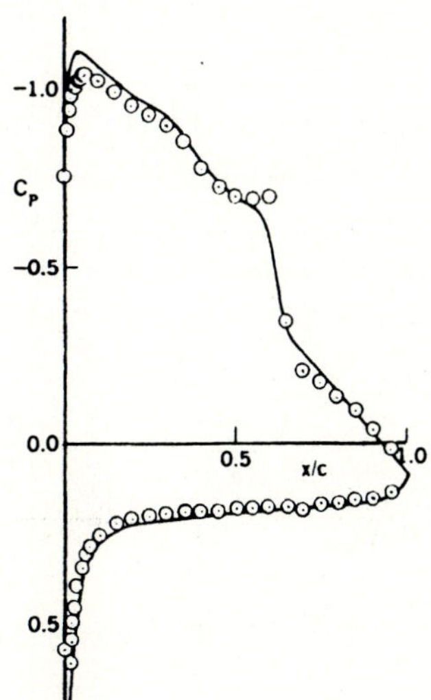

(c) α = 4°

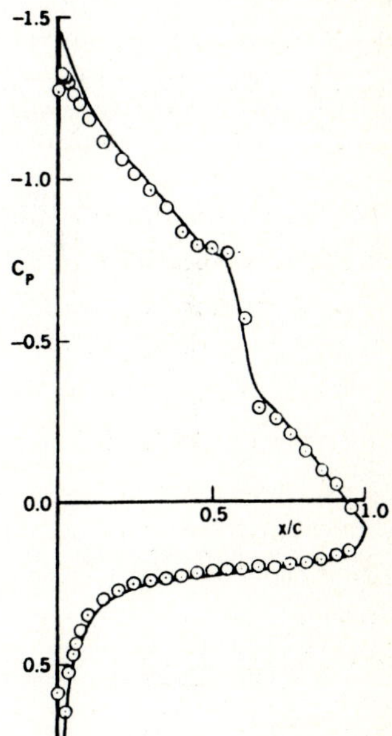

(d) α = 5°

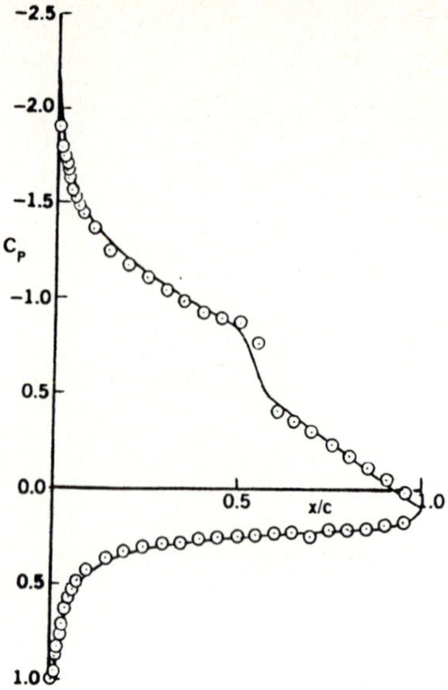

(e) α = 6°

Figure 3. Comparison of the calculated (—) and measured (o) distribution of pressure coefficients for different angles of attack.

Table 1. Experimental and Calculated Chordwise Separation (LS) and Reattachment (TR) and Transition Locations on the Upper Surface of the Eppler Airfoil

	Experiment			Calculated	
α, deg	LS	TR	$(\frac{x}{c})_{tr}$	LS	TR
0	0.48	0.74	0.68	0.52	0.75
2	0.43	0.67	0.63	0.48	0.70
4	0.40	0.62	0.57	0.43	0.64
5	0.38	0.59	0.53	0.41	0.60
6	0.37	0.55	0.48	0.39	0.55

Table 2. Measured and Calculated Values of Lift and Total Drag Coefficients

	Stuttgart Experiments		Langley Experiments		Eppler Code		Present Method	
α	c_ℓ	c_d	c_ℓ	c_d	c_ℓ	c_d	c_ℓ	c_d
0	0.330	0.0108	0.352	0.0105	0.393	0.0093	0.378	0.0095
2	0.520	0.0092	0.574	0.0118	0.613	0.0099	0.601	0.0104
4	0.710	0.0098	0.785	0.0133	0.832	0.0111	0.820	0.0116
5	0.800	0.0107	0.891	0.0138	0.940	0.0121	0.928	0.0122
6	0.880	0.0123	1.004	0.0141	1.032	0.0130	1.035	0.0127

the Eppler airfoil for a Reynolds number of 200 x 10^3 and angles of attack between 0 and 6 degrees. This calculation involved the determination of transition, which occurred in all cases within the separation bubble, and modeling of the separation induced transition region.

7. References

1. Proceedings of the Conference on Low Reynolds Number Airfoil Performance (ed. T. J. Mueller) UNDAS-CP-77B123, June 1985.

2. Proceedings of the International Conference on Aerodynamics at Low Reynolds Numbers. The Royal Aeronautical Society, Oct. 1986.

3. Cebeci, T., Clark, R.W., Chang, K.C., Halsey, N.D. and Lee, K.: Airfoils With Separation and the Resulting Wakes. J. Fluid Mech. 163, pp. 323-347, 1986.

4. Cebeci, T.: Essential Ingredients of a Method for Low Reynolds-Number Airfoil. To be published in the AIAA J., 1989.

5. Smith, A.M.O. and Gamberoni, N.: Transition, Pressure Gradient and Stability Theory. Proc. IX Int. Congress Appl. Mech., Brussels, 1956.

6. Van Ingen, J.L.: A Suggested Semi-Empirical Method for the Calculation of the Boundary-Layer Region. Rept. No. VTH71, VTH74, Delft, Holland, 1956.

7. McGhee, R.J., Jones, G.S. and Jouty, R.: Performance Characteristics from Wind-Tunnel Tests of a Low Reynolds-Number Airfoil. AIAA Paper 88-0607, Jan. 1988.

8. Veldman, A. E. P.: New Quasi-Simultaneous Method to Calculate Interacting Boundary Layers, AIAA J., 19, 1981, p. 769.

9. Cebeci, T. and Smith, A.M.O.: Analysis of Turbulent Boundary Layers, Academic Press, New York, 1974.

10. Chen, K.K. and Thyson, N.A.: Extension of Emmons' Spot Theory to Flows on Blunt Bodies. AIAA J., 9, 1971, pp. 821-825.

11. Chang, K. C., Bui, M. N., Cebeci, T. and Whitelaw, J. H.: The Calculation of Turbulent Wakes, AIAA J., 24, 1986, p. 200.

12. Cebeci, T. and Bradshaw, P.: Momentum Transfer in Boundary Layers. McGraw-Hill/Hemisphere Co., Wash. D.C., 1977.

13. Cebeci, T. and Egan, D.: Prediction of Transition Due to Isolated Roughness. To be published in AIAA J., 1989.

14. Mack, L. M.: Transition and Laminar Instability, Jet Prop. Lab. Publication 77-15, 1977.

THE INSTABILITY OF TWO-DIMENSIONAL LAMINAR SEPARATION

By Laura L. Pauley
The Pennsylvania State University, University Park, PA 16802, USA

Parviz Moin and William C. Reynolds
Stanford University, Stanford, CA 94305, USA

Abstract

The separation of a two-dimensional laminar boundary layer under the influence of an external adverse pressure gradient was studied. The unsteady, incompressible Navier-Stokes equations were solved using a fractional timestep method. Because the grid resolution did not allow the small scale structures found in turbulence to develop, these computations studied the development of large-scale, laminar structures in the boundary layer.

It was found that strong pressure gradients created periodic shedding from the separation. The Reynolds number and the strength of the adverse pressure gradient were varied independently to determine the characteristics of the unsteady separation. The frequency of shedding was influenced by the Reynolds number but not by the strength of the pressure gradient. The Strouhal number based on the local velocity and boundary layer thickness was independent of the Reynolds number and the pressure gradient. The vortex shedding was due to the inviscid instability of the separated laminar shear layer.

When the results were time-averaged, the unsteady separation produced a pressure gradient and streamline pattern similar to those found in laminar separation experiments. This suggests that the rapid reattachment found in experiments may be due to large-scale laminar structures.

---

1. Introduction

In many practical situations, a flow is three-dimensional. With separated flows the separation zone is complex and the characteristics of a separation structure may depend on whether the boundary layer is laminar or turbulent upstream of the separation. In order to understand unsteady separation it is important to know the conditions under which a separation occurs, the structural characteristics of the separated region, and the time required for the development or decay of the separation.

Many studies of boundary layer separation have been made. Often the work has concentrated on a steady structure in two or three dimensions (*eg.* Gaster, 1966, Peake and Tobak, 1982). Some have investigated the development of unsteady separation through flow visualization and selected measurements (*eg.* Despard and Miller, 1971,

Koromilas and Telionis, 1980). Due to limited data fields, these experimental studies do not characterize the entire flow. The complete flow fields from numerical computations can provide further insight into the structure of unsteady separation. The computational results describing a time-accurate unsteady separation allow us to understand the full structure of the separation and to test various theories which might account for this structure.

Two significant aspects of unsteady separation have been identified, these are the development of the separation region and the unsteady shedding of cross-stream vortices. Applications such as the enhancement of fighter aircraft performance drive the understanding of the conditions which induce a separation. Turbomachinery applications require an understanding of unsteady shedding which can create a significant modification of the efficiency and estimated life of turbine blades.

Koromilas and Telionis (1980) investigated the development of a separated region by impulsively bending a flat plate to produce an adverse pressure gradient on the boundary layer. The separated region started out quite thin but quickly grew and developed into a strong vortex near the reattachment point. From their photographs, it appears that this vortex is shed and a new vortex begins to develop. As the separated region becomes large, the lifted shear layer undergoes transition and the vortex structures disappear into the randomness of turbulence.

To understand the basic characteristics of boundary layer separation, many investigators have studied two-dimensional, steady, laminar separation. The typical structure of a laminar boundary layer separation bubble was described by Horton (1968). Figure 1 shows a sketch of the separation. Near the separation point, the reversed flow region contains slow moving fluid. The center of the reversed flow vortex lies near the reattachment point and causes the fluid to move faster in this region. At the point of separation, the dividing streamline is observed to rise slowly away from the wall. After the boundary layer undergoes transition to turbulent flow, the streamline quickly reattaches to the wall. Within the separated region, the wall pressure remains constant

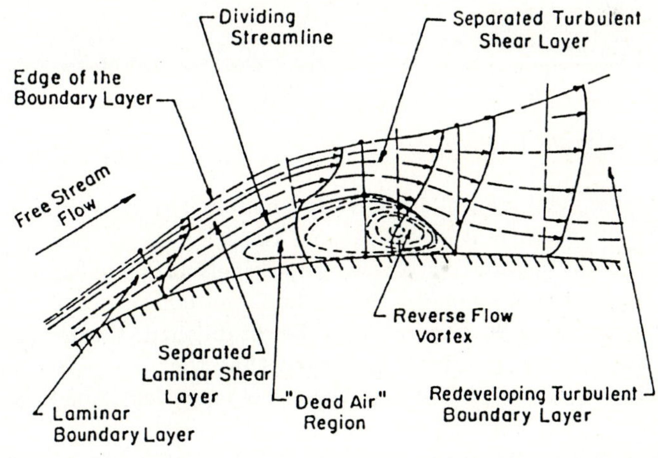

Fig. 1 The flow field in the vicinity of a transitional separation bubble. (Horton, 1968)

near the separation point while a steep pressure gradient accompanies the high velocity region near the reattachment point. Tani (1964) noted that the length of the separated region was reduced as the adverse pressure gradient increased. This was explained by the more rapid transition of the lifted laminar boundary layer. These characteristics of the laminar separation bubble have been confirmed by the results of many investigations.

2. Problem Definition

In our study a channel configuration was used to produce a known pressure distribution on a laminar boundary layer. A thin laminar boundary layer develops on one wall and a suction port on the opposite wall creates an adverse pressure gradient (see figure 2). For all results presented, distances are normalized by the channel height and the origin of the streamwise coordinate is at the leading edge of the flat plate. The velocity has been normalized by the inlet freestream velocity. In this paper, dimensional variables are denoted by lower case letters while nondimensional variables use upper case letters. The suction parameter, S, refers to the fraction of the entering flow which was removed through the suction port. Streamlines of the unsteady separated laminar boundary layer are shown. When separated, the laminar boundary layer of interest remained in the lower 1/5 of the channel–allowing us to study the effects of an externally applied adverse pressure gradient on a laminar boundary layer.

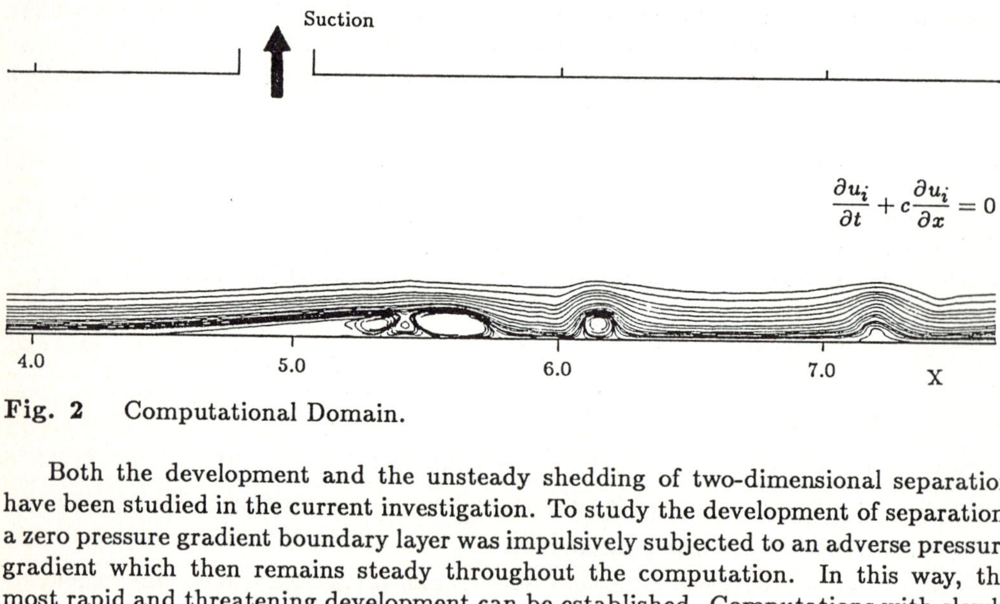

Fig. 2 Computational Domain.

Both the development and the unsteady shedding of two-dimensional separation have been studied in the current investigation. To study the development of separation, a zero pressure gradient boundary layer was impulsively subjected to an adverse pressure gradient which then remains steady throughout the computation. In this way, the most rapid and threatening development can be established. Computations with slowly increasing adverse pressure gradients indicated that the same structure develops but the rate of development was limited by the time-dependent function of the pressure gradient.

A fractional timestep method developed by Kim and Moin (1985) was used to solve the incompressible Navier-Stokes equations for this investigation. The numerical method is second order accurate in space and time. A no-slip boundary condition was used to develop a laminar boundary layer on the test wall. A no-stress condition on the suction wall reduces the required resolution in this region. Various exit boundary conditions were tested for this investigation (see Pauley, *et al.*, 1988). It was found that the convective exit boundary condition,

$$\frac{\partial u_i}{\partial t} + c\frac{\partial u_i}{\partial x} = 0, \qquad (1)$$

allowed the propagating vortex structure to exit the domain with minimum distortion. The propagation speed of the vortices within the computational domain was used as the value for c. It was found that the same computational results were obtained when the freestream velocity was used for c. The value of c was therefore not critical to the numerical solution.

All computations contained 256 points in the streamwise direction and 128 points in the normal direction. Half of the grid points were clustered in the boundary layer on the test wall using a hyperbolic tangent distribution. Even grid spacing was used in the streamwise direction. This created a solution which was grid independent. Because the grid resolution did not allow the small scale structures found in turbulence to develop, these computations studied the development of large-scale, laminar structures in the boundary layer.

The computational domain was set so that the separation solution was independent of inlet and outlet boundary effects. A Blasius boundary layer and uniform freestream were implemented at the inlet of the domain. All computations were time accurate. Computations began with a converged solution in the channel with no suction applied. At $T = 0.0$, a suction strength was impulsively applied.

3. Two-Dimensional Separation

Three different Reynolds numbers were used in this two-dimensional study. The parameters for these cases are listed below. The nominal Reynolds numbers listed are defined at the start of the suction port when no adverse pressure gradient is applied.

	Low Speed	Normal Speed	High Speed
Re_x	59,629	120,544	238,515
$Re_{\delta_{99}}$	1221	1737	2442
Re_θ	162	230	325

3.1 Steady Separation Structure and Stability

Computations were made for several suction strengths at each Reynolds number to determine the effects of adverse pressure gradient. For relatively weak adverse pressure gradients, the separated region developed to a steady, closed separation bubble. In each of these cases, the separated region was very thin.

For stronger adverse pressure gradients corresponding to increased suction, the separated region lengthened and small oscillations developed in the skin friction. The instantaneous skin friction and pressure coefficient curves for various suction strengths are shown in figure 3. The oscillations formed near the reattachment point and propagated

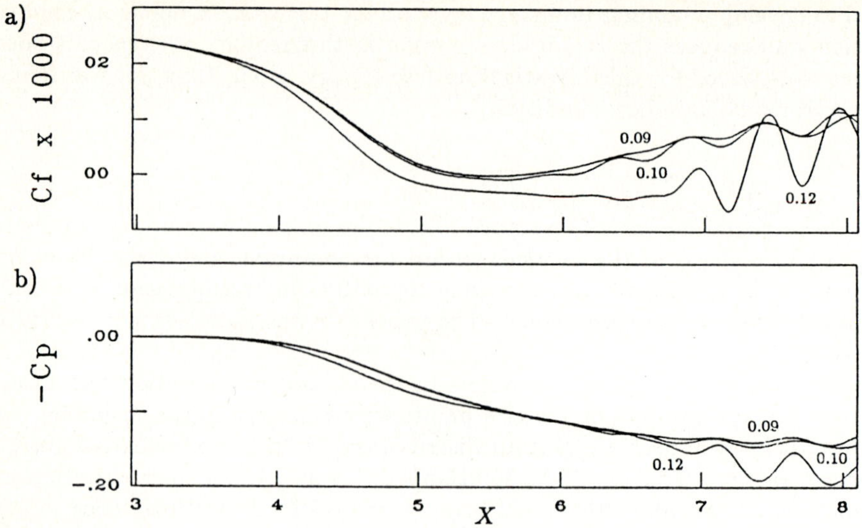

Fig. 3 Approaching the critical suction strength, $Re_x = 120,544$ and S = 0.09, 0.10, and 0.12. a) Skin friction b) Pressure coefficient.

downstream as "waves". These waves became more severe as the pressure gradient increased. The separation bubble length, however, remained steady until vortex shedding began at S=0.12.

The occurrence of shedding can be observed from the skin friction curves. Cases where the oscillations of the skin friction have a negative value are classed as "unsteady separation". Using this definition for unsteady separation, the critical suction strength was S = 0.12 for all Reynolds numbers tested. Above S = 0.12, discrete vortices are shed and additional freestream fluid is entrained into the separation.

3.2 Unsteady Separation Structure

As the adverse pressure gradient is increased beyond the critical condition, the strength of the vortices increases. These unsteady separated structures were examined in detail for the case of $Re_x = 120,544$ and S = 0.22. Under these conditions a strong separation was produced. The streamlines for the development of the unsteady separation event are shown in figure 4. Initially, the separation is symmetric, reminiscent of steady separation. As the separation develops, a recirculating region is formed which moves downstream and increases in strength. The separation then pinches into two distinct cells and a small region of counter-rotating fluid forms (indicated by an arrow in figure 4). Throughout the development, the upstream portion of the separation remains steady and quiescent. This separation development is very similar to that seen by Koromilas and Telionis (1980).

In this simulation, a steady limit cycle is reached after the initial transient. Figure 5 shows the streamline structure at six equally spaced intervals in the limit cycle. The flow near the separation point is virtually steady. A slight fluctuation in the height of the recirculating region indicates that boundary layer fluid is entrained to allow the

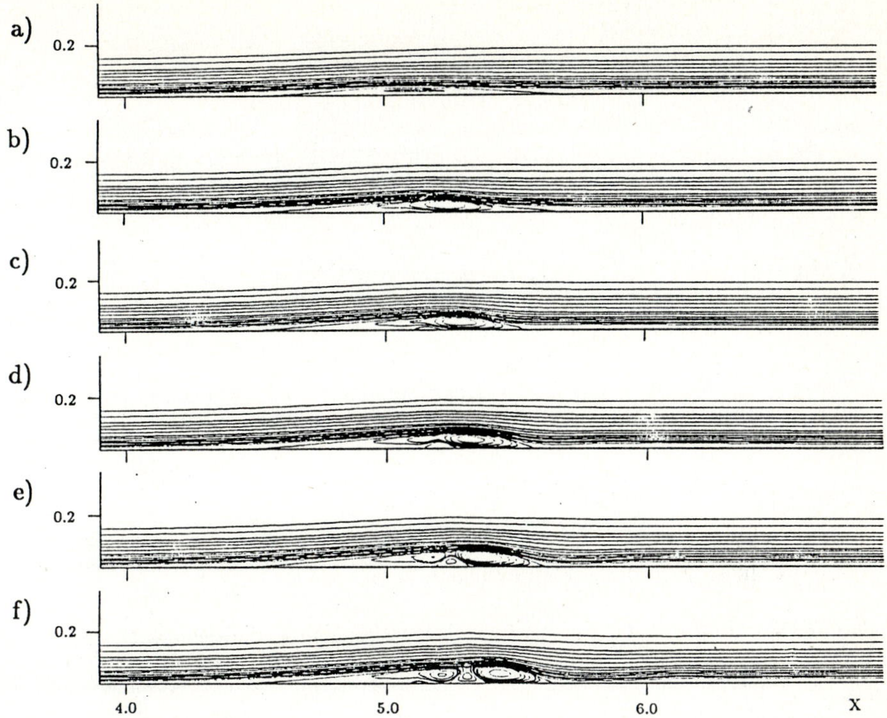

Fig. 4 Development of separation, $Re_x = 120,544$ and $S = 0.22$. a) $T = 2.4$ b) $T = 3.6$ c) $T = 4.0$ d) $T = 4.4$ e) $T = 4.8$ f) $T = 5.2$.

continuation of the shedding process. In the limit cycle, the shed vortices are considerably stronger than the first vortex shed after suction is initiated (compare figure 5 to figure 4). The shed vortices develop an oval shape due to the secondary counter-rotating vortices which cause the primary vortices to be pushed downstream near the wall and pulled upstream away from the wall. As the primary vortices propagate downstream and the secondary vortices lose strength, their shape becomes rounded and the vortices propagated at approximately 65% of the local freestream velocity. Strong viscous effects in the boundary layer cause the vortices to decay quickly.

To examine the development of the separation, streamwise velocity histories were recorded in the region of vortex shedding at four heights above the wall (see figure 6). The sampling points were at various Y locations all below the center of the vortices. Hence the histories characterize the near-wall flow. After a brief initial adjustment period, the boundary layer begins to shed periodically at a constant characteristic frequency. This shedding frequency was determined from an FFT of the steady state oscillations of the velocity. The velocity magnitudes settle to within 5% of their limit cycle magnitudes after 10 oscillations.

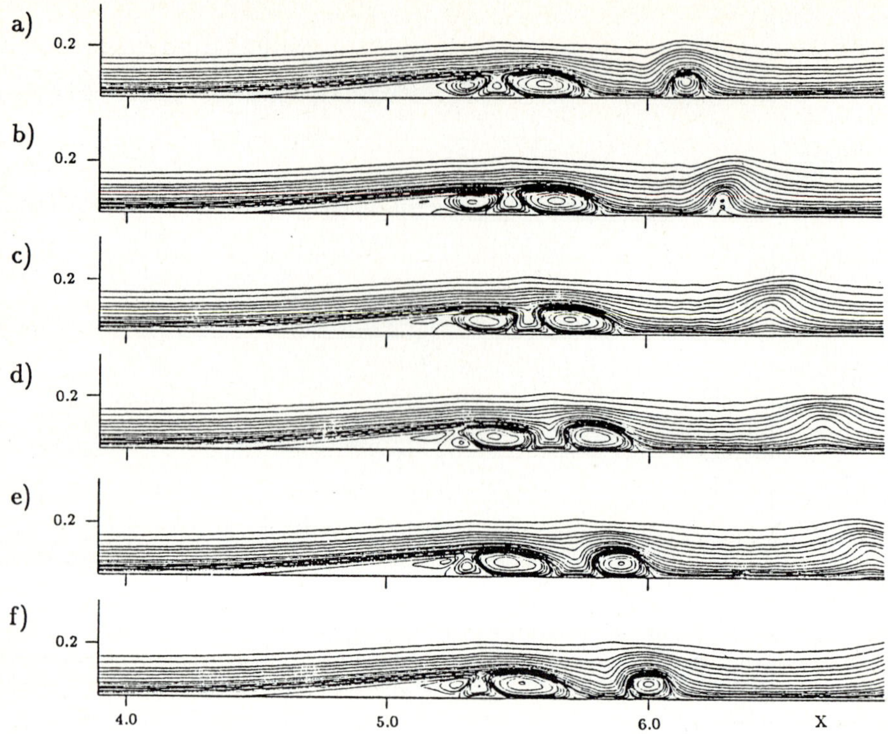

Fig. 5 Streamlines for limit cycle shedding, $Re_x = 120,544$ and $S = 0.22$. a) $T = 43.46$ b) $T = 43.72$ c) $T = 43.99$ d) $T = 44.25$ e) $T = 44.51$ f) $T = 44.78$.

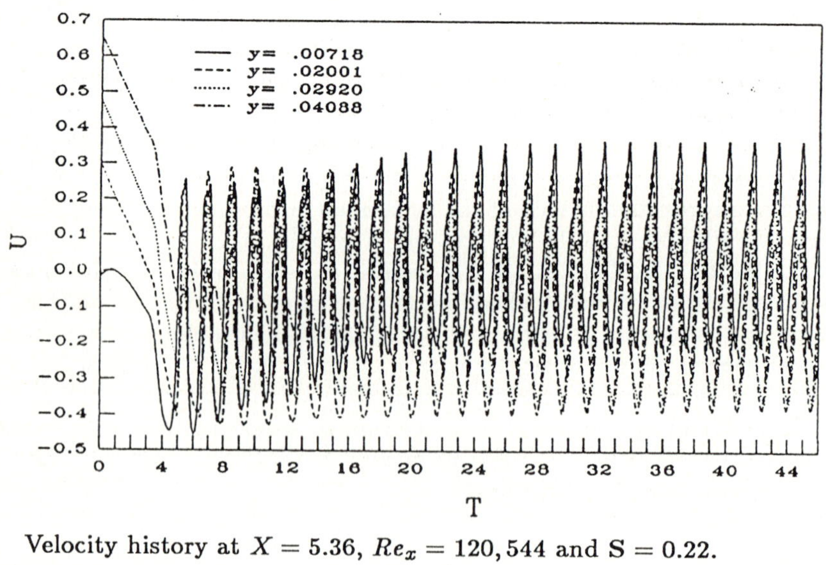

Fig. 6 Velocity history at $X = 5.36$, $Re_x = 120,544$ and $S = 0.22$.

3.3 Shedding Frequency

The parameters controlling the shedding frequency were studied by varying the suction strength and Reynolds number. At fixed Reynolds number, the shedding frequency was found to be independent of the suction strength. For example at $Re_x = 120,544$ the same frequency was observed for $0.12 < \mathbf{S} < 0.22$. The shedding frequency, however, was found to increase with increased Reynolds number.

The shedding frequency can be nondimensionalized with the boundary layer momentum thickness at separation and the local freestream velocity to produce a Strouhal number,

$$St_\theta = \frac{f\ \theta_{sep}}{(u_e)_{sep}}. \tag{2}$$

For the three Reynolds numbers examined, the Strouhal numbers based on freestream velocity and momentum thickness at separation collapse very well to

$$St_\theta = 0.00686 \pm 0.6\%. \tag{3}$$

This allows the frequency of shedding to be predicted from the momentum thickness at separation.

3.4 Vorticity and Inviscid Linear Stability

In figure 7 it is seen that the vorticity distribution found in the computations is similar to that in a free shear layer. Shedding occurred from the shear layer and the path of the shed vortex continued in the same direction as the center of the free shear layer. The presence of the adjacent wall appears to exert little influence on the propagating vortex. This suggests that the separation should be analyzed as a free shear layer, and that the vortex shedding is dominated by an inviscid instability. The stability of an unsteady separation was investigated by applying the results of an inviscid linear stability analysis for a mixing layer profile.

The results of Michalke (1964) for a free shear layer were used to predict the location of shedding. Considering a free shear layer between two freestreams, u_1 and u_2, the velocity difference across the shear layer is $\Delta u = u_2 - u_1$ and the average velocity in the shear layer is $\bar{u} = \frac{1}{2}(u_1 + u_2)$. His analysis used the vorticity thickness, $\delta_\omega = \Delta u/(\partial u/\partial y)_{max}$, and the velocity ratio, $\lambda = (\Delta u)/(2\bar{u})$, to describe the shear layer. A nondimensional group for frequency was calculated utilizing the vorticity thickness and the average velocity across the mixing layer,

$$\omega^* = \frac{\delta_\omega}{4}\frac{(2\pi f)}{\bar{u}}. \tag{4}$$

From inviscid theory, the most amplified frequency is $\omega^* \approx 0.21$ and is not strongly dependent on the velocity ratio, λ.

The velocity field obtained from the computation was time-averaged in order to obtain values of δ_ω and $\bar{u}$ for use in comparing the observed frequencies f with those predicted by the stability analysis. Beyond the separation point, the separated region contained virtually stagnant fluid, corresponding to $\lambda = 1.0$. In the unsteady region of the separation, λ and the vorticity thickness changed significantly with streamwise distance. Based on the time-averaged velocity profiles as one moves downstream, the

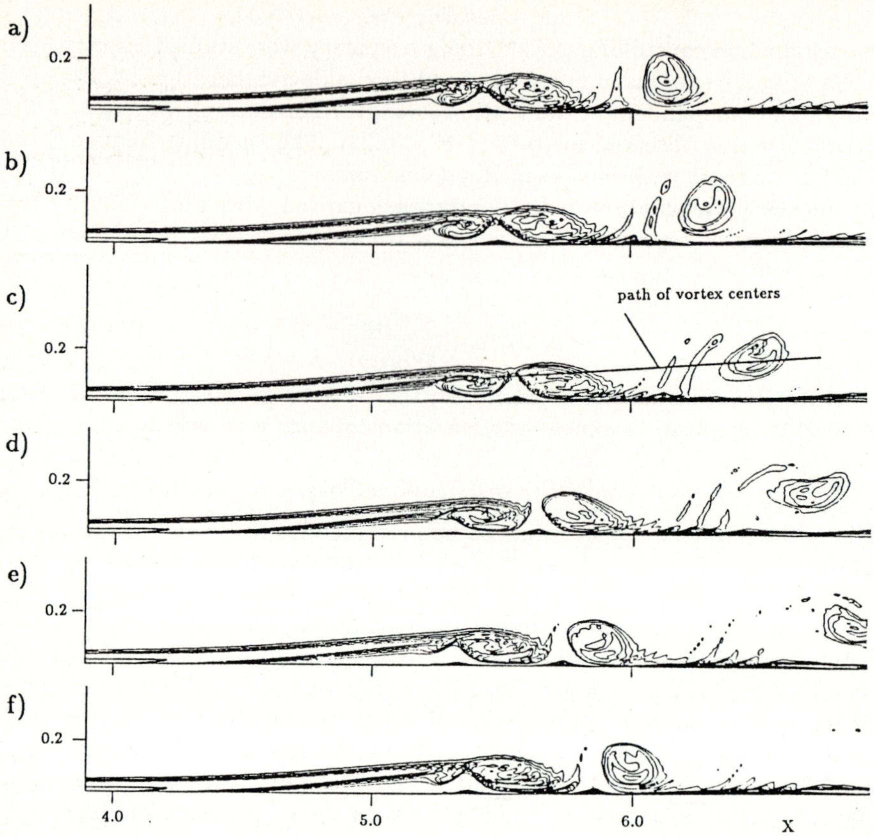

Fig. 7 Vorticity field for limit cycle shedding, $Re_x = 120,544$ and $\mathbf{S} = 0.22$. a) $T = 43.46$ b) $T = 43.72$ c) $T = 43.99$ d) $T = 44.25$ e) $T = 44.51$ f) $T = 44.78$.

shedding frequency, ω^*, varied rapidly near the vortex center. However, $\omega^* = 0.21$ was found at the streamwise location of the vortex center for all cases. Thus, this location, which is the location where a new vortex is produced, exhibits a shedding frequency that corresponds exactly with the most amplified frequency predicted by linear stability analysis.

3.5 Time-Averaged Separation Bubbles

When unsteady separation bubbles were time-averaged across the shedding cycle, the separation became a closed bubble. Shown in figure 8 are the time-averaged streamlines and pressure coefficient at $Re_x = 120,544$ and $\mathbf{S} = 0.22$. This separated region spanned the entire region of the inviscid adverse pressure gradient. Near the separation point, slow moving fluid is contained in the recirculating region and the pressure is nearly constant at the wall. Downstream, a strong recirculation is present in the region of strong pressure gradient. These characteristics were also seen by Horton (1968) in

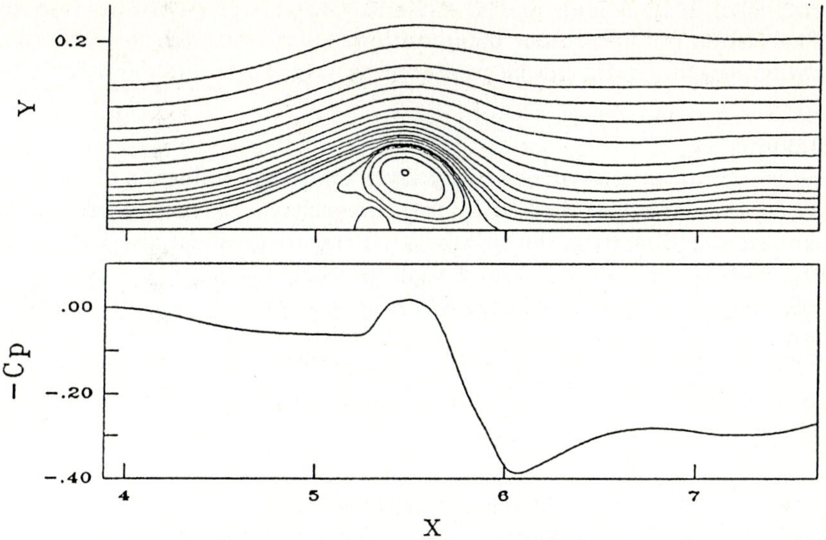

Fig. 8 Time-averaged streamlines and pressure coefficient, $Re_x = 120,544$ and $S = 0.22$.

his steady separation bubble study. Tani (1964), Gaster (1966), and others have also reported this same structure for steady separation bubbles.

The separation length, taken as the distance between separation and reattachment points of the time-averaged separation, was greatly affected by the presence of unsteady vortex shedding. For a given Reynolds number, the steady separation bubble increases in length when a stronger adverse pressure gradient is applied. After the shedding criterion is reached, the length of the time-averaged separation decreases when the pressure gradient is made more severe. The maximum time-averaged separation length therefore occurs at the onset of shedding.

Similar pressure distributions have been observed by McCullough and Gault (1949). A region of nearly constant pressure gradient was followed by a strong pressure gradient near reattachment. For separation bubbles with this pressure distribution, the length of the separation bubble decreased as the airfoil angle of attack increased. Gaster also found the separation length to decrease as the pressure gradient was increased. A strong similarity exists between the experimental results and the time-averaged unsteady separation from the computations.

Gaster observed a "bursting condition" where the length of the separation bubble suddenly changed. At pressure gradients above the bursting condition, he found short separation bubbles similar to the time-averaged bubbles found in this computation for unsteady separation. A constant pressure near separation was followed by a strong pressure gradient at reattachment. When the pressure gradient was reduced to the bursting condition, the bubble burst to a long bubble with no distinct strong adverse pressure gradient region. A similar pressure distribution was found for steady separation bubbles in the present computations.

The results of McCullough and Gault and Gaster were attributed to transition effects. The results of the present computations show that the pressure distribution and separation length results can be produced by large scale laminar structures.

4. Conclusions

The separation of a laminar boundary layer under the influence of an external adverse pressure gradient was studied in two dimensions. A strong pressure gradient created periodic shedding from the separation. Pressure gradient and Reynolds number effects were investigated. It was found that the Strouhal number based on the local velocity and boundary layer thickness was independent of Reynolds number and the pressure gradient. The unsteadiness in the separation was created by an inviscid instability of the shear layer which lifts from the wall. When the computational results were time-averaged, the closed separation bubble had a structure similar to laminar separation bubbles produced in experiments.

ACKNOWLEDGMENTS

This research has been supported by the Office of Naval Research under contract number N00014-84-K-0232. Computer facilities were provided by NASA-Ames Research Center. Three years of financial support for L.L. Pauley was provided by a fellowship from the National Science Foundation. These contributions are gratefully acknowledged.

REFERENCES

Despard, R.A., Miller, J.A., (1971), "Separation in Oscillating Boundary-Layer Flows", *J. Fluid Mech.*, 47: 21-31.

Gaster, M., (1966), "The Structure and Behavior of Laminar Separation Bubbles", AGARD CP-4, 813-854.

Horton, H.P. (1968), "Laminar Separation Bubbles in Two- and Three-Dimensional Incompressible Flow", Ph.D. Thesis, University of London.

Kim, J., Moin, P., (1985), "Application of a Fractional-Step Method to Incompressible Navier-Stokes Equations", *J. Comp. Phys.*, 59: 308-323.

Koromilas, C.A., Telionis, D.P., (1980), "Unsteady Laminar Separation: an Experimental Study", *J. Fluid Mech.*, 97: 347-384.

McCullough, G.B., and Gault, D.E., (1949), "Boundary-Layer and Stalling Characteristics of the NACA 64 A 006 Airfoil Section", NACA TN 1894.

Michalke, A., (1964), "On the Inviscid Instability of the Hyperbolic-Tangent Velocity Profile", *J. Fluid Mech.*, 19: 543- 556.

Pauley, L.L., Moin, P., Reynolds, W.C., (1988), "A Numerical Study of Unsteady Laminar Boundary Layer Separation", Report No. TF-34, Department of Mechanical Engineering, Stanford University, Stanford, CA.

Peake, D.T., Tobak, M., (1982), "Three-Dimensional Interactions and Vortical Flows with Emphasis on High Speeds", AGARDograph No. 252.

Tani, I., (1964), "Low-Speed Flows Involving Bubble Separations", *Prog. Aero. Sciences*, 5: 70-103.

BURSTING IN SEPARATING FLOW AND IN TRANSITION

Frank T. Smith
Department of Mathematics
University College London
Gower Street
London WC1E 6BT, U.K.

SUMMARY.

Three recent developments in the theory and associated computations of separating and transitional boundary layers are addressed. The three developments, which are very much inter-related, concern the following: a reversed-flow breakdown in the solution of the steady interacting boundary-layer equations; a finite-time break-up possible in any unsteady interactive boundary layer; and the absence of large-scale separation in turbulent flow. The aerodynamic implications for stall, intermittency, transition and turbulence effects are also emphasized.

§1. INTRODUCTION

Theoretical and computational work, from the late sixties to the present time, has led to almost complete accounts being available now to describe the separation process in steady laminar two-dimensional motion, in the subsonic, supersonic and to some extent the hypersonic ranges, for trailing-edge flows, corner flows and many others. See the reviews in Refs. 1-3 for example. The same applies to some degree in steady laminar three-dimensional motion, and for small- or large-scale separations, although to a much lesser degree for unsteady separation. Throughout viscous-inviscid interaction plays a key role.

The present contribution is intended to move on from the above aspects to the following recent developments, which are especially concerned with the inter-related issues of unsteadiness, stall, transition, turbulence effects, and their aerodynamic application. First (section 2), recent theory in Ref.4 proposes that a breakdown singularity can arise within the steady interactive boundary-layer framework whenever reversed flow is present. This provides a mechanism for stall to take place. Subsequent extensive computations for separated flow conducted to compare with and check on the theory are also described in section 2. The same ideas extend to three-dimensional and unsteady flows. In particular, a recent theoretical study (Ref.5) summarized in section 3 predicts that a break-up can occur in any unsteady separation controlled by the unsteady interactive boundary-layer equations. This is in effect an extension of previous special cases (see later references) exhibiting collapse processes in unsteady separation and it indicates a form of dynamic stall, with implications also for intermittency and other features in boundary-layer transition. Various unsteady-flow computations seem to be in line with the theoretical prediction, especially on the predicted large skin-friction, pressure-gradient and normal-velocity responses locally. The theory in sections 2,3 applies to both incompressible and compressible boundary layers, and further studies on extra compressibility effects are in progress. Finally, in section 4, recent work (Ref.6) on turbulent separation is described, based on Cebeci-Smith-like turbulence models. The major inference drawn there, again from analysis combined with computation, is in contrast with previous theories (e.g. Ref.7); it is suggested that large-scale separation need not take place at all in turbulent flow past thick airfoils or bluff bodies, at sufficiently high Reynolds numbers.

It should be remarked that, although there are several engineering criteria for separation-bursting (as well as for flat-plate transition, say) based on empiricism, rational theory on such bursting seems largely confined to Refs. 14,18, apart from the present more recent developments. Again, a review of the corresponding theory for transition, including other burstings and certain aspects of turbulent boundary-layer structures, is given in Ref.19.

The typical global Reynolds number Re is taken to be large and nondimensional velocities u,v, pressure p and streamwise and normal coordinates x,y are defined in the usual way, such that the free stream has $u = 1$, $v = 0$ and the airfoil chord is unity. The asymptotic theory aims among other things to provide comparisons, checks and guidelines for direct computations and experiments, as well as modelling and indicating more accurate computational approaches for high Reynolds numbers. External boundary layers are addressed mostly here, although the implications cross over to internal flows also. Further connections between the work in sections 2,3 and that on turbulent flows in section 4 are discussed in Ref.8, e.g. including a theoretical reasoning for the turbulent wall-layer scale $O(u_\tau^{-1}Re^{-1})$.

§2. STEADY LAMINAR SEPARATION AND STALL

Many interesting aerodynamic computations of laminar separating flows have been performed based on the interacting boundary-layer and similar approaches at finite or asymptotic values of the Reynolds number Re. The fairly typical situation is governed by the incompressible two-dimensional boundary-layer equations in scaled form

$$u = \frac{\partial\psi}{\partial y}, \quad v = -\frac{\partial\psi}{\partial x}, \tag{2.1a}$$

$$u\frac{\partial u}{\partial x} + v\frac{\partial u}{\partial y} = -p'(x) + \frac{\partial^2 u}{\partial y^2}, \tag{2.1b}$$

subject to no slip at the boundary and to an appropriate edge condition,

$$u = \psi = 0 \quad \text{at} \quad y = 0 , \tag{2.1c}$$

$$u \sim y + A(x) + \alpha f(x) \quad \text{as} \quad y \to \infty , \tag{2.1d}$$

respectively, but with the pressure $p(x)$ being unknown and related, via interaction, to the unknown displacement function $-A(x)$ or with $A(x)$ prescribed as in an inverse boundary-layer mode. Examples of the pressure-displacement interactions possible are numerous, e.g.

$$p(x) = \frac{1}{\pi}\int_{-\infty}^{\infty} \frac{dA}{d\xi}\,\frac{d\xi}{(x-\xi)} , \quad p = -\frac{dA}{dx} , \quad p = \pm A, \; p = -\frac{d^2A}{dx^2} , \tag{2.1e}$$

for subsonic, supersonic and hypersonic boundary layers, liquid-layer flows and jet or channel flows, in turn, for all of which the viscous-inviscid interaction structure is of triple-deck character or similar. Another common form of interaction used at finite Re has (2.1d,e) replaced by

$$u \to u_e(x) , \quad \psi \sim u_e(x)(y - \delta(x) + \alpha f(x)) \quad \text{as} \quad y \to \infty \tag{2.2a,b}$$

$$u_e u_e' = -p' , \quad u_e(x) = 1 - \frac{Re^{-1/2}}{\pi}\int_{-\infty}^{\infty} (u_e\delta)'(\xi)\,\frac{d\xi}{(x-\xi)} . \tag{2.2c,d}$$

Suitable constraints are imposed far upstream and downstream. Again, a Prandtl shift is applied above so that in (2.1d) $f(x)$ denotes the given wall shape, with an $O(1)$ scaling factor α . The usual task then is to solve (2.1a-c) with (2.1d,e) or (2.2) for various values of the parameter α, with regular separation/flow reversal occurring for sufficiently large α values.

A major issue is what happens to the flow solution as α is increased and the motion becomes increasingly separated. Based on a variety of evidence accumulating, a recent theoretical proposal (Ref.4) is that in general a singularity is encountered within the reversed-flow region $(u < 0)$ at a finite value $\alpha = \alpha_s$, i.e. the interactive approach breaks down then. Two main types of breakdown are proposed for $\alpha_s \to \alpha -$, namely

$$K = 3/2 \quad \text{and} \quad K \to 1 , \tag{2.3}$$

where for $\alpha = \alpha_s$ the local streamfunction profile ψ at the breakdown station, $x = x_s$

say, in the reversed flow, develops the form $\psi \propto -y^K$ as $y \to 0+$. Of the two types in (2.3) the second option which involves mainly inviscid dynamics seems to be favoured on theoretical grounds and is summarized below.

The case $K \to 1$ is associated with a short-scale region in which $x_s - x = \Delta X$, with Δ small and $\Delta \to 0$ as $\alpha \to \alpha_s-$, and u, ψ, p all remain $O(1)$ across most of the boundary layer. So the inviscid version of the boundary-layer equations holds,

$$u = \frac{\partial \psi}{\partial y} \quad , \quad V = -\frac{\partial \psi}{\partial X} \quad , \tag{2.4a}$$

$$u\frac{\partial u}{\partial X} + V\frac{\partial u}{\partial y} = -\frac{dp}{dX} \tag{2.4b}$$

with the upstream and downstream profiles of ψ, u and the pressures p being different as $X \to \pm\infty$, and A is replaced by $A(x_s)$, constant, in effect. The system can be treated analytically by consideration of vorticity conservation and Bernoulli's relation, indicating that the pressure $p(X)$ can increase from a constant value π_1 at $X = -\infty$ to another constant, π_2, at $X = \infty$ This forces deceleration of the motion and a turning back of much of the eddy motion present at $-\infty$, in readiness for the reattachment further downstream. Also, the profiles $u(\pm\infty)$ must both contain some reversed flow, for self-consistency in the viscous sublayer where y is $O(\Delta^{1/2})$. A Blasius-like response is induced for X large and negative and the local skin-friction (τ_w)minimum occurs at an $O(1)$ value of X, giving

$$\min(\tau_w) = -O(\Delta^{-1/2}) \ , \tag{2.5a}$$

where $\tau_w \equiv \partial u/\partial y\,(x,0)$. Hence shock-like discontinuities are implied in the velocity u and pressure p on either side of $x = x_s$, e.g.

$$p(x_s-) \neq p(x_s+) \ , \tag{2.5b}$$

and the maximum pressure gradient is predicted to occur at an $O(1)$ value of X, so that

$$\max\left(\frac{dp}{dx}\right) = +O(\Delta^{-1}) \ . \tag{2.5c}$$

Further details are presented in Ref.4. Again, the breakdown above, like that in the next section, can arise in virtually all the kinds of separating flows possible, including the supersonic, subsonic and hypersonic ranges, two- and three- dimensional motions, external or internal conditions and so on.

As a check on the above theoretical breakdown prediction new computational solutions have been obtained in the particular case of the supersonic ramp flows, where the interaction law is $p = -dA/dx$ and the wall shape has $f(x) = 0$ for $x < 0$, $f(x) = x$ for $x > 0$, with $p(\infty) = \alpha$. The numerical method used is essentially a global Newton iteration procedure, with finite differencing, in which the whole flowfield solution is guessed, a linearized system is solved for the unknown increments throughout the flowfield by repeated marching forward and/or backward depending on whether the local motion is forward or reversed, the increments are used to update the global guess, and the iterative scheme is then continued. The results obtained by this new method have gone further into separated flows than any other previous attempts as far as we know. The results agree first with Ref.9 at lower α, separation first occurring at $\alpha \approx 1.57$, and second with results obtained by an alternative method described in Ref.10 for somewhat increased α, but the present numerical solutions have then been continued up to much higher α with accuracy preserved with respect to grid effects. There the maximum pressure gradient increases dramatically, as does the negative minimum in the skin friction, and the computed results appear to be well in line with the theoretical prediction above. In particular, the behaviour

$$\min(\tau_w) \to -\infty\ , \quad \max(dp/dx) \to +\infty \quad , \tag{2.6a}$$

$$\text{with} \quad \frac{\max(dp/dx)}{[\min(\tau_w)]^2} \to O(1)\ , \quad \text{as} \quad \alpha \to \alpha_s -\ , \tag{2.6b}$$

is found to agree very well with the computational trends at the higher α values, within the reversed-flow region.

The main implications are that a stall, in the sense of a substantial change in the flow structure, takes place at $\alpha = \alpha_s$, and that new physical effects must come to the fore in the neighbourhood of $x = x_s$, principally a non-zero normal pressure gradient $\partial p/\partial y$ due to the enhanced normal velocities which are induced locally when the breakdown in (2.5) occurs. These and other features, including comparisons with previous computations, are discussed in Ref.4.

§3. UNSTEADY SEPARATION, STALL & TRANSITION

Much work has been done theoretically and computationally on unsteady separation, in both the interactive [p(x,t) unknown] and the non-interactive [p(x,t) prescribed] regimes. The present work concerns the interactive case, which is the more interesting and useful, and has application to boundary-layer intermittency and transition as well as to unsteady separation.

The typical governing equations here are the appropriate unsteady counterparts of those in section 2, so that (2.1b) is replaced by

$$\frac{\partial u}{\partial t} + u\frac{\partial u}{\partial x} + v\frac{\partial u}{\partial y} = -\frac{\partial p}{\partial x}(x,t) + \frac{\partial^2 u}{\partial y^2} \qquad (3.1)$$

and $(f,A,p)(x) \to (f,A,p)(x,t)$ in (2.1c-e) in effect. Many aerodynamic computations have been performed for the interactive system with (3.1), to yield unsteady separation; see e.g. Refs. 11, 12 and references therein. The present contribution, which like that in section 2 covers a very wide range of conditions (subsonic, supersonic and so on), suggests that a localised singularity or break-up is likely to arise in the unsteady motion at a finite time, say as $t \to t_0-$, and the break-up can take two dominant forms, as described below (based on Ref. 5).

The first type is a 'moderate' break-up, in which the length scale

$$x - x_0 = c(t - t_0) + (t_0 - t)^{3/2}\xi, \qquad (3.2)$$

operates near the break-up position $x = x_0$, where ξ is of order unity, as $t \to t_0-$, and the pressure takes the form

$$p - p_0 = (t_0 - t)^{1/2}p_1(\xi) + (t_0 - t)^{3/4}p_2(\xi) + \ldots \qquad (3.3)$$

where $p_0 \equiv p(x_0)$ is constant and p_1, p_2, ... are to be determined. The unsteady flow solution then sub-divides into three zones in the y direction, a main inviscid zone in which y is $O(1)$ and the streamwise velocity has the expansion

$$u = u_0(y) + (t_0 - t)^{1/2}u_1(\xi, y) + \ldots, \qquad (3.4)$$

a thin viscous wall layer where y is $O(t_0 - t)^{3/4}$, and a thin critical layer at the position where $u_0(y)$ is equal to the constant phase speed c. Analysis of the governing equations (3.1), etc., under the expansions such as (3.3), (3.4), then leads to the equation

$$p_1p_1' = b_1(p_1 - 3\xi p_1') \qquad (3.5a)$$

for the unknown pressure contribution $p_1(\xi)$, with b_1 being a constant. The solution therefore takes the implicit form

$$\xi = -b_2p_1 - b_3p_1^3, \qquad (3.5b)$$

where b_2, b_3 are constants having the same sign. This shows that the $p(\xi)$ solution is single-valued as required, and monotonic in ξ. Further, $|p_1| \propto |\xi|^{1/3}$ as $\xi \to \pm\infty$, which matches with the flow solution away from the break-up station $x = x_0$, giving in particular the behaviour

$$p - p_0 \propto |x - x_0|^{1/3} \text{ as } x \to x_0 \pm . \tag{3.6}$$

Hence a singularity in the pressure gradient, analogous to that arising in inviscid gas dynamics, is predicted at the break-up time $t = t_0$.

The second type of break-up is the 'severe' type, involving much stronger singularities. Here the principal middle zone of the flow at $t = t_0-$ is governed by the expansions

$$p = \bar{p}(\bar{x}) + \ldots , \quad u = \bar{u}(\bar{x}) + \ldots \tag{3.7a,b}$$

where the $O(1)$ local coordinate $\bar{x}$ is defined by

$$x - x_0 = (t_0 - t)\bar{x} . \tag{3.7c}$$

As a result the inviscid similarity equations

$$\bar{u} = \frac{\partial\bar{\psi}}{\partial y} , \quad (\bar{x} + \bar{u}) \frac{\partial\bar{u}}{\partial\bar{x}} - \frac{\partial\bar{\psi}}{\partial\bar{x}} \frac{\partial\bar{u}}{\partial y} = - \bar{p}'(\bar{x}) \tag{3.8a,b}$$

hold from (3.1), etc., subject to the constraints

$$\bar{\psi} = 0 \text{ at } y = 0+ , \tag{3.8c}$$

$$\bar{u} \sim y + \beta_0 \text{ as } y \to \infty \tag{3.8d}$$

where $\beta_0 \equiv A(x_0, t_0) + \alpha f(x_0, t_0)$ is constant. A viscous wall layer of thickness $O(t_0 - t)$ is also present. The primary task however is to solve (3.8a-d), and an essential feature of the solution is that the end conditions in terms of $\bar{x}$ are reached at finite distances, so that

$$\bar{u} \to \bar{u}_1(y) \text{ as } \bar{x} \to d_1 + , \tag{3.9a}$$

$$\bar{u} \to \bar{u}_2(y) \text{ as } \bar{x} \to d_2 - , \tag{3.9b}$$

where $d_1 < d_2$. Here the end velocity profiles $\bar{u}_1$, $\bar{u}_2$ are inter-related and d_1, d_2

act effectively as negative phase speeds. Solution properties and further details of the break-up structure are given in Ref.5. It is noteworthy that this severe case yields a discontinuity in the pressure itself at the break-up time, with

$$p(x_0-) \neq p(x_0+) , \tag{3.10}$$

in contrast with the moderate case where the discontinuity is only in the pressure gradient $\partial p/\partial x$, as in (3.6). Further, there is a whole range of possible singular forms stretching between the two particular forms described above and yielding increasingly discontinuous pressure gradients which culminate in the limiting severe case where (3.10) holds.

The theoretical predictions above for break-up in unsteady separation appear to agree well (Refs. 5, 12 and references therein) with the many previous computations done for unsteady interacting boundary layers. There is qualitative agreement also with unsteady Navier-Stokes computations (e.g. Ref.13). The focussed break-up involved at $x = x_0$ implies that locally extra short time and length scales must come into operation, analogous to the previous section, in particular bringing in nonzero $\partial p/\partial y$ contributions, and an unsteady Euler stage then takes effect to control the local concentration of high vorticity and enhanced pressure gradients. The break-up has ramifications for dynamic stall, similar to some extent to the more limited cases considered in Ref.14 for example, and it has potential application also with regard to intermittency in boundary-layer transition: again see Ref.5.

§4. ON TURBULENT SEPARATION

The theoretical study described below is from Ref.6's work on turbulence-modelled separation and follows on from that in Ref.15 on turbulent wake flow. The aim is to understand in an asymptotic sense the nature of such separation when a representative and much-used turbulence model is taken throughout, for quasi-steady turbulent flow, and to judge whether the separation properties thus predicted are sensible or not. There are signficant links also with the theoretical work in progress on transition (including that referred to in section 3) and turbulence dynamics (e.g. Ref.8).

The specific turbulence model chosen is the Cebeci-Smith one (Ref.16). With that model, and introducing for the moment an artificial "turbulence-level" parameter Tu , we address first the classical non-interactive boundary layer assumed on a thick airfoil, i.e. the composite scaled system consisting of (2.1a) combined with

$$u\frac{\partial u}{\partial x} + v\frac{\partial u}{\partial y} = u_e u_e' + Tu\frac{\partial}{\partial y}\left(y^2 B\left(\frac{\partial u}{\partial y}\right)^2\right) + \frac{\partial^2 u}{\partial y^2} \tag{4.1}$$

and $u = v = 0$ at $y = 0$, $u \to u_e(x)$ as $y \to \infty$. Here $u_e(x)$ is the prescribed edge velocity given by the assumed attached inviscid solution, y is scaled on $Re^{-1/2}$, B is unity except very near the wall, the term $\propto Tu$ is replaced by a diffusive term in the usual way at large y, and the parameter Tu is taken to be $O(1)$ at first, although it is raised to its correct large size subsequently. For the specific case of flow past a circular cylinder, where $u_e(x) \propto \sin x$, $0 \le x \le \pi$, computational solutions are given in Ref.6 for various values of Tu and the dependence of the Goldstein singular point $x = x_1$ on Tu is tracked. At zero Tu, x lies just beyond $\pi/2$ but as Tu is increased x_1 increases monotonically. An analytical investigation of (4.1) for large Tu then indicates that

$$x_1 \to x_{TE} \quad \text{as} \quad Tu \to \infty , \tag{4.2}$$

where $x_{TE} = \pi$ is the trailing-edge (rear stagnation-point) position, and the flow solution becomes multi-structured. For most x stations two main tiers in y occur, approximately where $y \sim Tu$ (outer small-deficit region) and $y \sim Tu^{-1}$ (inner stress-dominated region), but as $x \to x_{TE}$ the outer thickness increases like $(x_{TE} - x)^{-1}$ to a first approximation. That then leads to a localised zone near the trailing edge where $(x - x_{TE})$ is of order $[\ell n(Tu)]^{-1/2}$ approximately and the velocity deficit in the outer region rises to become comparable with u_e. The associated increase in the outer vorticity forces the nonlinear solution there, in the outer region to hit a zero-velocity condition at some station eventually, just ahead of x_{TE}, and it is found from analysis of the corresponding inner regions that the Goldstein point lies very close to that zero-velocity station. In particular the asymptotes

$$x_1 - x_{TE} = O[(\ell n\ Tu)^{-1/2}], \text{ thickness} \sim Re^{-1/2}(\ell n\ Tu)^{1/2} \tag{4.3}$$

are obtained as a result, to first approximation. These and other features are in keeping with the computations of (4.1) for increasing Tu.

The main implication from the above is that in the correct physical regime where Tu is large, $O(Re^{1/2})$, the Goldstein singular point is pushed very close to the trailing edge and therefore is not encountered in the two-tiered turbulent boundary layer. This suggests in formal terms that separation can be suppressed substantially.

A similar conclusion arises from a second approach, where large-scale separation is assumed (in contrast with (4.1)) at the start for laminar flow ($Tu = 0$), say past a thick airfoil or circular cylinder again, and the effects of increasing Tu from zero are examined. It is found that as Tu becomes large the separation position, based on extended Kirchhoff free-streamline theory, moves progressively downstream and the eddy size shrinks, tending ultimately to produce only an asymptotically small reversed-flow eddy concentrated near the trailing edge. This overall behaviour is due mainly to the strong increase in the skin friction τ_w in the boundary layer upstream of separation as Tu increases.

The flow structure implied, then, for a fully turbulent motion past a thick airfoil, say the circular cylinder, is essentially that of attached non-interactive flow. Large-scale separation need not occur at all, at least according to the assumed turbulence model. Given the attached inviscid solution, the two-tiered turbulent boundary layer remains attached right from the leading edge to the onset of the trailing edge. Its small-deficit outer thickness is $O(u_\tau)$, the inner stress sublayer has thickness $O(u_\tau^{-1}Re^{-1})$, with the friction velocity u_τ being of order $(\ell n Re)^{-1}$, and the resultant skin friction $\tau_w \propto u_e^2$ stays positive (see also Ref.17). Separation is confined to the vicinity of the trailing edge x_{TE}, specifically to within a distance of approximately $O(u_\tau^{1/2})$ of x_{TE}. The latter scale is due to the expansion $\propto(x_{TE} - x)^{-1}$ of the outer small-deficit tier of the boundary layer, much as in (4.3) and implying a non-slender scale where $u_\tau(x_{TE} - x)^{-1}$ becomes comparable with $(x_{TE} - x)$, that is, at distances of order $u_\tau^{1/2}$, and significant rotational effects occur. The governing equations there are the Euler equations plus the turbulent stress terms, analogous to the first situation addressed in this section.

This theoretical picture may seem bizarre at first sight but it is felt to be not inconsistent with computational and experimental findings. Some computed results for turbulent separation do indeed show the separated flow zone contracting substantially as Re is increased and the same is possibly true in experiments, although measurements then appear more difficult. Further, u_τ in practice is often about 0.1 - 0.2, and so the separation distance $u_\tau^{1/2} \approx 0.3 - 0.5$, which corresponds to a considerable portion of the airfoil chord.

REFERENCES

1. A F Messiter, "Boundary-layer interaction theory", Trans. A.S.M.E. Jnl. Appl. Mech., Vol 50 (1983), 1104-1113.
2. F T Smith, "Steady and unsteady boundary layer separation", Annual Revs. of Fluid Mechs., Vol 18 (1986), 197-220.
3. K Stewartson, "D'Alembert's paradox", S.I.A.M. Review, Vol 23 (1981), 308-343.
4. F T Smith, "A reversed-flow singularity in interacting boundary layers", Proc. Roy. Soc., Vol A420 (1988), 21-52.
5. F T Smith, "Finite-time break-up can occur in any unsteady interactive boundary layer", Mathematika, Vol 35 (1988), 256-273.
6. A J Neish, Ph.D. thesis, Univ. of London (1989); and A J Neish and F T Smith, theory on turbulent separation, in preparation (1989).
7. V V Sychev & Vik V Sychev, "On turbulent separation", Zh. Vyssh. Mat. Mekh. Fiz., Vol 20 (1980), 1500-1512.
8. F T Smith, D J Doorly and A P Rothmayer, "On displacement-thickness, wall-layer and mid-flow scales in turbulent boundary layers, and slugs of vorticity in channel and pipe flows", J. Fluid Mech., in press (1989).
9. D P Rizzetta, O R Burggraf and R Jenson, "Triple-deck solutions for viscous supersonic and hypersonic flow past corners", J. Fluid Mech., Vol 89 (1978), 535-552.
10. F T Smith, "Separating flow: small-scale, large-scale, and nonlinear steady effects", in Boundary-Layer Separation (eds. F T Smith and S N Brown), Springer-Verlag, Heidelberg (1987).
11. J D A Walker, V J Peridier and F T Smith, "Methods for the calculation of unsteady separation", A.I.A.A. 26th Aerospace Sci. Mtg., January (1988) Reno, Nevada, paper 88-0604.
12. F T Smith, "Steady and unsteady 3D interactive boundary layers", R T Davis Mem. Symp. (1987) and Computers & Fluids, to appear (1989).

13. H Fasel, in Turbulence and Chaotic Phenomena in Fluids, (ed. T Tatsumi) Elsevier Sci. Pub. B.V. (1984).
14. J W Elliott and F T Smith, "Dynamic stall due to unsteady marginal separation", J. Fluid Mech., Vol 179 (1987), 489-512; and F T Smith, "Concerning dynamic stall", Aeron.Quarterly, November (1982), 331-352.
15. A J Neish and F T Smith, "The turbulent boundary layer and wake of an aligned flat plate", J. Eng. Maths., Vol 22 (1988), 15-42.
16. T Cebeci and A M O Smith, Analysis of Turbulent Boundary Layers, Academic Press, New York (1974).
17. A A Townsend, The Structure of Turbulent Shear Flows, second ed. (1976).
18. F T Smith, Nonlinear effects and non-parallel flows: the collapse of separating motion, Utd. Tech. Rept. UTRC 85-55; and S N Brown, H K Cheng and F T Smith, J. Fluid Mech., Vol. 193 (1988) 191-216.
19. F T Smith, Interactions in boundary-layer transition, Invited Lecture, ICTAM Congress, Grenoble, 1988, published in "Theoretical and Applied Mechanics", (eds P. Germain, M.Piau, D. Caillerie), Elsevier Sci. Pub. B.V. (1989).

ACKNOWLEDGEMENTS

Thanks are due to the Royal Aircraft Establishment, Farnborough, the Science and Engineering Research Council, the University of London Computing Centre, and the United Technologies Independent Research Program, for support of various parts of this research, and to the author's colleagues Dr. A.J. Neish and Dr. D.J. Doorly for their collaboration in, and numerous discussions on, certain aspects of this work.

A REVIEW OF LOW REYNOLDS NUMBER AERODYNAMIC RESEARCH AT THE UNIVERSITY OF GLASGOW

R.A.MCD. Galbraith and F.N. Coton
Department of Aerospace Engineering, University of Glasgow,
Glasgow G.12 8QQ

SUMMARY

The paper gives the succinct overview of the Low Reynolds number work carried out at Glasgow University, together with some of the salient and interesting results observed. The work covers both experimental and computational aspects, with the emphasis being on a practical knowledge of aerofoil performance and prediction. The measured performance of four aerofoils is presented along with a discussion of specific aspects of the flow observed. A low-speed performance prediction scheme is outlined and its range of applicability highlighted by some particular examples. Finally, the main conclusions of the research programme are discussed.

INTRODUCTION.

The contemporary Low Reynolds number work at Glasgow has its origin in the high-performance aerofoil family of Nonweiler [1] developed for man-powered flight. One of these sections, GU 25-5(11)8. was wind tunnel tested by Kelling [2] and achieved its desired performance targets. Its low-speed characteristics rendered it most suitable for the canard wings of many microlight aircraft but, after some experience, it received a bad press [3]. A number of factors contributed to this including sensitivity to build tolerance and reduced performance in rain. It is interesting to note that the latter effect was primarily a result of water droplet formation on the polished wing surfaces. If a matt paint finish had been applied, the associated modification in surface tension effects would have ensured a more even distribution of water over the surface so retaining its original shape and hence performance [4]. As a result of these problems, however, the GA(W)-1 aerofoil and its derivatives began to replace the GU 25-5(11)8 in most designs and claims of significant performance improvement were made. For this reason, a comprehensive wind tunnel study, at Reynolds numbers from 50,000 to 600,000, was initiated to assess the relative performance of the two aerofoils [5, 6]. Subsequently, two other aerofoil sections were tested in a similar manner for wind turbine applications [7].

Coincident with the above, was the development of a low speed aerofoil design methodology [8], the first requirement of which was the acquisition of a high quality aerofoil performance predictor [9]. The incentive here was the very real market in Remotely Piloted Vehicles and high altitude compressor blades. The code developed was very general and became of interest to other sectors such as wind energy and also interfaced with research topics in helicopter aerodynamics, including dynamic stall [10, 11] and blade vortex interaction [12].

Aerofoil tests conducted to date have all been of a comparative nature, being similar in both the geometric and dynamic scales. It is therefore hoped that the relative performance of each section is well represented by the results obtained. Of the four aerofoils tested to date, it is interesting to note that the most insensitive to the variation in Reynolds number was the poorest performer in terms of lift and vice-versa. It is also apparent, from the data, that no simple extrapolation of performance characteristics from higher speeds is possible, due to the varied response of each aerofoil at lower Reynolds numbers. Consequently, identification of the relevant flow regions becomes significant and, in this case, was facilitated by both smoke and oil flow visualisation experiments.

Empirical data have proved valuable in assessing the applicable range of the developed performance code and, in some cases, have highlighted physical modelling limitations. Whilst prediction over the entire low Reynolds number range is not envisaged, the present technique does warn of the presence of analytically intractable flow phenomena. Since, also, the analysis method currently forms part of a design methodology, it is essential that reliable predictions are forthcoming from it.

EXPERIMENTS.

As reported in [5, 6, 7] , the four test aerofoils were pressure tapped along their mid-span and the associated surface static pressures were recorded in the Glasgow University 3' x 3' closed return tunnel. These pressures were then integrated to obtain the normal aerodynamic coefficients. No corrections were applied unless, as in [7], comparisons were made with other published data.

Figure 1 illustrates the obtained lift coefficients with respect to incidence and Reynolds numbers. It is to be anticipated that, to a first order, the general appearance of each set of contours will be insensitive to the extraneous disturbance environment. The maximum lift coefficient would, however, possibly be influenced by extraneous effects and the overall pattern may be displaced up or down the Reynolds number scale. The amount by which this occurs will vary from section to section but the more sensitive GU 25-5(11)8 section is liable to be significantly affected.

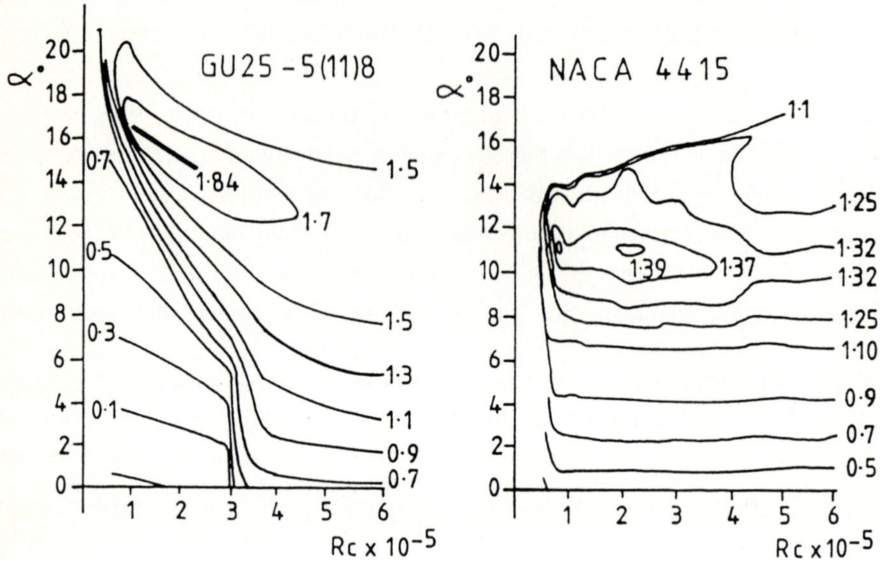

FIGURE 1.

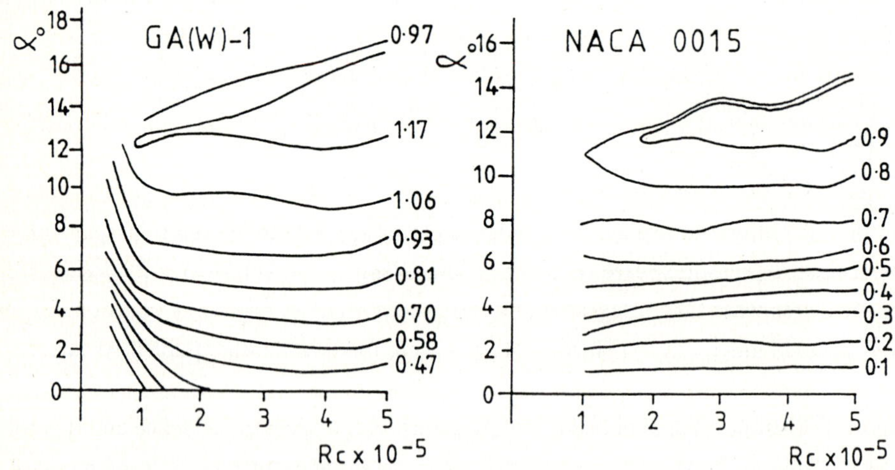

With the above in mind, there are several salient features worthy of note. First, the very striking difference between each aerofoil; this is particularly noticeable between the GU 25-5(11)8 and the NACA 0015. Comparing these two aerofoils, it may be observed that they appear to be at opposite extremes of Reynolds number sensitivity. The NACA 0015 shows hardly any change at all, as evidenced by the near horizontal contours, and the lift coefficient at $a = 4^0$ is always about 0.4. In contrast to this, the GU 25-5(11)8 lift coefficient, for the same incidence, varies from 0.1 to around 1.1 over the Reynolds number range; a tenfold increase. Of the two remaining aerofoils, the NACA 4415 has a constant Cl of 0.9 at 4^0 incidence and the corresponding GA(W)-1 value is 0.7 above a Reynolds number of 200,000; there is a marked decrease below this.

The obtained maximum lift coefficients for the GA(W)-1 section are noticeably well below those reported by McGhee and Beasley [13], where Cl max. = 1.63 at Re = 2×10^6 and Cl max. = 2.12 at Re = 12×10^6. These Reynolds numbers are obviously much higher than in Fig. 1 and the low Cl values presented here are most likely a result of the influence of significant, well formed separation bubbles.

A similar differential exists for the NACA 4415 maximum lift when compared with the higher Reynolds number data of Jacobs and Sherman [14]. Figure 2 highlights the development of the upper surface bubble observed in the present work by depicting the evolution of the static pressure distribution as incidence is increased. The extent of the bubble was obtained from a consideration of these profiles and oil flow visualisation results. It is evident that the bubble is

FIGURE 2.

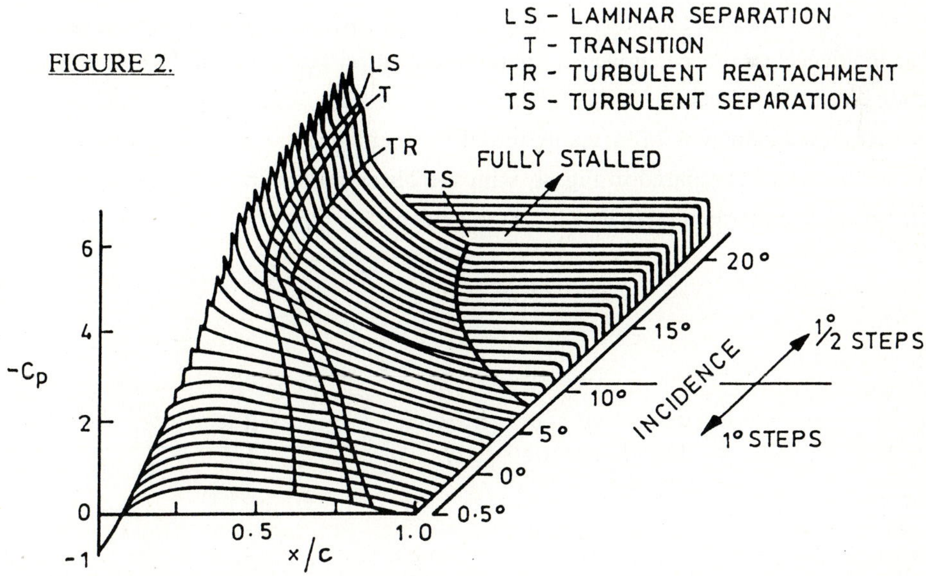

initially formed to the rear of the surface and migrates to the short classical form as the incidence increases. This continues until the combined stall condition occurs. The static pressure development on the upper surface is therefore both obvious and well behaved. It is common to be less concerned with the lower surface, unless something peculiar is observed. Figure 3 illustrates such a peculiarity (the kink) in the lift and pitching moment curves which was evident over the entire Reynolds number range considered. As reported in [15] this appears to be a relatively common effect and, from the discussions of Fig. 2, it is unlikely that the cause is associated with the upper surface flow development; so the lower surface becomes of interest.

FIGURE 4.

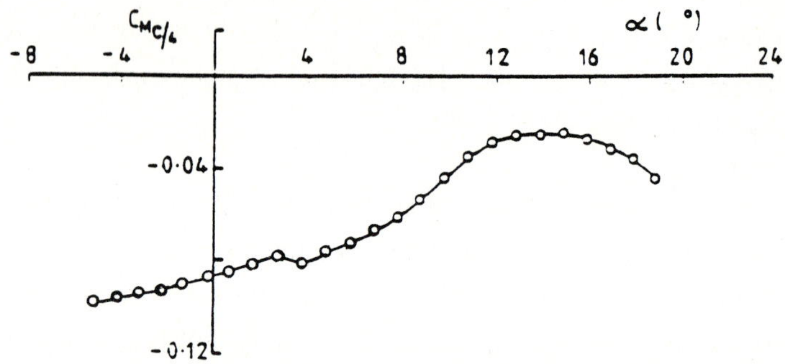

The performance prediction code [9] discussed below indicated, at low incidence, the likely existence of a lower surface laminar separation bubble, followed by a turbulent boundary layer. As the incidence increased, however, natural transition replaced the bubble and the transition location moved towards the trailing edge. The progression to fully laminar flow was an exceedingly rapid phenomenon and high resolution of the incidence at which this occurred was obtained. ,These results are compared in Fig. 4 with the observed occurrence incidence of the kink. The correspondence between these data is all too evident.

FIGURE 3.

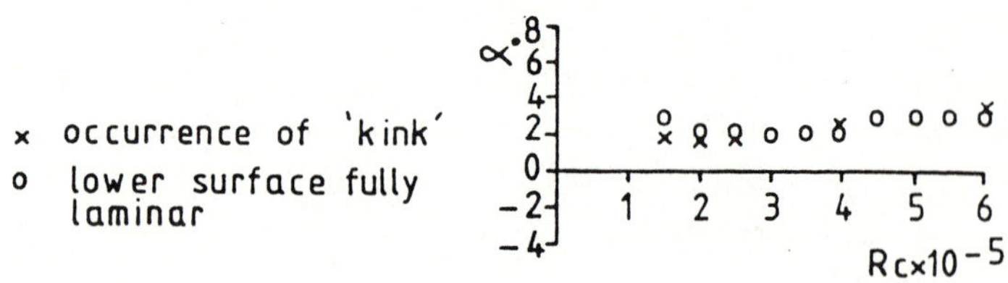

To assess the validity of predictions, a series of lower surface oil flow tests [16] were performed at the highest Reynolds number. Even although the surface stress levels were low, the resulting low incidence patterns clearly indicated the existence of a bubble followed by a turbulent boundary layer. At the same incidence as the occurrence of the kink, laminar flow was indicated over the entire surface.

The data obtained and discussed have been most useful, and will continue to be, in assessing the predictive capability of performance codes for low Reynolds number cases. It is also encouraging to note that, for the NACA0015 and 4415, there is good general agreement with the lift coefficient results of Jacobs and Sherman [14] at similar Reynolds numbers. Whilst their data were obtained at a much higher (2.0%) turbulence intensity than the present results (<0.5%) comparison of the two provides insight into the gross effects of high turbulence levels at low Reynolds numbers.

COMPUTATIONAL DEVELOPMENTS

Development [9] of a low speed two dimensional viscous-inviscid aerofoil performance prediction scheme was initiated to compliment the ongoing aerofoil testing programme. It was envisaged that the method would predict both turbulent and gross laminar separations and would provide a basis for the estimation of separation bubble effects on overall aerofoil performance. Currently, this scheme forms part of a comprehensive aerofoil design package where it is used to provide estimates of the performance of designed sections.

The viscous-inviscid analysis scheme is of the direct type with the external potential flow solution providing the velocity distribution input to the boundary-layer method. During calculation, the viscous solution yields new boundary conditions for the potential flow computation and so an iterative scheme is set up. This process is repeated until some predetermined convergence criterion is satisfied.

The external potential flow solution is provided by a linearly varying vortex panel method with incorporated wake model [17]. The model assumes that the wake region is bounded by free shear layers which have no significant thickness and are also streamlines of the flow. In the panel method, these shear layers are modelled by a series of uniform strength vorticity panels. During a viscous-inviscid calculation, the wake shape is iteratively determined from an initial estimate which is dependent on the aerofoil geometry and, where laminar separation exists, the boundary layer state at separation [18].

Solution of the laminar and turbulent boundary layer problems is achieved by application of an integral technique based on the method of Le Foll [19]. As this particular scheme is well behaved in regions of low skin friction, it is well suited to incorporation in a viscous-inviscid interaction scheme where accurate separation predictions are necessary. The location of boundary layer transition may either be fixed, calculated or, if laminar separation precedes transition, via Horton's [20] empirical separation bubble model.

The ability of the method to model gross laminar separation is demonstrated in Fig. 5 where low Reynolds number predictions for the GU 25-5(11)8 and FX 63-137 aerofoils are compared with empirical data. Since the inviscid wake shape correlation for laminar separation [18] was developed

FIGURE 5.

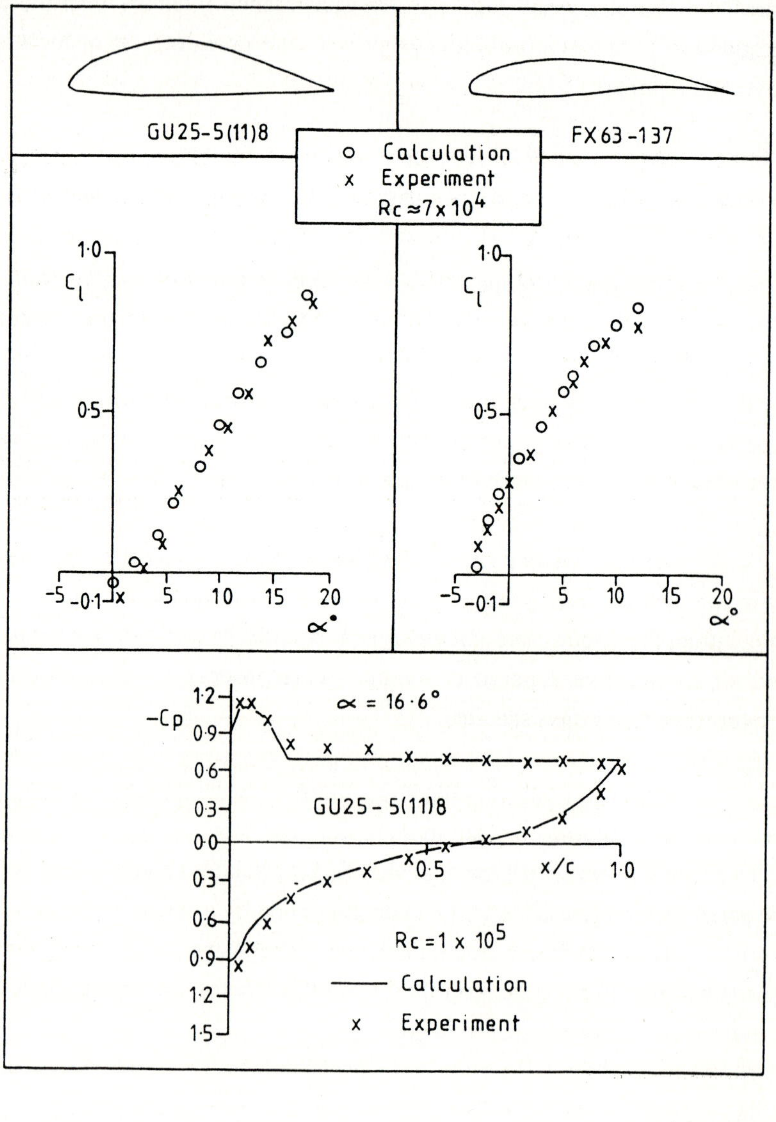

from a study of the GU 25-5(11)8 aerofoil, the quality of agreement in that case is not surprising. More significant, however, is the good prediction obtained for the FX 63-137 aerofoil when compared with data obtained at a much lower freestream turbulence level than that for which the original correlation was derived. For both aerofoils, successful predictions were obtained for cases with up to 95% upper surface separation. The low Reynolds number range of the method is, however, limited by its inability to account for simultaneous separation on both surfaces.

The viscous-inviscid interaction scheme has also been demonstrated capable of predicting flows with turbulent trailing edge separation. Figure 6 illustrates the agreement obtained with wind tunnel data for various aerofoils at low Reynolds numbers. In most cases, a satisfactory prediction was obtained by approximating the effect of a separation bubble with fixed transition at the predicted laminar separation point. This approach, however, produced some anomalies which appeared to be a consequence of inappropriate turbulent boundary layer starting conditions.

Of particular interest was the difference between the measured and predicted lift coefficients for the NACA 4415 aerofoil around stall. In this case, the calculation indicated the presence of a leading edge laminar separation bubble but, due to the simplified transition mechanism, did not account for its effect on the subsequent turbulent boundary layer development. For this reason, a study was conducted [21] to assess the possible influence of such a bubble formation on turbulent boundary-layer development and, so, aerofoil performance.

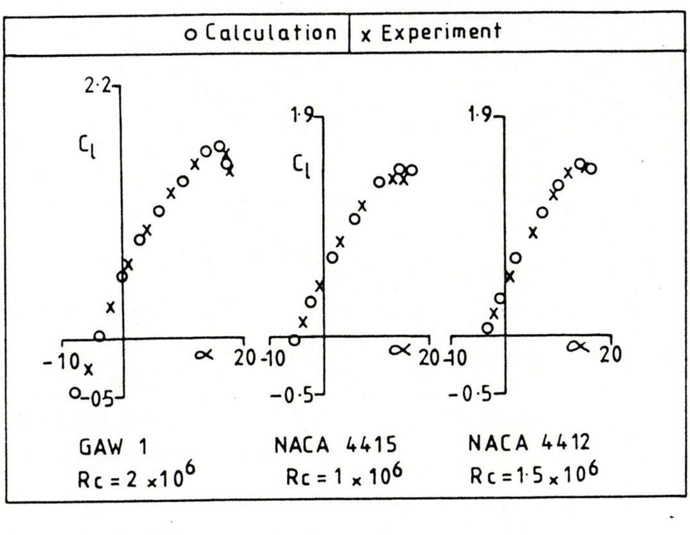

FIGURE 6.

A direct boundary-layer calculation scheme, capable of predicting short regions of separated laminar and turbulent flow, was developed to enable prediction of the growth within a separation bubble. In the method, the form factor and skin friction terms do not appear explicitly and instead semi-empirical functions which are well behaved around separation are the dependent variables. As a result, no difficulty is encountered when passing through separation and reattachment points.

In the study, it was found that the pressure gradient around the free shear layer transition location was particularly influential. Initially, a standard Horton type external velocity distribution was prescribed which resulted in a thinner turbulent boundary layer and delayed stall. Relaxation of the sharp transition corner resulted in substantial shear layer growth within the bubble and a subsequent early stall with a reduced CL max. Figure 7 illustrates these results and highlights the improved agreement between the prediction and measured data achieved by relaxation of the sharp corner.

FIGURE 7.

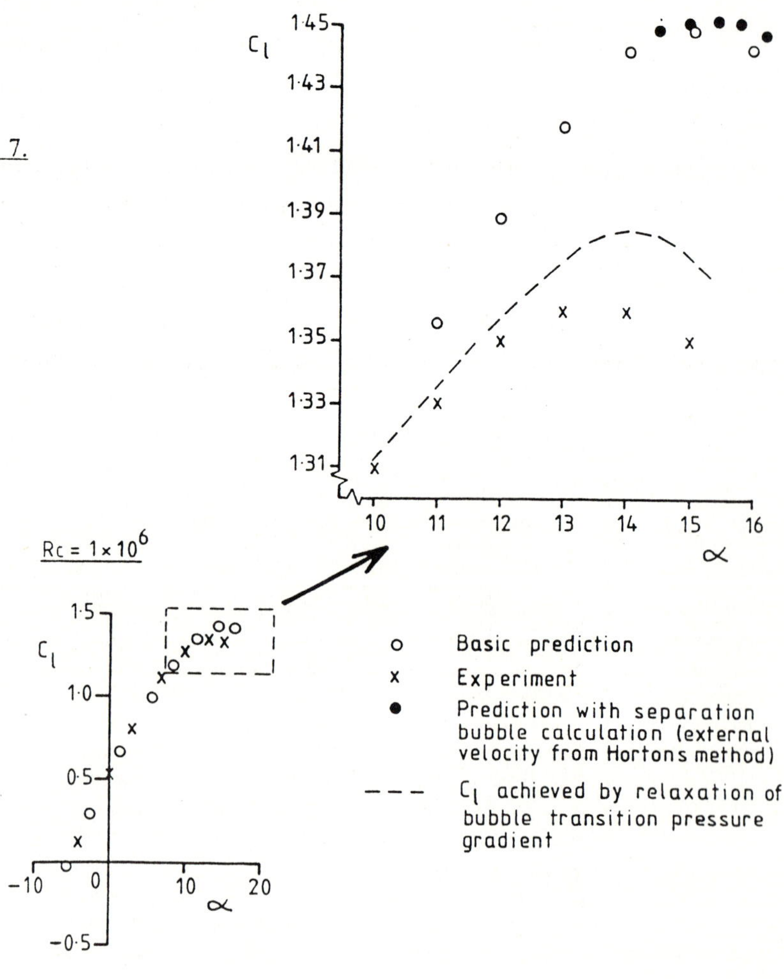

The implications of this study for semi inverse or fully inverse models, is that the accuracy with which they may be applied is strongly dependent on the boundary layer transition model employed. Of particular relevance, is the length of the transition region and the transition mechanism itself.

More recently, an aerofoil design package [8] incorporating inverse potential flow and geometric modification routines has been coupled to the performance prediction scheme. This facility permits intermediate designed sections to be tested analytically over a Reynolds number range prior to acceptance of a final aerofoil shape. The package also incorporates several utility elements including NACA section generation modules and a profile storage library. It is considered that the interactive nature of the package provides a flexible and powerful design tool.

CONCLUDING REMARKS.

The low Reynolds number research at Glasgow University was initiated in the late 1960's. Since then, a comparative aerofoil testing programme has been conducted to assess the relative merits of four aerofoil sections for specific applications. Throughout the programme, the data obtained have been supplemented by smoke and oil flow visualisation tests to clarify and identify the phenomena present. Further enhancement of the programme has been the concurrent development of an aerofoil design methodology incorporating a low-speed performance analysis code.

Future aspirations include interfacing the design code to general performance codes for wind turbines, remotely piloted vehicles and wind tunnel modelling and including unsteady effects. Further, tests are planned to secure the interpretation that we have placed on the observed phenomena and we propose modifying the turbulence levels in the tunnel to provide the more comprehensive tunnel database. It should be noted, that only one of the Glasgow University aerofoil sections has been wind tunnel tested and so it is our intention to consider others from that family.

REFERENCES.

1. Nonweiler, T. A New Series of Low Drag Aerofoils, Glasgow University Aero Report No. 6801, 1968.
2. Kelling, G.H. Experimental Investigation of a High-Lift Low-Drag Aerofoil, Glasgow University Report No. 6802, 1968.
3. Article on Q200 Microlight, Sport Aviation, March 1984.

4. Too, H.L. Flight Characteristics of a Canard Microlight Aircraft, G.U. Aero Final Year Project, 1982.
5. Galbraith, R.A.McD. The Aerodynamic Characteristics of a GU 25 5(11)8 Aerofoil for Low Reynolds Numbers, Experiments in Fluids, Vol. 3, pp 253-256, 1985.
6. Kokkodis, G. Low Reynolds Number Performance of a NACA 0015 and a GA(W)-1 Aerofoil, MSc. Thesis, University of Glasgow, 1987.
7. Saliveros, E. The Performance of a NACA 4415 Aerofoil Section at Low Reynolds Numbers, MSc. Thesis, University of Glasgow, 1989.
8. Coton, F.N., Galbraith, R.A.McD. An Aerofoil Design Methodology for Low Speed Aerofoils, 10th British Wind Energy Conference, London, 1988.
9. Coton, F.N., Galbraith, R.A.McD. A Direct Viscid-Inviscid Interaction Scheme for the Prediction of Two-Dimensional Aerofoil Lift and Pitching Moment in Incompressible Flow, Aeronautical Journal, April, 1989.
10. Niven, A.J., Galbraith, R.A.McD. The Effect of Pitch Rate on the Dynamic Stall of a Modified NACA 23012 Aerofoil and Comparison with the Unmodified Case, Vertica Vol. 11, No. 4, pp 751-759, 1987.
11. Vezza, M., Galbraith, R.A.McD. A Method for Predicting Unsteady Potential Flow about an Aerofoil, International Journal for Numerical Methods in Fluids, Vol. 5, No. 4, 1985.
12. Kokkalis, A., Galbraith, R.A.McD. Description of , and Preliminary Results from, a new Blade Vortex Interaction Test Facility, 12th European Rotocraft Forum, 1986.
13. McGhee, R.J., Beasley, W.D. Low-Speed Aerodynamic Characteristics of a 17-Percent-Thick Airfoil Section Designed for General Aviation Applications, NASA TN D-7428, 1973.
14. Jacobs, E.N., Sherman, A. Airfoil Section Characteristics as Affected by Variations of the Reynolds Number, NACA T.R. No. 586, 1937.
15. Saliveros, E., Galbraith, R.A.McD., Coton, F.N. The Low Reynolds Number Performance of a NACA 4415 Aerofoil, 10th British Wind Energy Conference, London, 1988.
16. Hamilton, S.M. Flow Visualisation of the Lower Surface of a NACA 4415 Aerofoil Section , Glasgow University Aero Final Year Project, 1989.
17. Leishman, J.G. Galbraith, R.A.McD., Hanna, J. Modelling of Trailing Edge Separation on Arbitrary Two-Dimensional Aerofoils in Incompressible Flow Using an Inviscid Flow Algorithm, Glasgow University Aero Report No. 8202, 1982.
18. Coton, F.N. Contributions to the Prediction of Low Reynolds Number Aerofoil Performance, PhD. Thesis, University of Glasgow, 1987.
19. LeFoll, J. A Theory of Representation of the Properties of Boundary Layers on a Plane, Proc. Seminar on Advanced Problems in Turbomachinery, V.K.I., 1965.
20. Horton, H.P. A Semi-Empirical Theory for the Growth and Bursting of Laminar Separation Bubbles, A.R.C. C.P. No. 1073, 1969.
21. Coton F.N., Galbraith, R.A.McD. A Direct Aerofoil Performance Code Incorporating Laminar Separation Bubble Effects, International Congress for the Aerospace Sciences, Israel, 1988.

Experimental aerodynamic characteristics of the airfoils LA 5055 and DU 86-084/18 at low Reynolds numbers

L.M.M. Boermans
F.J. Donker Duyvis
J.L. van Ingen
Faculty of Aerospace Engineering
Delft University of Technology

W.A. Timmer
Institute for Windenergy
Delft University of Technology

Abstract

In the Low-Speed Low-Turbulence Wind Tunnel (LTT) of the Low Speed Laboratory (LSL) at Delft University of Technology experimental results have been obtained for two low Reynolds number airfoils with different means of laminar separation- and transition control: airfoil LA 5055 designed by Liebeck and airfoil DU 86-084/18 designed at LSL.

Lift and pitching-moment data were obtained from pressure measurements (LA 5055) and balance measurements (DU 86-084/18) and drag data from wake surveys. The fluorescent oil flow technique was used to visualize the flow behaviour.

Design features of the LA 5055 airfoil are a distinct destabilizing region between 27% and 47% chord followed by a Stratford type pressure recovery distribution. To prevent early separation near the tunnel walls, suction was applied in a small region on the walls along the airfoil contour. In spite of the destabilizing region, drag producing laminar separation bubbles were present, and drag reductions up to 33% were measured by applying a zig-zag tape turbulator for transition control. The maximum lift coefficient, being governed by the Stratford limiting pressure distribution, could be raised by 38% with semi triangular vortex generators positioned at 20% chord.

The DU 86-084/18 airfoil with 18% chord camber changing flap has been designed for application in radio-controlled model sailplanes of the F3B (multi-task) class. Similar to the design of modern sailplane wing airfoils, artificial transition control as well as flap deflections were integrated from the start into the design. Due to the presence of long laminar flow regions on the upper and lower surface, the effectiveness of zig-zag tape turbulators, and the camber changing flap, low drag values were obtained at the C_L- and Re-range of interest for this class of model airplanes. In addition, stalling behaviour was gradual at all flap deflections.

1. <u>Airfoils</u>

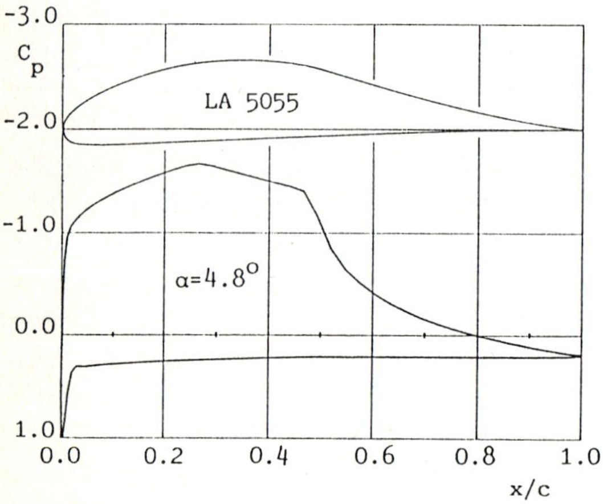

Fig. 1 Airfoil LA 5055

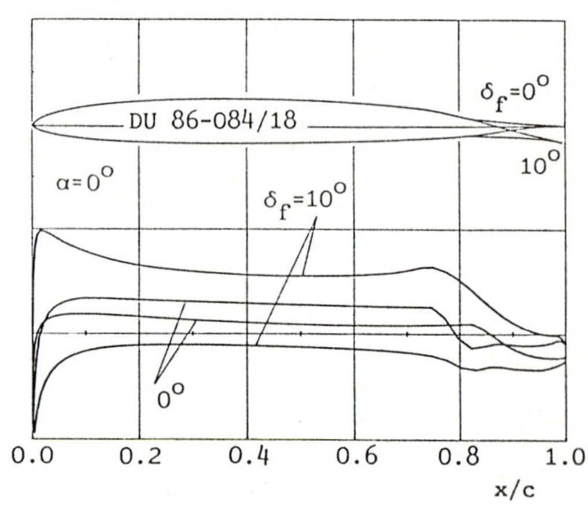

Fig. 2 Airfoil DU 86-084/18

The LA 5055 airfoil, designed by R.H. Liebeck, is characterized by an upper surface pressure distribution consisting of a rooftop with destabilizing region and a Stratford type pressure recovery distribution, Fig. 1. The destabalizing region is relatively long, from 27% chord to 47% chord, in the interest of promoting natural transition. The pressure recovery distribution was designed to take advantage of the conservatism in the Stratford distribution at low Reynolds numbers, as noted by Liebeck from windtunnel tests. The design conditions for LA 5055 were $C_\ell = 1.04$ at $\alpha = 4.82°$ and $Re_c=0.6*10^6$. The airfoil thickness is 15% chord.
The airfoil DU 86-084/18, Fig. 2, has been designed at LSL for application in radio controlled model sailplanes of the FAI F3B Class.

Competition in this class is focussed on speed (i.e. the time needed to fly four laps of 150 m and three turns), distance (i.e. the number of 150 m laps flown in four minutes) and duration (staying up and landing between marks after 6 minutes). A special winch launch technique is used to maximize altitude: after the initial climb phase a dive toward the winch is performed where maximum winch power and elasticity of the nylon line are converted to speed, and after disconnection a steep pull-up manoeuvre is used to convert this kinetic energy to potential energy for maximum altitude. The wings are built of carbon fiber material to withstand the high g-loadings during the pull-up manoeuvre.

Performance analysis shows that the range of operation for the airfoil is between lift coefficients of about 1 and 0 at corresponding Reynolds numbers of about 10^5 to 10^6. Low drag is emphasized in particular at lift coefficients below 0.5 (start, speed task and distance task), and a gradual stalling behaviour is required in view of the handling qualities in thermal flight conditions (duration task).

The LSL airfoil analysis and design computer program, Ref. 1,2,3, was used to design the airfoil. As shown in Fig. 2, the airfoil is relatively thin (8.4%c), has very long laminar flow regions on upper and lower surface, and a camber changing flap of 18%c. Similar to the design of modern sailplane wing airfoils, flap deflections and artificial transition control are integrated from the start into the design.

As indicated in the figure, the lower surface has been designed at the zero degree high speed flap deflection for laminar flow up to the flap, and the upper surface has been designed at the ten degree low speed flap deflection aiming at a long laminar flow region and a limited growth of the turbulent separated area at lift coefficients above the low drag bucket. The idea of the latter constraint is to avoid a dip in the lift versus angle of attack curve and accompanying bad handling and climbing qualities in thermal flight conditions; however, this sets a limit to the pressure gradient on the rear of the airfoil and thus the length of the laminar flow region.

As clarified in Ref. 4, artificial transition control to eliminate detrimental laminar separation bubbles, is preferable to the destabilizing region concept if an airfoil is required with a wide operating range of Reynolds numbers, lift coefficients and flap deflections, as in the present case. On the upper surface fixation of the laminar separation point, needed for the application of turbulators, is easily obtained by the change in pressure gradient at 75%c. On the lower surface this is the case at 82%c for the zero degree flap deflection only; the pressures induced by downward flap deflections will cause the laminar separation point to move forward, thus making mechanical turbulators ineffective when they are submerged into the separated flow of the bubble. However, extensive tests with mechanical and pneumatic turbulators on a sailplane airfoil with similar lower surface pressure distribution (Ref. 3) did not reveal a drag increase due to this type of turbulator malfunction. Probably the additional pressures due to the bubble, acting over the corner in the contour at the flap hinge have no component in flow direction and hence no additional pressure drag exists.

Both the LA 5055 and the DU 86-084/18 airfoil were tested with zig-zag tape turbulators, which are the thinnest effective mechanical turbulators known to the authors, Ref 3.

2. Experimental Arrangement

Models

The LA 5055 airfoil model was cast in 3 equal, spanwise parts using a movable, accurate mould; casting material was the synthetic resin Araldit. The model has a chord length of 360 mm and is positioned vertically, spanning the 1.25 m height of the windtunnel test section. A total of 106 pressure orifices with diameter 0.4 mm and drilled perpendicular to the surface was located in staggered formation at the midspan station. In addition, 25 pressure orifices were located at 40 mm from each model end, to detect any wall-induced separation. Accuracy of the model contour was not measured, however, previous models made by this casting technique showed tolerances generally within 0.1 mm.

The DU 86-084/18 airfoil model consisted of an aluminum core surrounded by epoxy resin and thin layers of carbon and glass fiber fabric with a gel topcoat forming the aerodynamic surface. It was built in female mould halves for the upper and lower surface, which in turn were carefully produced with the help of machined templates. After the model was built in one piece, the flap was cut out and provided with a tubular nose and five center-hinges. The model has a chord length of 235 mm and is positioned vertically between reflection plates which are 0.60 m apart. Contour accuracy has not been measured yet, but inspection with the help of machined templates indicate that tolerance is tight.

Wind Tunnel and Instrumentation

The Low-Speed Low-Turbulence Wind Tunnel is of the closed return type with a contraction ratio of 17.9. The free-stream turbulence level in the test section varies from about 0.015% at 10 m/s to 0.045% at 60 m/s. The interchangeable octagonal test section is 1.80 m wide, 1.25 m high and 2.70 m long.

The LA 5055 model was attached to mechanically actuated turntables which are flush with the test section top and bottom wall, Fig. 3. Suction boxes with perforated plates were integrated into the turntables, providing attachment for the model as well.

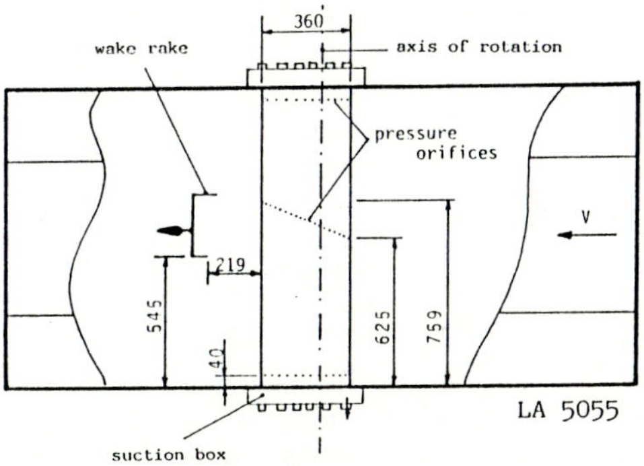

Fig. 3 LA 5055 test set-up

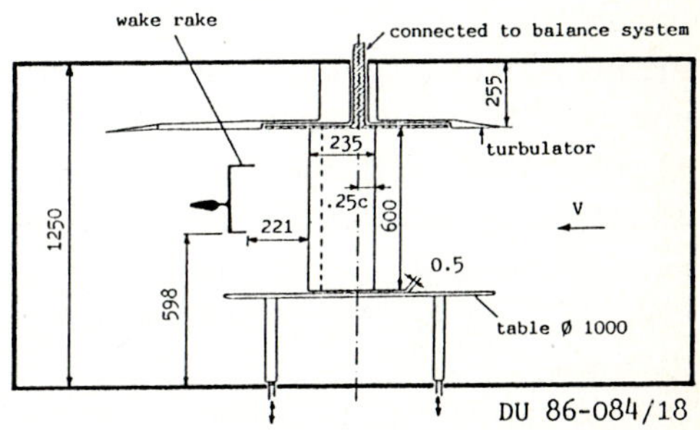

Fig. 4 DU 86-084/18 test set-up

The DU 86-084/18 model was attached vertically with one tip to a turntable, which is flush with a large reflection plate near the top wall; model and turntable are suspended to the windtunnel six component balance system, Fig. 4. A lower reflexion plate was present which left a small gap (0.5 mm) between the plate and the model tip.

In both cases, a wake rake, mounted on a cross beam and utilizing 50 total pressure tubes and 12 static pressure tubes, was used. Wakes were measured at 60%c and 95%c behind the trailing edge of LA 5055 and DU 86-084/18 respectively. All pressures were recorded by an automatically reading multi-tube liquid manometer (200 tubes). Pressure and balance data were on-line reduced using the Hp-21 MX-E laboratory computer.

Tests and methods

The LA 5055 model was tested in smooth condition at Reynolds numbers based on airfoil chord of $0.5*10^6$, $1*10^6$ and $1.5*10^6$. Strips of turbulator tape were used at Reynolds numbers of $0.5*10^6$ and $1*10^6$. The tape was 0.25 mm thick and 11 mm in width, and either unchanged, or cut in zig-zag form, Fig. 5.

To prevent early separation near the tunnel walls, suction of the tunnel wall boundary layer in the vicinity of the model was tested.

For that purpose the perforated plates were covered with adhesive plastic film except for the areas where suction was wanted.

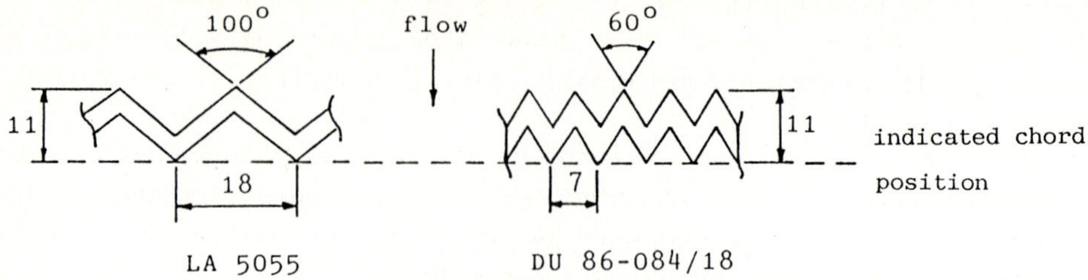

Fig. 5 Zig-zag tape turbulators

Finally, tests were performed with semi-triangular counter-rotating vortex generators positioned at 20%c, 30%c and 40%c. The vortex generators were similar to the devices used in Ref. 5, but the size was adapted for present purpose according to the recommendations of Ref 6. As shown in Fig. 6, the vortex generators were easily made by cutting and folding a thin brass strip.

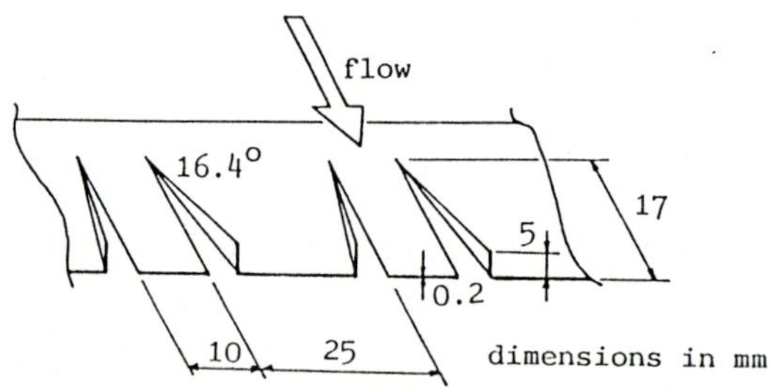

Fig. 6 Semi-triangular vortex generators

The Du 86-084/18 model was tested at Reynolds numbers from $0.2*10^6$ to $1*10^6$ and flap deflections from -5° to +15° both in smooth condition and with zig-zag tape turbulators, Fig. 5, on the upper and lower surface. On the upper surface 0.5 mm thick zig-zag turbulators were tested at positions between 67%c and 75%c, and on the lower surface 0.4 mm and 0.5 mm thick turbulators were tested at positions between 75%c and 79%c. The slots of the flap were sealed with tape of 0.04 mm thickness.

Laminar separation bubbles, turbulent separation and the effect of turbulators were visualized using the fluorescent oil-flow technique.

The static pressure measurements on the LA 5055 model surface were reduced to standard pressure coefficients and numerically integrated to obtain section normal force and pitching moment coefficients. For

the DU 86-084/18 model section normal force coefficients were obtained from the balance measurements taking the drag component of the turn-table (without model) into account. In both cases, profile drag coefficients were computed from the wake rake total and static pressures by the method of Jones, ref. 7.
In case of the LA 5055 measurements, standard low speed wind tunnel boundary corrections (Ref. 8) composed of solid and wake blockage, lift interference and wake-buoyancy, were applied to the section characteristics and pressure distributions. As an indication; the correction on the coefficients amount to less than 2% and on the angle of attack less than 0.2°. In case of the DU 86-084/18 measurements, no corrections were applied to the data because their magnitude is negligible. In both cases, the uncorrected coefficients are referred to the apparent dynamic pressure as measured without model at the center of the test section (standard procedure).

3. Experimental Results

LA 5055

The pressure distributions near the tunnel walls without suction showed early wall induced separation; this separation expanded to the midspan section and degraded the airfoil's performance at higher angle of attack, as shown in Fig. 7. Attempting to obtain instantaneous flow separation on the complete model, suction was applied in various

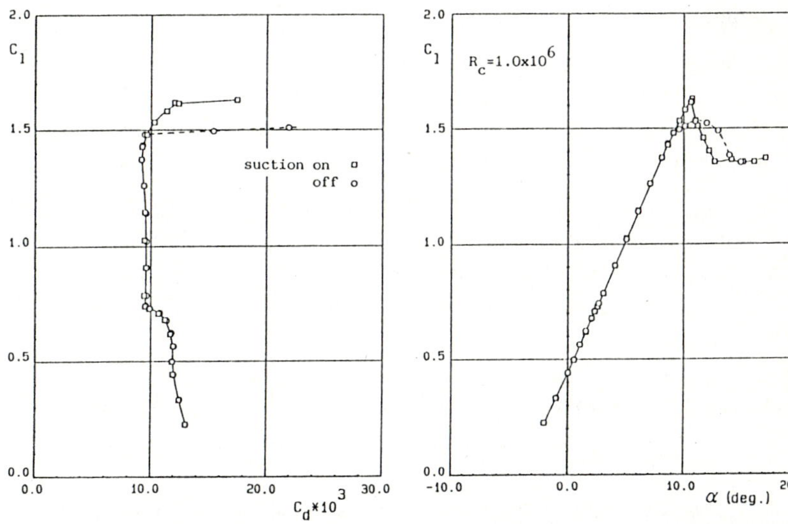

Fig. 7 Measured characteristics of LA 5055 with and without wall boundary layer suction

regions on the walls in front of the model as well as along the airfoil contour. The highest $c_{\ell_{max}}$ at the midspan section was obtained at every Re-number by suction in a 5 mm wide region along the airfoil upper surface, however, a perfect 2-D stall could not be obtaind since separation started near the upper wall and rapidly progressed toward the lower wall. Stall near the upper wall could be postponed with a pair of counter-rotating vortex generators at 30%c, however no differences at the midspan section were detected.
At five different angles of attack ($Re=0.5*10^6$) spanwise drag surveys were made to check the two-dimensionality of the flow and to find a wake rake position where the drag represents a mean value. Some

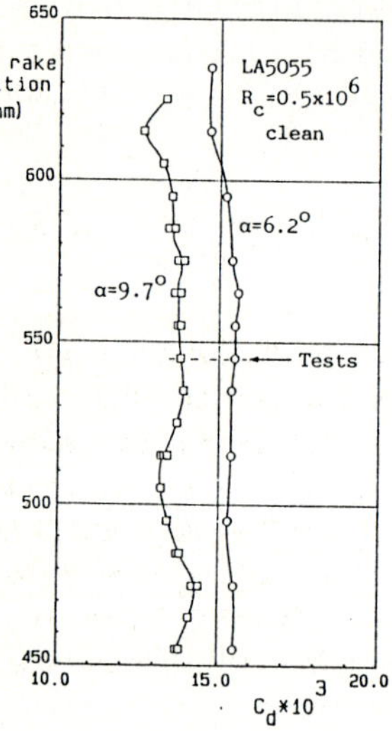

Fig. 8 Spanwise drag traverse measurements, LA 5055

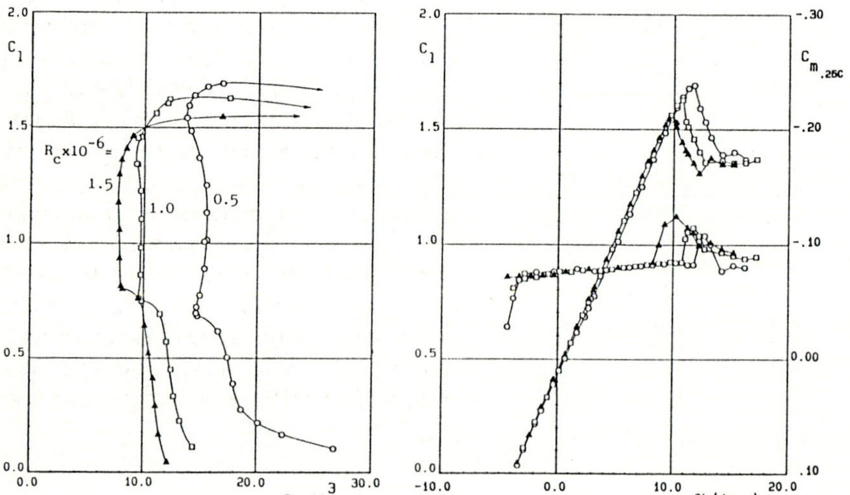

Fig. 9 Measured characteristics of LA 5055

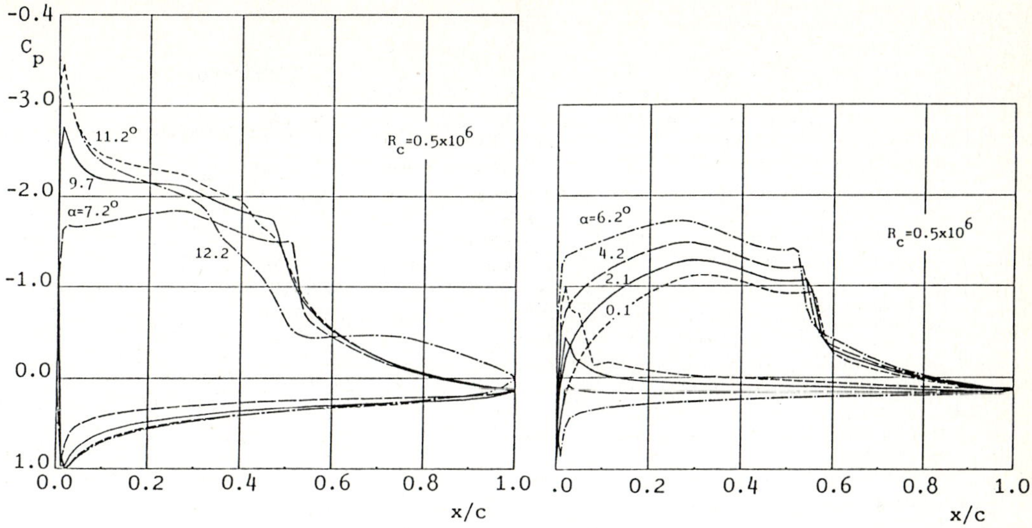

Fig. 10 Measured pressure distributions of LA 5055

typical results are shown in Fig. 8; spanwise variations in drag coefficient are generally less than 0.002. The wake rake was set at the position indicated in the figure.

Fig. 9 shows the measured aerodynamic characteristics at $Re=0.5*10^6$, $1*10^6$ and $1.5*10^6$. The flow behaviour at $Re=0.5*10^6$ is illustrated by the pressure distributions in Fig. 10. The flow on the lower surface is laminar up to the trailing edge at angles of attack higher than 2° i.e. the lower end of the low drag bucket, while at lower angles of attack there is a laminar separation bubble near the leading edge caused by the steep adverse pressure gradient after the peak on the nose. On the upper surface a laminar separation bubble is present which slowly moves forward with increasing angle of attack. At $\alpha < 10^\circ$ the laminar boundary layer separates in the destabilizing region before transition occurs and reattachment is in the steep part of the pressure recovery region. At $\alpha = 10^\circ$ transition is at the onset of the pressure recovery region and the drag is a minimum. Beyond $\alpha = 10^\circ$ the bubble moves forward in the destabilizing region, and at $\alpha = 11.7^\circ$ the maximum lift coefficient is reached; beyond that angle complete separation of the pressure recovery region occurs.

By increasing the Reynolds number the boundary layer becomes thinner and transition moves forward. The decrease of boundary layer thickness causes the change in effective airfoil contour indicated by the pressure distributions in Fig. 11 and corresponding slight lift increase in Fig. 9. Due to earlier transition the length of the laminar separation bubble decreases, Fig. 11, but so does the width of the low drag bucket and the maximum lift coefficient, Fig. 9.

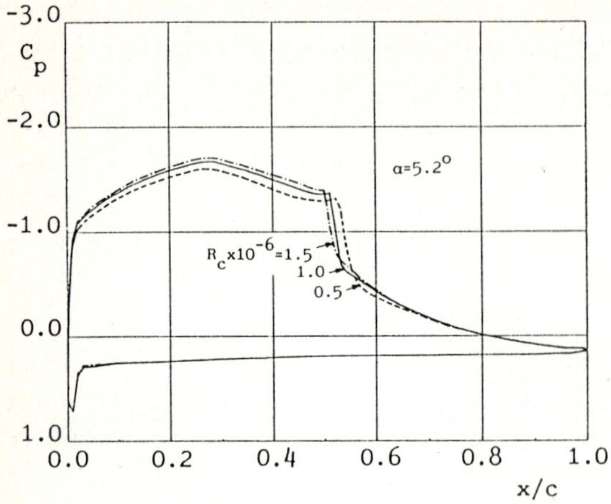

Fig. 11 Measured pressure distributions of LA 5055

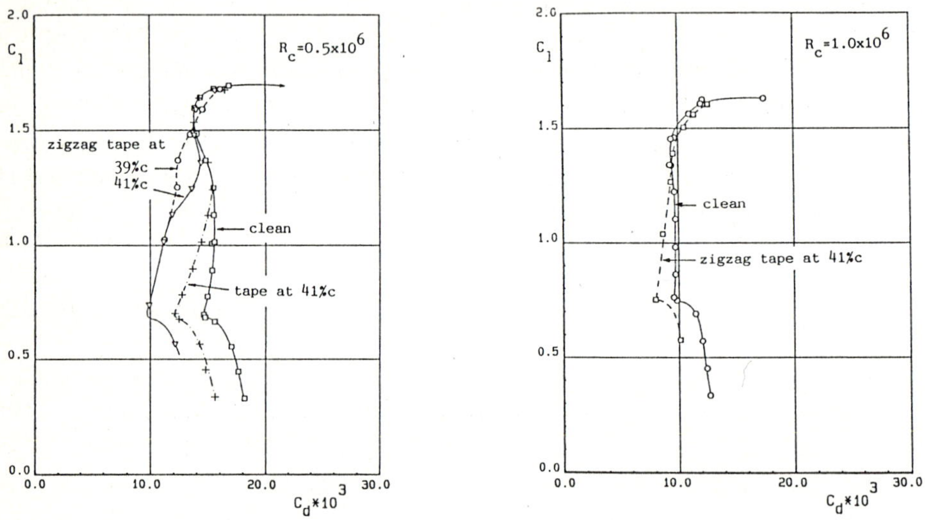

Fig. 12 Measured characteristics of LA 5055 with and without turbulators

Fig. 12 shows results of tests with <u>turbulators</u> on LA 5055. A strip of tape positioned at 41%c shows at Re=0.5*10^6 a drag reduction at lift coefficients below 1.25; at higher lift coefficients the tape is submerged in the laminar separation bubble. A zig-zag form, cut from the same tape and applied at the same chord position, is considerably more effective and produces a drag reduction up to 33%, but also submerges in the bubble at higher lift coefficients. Shifting the zig-zag tape 2%c forward makes it effective at all lift coefficients, with a drag penalty at lift coefficients near the upper end of the low drag bucket (probably transition occurs too early). Similar results are obtained at Re=1*10^6 and the zig-zag tape at 41%c.

The measured results illustrate both the difficulty of designing and the negative effects of this type of destabilizing region where, ideally, the laminar boundary layer does not separate and transition occurs just at the end of the region. As explained in Ref. 4, this situation can only be obtained at one particular angle of attack and Reynolds number and deviations will cause transition to move forward on the destabilizing region or rearward, forming a bubble; either of which costs extra drag.

As illustrated in Ref. 9 and 10, it is favourable for the upper surface of an airfoil designed for sailplane application to design a pressure distribution which is rounded off in the transition region and, given the decrease of flight speed with increase of angle of attack, to exploit the destabilizing effect of the pressure gradient and at the same time the stabilizing effect of the Reynolds number to avoid the occurrence of a detrimental laminar separation bubble. On the lower surface the change in pressure gradient and Reynolds number have both stabilizing effects, and a rather long destabilizing region would be needed to avoid bubbles at the lower Reynolds number at the cost of a drag increase due to forward movement of transition at the higher Reynolds number. Here, and even more in case of airfoils designed for lower Reynolds numbers, artificial transition control provides a solution. The measurements on LA 5055 illustrate that for an effective application of artificial transition devices, the position of laminar separation should be fixed.

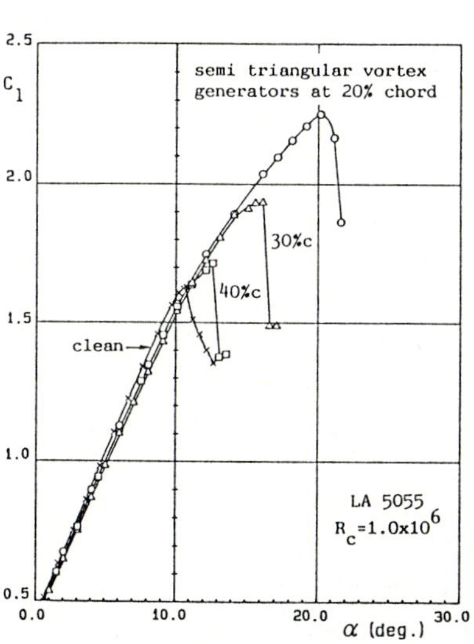

Fig. 13 Measured lift of LA 5055 with vortex generators

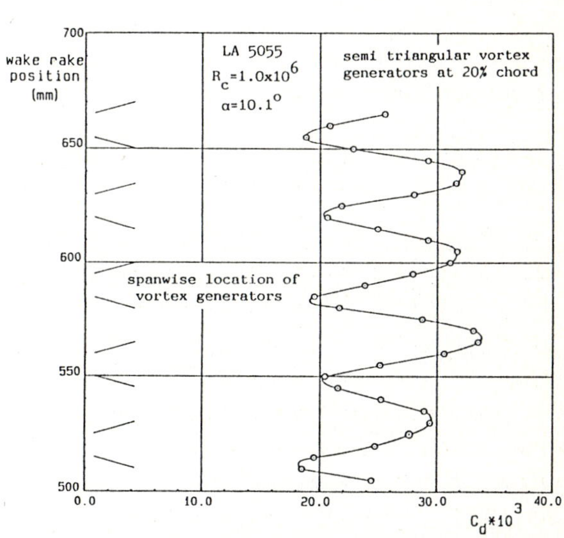

Fig. 14 Spanwise wake traverse measurements, LA 5055 with vortex generators

Tests with <u>vortex generators</u> on LA 5055 were performed in order to study the possibility of forced mixing in case of an imminent separation pressure recovery distribution, as suggested in Ref 11.

As shown by the results in Fig. 13 the maximum lift coefficient increases when the vortex generators are shifted forward. With the vortex generators at 20%c the maximum lift coefficient is raised by 38% with respect to the clean airfoil case. In all cases the turbulent boundary layer eventually separates at approximately 55%c due to the steep local pressure gradient. It is concluded that vortex generators postpone flow separation in case of Stratford type pressure recovery distributions provided that they are positioned some distance ahead of the separation point without generators. This conclusion is in accordance with the advice given by Taylor (Ref. 6) to locate the generators some 10 to 30 times the local boundary layer thickness upstream of the separation position without generators.

Fig. 14 shows results of spanwise wake traverse measurements. The periodicity of the results clearly shows the effects of the streamwise vortices emerging from the generators. The counter-rotating vortices of each pair of generators cause a decrease of boundary layer thickness at their centerline downstream, while the co-rotating vortices cause a thickening of the boundary layer between each pair of generators. The question arises how accurate the wake drag results are when vortices are present in the wake; comparison with balance measurements, as described in Ref. 12, would provide an answer. Anyhow, the results show again the necessity to do wake traverse measurements, which are very time consuming. Therefore, a computer controlled traversing wake rake system, using the 3-axis traversing system of the laboratory and applying ESP pressure scanners, is under development at LSL and will soon be available.

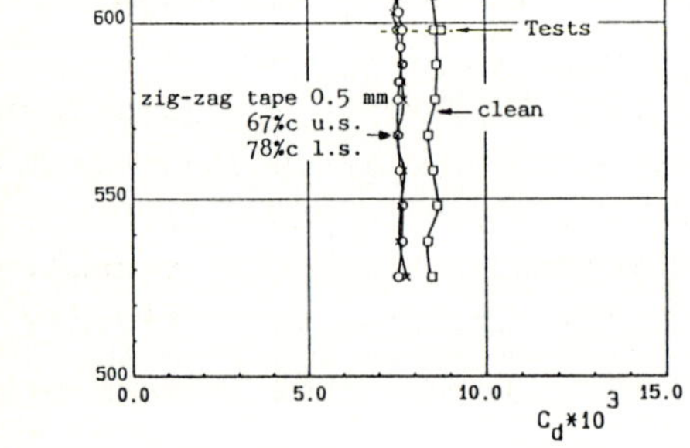

Fig. 15 Spanwise drag traverse measurements, DU 86-084/18

In previous tests with different airfoils (Ref. 13) it was found that the lift measured with the balance system is independent of the width of the gap between the tip and lower reflection plate as long as this gap is smaller than 1 mm. This result was confirmed by tests with DU 86-084/18 where the gap was varied between 0.08 mm and 3 mm; the present results were obtained with the usual gap width of 0.5 mm. The model forces could be measured with negligible scatter by the balance system at Re = 100.000, however, the small pressure differences in the wake limited the drag measurements with the automatically reading multi-tube liquid manometer to Re = 200.000.

Spanwise drag surveys were made at several angles of attack and Reynolds numbers, both at zero and at ten degree flap deflection as well as without and with zig-zag tape. While the results without zig-zag tape show spanwise variations in drag coefficient up to 0.003 due to laminar separation bubbles on both sides of the airfoil, the results with zig-zag tape are nearly constant as illustrated in Fig. 15.

To find the best location and thickness of the zig-zag tape turbulators on the airfoil upper and lower surface, extensive systematic tests were performed at zero and ten degrees flap deflection and various Reynolds numbers. The final selection, being zig-zag tape of 0.5 mm thickness at 67%c upper surface and 78%c lower surface, produces the least drag at Reynolds numbers below 600.000. At higher Reynolds numbers there is a drag penalty (up to $\Delta C_d = 0.0005$ at $Re=1*10^6$) with respect to the lowest drag which could be obtained, being the price which has to be paid for these mechanical turbulators to be effective (not submerged) at low Reynolds numbers. Note that the Reynolds number for this airfoil varies with a factor 10.

Fig. 16 shows measured airfoil characteristics at two typical Reynolds numbers and flap deflections, without and with turbulators. The necessity and effectiveness of the turbulators is evident. At zero degree flap deflection the drag reduction is mainly due to the elimination of the bubble on the upper surface. At ten degrees flap deflection the turbulator on the lower surface is submerged in the bubble, as shown by oil flow studies, hence the large drag reduction is entirely due to the elimination of the bubble on the upper surface. No further drag reduction could be obtained at this flap deflection by changing the turbulator thickness and position on the lower surface; the drag already equals the drag at zero flap deflection as shown in Fig. 15. Hence, the presupposition in designing the airfoil that turbulator malfunction on the lower surface due to downward flap deflection would be harmless, has been substantiated.

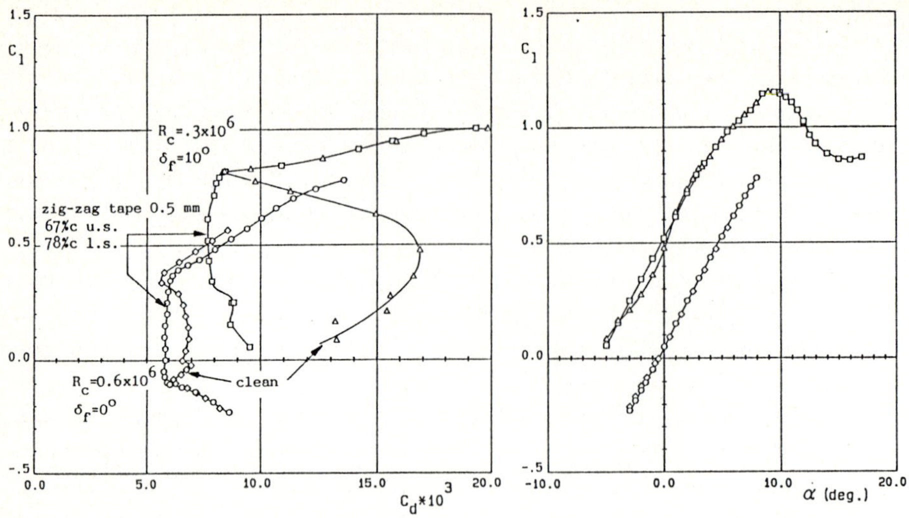

Fig. 16 Measured characteristics of DU 86-084/18 with and without zig-zag tape turbulators

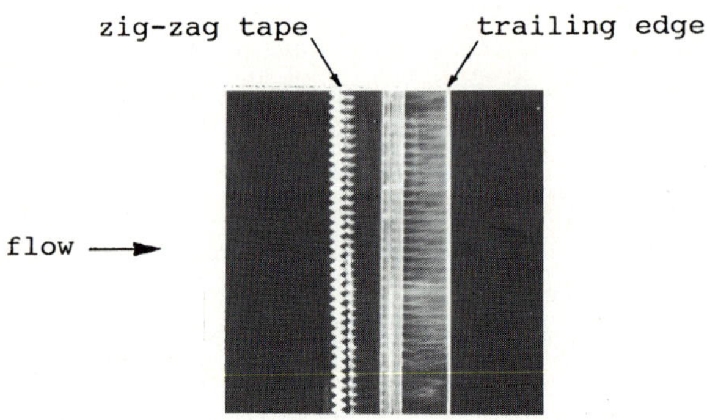

Fig. 17 Fluorescent oil flow pattern

The traces of the vortices produced by the zig-zag tape are clearly visible in the fluorescent oil flow pattern, an example of which is shown in Fig. 17.

Fig. 18 presents the characteristics for zero and ten degree flap deflection at relevant Reynolds numbers. The reproduction of the drag values at Re = 300.000 and higher is excellent, however the data at Re = 200.000 show scatter for reasons mentioned before. The gradient of the lift curves change when transition is moving forward and the drag increases. At ten degrees flap deflection the lift curve shows a small irregularity just before the maximum lift coefficient is reached; flow patterns revealed that this is due to a short bubble on the airfoil's nose which turns into a long bubble at increasing angle of attack. This also happens at the zero flap setting although it is not noticeable in the lift curve. The stalling behaviour of the airfoil is gradual and no hysteresis loops were detected by increasing and decreasing the angle of attack.

Fig. 19 shows the effects of flap deflection at Re = 300.000. At δ = 15° the upper limit of the low drag bucket is near C_ℓ = 1.0 where the lift curve shows a small horizontal step and no dip as intended. At α = 7° the short bubble on the airfoil's nose changes into a long

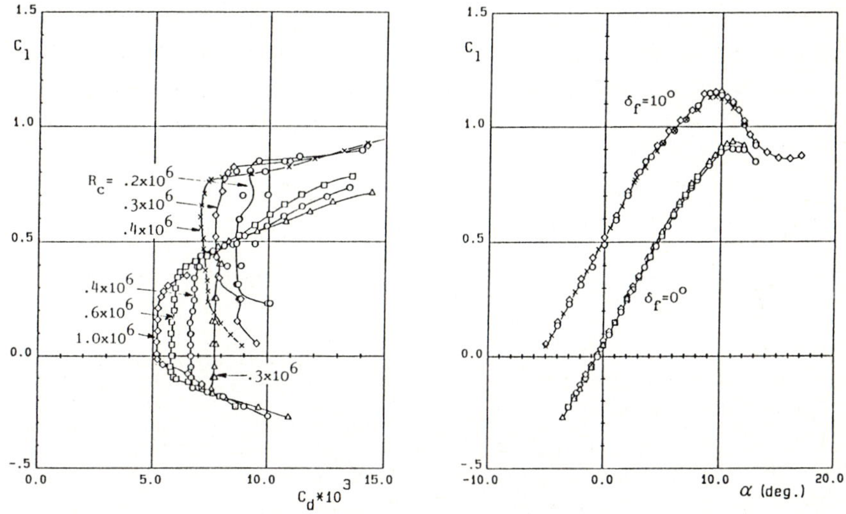

Fig. 18 Measured characteristics of DU 86-084/18 with 0.5 mm zig-zag tape at 67%c u.s. and 78%c l.s.

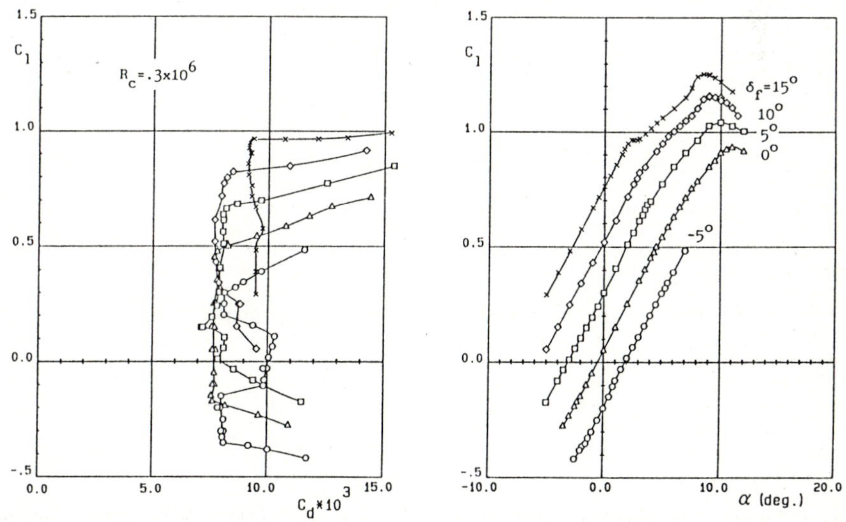

Fig. 19 Measured characteristics of DU 86-084/18 with 0.5 mm zig-zag tape at 67%c u.s. and 78%c l.s.

bubble. At $\delta = -5^0$ there is turbulator malfunction on the upper surface; the pressures induced by this negative flap deflection cause laminar separation ahead of the turbulator which in turn submerges in the separated bubble flow. Contrary to the situation on the lower surface at positive flap deflections, there is a drag increase, which can be eliminated by shifting the turbulator forward or increase its thickness, but both measures have a drag penalty at higher Reynolds numbers. Pneumatic turbulators (blowing a small amount of air through orifices periodically spaced in spanwise direction) have the advantage that they are still active in the region behind laminar separation because the air jets pass through this region and disturb the laminar outer flow (Ref. 4). In view of this self-adjusting capability, it is intended to test the DU 86-084/18 airfoil with pneumatic turbulators, located further backward than the zig-zag tape turbulators, in order to see if a further drag reduction is possible.

4. References

1. Ingen, J.L. van, Boermans, L.M.M. and Blom, J.J.H.: Low speed airfoil section research at Delft University of Technology. ICAS-80-10.1, Munich, 1980.
2. Ingen, J.L. van: On the analysis and design of low speed airfoils using potential flow methods and boundary layer theory. Report LR-365, Department of Aerospace Engineering, Delft University of Technology, 1982.
3. Ingen, J.L. van, Boermans, L.M.M.: Aerodynamics at low Reynolds numbers: A review of theoretical and experimental research at Delft University of Technology. Paper no. 1 in Proceedings of the International Conference on Aerodynamics at low Reynolds numbers, October 1986, R.A.S. London.
4. Horstmann, K.H., Quast, A. and Boermans, L.M.M.: Pneumatic turbulators - a device for drag reduction at Reynolds numbers below 5×10^6. Paper no. 20 in AGARD CP-365, Brussels, 1984.
5. Wentz, W.H. and Seetharam, H.C.: Development of a fowler flap system for a high performance general aviation airfoil. NASA CR-2443, 1974.
6. Taylor, H.D.: Summary report on vortex generators. Report R-052809, United Aircraft Corporation, Research Dept., 1950.
7. Jones, B.M.: Measurement of profile drag by the pitot-traverse method. R&M no. 1688, Brit. A.R.C., 1936.
8. Allen, H.J., Vincenti, W.G.: Wall interference in a two-dimensional flow windtunnel with consideration of the effect of compressibility. NACA Report 782, 1944.
9. Boermans, L.M.M., Selen, H.J.W.: Design and tests of airfoils for sailplanes with an application to the ASW-19B. ICAS-paper 82-5.5.2, 1982.
10. Boermans, L.M.M., Waibel, G.: Aerodynamic and structural design of the Standard Class sailplane ASW-24. ICAS-paper 88-2.7.2, 1988.
11. Schubauer, G.B., Spangenberg, W.G.: Forced mixing in boundary layers. Journal of Fluid Mechanics, Vol. 8, 1960.
12. Boermans, L.M.M., Borne, P.C.M. van den: Design and tests of a flexible sailwing airfoil for lightweight aircraft. OSTIV Publication XVIII, 1985.
13. Volkers, D.F.: Preliminary results of wind tunnel measurements on some airfoil sections at Reynolds numbers between $0.6*10^5$ and $5.0*10^5$. Memorandum M-276, Faculty of Aerospace Engineering, Delft University of Technology, 1977.

PERFORMANCE MEASUREMENTS OF AN AIRFOIL AT LOW REYNOLDS NUMBERS

Robert J. McGhee and Betty S. Walker
NASA Langley Research Center
Mail Stop 339, Hampton, Virginia 23665

Abstract

Performance characteristics of an Eppler 387 airfoil using both direct (force) and indirect (pressure) measurement techniques have been obtained at Reynolds numbers from 60,000 to 460,000 in the Langley Low-Turbulence Pressure Tunnel. Lift, drag, and pitching-moment data were obtained from two internally-mounted strain-gage balances specifically designed for small aerodynamic loads. Comparisons of these results with data from a pressure model of an Eppler 387 airfoil are included. Drag data for both models using the wake traverse method are compared with the balance data. Oil flow visualization and surface mounted hot-film sensors were used to determine laminar-separation and turbulent-reattachment locations. Problems associated with obtaining accurate wind-tunnel data at low Reynolds numbers are discussed.

Symbol List

c	airfoil chord
c_d	section drag coefficient
c_l	section lift coefficient
c_m	section pitching-moment coefficient about quarter-chord point
l/d	section lift-drag ratio
p_t	total pressure
q	dynamic pressure
R	Reynolds number based on free-stream conditions and airfoil chord
x	airfoil abscissa
W	uncertainty in measurement
α	angle of attack

Abbreviations:

EDM	electric discharge machining
LTPT	Low-Turbulence Pressure Tunnel
rms	root mean square

Introduction

Recent interest in low Reynolds number aerodynamics has increased for both military and civil applications with emphasis on providing better vehicle performance [1]. Reynolds numbers below 500,000 are usually identified as being in this classification. Applications are varied and include remotely piloted vehicles, ultra-light human-powered vehicles, wind turbines, and propellers.

Experimental results obtained on various airfoils at low Reynolds numbers in different wind tunnels have shown large differences in airfoil performance (refs. 1 and 2). This is not surprising because of the sensitivity of the airfoil boundary layer to free-stream disturbances, model contour accuracy, and model surface roughness. Also, the model forces and pressure differences are small and difficult to measure accurately. Transition and separation play a critical role in determining the development of the boundary layer which in turn has a large affect on airfoil performance.

The Fluid Dynamics Branch at the NASA Langley Research Center has initiated a research program to determine performance characteristics of airfoils at low Reynolds numbers. This effort includes evaluation of existing airfoil analysis codes and development of test techniques to obtain accurate wind tunnel data. The experimental program uses the Langley Low-Turbulence Pressure Tunnel (LTPT) and consists of performance measurements of an Eppler 387 airfoil using both direct (force) and indirect (pressure) measurement techniques. Oil-flow visualization and surface-mounted thin-film gages were used to determine laminar-separation and turbulent-reattachment locations.

This paper presents experimental results obtained on a force model of an Eppler 387 airfoil. Low-speed results were obtained in the LTPT for chord Reynolds numbers from 60,000 to 400,000. Lift, drag, and pitching-moment data were obtained from two internally-mounted strain-gage balances specifically designed for small aerodynamic loads. Comparisons of these results with data [2] from a pressure model of an Eppler 387 airfoil tested in the same wind tunnel are included. Drag data for both models were also obtained using the wake traverse method.

Experimental Arrangement

Wind Tunnel

The test was conducted in the NASA Langley Low-Turbulence Pressure Tunnel. This facility is described in detail in reference 3 and dynamic flow quality measurements are reported in reference 4. The LTPT is a pressurized, closed-circuit, continuous-flow wind tunnel with an operating pressure from approximately 0.10 to 10 atmospheres. The test section was designed for two-dimensional testing of airfoil sections and is 7.5 feet high, 7.5 feet long, and 3 feet wide. The contraction ratio is 17.6 to 1 and nine anti-turbulence screens are installed in the settling chamber. This facility was selected to develop test techniques for low Reynolds number aerodynamics because of its good flow quality, precision pressure instrumentation, and variable pressure capability. The variable pressure capability of the LTPT allows one to enhance the resolution of model forces and pressure differences by operating at higher dynamic pressures compared to an atmospheric wind tunnel (for a constant chord length and Reynolds number).

Model and Balance

A sketch of the model and balance arrangement is illustrated in figure 1. The airfoil model was made in three sections with a metric center section "floating" between the two non-metric sections. The center section is attached to the non-metric panels by two internally mounted strain-gage balances. Gaps of 0.006 inch were provided between the

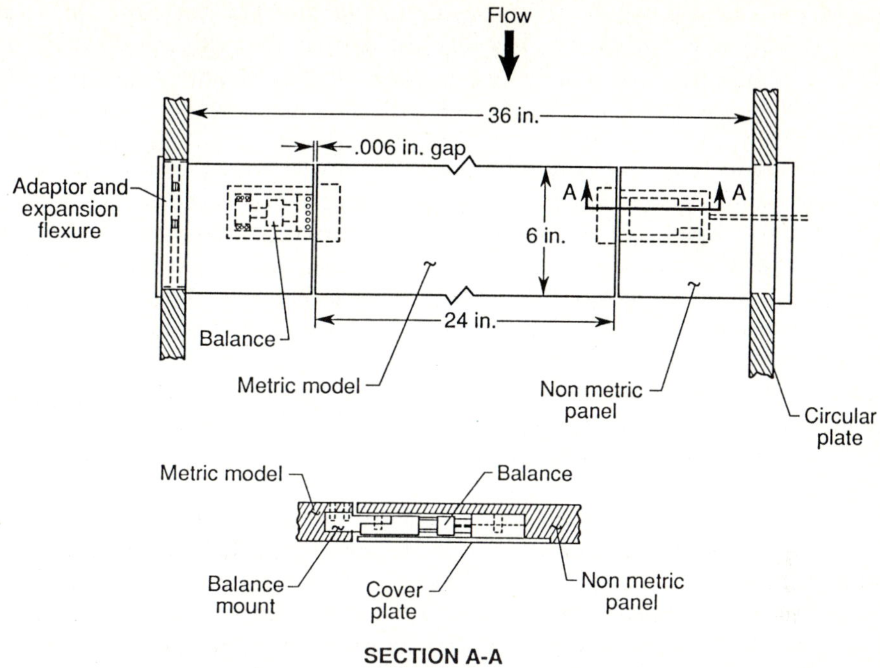

Figure 1. Model-balance arrangement.

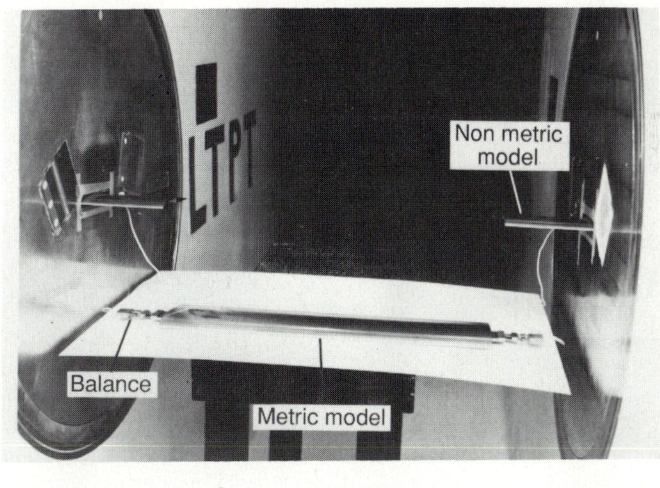

Figure 2. Photograph of model-balance arrangement in LTPT.

metric and non-metric sections to minimize airflow and provide clearance for balance measurements.

The metric model was machined from an aluminum plate using the Electric Discharge Machining (EDM) wire technique. The forward, center, and aft regions of the model were hollowed out to provide a light-weight model of approximately 2.75 pounds. An aluminum balance mount plate was bonded and pinned at each end of the model to provide balance attachment between the metric and non-metric sections. The final exterior of the model was machined using the EDM wire technique to a contour accuracy of ±0.001 inch. The metric model had a chord length of 6 inches and a span of 24 inches. The non-metric airfoil panels were machined from stainless steel with a contour accuracy of ±0.002 inch. They were attached to the tunnel sidewall circular plates using adaptor plates. A specifically designed thermal expansion flexure was employed at the left attachment plate. This flexure provides for any lateral expansion/contraction of the airfoil model due to temperature effects. This flexure is an important part of the model/balance arrangement as it eliminated zero shifts and preloads on the very sensitive axial-force strain gages. The photograph of figure 2 illustrates the model/balance arrangement in the LTPT.

Lift, drag, and pitching-moment data were obtained from two three-component, internally-mounted, strain-gage balances specifically designed for small aerodynamic loads. The balances were designed by Paul W. Roberts of the Langley Research Center. The design loads are 4 pounds in normal force, 0.25 pound in axial force, and 8 inch-pounds in pitching-moment. The balances were calibrated installed in the model to account for first and second order interactions and had a force/moment resolution of: normal force, ±0.04 pounds; axial force, ±0.003 pounds; and pitching moment, ±0.08 inch-pounds. The largest error occurred in axial force (about 2%) when the other two components were loaded to their full design load.

Wake Survey Rake

Figure 3. Photograph of wake rake in LTPT.

The computer-controlled wake survey rake (fig. 3) was mounted on a vertical traverse mechanism. Vertical position is achieved by a stepper motor that drives a 0.75 inch diameter ball screw. Vertical range is 50 inches and position accuracy is ±0.001 inch. Position tracking is accomplished by an optical shaft encoder in a closed loop configuration. The rake consisted of three five-hole probes located on the tunnel centerline and at spanwise stations of 4 and 8 inches. Each probe has four static pressure holes located at 45° to the probe centerline and 90° apart with a total pressure hole at the center. All orifices had an internal diameter of 0.020 inch. The five-hole probes were calibrated in the LTPT prior to the airfoil model tests. For the calibration, the rake head was rotated in the vertical plane (±10°) by a gear motor and ball screw arrangement. A precision potentiometer provided measurement of the probe pitch angle. The rake was located 1.5 chords behind the trailing edge of the airfoil. A typical wake survey consisted of selecting a starting position, stopping position, and incremental distance moved by the rake. For a wake traverse of about 2 inches an increment of 0.03 inches was used, giving 67 positions in the airfoil wake. At each position 4 samples of the pressures were taken, then averaged.

Instrumentation

Lift, drag, and pitching-moment data were obtained from the two strain-gage balances. Measurements of the wake-rake pressures and basic tunnel pressures were made with variable-capacitance precision transducers. These transducers have an accuracy of ±0.25% of reading. A thermocouple was installed in the model in order to monitor the balance temperature. The model temperature was maintained at approximately 70° F, the balance calibration temperature. Model angle of attack was measured by a calibrated digital shaft encoder and checked using a precision electric inclinometer. Data were obtained by a high-speed data acquisition system and recorded on magnetic tape. Real time data displays were available for tunnel parameters, balance loads, and wake profiles.

Tests and Methods

The force model was tested at Reynolds numbers based on airfoil chord from approximately 60,000 to 400,000 and angles of attack of 0, 2, 4, and 8 degrees. All data were obtained with the model in an aerodynamically smooth condition.

Laminar-separation and turbulent-reattachment locations were determined using the oil flow technique reported in reference 5. These oil flow data are compared with data from the multielement hot-film sensor technique [6] developed by John P. Stack of the Langley Research Center and typical results are shown in figure 4. Good agreement between the oil flow and hot-film data are shown.

Standard low-speed wind tunnel boundary corrections [7] have been applied to the section data. Corrections were applied to the free-stream dynamic pressure because of solid and wake blockage and applied to lift, pitching-moment, and angle of attack because of the effects of floor and ceiling constraints on streamline curvature. Blockage corrections were also applied to the profile-drag coefficients determined from the wake-rake total and static pressures. The magnitude of these corrections are negligible for a 6 inch chord model placed in the LTPT.

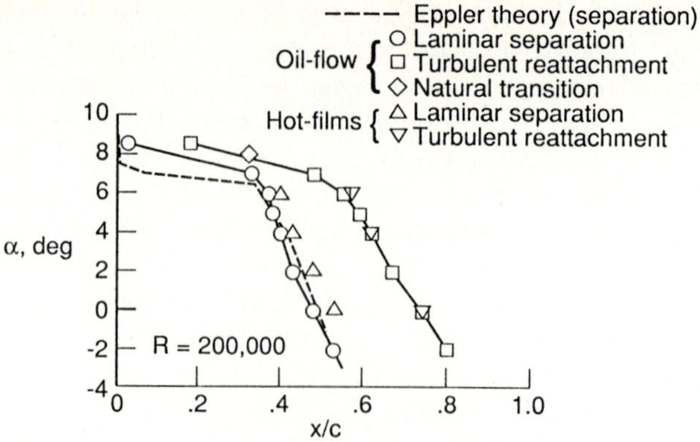

Figure 4. Separation and reattachment locations at R = 200,000.

It is important when measuring performance characteristics of airfoils to provide some indication of the data accuracy. Following is a list of areas in two-dimensional testing at low Reynolds numbers that may contribute to the overall uncertainty of the results:

- o Tunnel flow quality
- o Model contour accuracy
- o Model surface roughness
- o Confinement effect of wind-tunnel walls
- o Instrumentation accuracy
- o Tunnel sidewall boundary-layer interaction
- o Unsteady wakes

The present tests were conducted in a low disturbance wind-tunnel employing a model with a contour accuracy of ±0.001 inch and with an aerodynamically smooth surface (64 microinch, rms). The confinement effect of the wind-tunnel floor and ceiling was minimized by testing a model with a chord-to-tunnel height ratio of about 0.07. The degree of uncertainty associated with the instrumentation accuracy was minimized by using precision pressure transducers and carefully calibrated strain-gage balances.

In order to minimize the sidewall boundary-layer interference effects on the balance measurements the model was made in 3 sections with a center section "floating" between two end panels. A gap of 0.006 inch was provided between the metric and non-metric panels to minimize airflow and provide clearance for balance measurements. Thus the airfoil center section was isolated from the thick tunnel sidewall boundary layers which occur at low Reynolds numbers. The effect of the gaps on the section data are illustrated in figure 5 for $\alpha = 0°$. Tests were performed with the gaps both unsealed and sealed with petroleum jelly for Reynolds numbers from 200,000 to 400,000. Essentially no effect on the section data occurred due to the gap.

The wake-rake traverse method has been used successfully for many years to determine the profile-drag of airfoils. However, at low Reynolds numbers where laminar separation bubbles are present, the wakes are composed of large-scale vortices which

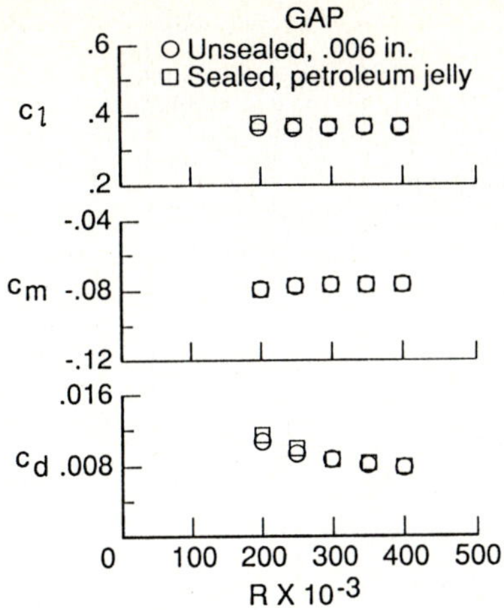

Figure 5. Effect of gap on section data at $\alpha = 0°$.

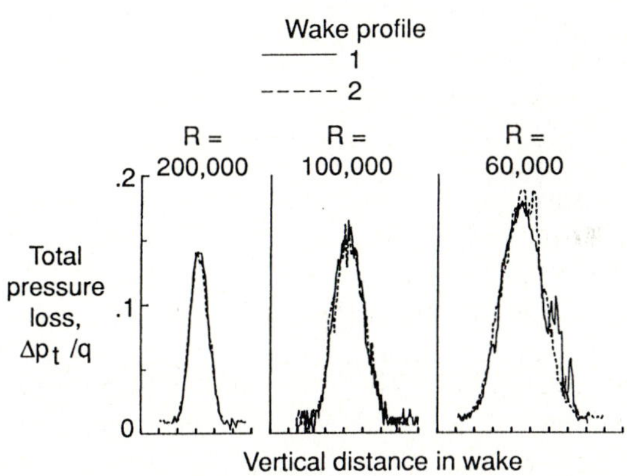

Figure 6. Wake profiles for Eppler 387 airfoil at $\alpha = 4°$.

produce unsteady wakes. Figure 6 illustrates this flow environment for R $\leq$ 100,000 for the Eppler 387 airfoil [2]. Thus, the wake traverse method is subject to errors related to

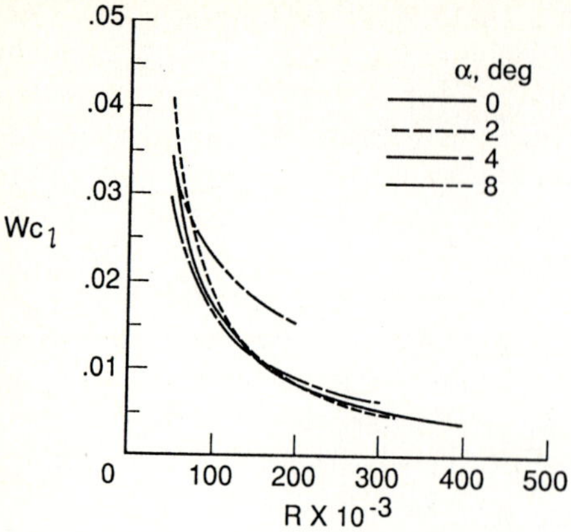

(a) Lift coefficient.

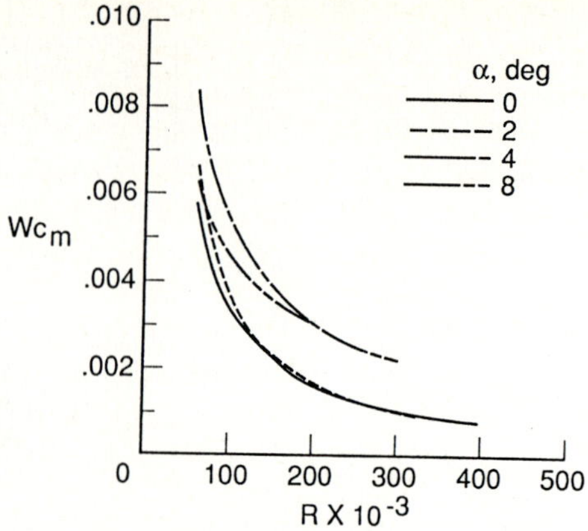

(b) Pitching-moment coefficient.

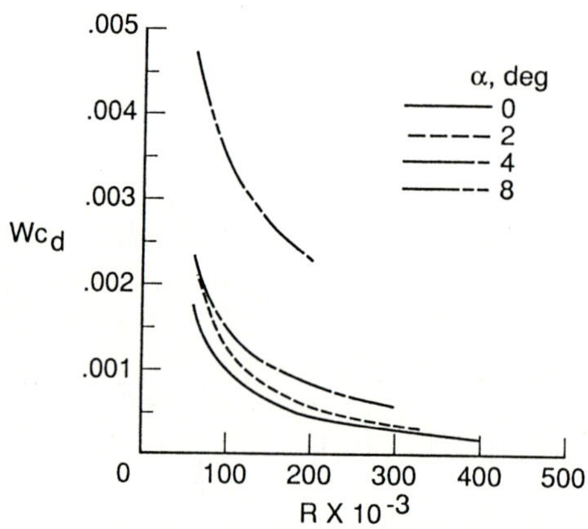

(c) Drag coefficient.

Figure 7. Uncertainty in section data.

the changing flow directions in the unsteady wakes. Therefore, the strain-gage balance method was used for the present test.

An estimate of the uncertainties for the force/moment measurements were determined using the technique of reference 8. These results are illustrated in figure 7 for the angles of attack and Reynolds number tested. The uncertainties increase with decreasing

Reynolds numbers. Generally the uncertainty in the lift and pitching-moment coefficients are acceptable for low Reynolds number testing. For example, at R = 100,000 the maximum uncertainty in the lift and pitching-moment coefficients are about ±0.025 and ±0.006, respectively. The uncertainty in drag coefficient is considered acceptable for angles of attack of 0, 2, and 4 degrees. For $\alpha = 8°$, the uncertainty in drag coefficient is greater than ±0.0025 and these drag data are not included in the paper. The large uncertainties in drag coefficient for $\alpha = 8°$ are attributed to the large aerodynamic loads on the axial-force gages requiring the data to be obtained at reduced dynamic pressures. Induced effects may also be present due to the complex three-dimensional secondary flows that occur in the vicinity of the non-metric panels and tunnel sidewall boundary layers. A repeat run of the balance data was obtained for an angle of attack of 0° and the results are shown in figure 8. Excellent repeatability is shown for both the lift and pitching-moment data. The drag data could be repeated within 10 drag counts ($c_d = 0.0010$).

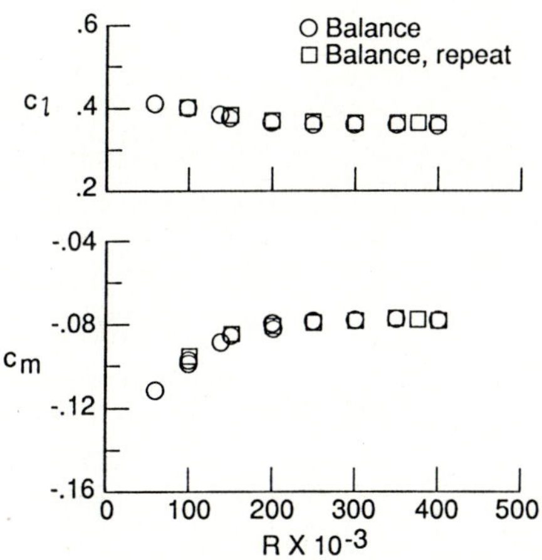

(a) Lift and pitch data.

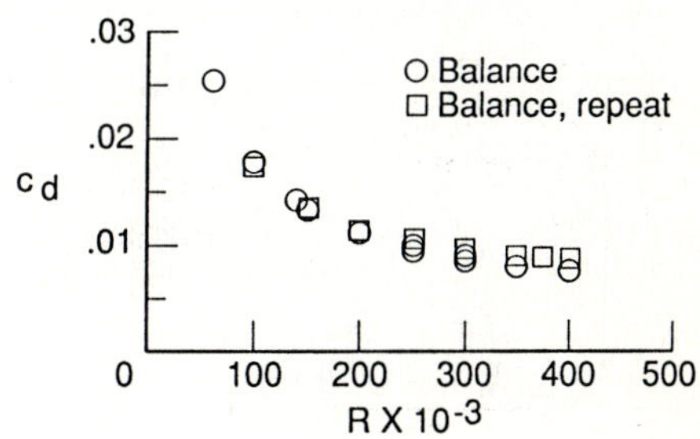

(b) Drag data.

Figure 8. Balance repeatability at $\alpha = 0°$.

Discussion of Results

The results of the present force model measurements are compared with the data obtained on a pressure model of the Eppler 387 airfoil tested in the LTPT (ref. 2) and are illustrated in figures 9 and 10. The lift and pitching-moment data obtained from the balance generally agree well with the results obtained from integration of the pressure data. The largest differences in the lift and pitch data between the two techniques occur at R = 60,000 and angles of attack less than about 4°. The largest uncertainties in the force and moment coefficients occur at this Reynolds number as illustrated by the data in figure 7. Also, for an angle of attack of 4° and R = 60,000 (fig. 9c), the results of reference 2 indicate that two flow phenomena (laminar separation with and without turbulent reattachment) were possible. The determination of lift and pitching-moment data from either the direct (force) or indirect (pressure) measurement technique is considered acceptable for low Reynolds number testing.

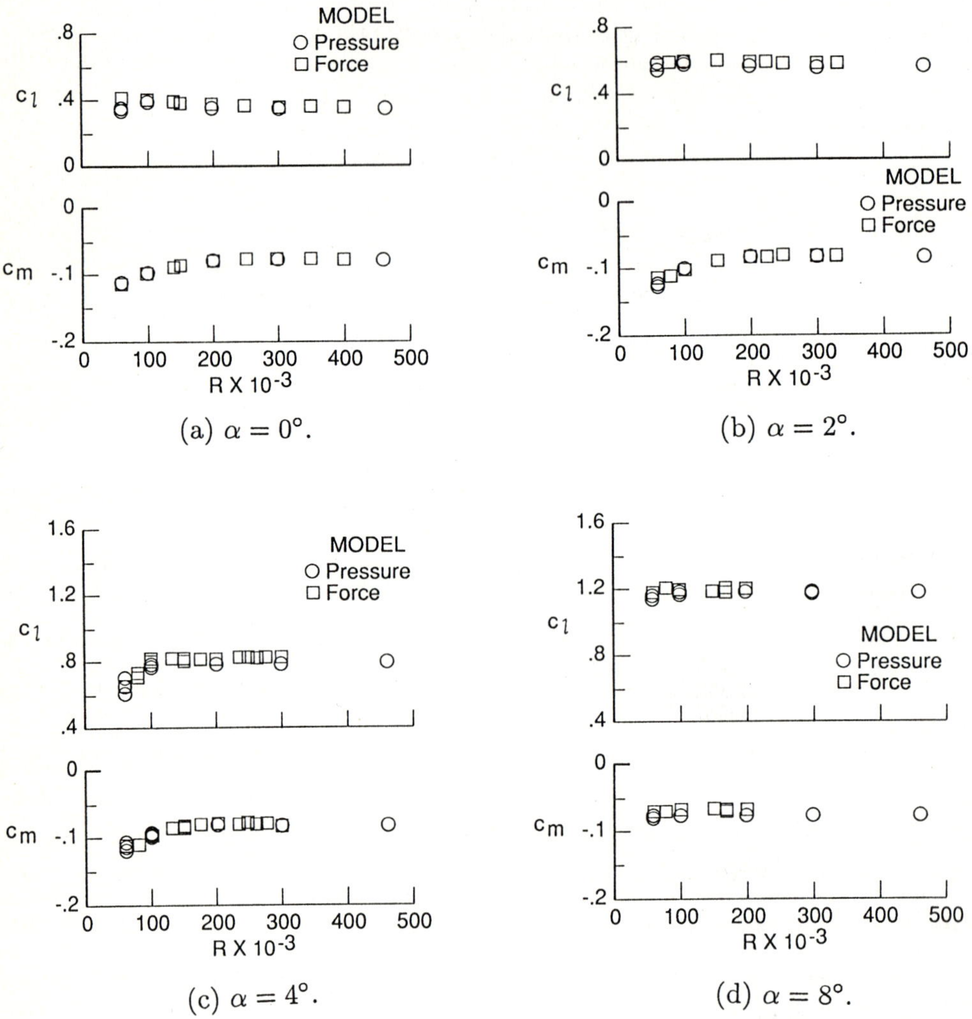

(a) $\alpha = 0°$.

(b) $\alpha = 2°$.

(c) $\alpha = 4°$.

(d) $\alpha = 8°$.

Figure 9. Lift and pitch data.

Drag data obtained from the balance measurements are compared with the drag data from the wake traverse method in figure 10. Wake-rake data are shown for both the force and pressure models. Good agreement between the balance measurements and wake traverse measurements for both models are shown for Reynolds numbers greater than 100,000 and angles of attack of 0, 2, and 4 degrees.

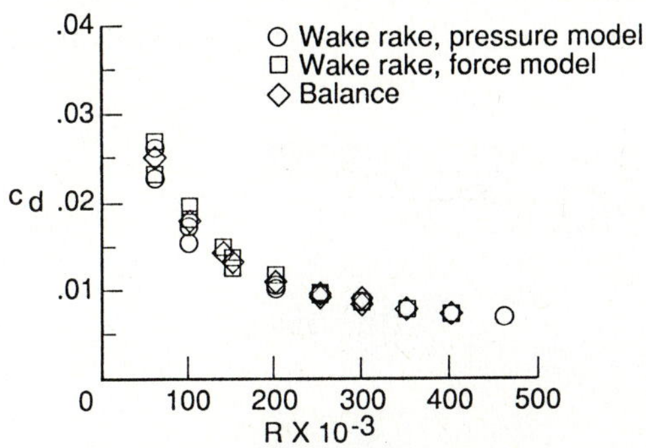

(a) $\alpha = 0°$.

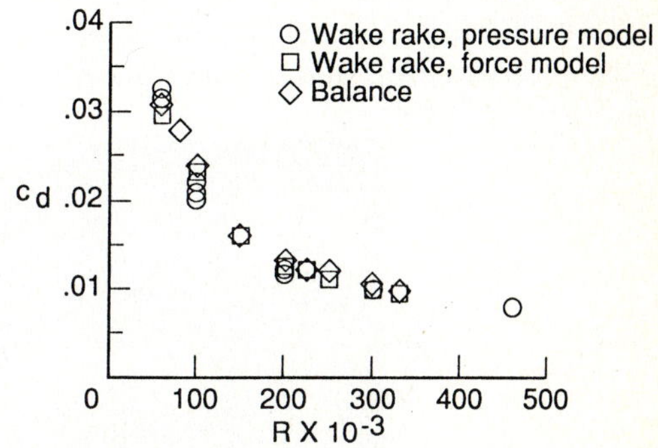

(b) $\alpha = 2°$.

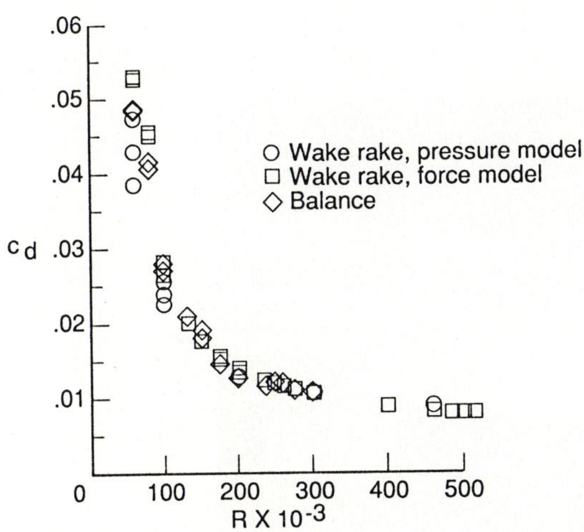

(c) $\alpha = 4°$.

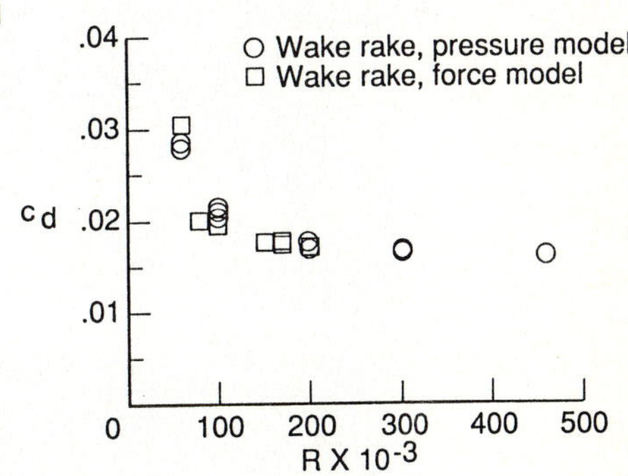

(d) $\alpha = 8°$.

Figure 10. Drag data.

At a Reynolds number of 100,000 or less, the flow phenomena are dominated by the presence of large laminar separation bubbles which result in unsteady wakes and significant spanwise flow structure. An excellent discussion of this flow environment is reported in reference 1. The unsteady wake profiles (fig. 6), oil-flow results, and spanwise drag data clearly illustrate the nature of this complex flow environment. At a Reynolds number of 300,000 and an angle of attack of 4° (fig. 11a), the oil flow results show that the turbulent reattachment line is uniform across a span distance of about 10 inches. However, for a Reynolds number of 100,000 the photograph (fig. 11b) shows a non-uniform turbulent reattachment line and turbulent trailing-edge separation of about 5% of the airfoil chord. The spanwise wake-rake drag data (figs. 12 and 13) indicate significant variations in spanwise drag for Reynolds numbers of 100,000 or less for both the force and pressure models.

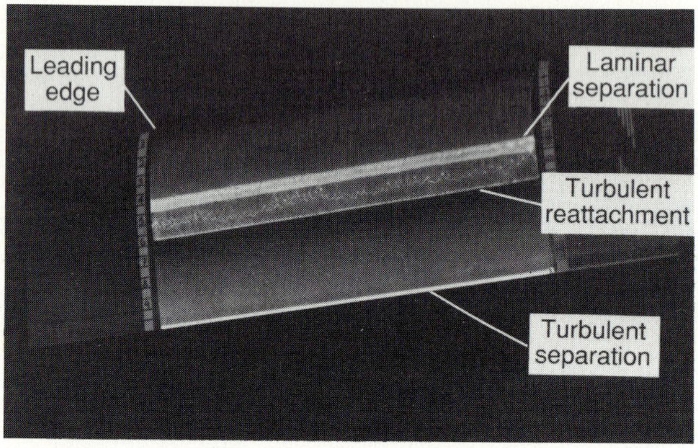

(a) R = 300,000.

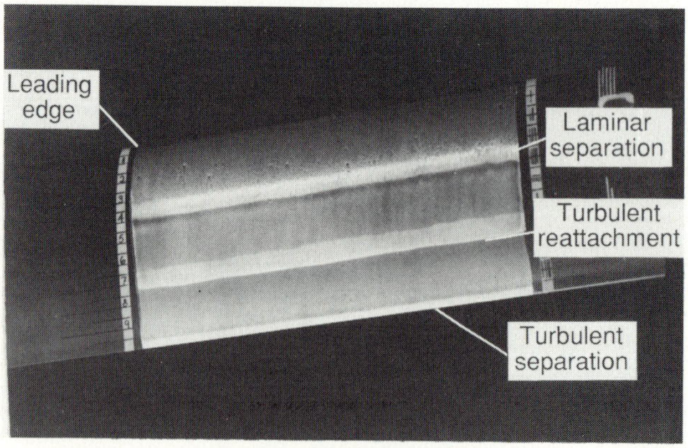

(b) R = 100,000.

Figure 11. Oil flow photographs at $\alpha = 4°$.

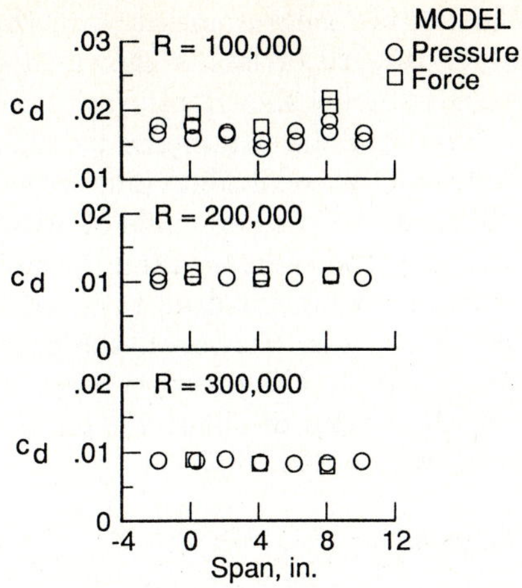

Figure 12. Spanwise wake-rake drag data at $\alpha = 0°$.

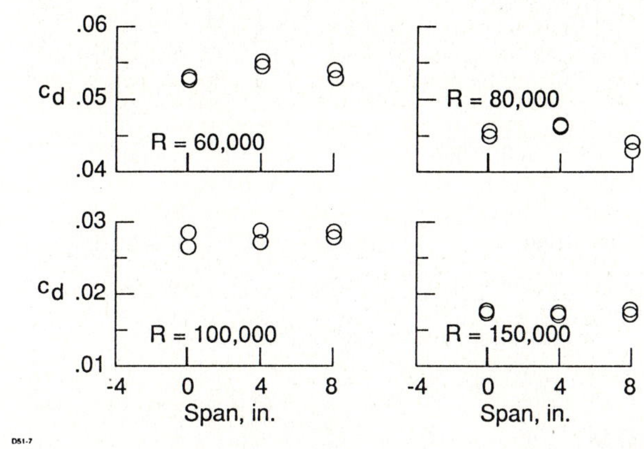

Figure 13. Spanwise wake-rake drag data for force model at $\alpha = 4°$.

A comparison of the balance drag measurements with the wake traverse measurements for Reynolds numbers of 100,000 or less is shown in figure 10. The wake-rake data for both the force or pressure model show large variations in drag coefficient characteristic of unsteady flows. The drag data from the balance measurements generally indicate a mean or higher value of drag coefficient compared to the wake traverse data.

A comparison of section lift-drag ratio at an angle of attack of 4° between the force data and pressure data (lift from pressure integration, drag from wake rake) is shown in figure 14(a). Generally lower values of lift-drag ratio are shown for the force data for Reynolds numbers of 100,000 or less. However, the uncertainty in lift-drag ratio is larger for the force data compared to the pressure data. For example, figure 14(b) indicates an uncertainty in lift-drag ratio of about 7% for the force data compared to about 2% for the pressure data at R = 100,000.

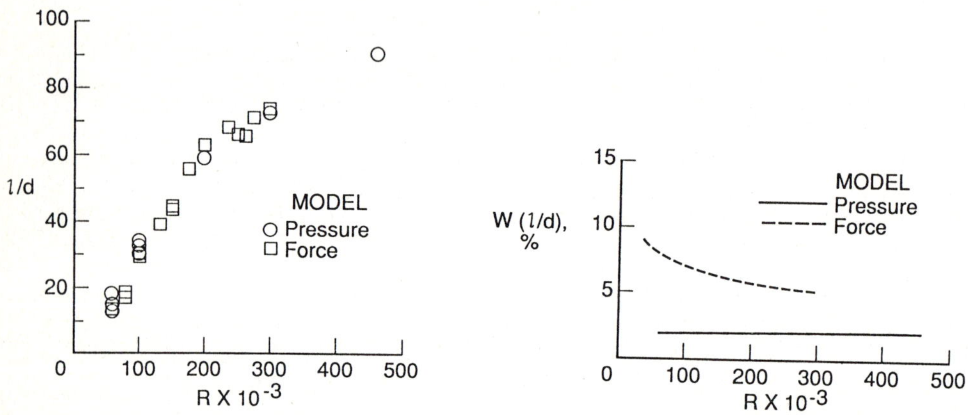

(a) Lift-drag ratio.

(b) Uncertainty in lift-drag ratio.

Figure 14. Lift-drag data at $\alpha = 4°$.

Concluding Remarks

Performance characteristics of an Eppler 387 airfoil using both the direct (force) and indirect (pressure) measurement technique have been obtained at Reynolds numbers from 60,000 to 460,000 in the Langley Low Turbulence Pressure Tunnel. The tests were conducted in a manner as to minimize both experimental apparatus and instrumentation uncertainties. As a result of this experiment, the following conclusions and remarks are considered appropriate:

(1) Comparison of the lift and pitching-moment data obtained from the balance measurements with the results obtained from integration of the pressure data showed good agreement.

(2) Comparison of the drag data obtained from the balance measurements with the results obtained using the wake traverse method showed good agreement for Reynolds numbers greater than 100,000 and angles of attack of 0, 2, and 4 degrees.
(3) At Reynolds numbers of 100,000 or less, where the flow phenomena are dominated by laminar separation bubbles, the wake traverse method for drag determination is suspect. Unsteady wakes and significant spanwise drag variations preclude accurate measurement. The direct measurement technique offers potential for improved drag determination.
(4) The use of the strain-gage balance for the measurement of small drag forces requires a careful examination of the design, calibration, model-balance arrangement, and evaluation of flow-induced interference effects in order to obtain accurate measurements.

References

1. Mueller, Thomas J., *Low Reynolds Number Vehicles.*, AGARD-AG-288 (Feb. 1985).
2. McGhee, Robert J.; Walker, Betty S.; and Millard, Betty F., *Experimental Results for the Eppler 387 Airfoil in the Langley Low-Turbulence Pressure Tunnel.*, NASA TM-4062 (Oct. 1988).
3. McGhee, Robert J.; Beasley, William D.; and Foster, Jean M., *Recent Modifications and Calibration of the Langley Low-Turbulence Pressure Tunnel.*, NASA TP-2328 (1984).
4. Stainback, P. Calvin; and Owen, F. Kevin, *Dynamic Flow Quality Measurements in the Langley Low Turbulence Pressure Tunnel.*, Technical Papers – 13th Aerodynamic Testing Conference, AIAA (1984), 257–265.
5. Loving, Donald L.; and Katzoff, S., *The Fluorescent-Oil Film Method and Other Techniques for Boundary-Layer Flow Visualization.*, NASA Memo 3-17-59L (1959).
6. Sewall, W. G.; Stack, J. P.; and McGhee, R. J., *A New Multipoint Thin-Film Diagnostic Technique for Fluid Dynamic Studies.*, SAE Paper No.881453 (Oct. 1988).
7. Pankhurst, R. C.; and Holder, D. W., "Wind-Tunnel Techniques.," Sir Isaac Pitman and Sons, Ltd., London, 1965.
8. Kline, S. J.; and McClintock, F. A., *Describing Uncertainties in Single-Sample Experiments.*, Mech. Eng. **75**, **No. 1** (Jan. 1953), 3–8.

CORRELATION OF THEORY TO WIND-TUNNEL DATA AT REYNOLDS NUMBERS BELOW 500,000

Raquel Evangelista
Analytical Services & Materials, Inc.
107 Research Drive, Hampton, Virginia 23666

Robert J. McGhee and Betty S. Walker
NASA Langley Research Center
Mail Stop 339, Hampton, Virginia 23665

Abstract

This paper presents results obtained from two airfoil analysis methods compared with previously published wind tunnel test data at chord Reynolds numbers below 500,000. The analysis methods are from the Eppler-Somers airfoil design/analysis code and from ISES, the Drela-Giles airfoil design/analysis code. The experimental data are from recent tests of the Eppler 387 airfoil in the NASA Langley Low Turbulence Pressure Tunnel. For R $\geq$ 200,000, lift and pitching moment predictions from both theories compare well with experiment. Drag predictions from both theories also agree with experiment, although to different degrees. The Drela-Giles code also predicts pressure distributions which compare well with the experimental pressure distributions, including those cases with laminar separation bubbles. For Reynolds numbers of 60,000 and 100,000, lift and pitching moment predictions from both theories are in fair agreement with experiment. However, most of the drag predictions from the Eppler-Somers code are accompanied with separation bubble warnings which indicate that the drag predictions are too low. With the Drela-Giles code, there is a large discrepancy between the computed and experimental pressure distributions in cases with laminar separation bubbles, although the drag polar predictions are similar in trend to experiment.

Symbol List

c	airfoil chord
c_d	airfoil section drag coefficient
c_l	airfoil section lift coefficient
$c_{l,max}$	maximum airfoil section lift coefficient
c_m	airfoil section quarter-chord pitching moment coefficient
C_p	pressure coefficient
n	critical disturbance amplification ratio (transition criterion in Drela-Giles code)
R	Reynolds number based on airfoil chord and freestream conditions
x	airfoil abscissa
$(x/c)_{Tr}$	theoretical transition location and experimental turbulent reattachment location
α	angle of attack relative to chord line, degrees

Introduction

The Fluid Dynamics Branch at NASA Langley Research Center has initiated a research program to determine airfoil performance at low Reynolds numbers. Part of this effort consists of an evaluation of existing airfoil analysis codes. As a result of this effort, this paper presents the results obtained from two airfoil analysis methods compared with wind tunnel test data at chord Reynolds numbers below 500,000. The two airfoil analysis methods are from the Eppler-Somers code [1,2] and ISES, the Drela-Giles code [3,4]. The airfoil analyzed is the Eppler 387 airfoil, and the wind tunnel data for this airfoil are from recent tests [5,6] in Langley's Low Turbulence Pressure Tunnel (LTPT) for Reynolds numbers between 60,000 and 460,000.

Description of the Codes

The Eppler-Somers and Drela-Giles codes are documented in detail by their developers in References [1,2] and [3,4] respectively. The following descriptions of the two codes are included only to outline the similarities and differences between them.

The Eppler-Somers Code

The Eppler-Somers low-speed airfoil design/analysis code has been widely used since it became available in 1980. The airfoil analysis uses a panel method to compute potential flow velocities and an integral method to calculate the boundary-layer characteristics. Airfoil lift and pitching moment are calculated from the inviscid velocity distribution which is corrected for viscous effects determined by the integral boundary layer calculations. Drag is calculated using a modified version of the Squire-Young formula which is applied to the boundary layer characteristics at the trailing edge. The prediction of laminar or turbulent separation is determined by the shape factor based on energy and momentum thicknesses. Transition is assumed upon the prediction of laminar separation. The prediction of transition is based on an empirical criterion which contains the Reynolds number (based on local conditions and momentum thickness) and the shape factor. The code predicts the existence of significant laminar separation bubble formation by an empirical criterion based on the velocity decrease across the empirical bubble length. In this case, the code issues a separation bubble warning to indicate that the predicted drag coefficient may be too low because the code does not account quantitatively for the influence of the separation bubble on drag.

The Drela-Giles Code

The Drela-Giles airfoil design-analysis code was recently developed at MIT. The airfoil analysis consists of a simultaneous solution of the discretized Euler and integral boundary-layer equations, typically on 96x16 or 132x32 size grids. This direct coupling of the viscous and inviscid parts of the analysis is intended to handle flows with strong viscous-inviscid interactions such as at low Reynolds numbers.

Laminar and turbulent separation are determined by local kinematic shape parameter criteria. Unlike the Eppler code, the Drela code can operate in an inverse mode to calculate across separated regions. In the case of laminar separation, the code continues in the inverse mode until transition is predicted. Transition prediction is based on the e^n method and the Orr-Sommerfeld equation of linear compressible stability theory for spatial disturbance amplification; the user specifies a value for the critical disturbance amplification ratio, n, as the transition criterion. The code does not actually perform stability calculations; instead, it uses the results from stability analyses of Falkner-Skan velocity profiles to formulate the equations for predicting the transition locations on the airfoil. Since n is an input variable, different values for n yield different results. However, it has been found that the choice for n is not arbitrary. In [7], it was found that using n values from 13 to 15 in the Drela code produced results that correlated well with experiment for cases with laminar separation bubbles at $R \geq 200{,}000$, whereas using lower values for n (from 7 to 11) produced better results for cases without laminar separation bubbles. To explain this difference, Pfenninger and Vemuru [8] have suggested that when a laminar separation bubble causes transition, the disturbance amplification ratio increases much more rapidly before transition than in the case in which natural transition occurs.

Since this code can calculate across a separation bubble, it has the potential to predict the effects of separation on an airfoil's section characteristics. The bubble effects are incorporated in the computation of the airfoil pressure distribution, which automatically includes the bubble effects on the airfoil section lift and pitching moment. Section drag is obtained from a wake momentum balance integration.

The Drela code requires more user interaction than the Eppler code because of its iterative solution procedure, especially in cases involving separation bubbles or considerable turbulent separation [7]. Compared to the Eppler code, the Drela code takes much more run time, both in terms of real and cpu time. Obtaining a drag polar from the Drela code over a ten-degree range of α can take at least 1600 cpu seconds on a CONVEX C210 machine whereas a drag polar for the same range of α from the Eppler code takes only about 40 cpu seconds on a CDC Cyber 170/180 machine.

Experimental Data

The experimental data for the Eppler 387 airfoil used in the present study were obtained from the Low-Turbulence Pressure Tunnel at NASA Langley. Pressure and force models of the Eppler 387 airfoil were tested at Reynolds numbers between 60,000 and 460,000. Results from these tests are documented in References [5] and [6] respectively.

At $R \geq 200{,}000$, both pressure integration and force measurement techniques from [5] and [6] produce very similar data, indicating that both techniques work well at these Reynolds numbers. However, when the test Reynolds numbers are 100,000 or less, unsteady flow effects and significant laminar separation with and without turbulent reattachment complicate the flow, resulting in differences in the data obtained with the two techniques. An example of the differences in data at these low Reynolds numbers is shown in fig. 1 (from Ref. 6), which compares the drag measurements using the wake traverse method with force measurements for different Reynolds numbers. At Reynolds numbers of 100,000 or less, unsteady wake flow precludes accurate measurements by the wake traverse method. At the same time, the uncertainty in the force measurements increases at these Reynolds numbers because of the small magnitude of the forces being measured. The discussion of results at these lower Reynolds numbers is thus separate from the discussion of results for $R \geq 200{,}000$.

Discussion of Results

All experimental data shown in the following discussion are from the pressure model test [5] of the Eppler 387 airfoil in the LTPT. All Drela analytical results shown were obtained using a 132x32 grid.

200,000 $\leq$ R $<$ 500,000

Results from analyses of the Eppler 387 airfoil using both the Eppler and Drela codes are compared to experimental data in figs. 2, 3, and 4, for R = 200,000; 300,000; and 460,000 respectively. The theoretically computed and experimentally obtained airfoil data are also tabulated in Tables 1, 2, and 3 for these Reynolds numbers. The Drela results discussed in this section were obtained using n = 14. (A supplement to the discussion in [7] of the effects of n on the Drela code's analytical results is provided in Appendix A.)

Fig. 2a compares the airfoil section characteristics from the Eppler analysis with the experimental data at $R = 200,000$. The experimental and theoretical lift and pitching moment curves agree well. A discrepancy exists between theory and experiment in the drag polars. There are no separation bubble warnings from the Eppler code except near $c_{l,max}$ on the upper surface, and at a lift coefficient of about 0.1 on the lower surface. However, experimental pressure distributions and flow visualization in [5] have shown that laminar separation bubbles exist on the Eppler 387 airfoil at $R = 200,000$ for angles of attack between -2 and 7 degrees. Since the Eppler code does not account for the drag due to laminar separation bubbles, the code underpredicts drag by as much as 17% in the lift coefficient range between approximately 0.3 and 1.0. Eppler has reported in [9] that the bubble analogy in the code is not conservative enough. With Eppler's modification of the bubble analogy in [9], more bubble warnings are issued for the Eppler 387 at $R = 200,000$ than prior to his modification.

Fig.2b compares the Drela predictions with experiment. As in the case with the Eppler code, the Drela code's lift and pitching moment curves agree well with experiment. The Drela code's drag predictions agree well with experimental data except near $c_{l,max}$ ($c_{l,max} \sim 1.2$). The difference between Drela-computed and experimental drag values in the c_l range between 0.2 and 1.0 is at most 2%.

Fig. 3a compares the Eppler theory with experiment at $R = 300,000$. As in the previous case, the lift and pitching moment curves for both theory and experiment are in good agreement. However, the Eppler code still underpredicts drag by as much as 13.5% in the lift coefficient range between 0.3 and 1.0.

Fig.3b compares the Drela theory with experiment at $R = 300,000$. The lift and pitching moment curves for both theory and experiment are in good agreement. In addition, the Drela drag predictions at $R = 300,000$ agree within 3% of experimental drag data for c_l between 0.3 and 1.0.

Figs. 4a and 4b show the comparisons between theory and experiment at $R = 460,000$. At this higher Reynolds number, the separation bubble effects are less significant than in the two previous cases; therefore, both the Drela and Eppler codes produce results in very good agreement with experiment. The difference in drag predictions between the two theories is at most 4% in the range of lift coefficients between 0.3 and 1.0.

Unlike the Eppler code which computes inviscid pressure distributions only, the Drela code computes airfoil pressure distributions which incorporate the effects of the viscous-inviscid interactions. Figs. 5a, 5b, and 5c are typical examples of how the Drela-computed pressure distributions compare to experiment. All three figures correspond to an experimental $\alpha = 4°$ at R = 200,000; 300,000; and 460,000 respectively. Also indicated in figs. 5a and 5b are the regions where laminar separation bubbles were observed by oil flow visualization in the experiment. The Drela pressure distributions were obtained by matching experimental lift coefficients, so that the theoretical and experimental angles of attack are not exactly the same. The overall match in the pressure distributions between the Drela theory and experiment is good for all three cases; there are only slight discrepancies between the computations and experiment in the separation bubble region on the airfoil upper surface. This is not unreasonable since the Drela code can converge upon a solution which matches the experimental lift coefficient, but this solution may have a slightly different pressure distribution from that of the experiment due to small variations in separation bubble dimensions. In addition to the good correlation between the computed and experimental pressure distributions, the agreement between the Drela-computed and experimental section characteristics for the cases in fig. 5 is excellent.

Figs. 6a (R = 200,000) and 6b (R = 300,000) show the experimental turbulent reattachment location and the Drela-computed transition location on the upper surface of the airfoil where separation bubbles form at various lift coefficients. Turbulent reattachment was determined experimentally from oil flow visualization, and accuracy of the oil flow data was confirmed by hot film data [10] at $R = 200,000$. These figures illustrate the good agreement between the reattachment locations and predicted transition locations.

R < 200,000

The complex flow phenomena generated at these low Reynolds numbers (i.e., unsteady flow effects, and significant laminar separation with and without turbulent reattachment) proved to be a challenge for both the experimental techniques used in refs. [5] and [6] and the computational techniques of the present study. The Eppler and Drela predictions for these low Reynolds numbers are presented here with experimental data to illustrate how the codes handle such low Reynolds number cases.

Fig. 7a compares predictions from the Eppler code with experiment at $R = 100,000$. The theoretical lift curve falls almost exactly on the experimental curve, whereas the magnitudes of the pitching moments are underpredicted. The Eppler code does not account for laminar separation bubble drag as evidenced by the discrepancies between the theoretical and experimental drag polars. The Eppler code issues bubble warnings throughout the entire range of lift coefficients analyzed, except at $c_l = 1.06$ where the Eppler drag prediction agrees well with the experimental drag value.

Fig. 7b compares predictions from the Drela code, using $n = 14$, with experiment. There is good agreement in the lift and pitching moment characteristics between theory and experiment. However, the agreement between the theoretical and experimental drag polars is not as good as in the previous cases. The Drela code underpredicts drag in the c_l range between 0.3 and 0.9. A comparison of computed and experimental pressure distributions shows that the theory predicts shorter bubble lengths than those found in the experiment. An example of this is shown in fig. 8 for $R = 100,000$ and $c_l = .778$. In this figure, one can see the discrepancy between the pressure distributions in the region where the experimentally observed laminar separation bubble is indicated. For lift coefficients above 0.9, the Drela code overpredicts drag; it only matches the experimental drag near the lift coefficients ($c_l \sim 1.2$) in which the experiment showed that a very short bubble (less than 1% of the airfoil chord) existed.

In an attempt to improve the correlation between the Drela predictions and theory at $R = 100,000$, higher values for n, the transition criterion, were used as input to the Drela code. However, increasing n to 20 had only a negligible effect on the separation bubble length. Fig. 9 shows that the experimentally observed turbulent reattachment locations occurred much farther downstream than the Drela-predicted transition locations for both $n = 14$ and $n = 20$ cases. Further increases in the transition criterion were not considered since increasing n from 14 to 20 did little to move the transition location.

Although the code could not accurately predict the extent of the separation bubbles at $R = 100,000$, fig. 10 shows that the increase in transition criterion to $n = 20$ did move the drag predictions closer to experiment in the c_l range between 0.3 and 0.9. In the higher c_l range ($c_l > 0.9$), the drag values computed using $n = 14$ are closer to experimental values than when using $n = 20$. This can be seen by comparing fig. 7b to fig. 10.

The Eppler predictions and experimental data at $R = 60,000$ are shown in fig. 11a. The Eppler code issues bubble warnings for the entire lift coefficient range analyzed. Thus, there is little agreement in the pitching moment curves and drag polars between theory and experiment. However, there is still good agreement between the theoretical and experimental lift curves. The code predicts a break in the lift-curve slope corresponding to the experimentally observed laminar stall at a lift coefficient of about 0.6, and another break in the lift-curve slope corresponding to the experimentally observed flow reattachment at a higher lift coefficient.

The Drela predictions using $n = 14$ and $n = 20$ are compared to experiment at $R = 60,000$ in figs. 11b and 11c, respectively. Both Drela calculations also show breaks in the slopes of the lift curves similar to the experimental lift curves. As in the case of $R = 100,000$, the Drela analysis using $n = 20$ gave drag values that agreed better with experiment in the c_l range between 0.3 and 0.9, whereas at $c_l > 0.9$, using $n = 14$ gave results that were closer to experiment. Again, the Drela theory was able to predict drag polars similar in trend to the experimental drag polar. However, fig. 12 shows that the computed pressure distributions do not agree well with experiment for $R = 60,000$. In addition, the pressure data indicate that laminar separation with and without turbulent reattachment occurred at the same angle of attack in the experiment.

Conclusions

Airfoil performance predictions from the Eppler-Somers and the Drela-Giles codes have been compared with experimental data for the Eppler 387 airfoil at Reynolds numbers below 500,000. At $R \geq 200,000$, both theories predicted lift and pitching moment characteristics that compared well with experiment. Drag predictions from both theories also agreed with experiment, although to different degrees. For $.3 < c_l < 1.0$, the Eppler code underpredicted drag by as much as 13.5% at $R = 300,000$ and 17% at $R = 200,000$. On the other hand, the Drela-Giles code was able to predict drag to within 3% of experimental drag data at $R \geq 200,000$ for $.3 < c_l < 1.0$. The Drela-Giles code was also able to predict pressure distributions which compared well with the experimental pressure distributions, including those cases with laminar separation bubbles.

For Reynolds numbers of 60,000 and 100,000, unsteady flow effects and significant laminar separation with and without turbulent reattachment complicate the flow, but both codes were able to predict lift and pitching moment fairly well. Most of the Eppler-Somers drag predictions at these low Reynolds numbers were accompanied with separation bubble warnings which indicated that the drag predictions were too low. With the Drela-Giles code, there was a large discrepancy between the computed and the experimental pressure distributions in cases with significant laminar separation bubbles, although the code did predict drag polars that were similar in trend to the experimental drag polars.

Appendix A

The evaluation of the Drela-Giles code in [7] showed that different values for the transition criterion (i.e., the critical disturbance amplification ratio, n) yielded different results. However, a rule of thumb emerged from this evaluation for $R \geq 200,000$; that high values for n (13 to 15) yield results that correlate better with experiment for cases with moderate laminar separation bubble formation, whereas lower values for n (7 to 11) yield results that correlate better with experiment when transition occurs without laminar separation.

The Drela-computed section characteristics using $n = 9$ and $n = 14$ are compared to experimental section characteristics at $R = 300,000$ in fig. A1. The $n = 14$ case correlates better with experiment than the $n = 9$ case. For this Reynolds number, laminar separation bubble formation occurs in the c_l range between 0.1 and 1.0 [5]. In fig. A2, which illustrates the computed airfoil section characteristics at R=200,000 using $n = 9$ and $n = 14$ compared to the experimental section characteristics, the $n = 14$ case also correlates better with experiment than the $n = 9$ case. For this Reynolds number, laminar separation bubble formation occurs in the c_l range between 0.1 and 1.1 [5].

Fig. A3 compares the theoretical and experimental drag coefficients at various Reynolds numbers for $\alpha = 0$ and 4 degrees. For Reynolds numbers of 200,000 and 300,000, laminar separation bubbles exist at both these angles of attack [5]. These plots further illustrate that the higher n value produces results which correlate better with experiment than the lower n value.

Fig. A4 compares the experimentally-determined turbulent reattachment locations to the Drela-computed transition locations using $n = 9$ and $n = 14$ on the airfoil's upper surface for several lift coefficients at $R = 200,000$ and $R = 300,000$. These plots show that using $n = 9$ causes transition to occur earlier than $n = 14$, whereas the higher n value predicts transition locations that are very close to the experimental turbulent reattachment locations.

Table 1.- R = 200,000 (Drela code, n = 14)

α, deg	c_l, Drela	c_l, Eppler	c_l, Exp.	c_d, Drela	c_d, Eppler	c_d, Exp.	c_m, Drela	c_m, Eppler	c_m, Exp.
-2	0.193	0.173	0.156	0.0117	0.0118*	0.0133	-0.0850	-0.0822	-0.0814
-1	.289	.283	.249	.0101	.0090	.0105	-.0812	-.0832	-.0804
0	.397	.393	.350	.0105	.0093	.0106	-.0816	-.0842	-.0780
1	.506	.503	.466	.0111	.0097	.0113	-.0822	-.0853	-.0796
2	.615	.613	.574	.0119	.0099	.0118	-.0824	-.0864	-.0794
3	.722	.723	.680	.0125	.0105	.0127	-.0828	-.0875	-.0807
4	.827	.832	.785	.0133	.0111	.0133	-.0832	-.0885	-.0803
5	.932	.940	.891	.0139	.0121	.0138	-.0832	-.0892	-.0809
6	1.034	1.032	1.004	.0143	.0130	.0141	-.0828	-.0863	-.0809
7	1.131	1.098	1.103	.0139	.0190	.0145	-.0808	-.0783	-.0789
7.5	1.163	1.127	1.155	.0147	.0205*	.0152	-.0779	-.0740	-.0783
8	1.171	1.159	1.180	.0183	.0215*	.0175	-.0739	-.0707	-.0763

Table 2.- R = 300,000 (Drela code, n = 14)

α, deg	c_l, Drela	c_l, Eppler	c_l, Exp.	c_d, Drela	c_d, Eppler	c_d, Exp.	c_m, Drela	c_m, Eppler	c_m, Exp.
-2	0.186	0.173	0.146	0.0111	0.0107*	0.0118	-0.0810	-0.0822	-0.0788
-1	.282	.283	.254	.0083	.0078	.0101	-.0783	-.0832	-.0807
0	.393	.393	.352	.0087	.0081	.0087	-.0794	-.0842	-.0768
1	.504	.503	.465	.0091	.0084	.0093	-.0800	-.0853	-.0784
2	.614	.613	.573	.0097	.0086	.0099	-.0806	-.0864	-.0785
3	.722	.723	.685	.0101	.0090	.0104	-.0812	-.0876	-.0795
4	.830	.833	.792	.0107	.0095	.0109	-.0818	-.0887	-.0794
5	.936	.943	.901	.0111	.0104	.0114	-.0820	-.0898	-.0799
6	1.038	1.051	1.009	.0115	.0113	.0118	-.0820	-.0907	-.0799
7	1.130	1.112	1.106	.0119	.0173	.0129	-.0802	-.0810	-.0785
7.5	1.158	1.144		.0141	.0189*		-.0781	-.0771	
8	1.176	1.177	1.180	.0175	.0199*	.0168	-.0751	-.0737	-.0756
8.25	1.182	1.193		.0191	.0204*		-.0735	-.0721	

Table 3.- R = 460,000 (Drela code, n = 14)

α, deg	c_l, Drela	c_l, Eppler	c_l, Exp.	c_d, Drela	c_d, Eppler	c_d, Exp.	c_m, Drela	c_m, Eppler	c_m, Exp.
-2	0.181	0.173	0.153	0.0097	0.0097*	0.0103	-0.0785	-0.0822	-0.0774
-1	.281	.283	.263	.0073	.0068	.0090	-.0783	-.0832	-.0797
0	.391	.393	.356	.0070	.0070	.0073	-.0781	-.0842	-.0785
1	.503	.503	.470	.0075	.0073	.0076	-.0790	-.0853	-.0783
2	.614	.613	.580	.0079	.0075	.0078	-.0798	-.0864	-.0786
3	.723	.723	.693	.0083	.0078	.0084	-.0804	-.0876	-.0799
4	.831	.833	.803	.0087	.0083	.0090	-.0812	-.0887	-.0801
5	.938	.943	.914	.0091	.0090	.0093	-.0816	-.0900	-.0807
6	1.040	1.053	1.022	.0095	.0098	.0101	-.0818	-.0911	-.0807
7	1.130	1.120	1.109	.0109	.0157	.0120	-.0808	-.0826	-.0788
7.5	1.161	1.159		.0135	.0175*		-.0794	-.0799	
8	1.185	1.196	1.179	.0163	.0184*	.0161	-.0767	-.0771	-.0759

* Eppler code bubble warning

References

1. Eppler, R. and Somers, D. M., *A Computer Program for the Design and Analysis of Low-Speed Airfoils.*, NASA TM 80210 (1980).
2. Eppler,R. and Somers, D. M., *Supplement to: A Computer Program for the Design and Analysis of Low-Speed Airfoils.*, NASA TM 81862 (December 1980).
3. Drela, M., *Two-Dimensional Aerodynamic Design and Analysis Using the Euler Equations.*, Massachusetts Institute of Technology, Gas Turbine Laboratory Rept. 187 (February 1986).
4. Drela, M. and Giles, M. B., *Viscous-Inviscid Analysis of Transonic and Low Reynolds Number Airfoils.*, AIAA Journal **25, No. 10**, 1347–1355.
5. McGhee, R. J., Walker, B. S. and Millard, B. F., *Experimental Results for the Eppler 387 Airfoil at Low Reynolds Numbers in the Langley Low-Turbulence Pressure Tunnel.*, NASA TM 4062 (October 1988).
6. McGhee, R. J. and Walker, B. S., *Performance Measurements of an Airfoil at Low Reynolds Numbers.*, Proposed Paper for Conference on Low Reynolds Number Aerodynamics, University of Notre Dame, Indiana (June 5–7, 1989).
7. Evangelista, R. and Vemuru, C. S., *Evaluation of an Analysis Method for Low-Speed Airfoils by Comparison with Wind Tunnel Results.*, AIAA 89-0266, 27th Aerospace Sciences Meeting, Reno, Nevada (January 1989).
8. Pfenninger, W. and Vemuru, C. S., *Design of Low Reynolds Number Airfoils-I.*, AIAA 88-2572, 6th Applied Aerodynamics Conference, Williamsburg, Virginia (June 1988).
9. Eppler, R., *Recent Developments in Boundary-Layer Computation.*, Proceedings of Aerodynamics at Low Reynolds Numbers $10^4 < Re < 10^6$ International Conference Vol. II, London (October 15–18, 1986).
10. Sewall, W. G., Stack, J. P., McGhee, R. J., and Mangalam, S. M., *A New Multipoint Thin-Film Diagnostic Technique for Fluid Dynamic Studies.*, SAE Technical Paper 881453, Aerospace Technology Conference and Exposition, Anaheim, California (October 1988).

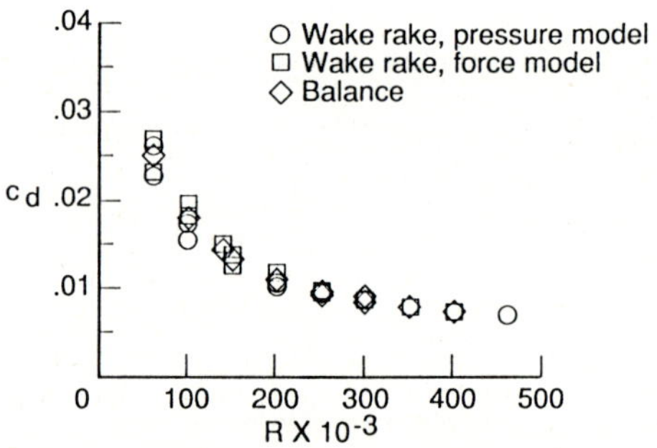

Figure 1. Drag measurements for Eppler 387 airfoil at $\alpha = 0°$.

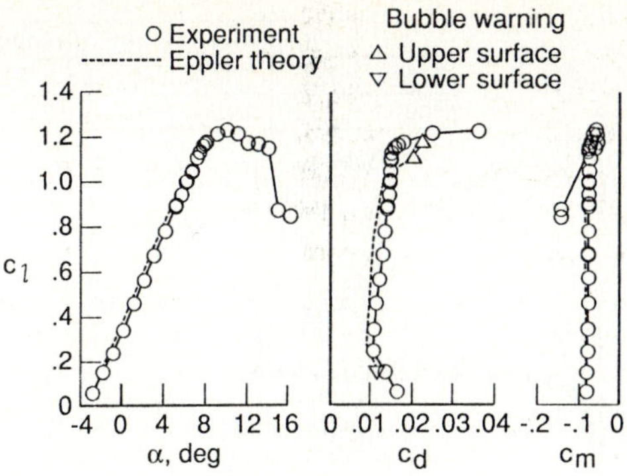

(a) Eppler code.

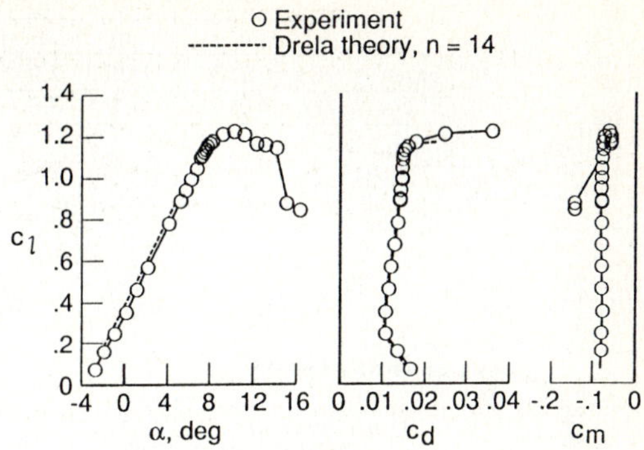

(b) Drela code.

Figure 2. Theory and experiment at R = 200,000.

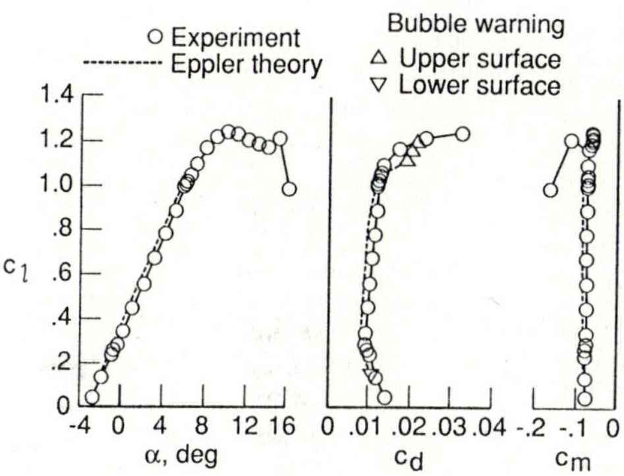

(a) Eppler code.

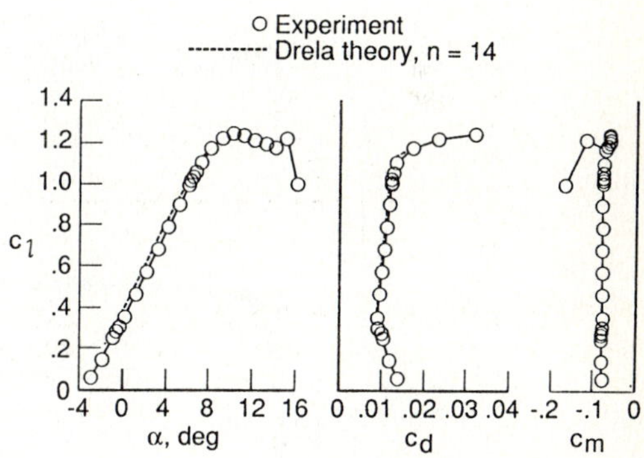

(b) Drela code.

Figure 3. Theory and experiment at R = 300,000.

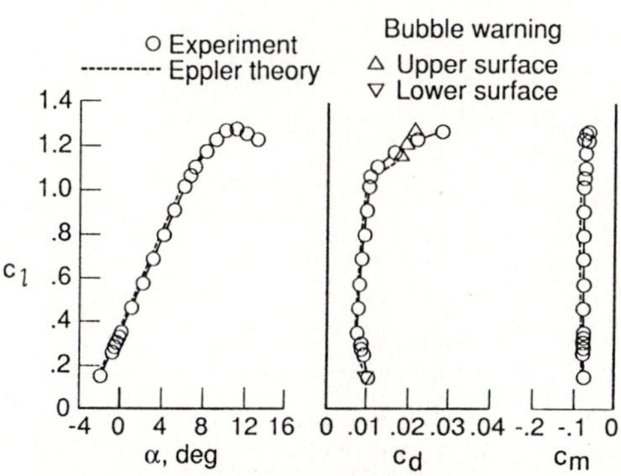

(a) Eppler code.

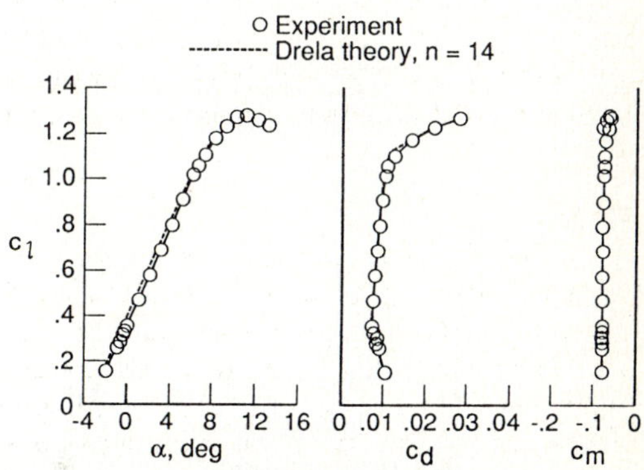

(b) Drela code.

Figure 4. Theory and experiment at R = 460,000.

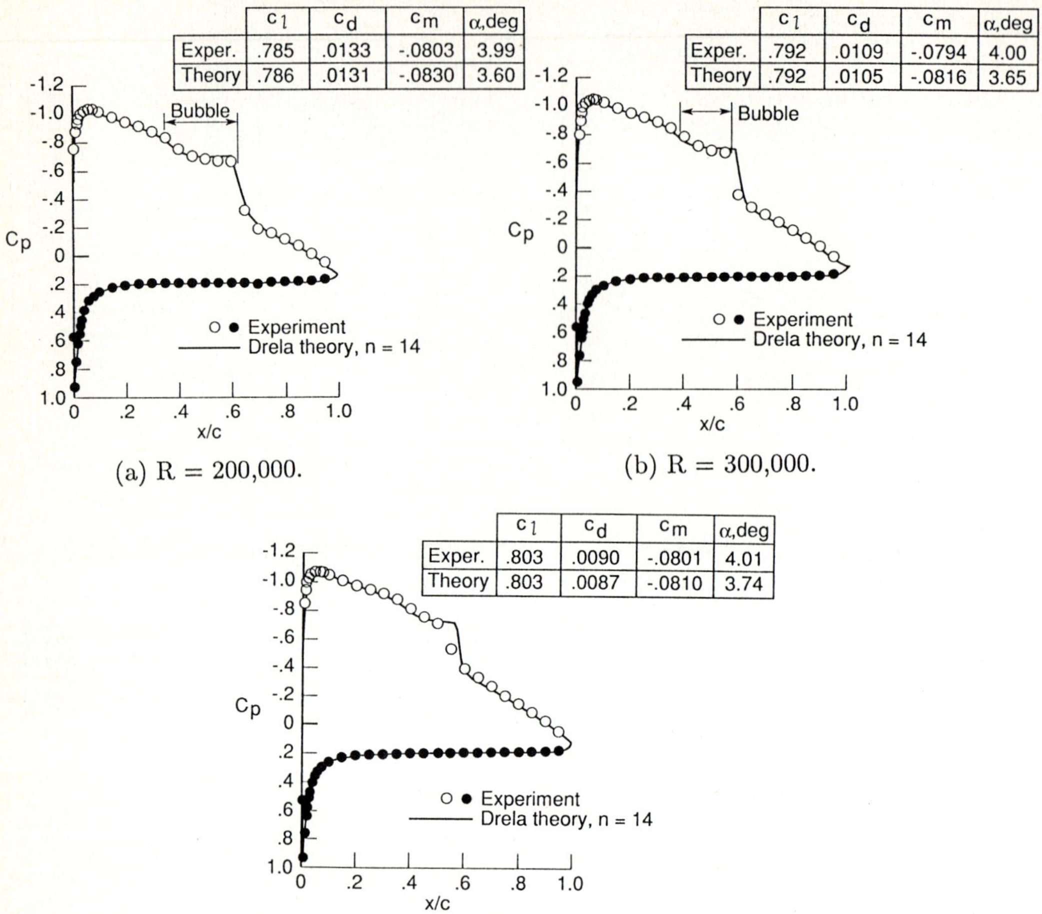

	c_l	c_d	c_m	α,deg
Exper.	.785	.0133	-.0803	3.99
Theory	.786	.0131	-.0830	3.60

(a) R = 200,000.

	c_l	c_d	c_m	α,deg
Exper.	.792	.0109	-.0794	4.00
Theory	.792	.0105	-.0816	3.65

(b) R = 300,000.

	c_l	c_d	c_m	α,deg
Exper.	.803	.0090	-.0801	4.01
Theory	.803	.0087	-.0810	3.74

(c) R = 460,000.

Figure 5. Experimental pressure data and predictions from Drela code.

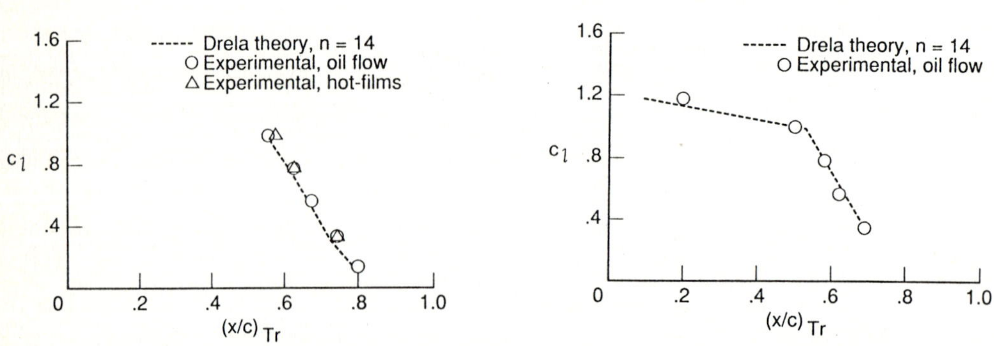

(a) R = 200,000.

(b) R = 300,000.

Figure 6. Experimentally observed turbulent reattachment locations and Drela code predicted transition locations for upper surface of Eppler 387 airfoil.

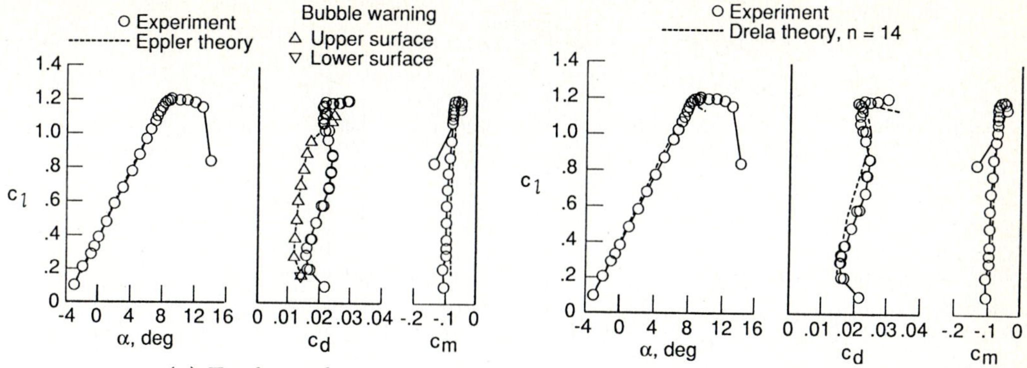

(a) Eppler code.

(b) Drela code.

Figure 7. Theory and experiment at R = 100,000.

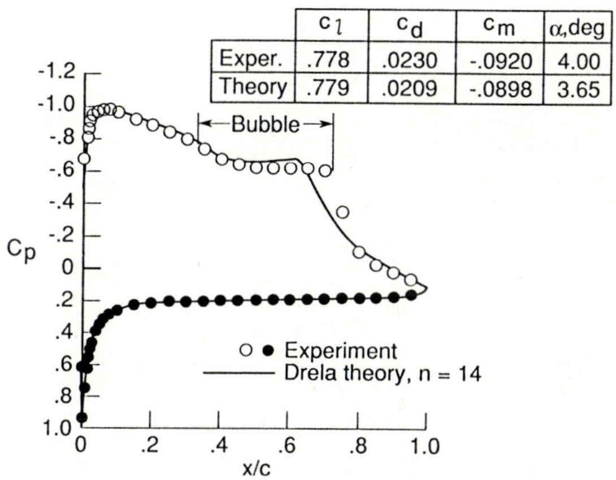

	c_l	c_d	c_m	α,deg
Exper.	.778	.0230	-.0920	4.00
Theory	.779	.0209	-.0898	3.65

Figure 8. Experimental pressure data and predictions from Drela code at R = 100,000.

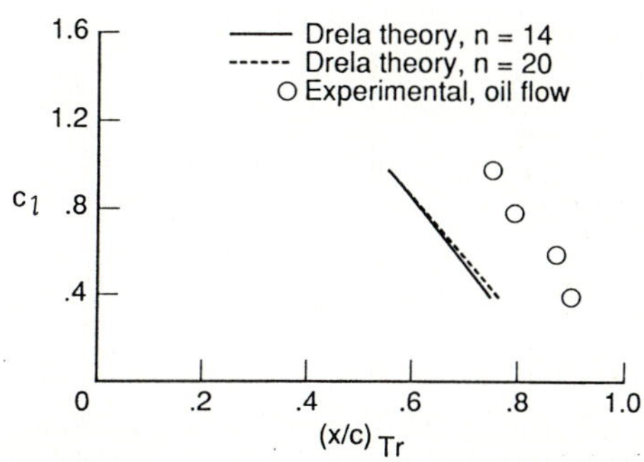

Figure 9. Experimentally observed turbulent reattachment locations and Drela code predicted transition locations for upper surface of Eppler 387 airfoil at R = 100,000.

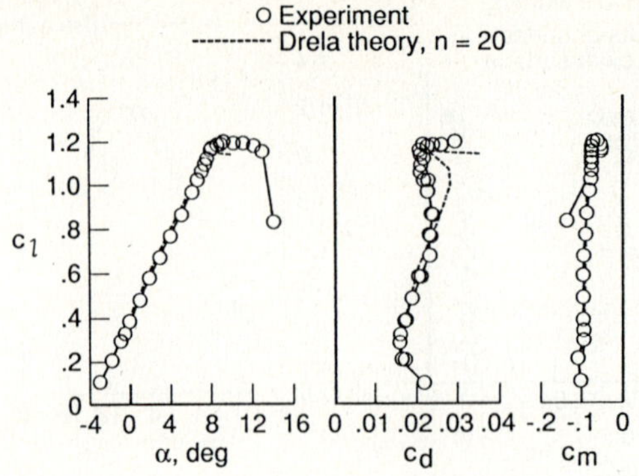

Figure 10. Experimental section data and Drela code predictions at R = 100,000 and $n = 20$.

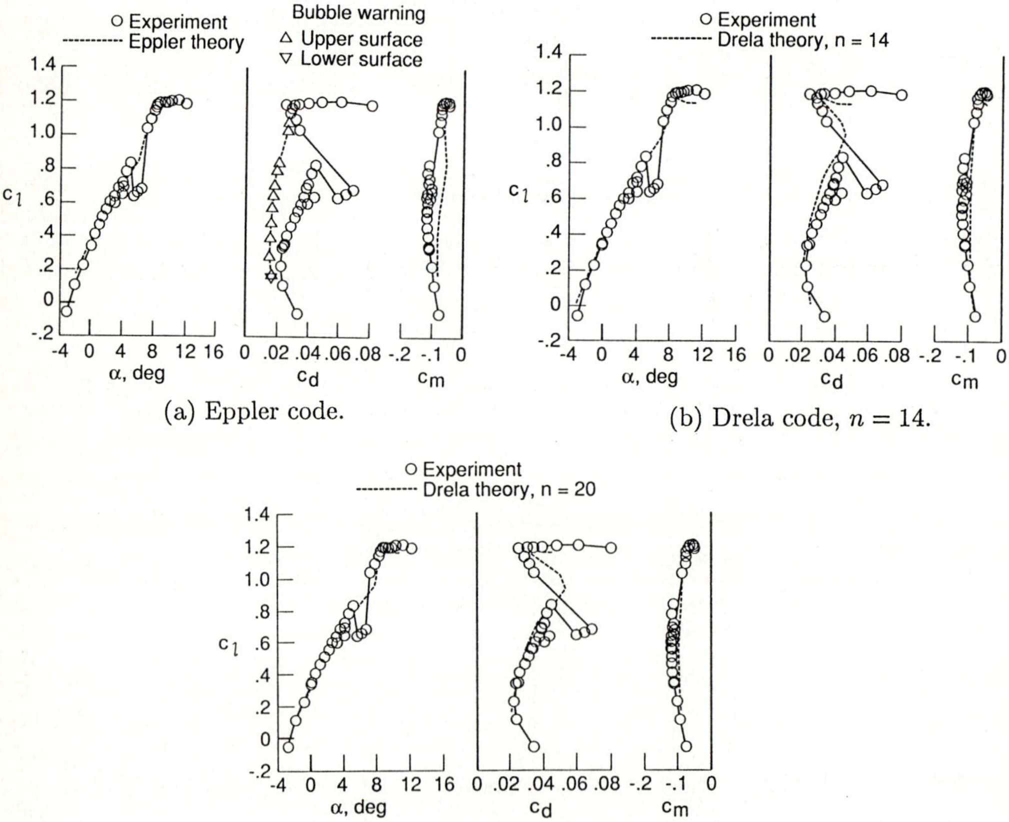

(a) Eppler code.

(b) Drela code, $n = 14$.

(c) Drela code, $n = 20$.

Figure 11. Theory and experiment at R = 60,000.

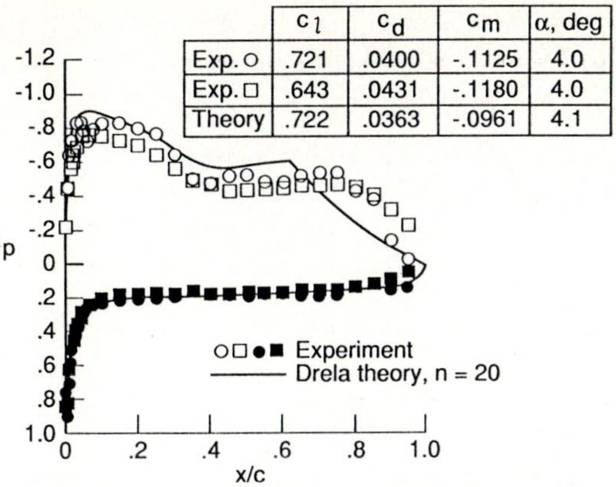

	c_l	c_d	c_m	α, deg
Exp. ○	.721	.0400	-.1125	4.0
Exp. □	.643	.0431	-.1180	4.0
Theory	.722	.0363	-.0961	4.1

Figure 12. Experimental pressure data and predictions from Drela code at R = 60,000.

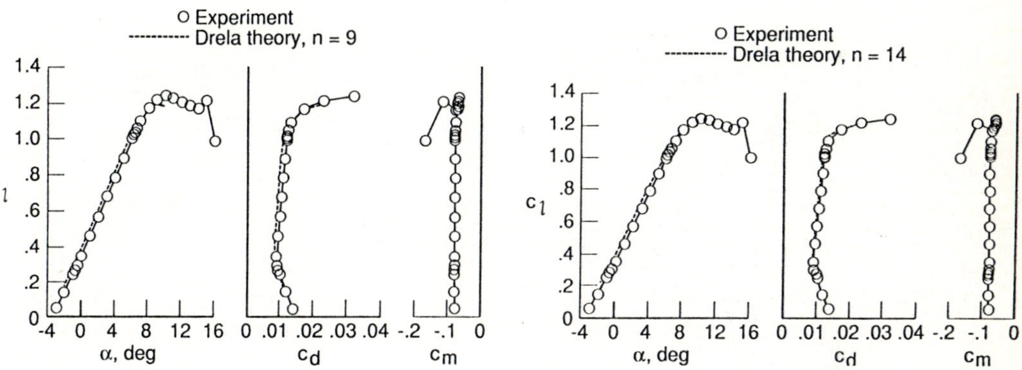

(a) $n = 9$.

(b) $n = 14$.

Figure A1. Experimental data and predictions from Drela code at R = 300,000.

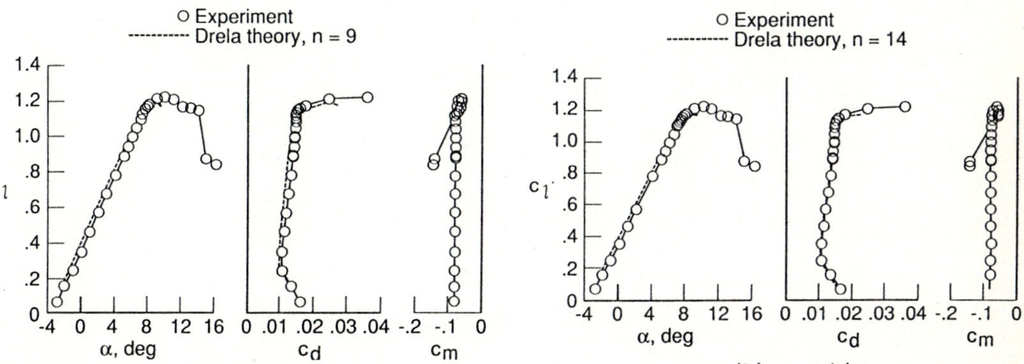

(a) $n = 9$.

(b) $n = 14$.

Figure A2. Experimental data and predictions from Drela code at R = 200,000.

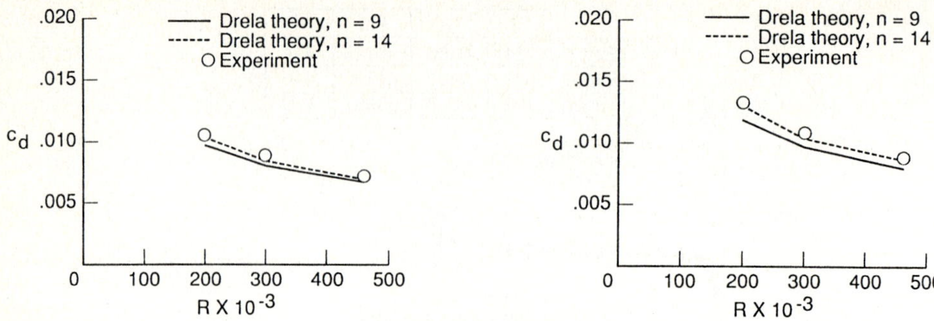

(a) $\alpha = 0°$.

(b) $\alpha = 4°$.

Figure A3. Experimental drag data and predictions from Drela code.

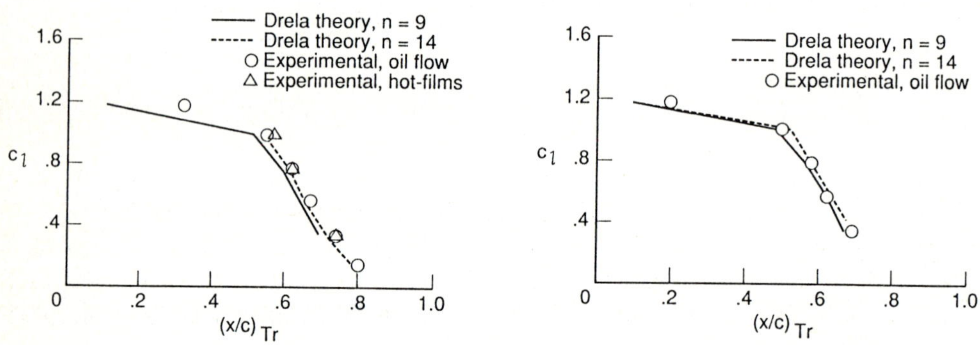

(a) R = 200,000.

(b) R = 300,000.

Figure A4. Experimental turbulent reattachment locations and predicted transition locations.

AN EXPERIMENTAL STUDY OF LOW-SPEED SINGLE-SURFACE AIRFOILS WITH FAIRED LEADING EDGES

James D. DeLaurier
University of Toronto Institute for Aerospace Studies
4925 Dufferin St., Downsview, Ontario, Canada M3H 5T6

ABSTRACT

The development of a large model ornithopter required the design of an aerodynamically efficient wing with torsional compliance. A solution for this was to use an essentially single-surface airfoil with a faired leading edge. This paper describes wind-tunnel tests on various candidate airfoils, with the goal of attaining the largest angle-of-attack range for attached flow. Also tested were several "benchmark" sections, such as single-surface airfoils with sharp leading edges and an Eppler 193 double-surface airfoil. The results showed no candidate as good as the Eppler design, but very acceptable performance was obtained from a section with a streamlined leading-edge fairing on the upper side of a 5% circular arc.

INTRODUCTION

The motivation for this work originated from the goal of designing an efficient flapping wing for a large model ornithopter. The original airfoil was similar to that used in many ultralight airplanes, consisting of a single surface of fabric stretched over cambered ribs and wrapped around a tubular leading-edge spar (Fig. 1). The fabric is then attached to the main sheet, aft some small percentage of the chord, thus forming a pocket fairing behind the leading-edge tube.

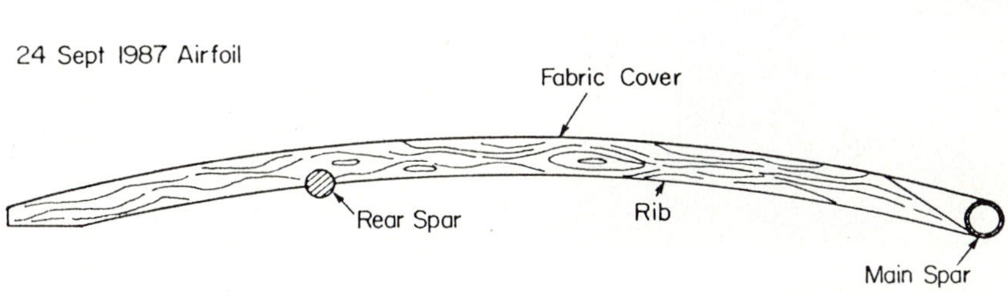

Figure 1. Original (24 Sept. 1987) Airfoil Design

Such an airfoil seemed to be particularly appropriate for the ornithopter wing which, structurally, consisted of a round leading-edge spar, about which the ribs could freely pivot to give the twisting required for efficient thrusting. The essentially single-surface covering provides the compliance necessary for this twisting, with the magnitude controlled primarily by the bending stiffness of the rear spar.

Flight tests, however, proved disappointing in that only "powered glides" were achieved (approximately 30:1 slope) and, in fact, with the wings fixed, the glide slope was approximately 4:1. Clearly, the airfoil required further study.

The UTIAS return-flow wind tunnel was used to perform static tests on a rectangular-planform rigid wing of Aspect Ratio=3.11 incorporating this airfoil (Fig. 2). The results, shown in Fig. 3, gave profile drag coefficients which were much higher than original estimations. Also, there was no angle where the flow was simutaneously attached on both the top and bottom, which is a particularly poor result considering the large angle-of-attack range in which an ornithopter airfoil typically operates. Therefore, it was decided to perform a detailed experimental study of alternative geometries for airfoils of this type, with the goal of obtaining the largest possible angle-of attack range for simultaneous attached flow on both the airfoil's upper and lower surfaces.

Figure 2. 24 Sept.1987 Aspect Ratio=3.11 Wing Mounted in the UTIAS Wind Tunnel

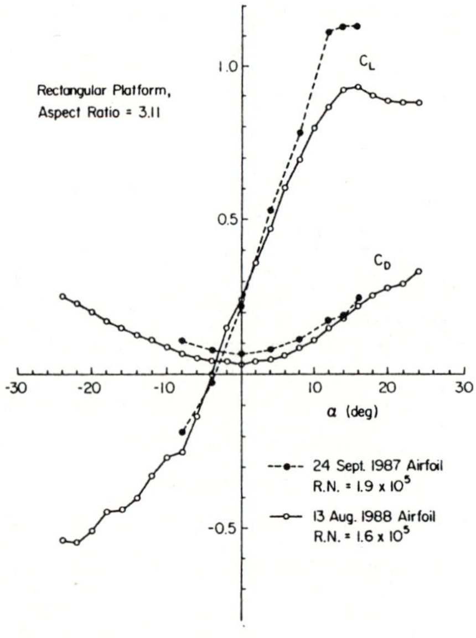

Figure 3. Lift and Drag Coefficient Results for Aspect Ratio=3.11 Wings

It was decided that the most expedient way to identify promising design directions was to perform flow-visualization tests (with tufts) on candidate profiles fitted, wall to wall, in a narrow test section. Therefore, the behavior is essentially that for a two-dimensional airfoil. Also, it is relatively easy to construct models of this span out of balsa and thin plywood.

Sixteen models were tested, which included two "pure" single-surface sections and an Eppler 193 double-surface section as baselines for comparing the behavior of the candidate airfoils. Among the candidates themselves, the main variables were camber and nose-fairing geometry, including leading-edge radius.

The results of this study, along with the selection and further testing of a suitable airfoil design, are described in the balance of this paper. Also presented are wind-tunnel results on the propulsive behavior of an ornithopter wing incorporating this airfoil.

EXPERIMENTAL METHOD

A diagram of the small wind tunnel used for the flow-visualization experiments is shown in Fig. 4. The test-section size is 16" long by 11.25" high by 1.6" (average) wide. The power source is a 1/8 hp , 3400 rpm, centrifugal blower from a Xerox machine, which gives a nominal test-section speed of 26 ft/s. To account for the speed increase caused by boundary-layer growth, the test section's width has a uniform 0.25° divergence, which results in a measured speed variation, along its length, of less than 1%.

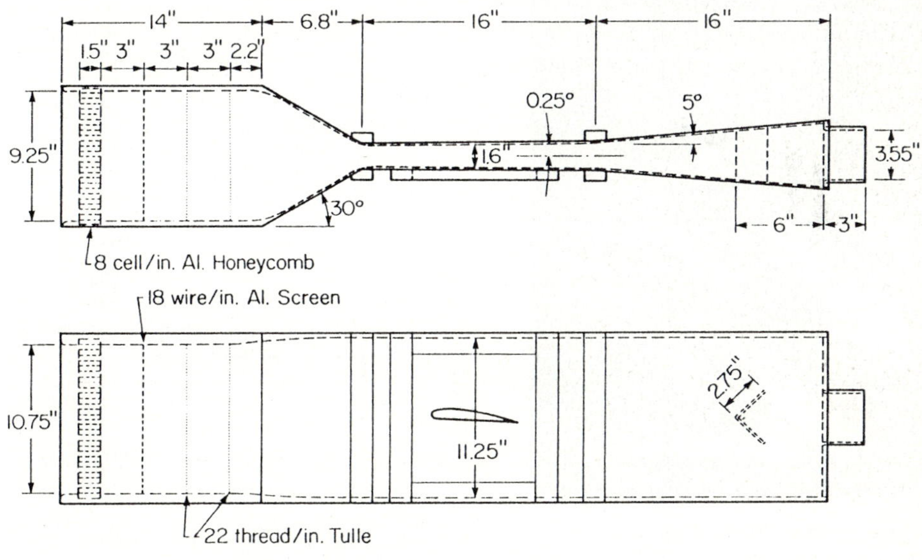

Figure 4. Schematic Drawing of Flow-Visualization Wind Tunnel

The incoming flow is conditioned by first passing through a 1.5 inch deep , 8 cells per inch, aluminum honeycomb. This is followed by an aluminum screen of 18 wires per inch and two screens of polyester tulle (a net fabric) of approximately 22 threads per inch. The flow is then contracted by a factor of 5.52, resulting in a measured test-section turbulence of no more than 0.08%.

Flow straightness was assessed with a long thread attached to a stiff wire, and all locations in the test section had satisfactory performance. The unusual divergence section, shown in Fig. 4, gave no undesirable nonuniformities or swirl.

The 16 airfoil models of 9" chord (shown in Fig. 5) were constructed primarily from balsa wood and mounted onto a 1/32" plywood backplate. A thick wire extended out behind the backplate, through a hole in the center of the wind tunnel's back wall, in order to provide a pivot point for changing the airfoil's angle of attack. The width of the airfoil was carefully measured so that it gave a tight wall-to-wall fit when the door was closed. This would "lock" the airfoil at the desired angle, as well as giving conditions for 2-dimensional flow.

The tufts used for the flow visualization were cut, 1 mm wide by 1.5" long, from 1/4-mil aluminized Mylar. These were mounted with small patches of adhesive tape in the pattern shown in Fig. 6. It was found that the tufts worked best if they had a small amount of "crinkle", rather than being absolutely smooth and straight. Smooth tufts would tend to stick flat to the airfoil's surface.

Figure 5. Flow-Visualization Wind-Tunnel Models

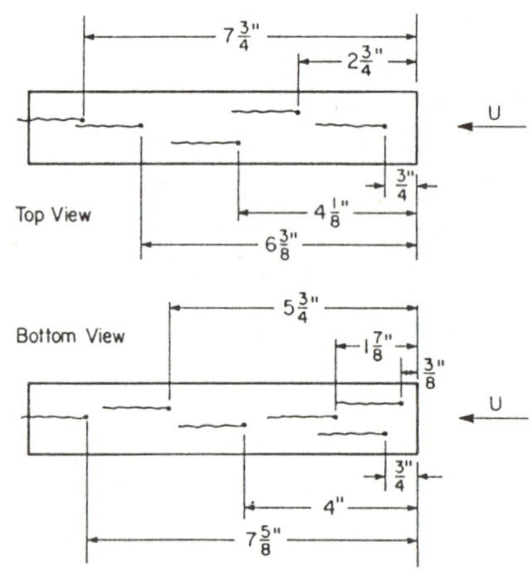

Figure 6. Tuft Pattern on Airfoil Models

For the flow-visualization tests, the airfoil would be set to a given angle in the test section while the blower was off. Then, after the door was closed and secured, the blower would be switched on and observations would begin when equilibrium was attained. For each tuft, a judgement was made on its flow condition in accordance with the following system:

"a", smooth attached flow,
"a*", attached flow with consistent twitching,
"s^{t}", separated, but generally waving in the mean-flow direction,
"s", totally separated, with reverse flow.

Note that due to the speed and size constraints of the flow-visualization wind tunnel, the Reynolds number for these tests was 1.2×10^5, which is less than the nominal flight value of 2.0×10^5. However, it was assumed that any design improvements discovered at the lower Reynolds number would also apply at the higher value. Also, this would be ultimately checked by testing the resulting candidate airfoil in a finite-wing test at a higher Reynolds number.

The finite-wing tests were performed in the UTIAS low-speed return-flow wind tunnel, which has a 3.67 ft high by 5.50 ft wide test section, and an average turbulence level of 1.3%. As seen in Fig. 2, the wing was mounted on a one-component parallelogram-beam strain-gage balance. Therefore, drag and lift measurements required separate tests, with different orientations of the balance. The electronic signal passed through an A to D converter in a 64K Tecmar computer, and the resulting digitized samples were averaged to give the values plotted in Fig. 3.

Also, as for the flow-visualization tests, the tunnel was switched off each time the angle of attack was changed. It is understood that this procedure could show somewhat different stall behavior than if the angle were changed in constant flow; but it has been assumed, for the relative-comparison purpose of this research, that this difference is small enough to justify the simpler angle-setting method.

RESULTS

The results from the flow-visualization tests are shown in Figs. 7 through 17. Note that although 16 airfoil models were tested, only 11 will be described, since these best illustrate the design evolution to the final section.

The Figures summarize separation behavior by plotting regions of the "a", "a*", "s^{t}", and "s" flow conditions vs. angle of attack and chordwise distance. Note that the 1.5" length of the tufts means that the identified behavior is an average over this distance. Thus, despite the clearly marked boundaries on the Figures, this experimental procedure actually cannot

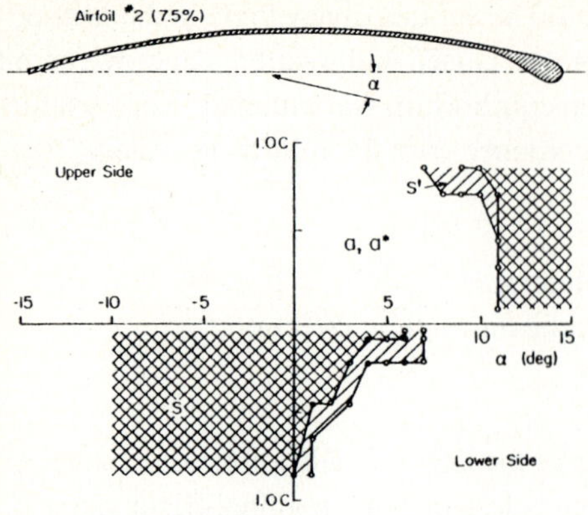

Figure 7. Fineness Ratio 3.17 Fairing

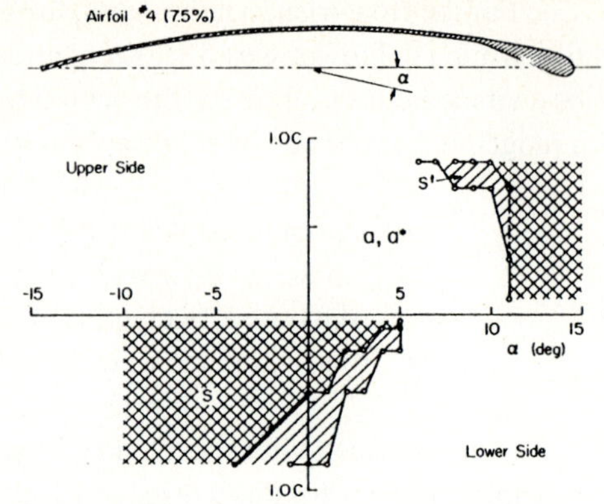

Figure 8. Fineness Ratio 4.50 Fairing

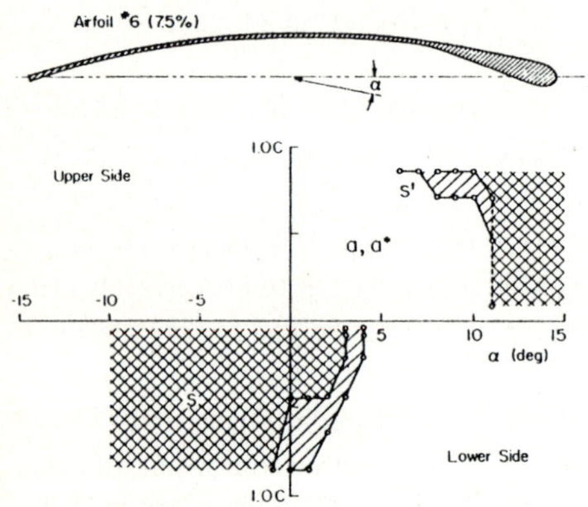

Figure 9. Fineness Ratio 5.83 Fairing

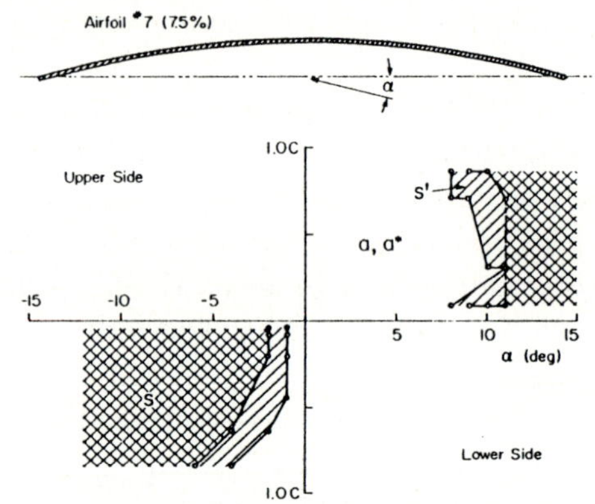

Figure 10. Thin 7.5% Arc

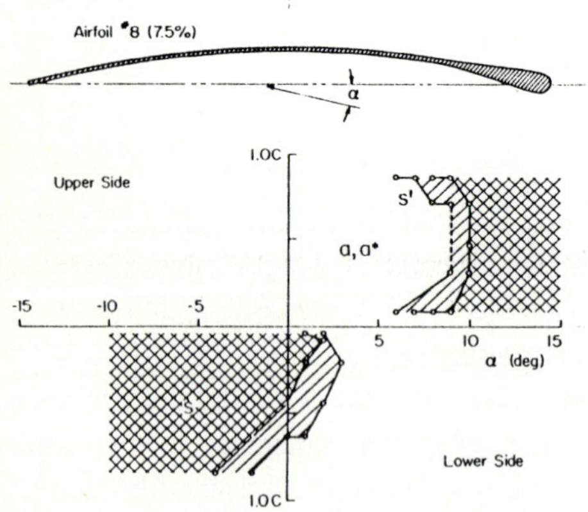

Figure 11. Camber-Centered Fairing

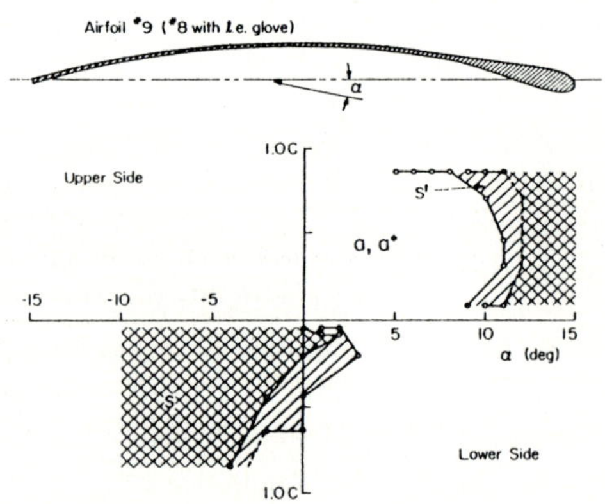

Figure 12. #8 with Reshaped L. E.

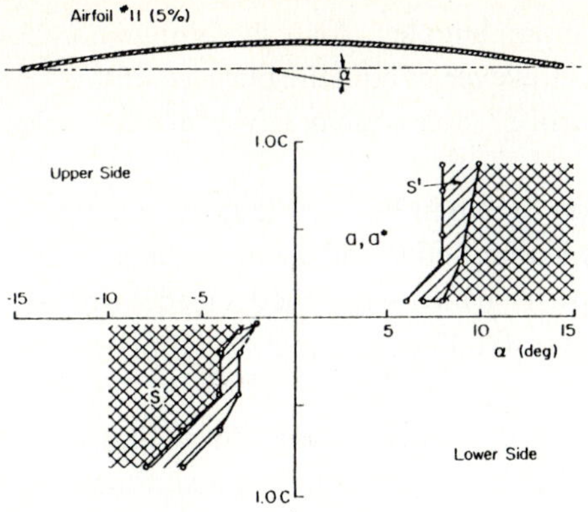

Figure 13. Thin 5% Arc

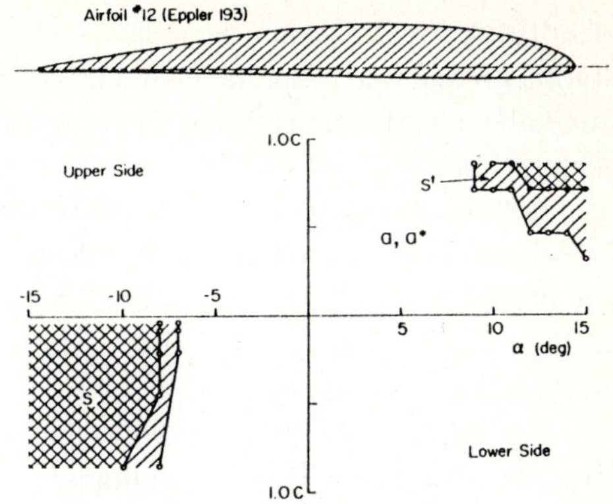

Figure 14. Eppler 193

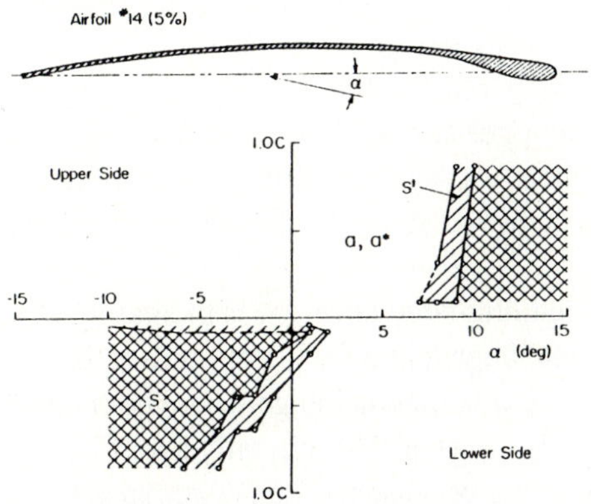

Figure 15. Smoothly-Shaped Fairing

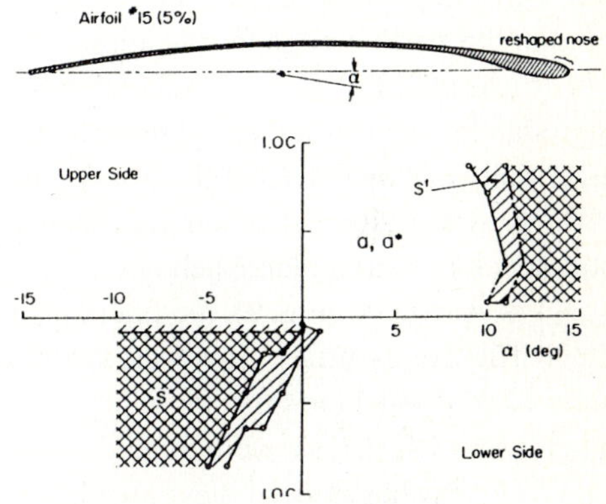

Figure 16. #14 with Reshaped L. E.

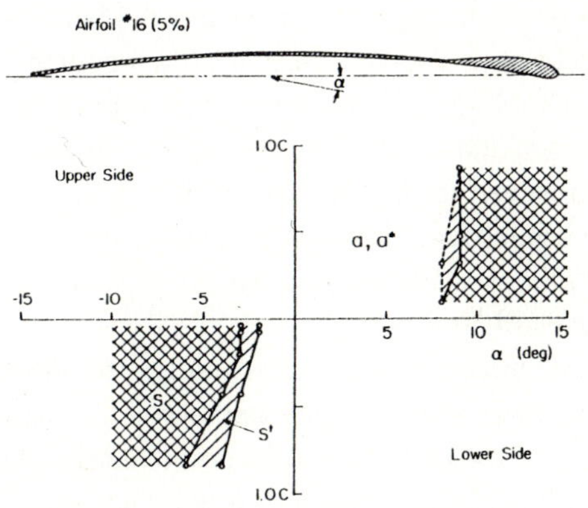

Figure 17. Upper-Surface Fairing

identify precise separation boundaries or small separation bubbles. Also, the limited number of tufts gives a fairly angular character to the plots. However, within the purpose of this study, the overall characteristics and trends of the airfoils' separation behavior are very adequately described.

Airfoil #2 (Fig. 7) represents the baseline section shown in Fig. 1, minus the rear spar (which was assumed to have negligible effect on the leading-edge and upper-surface flow). The basic form for the section was a 7.5% circular arc, to which was added a leading-edge fairing on the underside consisting of a circle (diameter=0.042 chord) and straight line, thus giving a fineness ratio of 3.17.

Fig. 7 shows similar behavior to that observed during the finite-wing test, in that there is no angle at which the flow is simultaneously attached to both upper and lower surfaces. Note that although the upper surface's flow seemed to follow a trailing-edge stall pattern, with total separation occurring at a reasonable value of 11°, the lower side gave very poor behavior, with complete attached flow only achieved above 7°. Therefore, increasing the fineness ratio of the leading-edge fairing appeared to be a promising design direction.

Airfoil #4 (Fig. 8) was identical to Airfoil #2, except that the fineness ratio of the leading-edge fairing was increased to 4.50. It is seen that upper-surface flow behavior remained essentially unchanged from that for Airfoil #2; and there was a small improvement in the lower-surface flow in that complete attachment was achieved above 5°. However, this was still not close to acceptable behavior.

For Airfoil #6 (Fig. 9), the fineness ratio of the leading-edge fairing was increased to 5.83. The upper-surface flow was, again, virtually unchanged; and the lower-surface flow showed additional incremental improvement. Note, though, that the range where both upper and lower-surface flow were totally and simultaneously attached was still only about 2°.

The poor behavior of these airfoils was disappointing because the 7.5% circular-arc section, on which these were based, was documented in Hoffman[1] and DeLaurier & Harris[2] as being an excellent low-Reynolds number design. Therefore, it was decided to test this as Airfoil #7 and, as seen in Fig. 10, the results indeed showed behavior which was much better than that for the previous sections. The upper-surface flow evidenced local leading-edge separation which, according to Schmitz,[3] is to be expected for Airfoil #7's small leading-edge radius. Also, the upper surface attained total stall at about the same angle (11°) as that for the previous airfoils. It was, however, the lower-surface flow which gave this profile its overall improved behavior, in that separation disappeared at angles above -1°. This resulted in a range of totally and simultaneously-attached flow from -1° to 8°, a great improvement over the range from 4° to 6° of Airfoil #6.

The results for Airfoil #7 clearly confirm the findings from Ref. 3, which showed the benefits of a "sharp" leading edge on thin highly-cambered low-Reynolds number airfoils. However, the requirement for a leading-edge spar compelled further studies of streamlined fairings. It was decided, with Airfoil #8 (Fig. 11), to try a design where a fairing such as on Airfoil #4 was centrally aligned with the circular-arc curve. That is, the mean-camber line of

Airfoil #8 remains a 7.5% arc, unlike the essentially "drooped" noses of Airfoils #2, #4, and #6.

As seen in Fig. 11, Airfoil #8 did show better performance than the previous faired airfoils, with a totally attached-flow range from 3° to 6°, but this was still very poor compared with the behavior of Airfoil #7. Also, note that the upper-surface separation behavior, though similar to that for Airfoil #7, occurred at a lower angle and also was inferior to that for the previous faired airfoils.

At this point, it was clear that a simple modification to the baseline airfoil of Fig. 1 was not going to give the desired performance improvement. In this respect, reconsideration was given to the original structural design where the ribs freely pivoted about the round leading-edge spar, with the rear spar controlling the torsional stiffness. Instead, an alternative design was postulated, where the ribs would be firmly attached to the leading-edge spar, which now provides the torsional stiffness as well. The advantages are, first, that since the fairing would be integral with the spar, it could be shaped into any appropriate aerodynamic configuration; and second, the rear spar could be eliminated, thus further reducing drag. The major disadvantage is that everything on the wing would now have to be built in, with no easy way to perform adjustments. With the previous design, a certain amount of "cut and try" was possible by changing the rear spar's stiffness.

In conjunction with this new wing design direction, a comprehensive computer analysis was developed which would predict the structural dynamic response and the resulting thrust and lift performance. It was found that considerable care and accuracy is required in the selection of the geometric and elastic parameters to give an efficiently performing flapping wing.

In this spirit, Airfoil #8 was modified with the addition of an elliptically-shaped leading edge shown in Fig. 12. Because this gives the smaller nose radius and smoothly-changing curvature typical of efficient low-Reynolds number shapes (Miley[4]), a performance improvement was expected. The resulting section, Airfoil #9, did show significantly better behavior in that, if the upper-surface's extreme-rearward separation is ignored, the flow was totally-attached from 3° to 8°. Although this still did not match the performance of Airfoil #7, it was felt that a very promising design feature was identified.

Now, it was observed that the major problem constraining the range of totally-attached flow was leading-edge separation on the underside. With the exception of Airfoil #7, none of the sections described, to this point, were even capable of totally-attached flow at zero angle of attack. Therefore, it was decided to reduce the fairly high camber of 7.5% to 5%. As a first step, a thin 5% section (Airfoil #11) was tested in order to find the baseline performance for this camber shape.

The results, shown in Fig. 13, demonstrate the same sort of excellent behavior as seen for Airfoil #7, in that a large range of totally-attached flow is achieved. Likewise, the upper surface flow experiences leading-edge separation, although note that this now occurs at 6° instead of 8°. The lower surface, however, retains attached flow down to -2°, so the overall range of totally-attached flow is only 1° less than that for Airfoil #7. Therefore, the 5% profile appeared to be a very acceptable shape upon which to continue the study of the addition of faired leading edges.

Before this work proceeded, though, it was decided to test a "benchmark" double-surface section so as to observe the behavior, for this particular experimental setup, of a recognized excellent airfoil design. The profile chosen was the Eppler 193, shown in Fig. 14 as Airfoil #12. Ref. 4 documented outstanding aerodynamic performance at this Reynolds number; and Fig. 14 clearly confirms this, in that totally-attached flow ranged from -7° to 9°. Even beyond this, the stall pattern was very benign. However, as discussed previously, such a design was not compatible with the torsionally-compliant structural requirements of the ornithopter wing, so it could only serve to demonstrate "ideal" behavior, relative to which the candidate airfoils could be compared. Also, note that this also served to give confidence to the experimental setup, in that the Eppler's separation behavior was in agreement with the increasingly nonlinear behavior of lift coefficient with angle plotted in Ref. 4.

Because separation on the underside of the candidate airfoils was the major obstacle to obtaining a wide range of totally-attached flow, it was decided, with Airfoil #14, to explore the effect of a smoothly-curved fairing on the underside of the leading edge. This shape, seen in Fig. 15, is quite different from the straight-line fairing on the previous airfoils, and did give the best lower separation angle (2°) among the candidate sections. However, the upper separation angle of 7° gave a totally-attached flow range which was no improvement over that achieved by Airfoil #9. Since Airfoil #9 had a smoothly-shaped nose, which gave it significantly better performance than that for its previous shape in Airfoil #8, it was decided to try this with Airfoil #14. The modified section, called Airfoil #15, also gave a significant performance improvement in that the attached-flow range now extended from 1° to 9°, as seen in Fig. 16. This was the best behavior yet among the candidate airfoils, confirming the importance of nose shape as well as the aft geometry of the fairing.

During a visit in March 1988 to Jeremy Harris (a Principal Research Engineer at Battelle Memorial Institute in Columbus, Ohio, and the author's partner in the ornithopter project), these results were discussed and it was decided to test a design with the fairing located on the upper surface of the 5% circular arc. It was reasoned that this would give a particularly smooth entry to the underside flow, further improving the lower-surface separation behavior. Also, the fairing would have a similar nose radius and fineness ratio to that for Airfoil #15.

The resulting design, shown as Airfoil #16 in Fig. 17, finally extended the under-surface flow-separation angle to the negative regime (-2°). That, along with the upper-surface flow-separation angle of 8°, gave an overall range for totally-attached flow of 10°.

At this point it was decided that Airfoil #16 was sufficiently promising to warrant wind-tunnel tests for lift and drag characteristics, as was done for the original section. Another test panel of Aspect Ratio=3.11 was built incorporating Airfoil #16, with its construction intended to be representative of that for the new ornithopter wing. A section is sketched in Fig. 18, showing how the ribs were now positioned on the upper side of the covering.

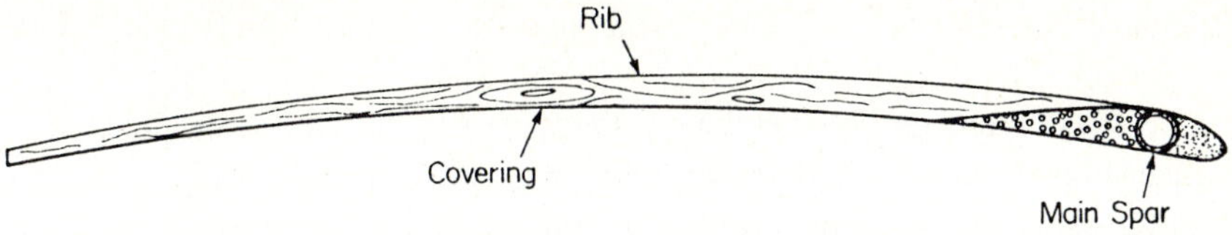

Figure 18. 13 August 1988 Airfoil Design

The results are given in Fig. 3 where it is seen that the drag has been greatly reduced in comparison with that for the original wing (nearly 50% at zero angle of attack). Also, relative to the airfoil data presented in Ref. 4, the overall behavior was reasonable for an airfoil of this type. Therefore, it was decided that Airfoil #16 would be incorporated into the new ornithopter wing.

CONCLUDING REMARKS

The major results from this investigation may be summarized as follows:

(1) The Eppler 193 gave the best behavior of all airfoils tested, with totally-attached flow from -7° to 9° angle of attack.

(2) The two single-surface sections did not behave as well as the Eppler, especially at the negative angles, but still gave a wider range of attached flow than the candidate airfoils with the nose fairing on the bottom.

(3) An airfoil with the fairing on top gave the best behavior among the candidates tested, including the two single-surface sections, with totally-attached flow from -2° to 8° angle of attack.

Subsequent force tests on this airfoil showed much improved performance over the original section, with profile drag reduced by nearly 50%. Therefore, this airfoil was selected to be incorporated into the new ornithopter wing.

As to the limitations of these experiments, note first that the 26 ft/s maximum speed of the flow-visualization wind tunnel meant that the airfoils had to be relatively large in order to achieve Reynolds numbers approaching those for the ornithopter. Therefore, wall effects could be significant, especially for high angles. Next, the boundary layer on the side walls, as discussed by Mueller, et al.[5], could influence the observations. However, all airfoils were tested in the same conditions, and compared relative to one another. Also, it was encouraging that the separation behavior of the Eppler airfoil appeared to agree with the force data in Ref. 4.

The finite wings were also built at the ornithopter's original full-scale chord of 9" in order to utilize actual construction methods, templates, and materials. In fact, the 1988 wing was structurally tested for the elastic properties required by the previously-mentioned flapping-wing performance analysis. However, the span was restricted in order to minimize wall effects for testing in the UTIAS wind tunnel, resulting in low Aspect Ratios of 3.11. Therefore, one could question whether such a test could yield accurate 2-D airfoil information; but, again, it was the relative comparison between the original and final wing that was important.

Finally, it would be interesting to show how the new ornithopter wing behaved. Its planform geometry, shown in Fig. 19, was determined by the performance analysis, which also gave an 11" maximum chord and a 41" semispan. The airfoils were scaled from Airfoil #16, with the exception that the camber was decreased on the outer sections. Also, a washin distribution was built into the wing to give zero twist at the nominal flight speed of 35 ft/s and 5° flapping axis angle of attack.

Figure 19. Mark-6 Ornithopter Wing Mounted in UTIAS Wind Tunnel

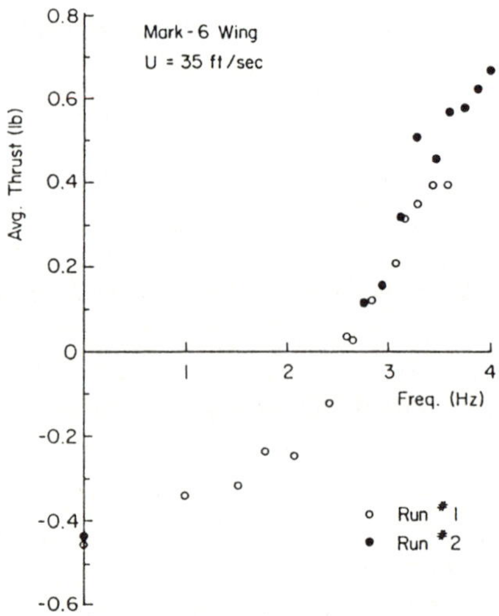

Figure 20. Thrust Results for Mark-6 Ornithopter Wing

The flapping tests were performed in the UTIAS wind tunnel, and the results are shown in Fig. 20. Note that the desired thrust was 0.5 lb which, from a previous test, is the value necessary to achieve sustained flight. This value is attained, for the half wing alone, at a flapping frequency of approximately 3.4 Hz; and for the full wing, sustained flight is promised at a very reasonable value of approximately 3.1 Hz (the original wings flapped at 5 Hz). Therefore, the new ornithopter wing incorporating this airfoil appears to have the performance required to give successful flight.

ACKNOWLEDGMENTS

During the course of this research, Jeremy Harris, the author's partner in the ornithopter project, gave continuing helpful insights and suggestions despite the difficulties caused by living 450 miles away. Further, Prof. Thomas Mueller, of the University of Notre Dame, suggested appropriate damping screens for the flow-visualization tunnel, and Lisa Kerr, of UTIAS, performed hot-wire measurements which showed what a good job the screens were doing. Also, the author very much appreciates help given by UTIAS Research Engineer Chris Hayball, who reviewed this paper and patiently provided instructions on word processing.

REFERENCES

[1] Hoffman, R. J., "Free-Flight Models", Model Aeronautics Made Painless, Published by the Author, 1955, pp. 21-22.

[2] DeLaurier, J. D. and Harris, J. M., "An Experimental Investigation of the Aerodynamic Characteristics of Stepped-Wedge Airfoils at Low Speeds", AIAA Paper 74-1015, Sept. 1974.

[3] Schmitz, F. W., "Aerodynamics of the Model Airplane. Part 1-Airfoil Measurements", NASA TM-X-60976, Nov. 1967.

[4] Miley, S. J., "A Catalog of Low Reynolds Number Airfoil Data for Wind Turbine Applications", Texas A&M University, Feb. 1982, distributed by the National Technical Information Service: DE82-021712.

[5] Mueller, T. J., Batill, S. M., Brendel, M., Perry, M. L., Bloch, D. R., Huber, A. F., Bastedo, W. G., O'Meara, M. M., Schmidt, G. S., "Low Reynolds Number Wind Tunnel Measurements: The Importance of Being Earnest", Proceedings of the International Conference on Aerodynamics at Low Reynolds Numbers $10^4 < R_e < 10^6$, London, 15th - 18th October, 1986.

A COMPUTATIONALLY EFFICIENT MODELLING OF LAMINAR SEPARATION BUBBLES

Paolo Dini and Mark D. Maughmer
Department of Aerospace Engineering
The Pennsylvania State University, University Park, Pennsylvania

Introduction

In order to predict the aerodynamic characteristics of airfoils operating at low Reynolds numbers ($R < 5.0 \times 10^5$), it is necessary to accurately account for the effects of laminar (transitional) separation bubbles.[1] In general, the greatest difficulty comes about when attempting to determine the increase in profile drag that results from the presence of separation bubbles. Because the drag on an airfoil depends on the trailing-edge value of the momentum thickness,[2] the increase in drag that accompanies separation bubbles is primarily due to the rapid increase in momentum thickness that occurs in a bubble compared to the much smaller growth that occurs with a natural transition from a laminar to a turbulent boundary layer. As a consequence, when a separation bubble is the actual mechanism of transition, methods that assume transition occurs at the laminar separation point and use the boundary-layer properties at that point as the initial conditions for the turbulent boundary-layer calculations usually underpredict the drag. Thus, to successfully determine the increase in drag on an airfoil due to separation bubbles, it is necessary to accurately model the development of the boundary layer through the bubble.

While a number of empirically based separation bubble models have been introduced in the past, some of which are detailed in Ref. 3, the majority of these assume that the bubble development is fully predictable from upstream conditions. More recently, much more accurate predictions have been made possible using viscous/inviscid interaction approaches such as those described in Refs. 4-6. By such means, the influence of the bubble on the entire velocity distribution over the airfoil is accounted for globally by iterating between the inviscid flow and boundary-layer solutions. While not of much concern in predicting the aerodynamic characteristics of a single airfoil, the amount of computational time required for such boundary-layer iteration methods becomes consequential in the case of airfoil design for which the number of analysis cases required can become very large.

One way of accounting for laminar separation bubbles in airfoil design is the bubble analog used in the design and analysis program of Eppler and Somers.[7,8] In this method, the designer is warned about the presence of separation bubbles which might unacceptably increase the drag over that which is predicted assuming that transition occurs at laminar separation. Although this approach has proven very useful in designing airfoils for low Reynolds number applications, it would be advantageous to have predictions of section properties which more fully account for the presence of laminar separation bubbles provided this can be done without significantly increasing the computational time. Toward this end, a locally interactive separation bubble model has been developed and incorporated into the Eppler and Somers program. Although unable to account for strong interactions such as the large reduction in suction peak sometimes caused by leading-edge bubbles, it is able to predict the increase in drag and the local alteration of the airfoil pressure distribution that is caused by bubbles occurring in the operational range which is of most interest.

To fully determine the behavior and influence of a laminar separation bubble, it is necessary to accurately predict the shear-layer development in the regions of the laminar separation bubble indicated in Fig. 1. The formation of a bubble is initiatated at point S, shown in the figure, by the laminar boundary layer separating from the airfoil surface. Using integral boundary-layer methods, this point can be determined with the accuracy needed for airfoil design work. Once separated, the free shear layer development must be tracked and the transition from laminar to turbulent flow, which occurs near the point T, predicted. As shown in Fig. 2, the separation bubble causes a plateau to form in the velocity distribution between the points corresponding to laminar separation and the end of the transition region. From this point, the turbulent part of the bubble encompasses a pressure recovery region which leads to the reattachment of the turbulent shear layer at point R. As an additional pressure recovery always occurs downstream of a reattachment point, the velocity distribution corresponding to the highly non-equilibrium, relaxing boundary layer downstream of reattachment "undershoots" the inviscid distribution. Eventually, the turbulent boundary layer reaches its fully-developed state and the undershoot region merges smoothly from below with the inviscid velocity distribution. Clearly it is possible, especially at low Reynolds numbers, that the turbulent boundary layer never reaches equilibrium before the trailing edge of the airfoil.

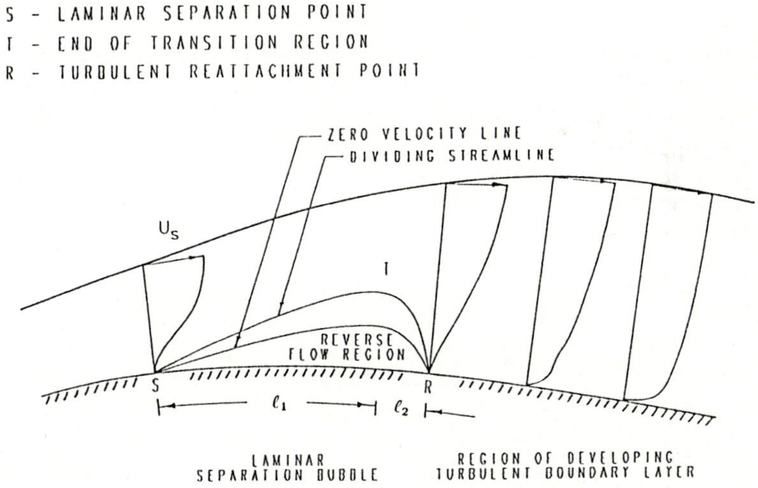

Fig. 1: Sectional view of a two-dimensional short laminar separation bubble.

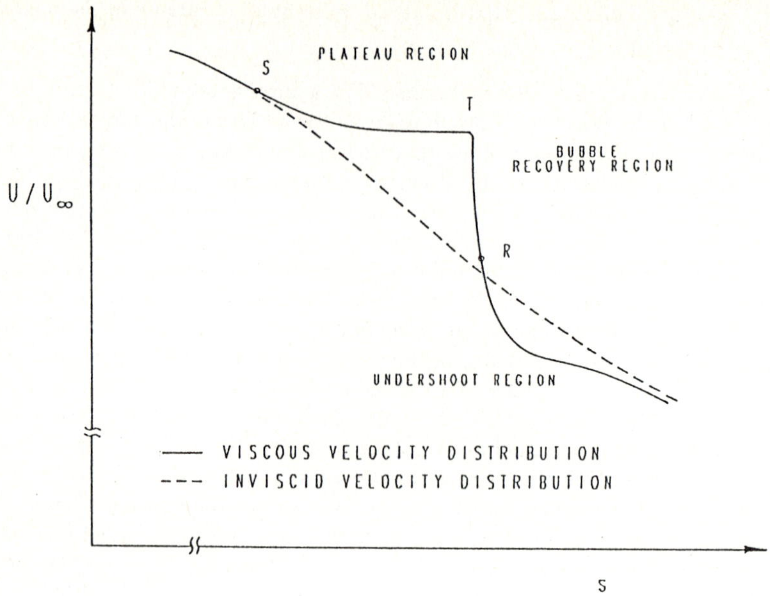

Fig. 2: Influence of a short laminar separation bubble on the velocity distribution over an airfoil.

Overview of Procedure for Modelling Laminar Separation Bubbles

Efforts to develop a method able to predict the effects of a laminar separation bubble which interacts weakly with the inviscid flow, the so-called short bubble, began with the incorporation of the classical empirical model of Horton,[9] modified according to the suggestions of Roberts[10] and Schmidt,[3] into the Eppler and Somers program. Because they are formulated in terms of integral boundary-layer properties, bubble models such as these are well suited to the integral boundary-layer analysis employed by Eppler. Specifically, this method makes use of the momentum and energy integral equations,

$$\frac{d\delta_2}{ds} = \frac{c_f}{2} - (H_{12} + 2)\frac{\delta_2}{U}\frac{dU}{ds} \tag{1}$$

$$\frac{d\delta_3}{ds} = C_D - 3\frac{\delta_3}{U}\frac{dU}{ds} \tag{2}$$

along with the appropriate closure relations.[11]

Using the empirical separation bubble model noted, the sensitivity of the boundary-layer development and drag prediction to various parts of the bubble was explored. As detailed in Refs. 12 and 13, and reported by other researchers as well,[14] it is found that the bubble development for such models is very sensitive to small variations in the governing parameters. Thus, although generally accurate for predicting features of the bubble to within twenty percent, empirical bubble models based only on upstream conditions are not able of providing acceptable drag predictions. Consequently, it was concluded that the accuracy desired could be achieved only by a model which accounts for the effects of the downstream portions of the bubble on those upstream. In particular, along with the chord Reynolds number and the upstream development of the laminar boundary layer, the development of the bubble is found to depend strongly on the total amount of pressure which is to be recovered along the bubble. These variable flow characteristics scale well with two dimensionless quantities, the Reynolds number based on momentum thickness at the laminar separation point, $(R_{\delta_2})_S$, and Gaster's pressure gradient parameter,[15]

$$P = \frac{[(\delta_2)_S]^2}{\nu}\frac{\Delta U}{\Delta s} = R\left[\frac{(\delta_2)_S}{c}\right]^2 \frac{\Delta(U/U_\infty)}{\Delta(s/c)} \tag{3}$$

It should be noted that, while $(R_{\delta_2})_S$ has been used in other models to account for decreasing transition length with increasing Reynolds number, Gaster's parameter has been used in the past only as a criterion of when reattachment of the short bubble is not possible and a long bubble forms. Justification that the value of P has a strong influence on the transition process and other aspects of the bubble development, such as the velocity plateau region of the bubble, is motivated by the unsteady laminar simulation presented in Ref. 16. In this research, the criterion for the "bursting" of short bubbles is found to correlate well with the boundary between steady and unsteady reattachment of a laminar boundary layer subjected to an imposed pressure gradient. The unsteadiness is due to a periodic vortex-shedding which increases in frequency and intensity as the value of Gaster's parameter increases. In view of the ellipticity of this recirculating flow field, it is not too surprising that this frequency is found to correspond to the most unstable frequency from a linear stability analysis of the laminar free shear layer. Thus, it is expected that transition in such a flow field would be strongly dependent on the shedding frequency as is characterized by the value of Gaster's parameter. Because the short separation bubble is analogous to this unsteady reattachment process, it is reasonable that, in addition to $(R_{\delta_2})_S$, the transition process should depend on the downstream reattachment of the turbulent shear layer and the value of Gaster's pressure gradient parameter.

Starting with the inviscid velocity distribution over an airfoil, the scheme used by the present method of predicting the development of laminar separation bubbles is summarized by the flow diagram shown in Fig. 3. As indicated, the bubble model is evoked when laminar separation is predicted. At that point $(R_{\delta_2})_S$ is determined and, based on the inviscid velocity gradient at the laminar separation point, an initial estimate of Gaster's parameter, P, is made. From the separation point, the velocity distribution in the plateau region is prescribed. The function which defines this distribution depends both on P and the matching of its slope to that of the inviscid velocity distribution at the laminar separation point. Using the prescribed velocity distribution over the laminar part of the bubble, the boundary-layer development is accomplished using the momentum and energy integral equations, Eqs. (1) and (2), along with appropriate closure relationships which, as will be described, were derived using the reverse-flow velocity profiles developed by Green.[17] Likewise, these profiles were used to develop the criterion used to predict transition in the free shear layer. This criterion is also a function of $(R_{\delta_2})_S$ and P. The turbulent part of the bubble and the undershoot region are determined by prescribing the distribution of H_{32} and solving the integral boundary-layer equations in the inverse mode. At this point, a local iteration is carried out to ensure that Horton's reattachment condition[9] is satisfied for a velocity distribution in the non-equilibrium region of the bubble that merges smoothly downstream with the inviscid velocity distribution. Once this iteration has converged, a new value of P is calculated from the velocity gradient along the bubble as determined using the points of laminar separation and the point where the bubble recovery first crosses the inviscid distribution. With this new value of P, an outer iteration repeats the calculations indicated until the overall length of the bubble no longer changes. It is through the two iteration schemes that the influence of the downstream conditions in the bubble are communicated upstream.

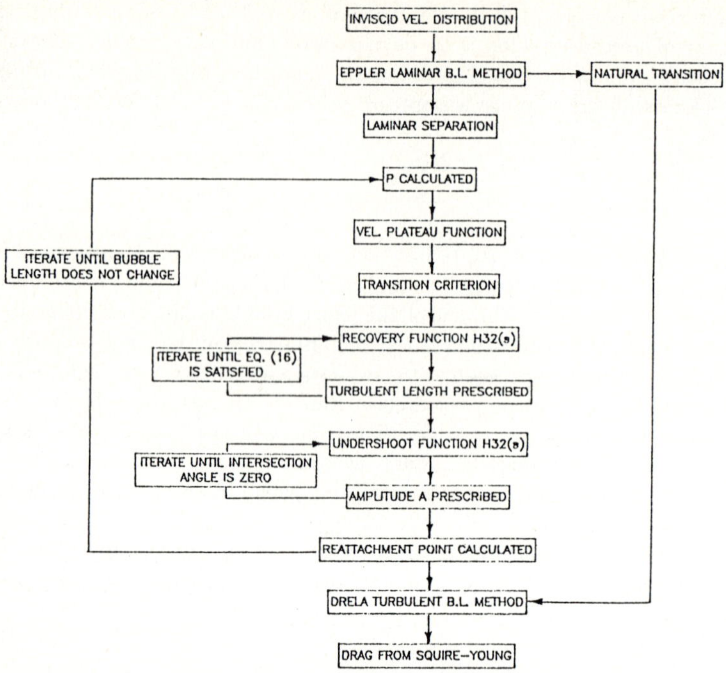

Fig. 3: Flow chart of computational scheme used to model laminar separation bubbles in the flow over airfoils.

Details of the Laminar Separation Bubble Model

In the present procedure for modelling laminar separation bubbles, the laminar closure correlations, the velocity distribution in the plateau region, transition, and the function describing the undershoot region depend only on $(R_{\delta_2})_S$ and P. All other features of the bubble are determined from the governing equations. The specific regions of the bubble model will now be described.

Laminar Part of the Bubble

The Eppler and Somers program uses a very reliable criterion to detect laminar separation, based on the value of the energy to momentum thickness shape factor,

$$(H_{32})_S = 1.515095 \tag{4}$$

This value is approached from above. Since the analysis method is formulated in the direct mode, a small error is introduced in the prediction of the laminar separation point by the presence of the Goldstein singularity. This causes the distribution of H_{32} to exhibit a very steep slope immediately upstream of separation, which is a similar behavior to that of the skin-friction coefficient. Furthermore, the direct formulation precludes any upstream influence of the bubble on the pressure distribution. In the present version of the bubble model this local interaction is neglected. Although possibly important for leading-edge bubbles, it should not have a major impact on the development of mid-chord bubbles.

Upon detection of laminar separation, the development of the separated laminar shear layer is calculated using the same governing equations, the momentum and energy integral equations, together with closure relations obtained from a family of reversed laminar velocity profiles. Instead of implementing this boundary-layer method in the inverse mode as it is usually done, the development of a family of pressure distributions in the laminar part of the bubble allows its calculation in the direct mode.

The function used to approximate the pressure distribution in the laminar part of the bubble is a generalization of that developed by van Ingen and Boermans and presented in Refs. 18 and 19,

$$\frac{U}{U_S} = .978 + .022\exp(-4.454\xi - 2.5\xi^2) \tag{5}$$

where

$$\xi = \frac{s - s_S}{(R_{\delta_2})_S(\delta_2)_S} \tag{6}$$

This distribution, unlike the constant-pressure plateau used by Horton, allows a slight pressure rise after laminar separation. Using detailed pressure distributions obtained from recent wind-tunnel tests of the NASA NLF(1)-1015 airfoil[20] in the Low-Turbulence Pressure Tunnel at the NASA Langley Research Center, the accuracy of Eq. (5) has been checked for several different conditions. It is found that, when the overall pressure gradient along the bubble decreases, the velocity distribution over the plateau region falls below that given by Eq. (5) while, when the pressure gradient steepens, it rises above. To account for this effect, Eq. (5) is rewritten as

$$\frac{U}{U_S} = (1 - DU) + DU\exp(-4.454\xi - 2.5\xi^2) \tag{7}$$

where DU is indicative of the amount of the pressure rise accomplished by the plateau velocity distribution over the laminar part of the bubble. It is found to be well represented as a function of the Gaster pressure gradient parameter, P. This functional relationship, shown in Fig. 4, was developed by extracting corresponding values of DU and P directly from the experimental pressure distributions of the NLF(1)-1015 and the Eppler E387 airfoils.[20,21] The solid line is a quadratic least-squares fit that has been included in the model,

$$DU = 0.0609691 + 0.304819P + 0.507176P^2 \tag{8}$$

It should be noted that the value of $DU = 0.022$ used by van Ingen and Boermans falls in the middle of the variation in DU shown in Fig. 4.

Upon examination of the velocity distributions obtained using Eqs. (6)-(8), it was noticed that unlike in the experimental distributions, a discontinuity in the velocity gradient at the point of laminar separation was present. Therefore, to match the gradients at laminar separation, an additional variable is introduced into the velocity distribution given by Eq. (7). Specifically, the product $[(R_{\delta_2})_S(\delta_2)_S]$ in Eq. (6) can be treated simply as a scaling factor between the physical variable, s, and the universal dimensionless variable, ξ. Consequently, the value of this scaling factor is determined as that which yields a velocity distribution in the laminar part of the bubble whose gradient is continuous with the gradient of the inviscid velocity distribution at separation.

Using the prescribed velocity distribution over the laminar part of the bubble, the development of the separated laminar shear layer is calculated by means of the same governing equations used for the rest of the boundary layer, the momentum and energy integral equations. Since δ_2 and δ_3 are obtained directly from the governing equations, the transition criterion is expressed in terms of $(H_{32})_T$ as a function of $(R_{\delta_2})_S$ and P. A family of velocity profiles must likewise be chosen to develop the closure correlations for $H_{12}(H_{32})$, $c_f(H_{32}, R_{\delta_2})$, and $C_D(H_{32}, R_{\delta_2})$. The correlations based on the reversed Falkner-Skan, or Stewartson,[22] profiles developed by Drela[5] were originally included in the model. As discussed by Fitzgerald and Mueller,[23] however, recent measurements inside the bubble show velocity profiles that are quite different from the Stewartson profiles and closer to the two-parameter profile family originally developed by Green.[17] As shown in Fig. 5, the two parameters, h/b and G, are linked to the geometrical characteristics of the profiles. (h/b) is the ratio of the distance of the shear layer from the centerline of the wake to the width of the shear layer and G is the amplitude of the Coles wake function. Since there is slip along the centerline of such a recirculating base flow, these profiles cannot be used to develop a correlation for c_f. In view of the characteristically small values of c_f in the laminar part of the bubble, however, this should not be a problem.

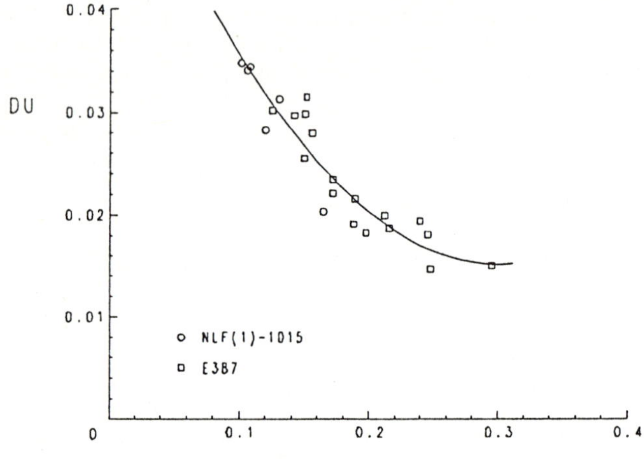

Fig. 4: Pressure recovery in the laminar part of the bubble as a function of Gaster's pressure gradient parameter. o : NLF(1)-1015 airfoil, NASA LaRC, June 1987; □ : E387 airfoil.[21]

By applying the definitions for the integral thicknesses of the boundary layer and the dissipation coefficient, relationships for H_{12}, H_{32}, and $R_{\delta_2} C_D$ as functions of h/b and G are obtained for the Green profiles. The results are shown in Figs. 6 and 7 where these new two-parameter correlations are compared to those developed by Drela from the Stewartson profiles. The solid lines utilize the fitted variations of G and (h/b). As both H_{12} and H_{32} increase monotonically between separation and transition, moving to greater values of the abscissa on these plots corresponds to moving downstream inside the bubble. Thus, both are similar to the Stewartson correlations near separation but can be quite different further downstream. It appears from the measurements that the back-flow, which is proportional to G, may be constant within each bubble although different for different bubbles. From Fig. 6, the values of shape factors actually measured, although different in absolute value, follow the same slope thus confirming a constant value of back-flow velocity. These considerations justify eliminating (h/b) between the expressions for the shape factors and expressing the closure relationships in terms of H_{32}, calculated

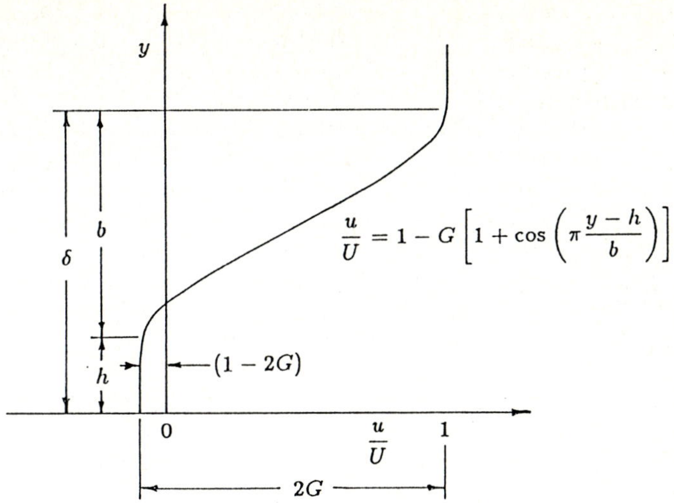

Fig. 5: Two-parameter reversed velocity profile of Green.[17]

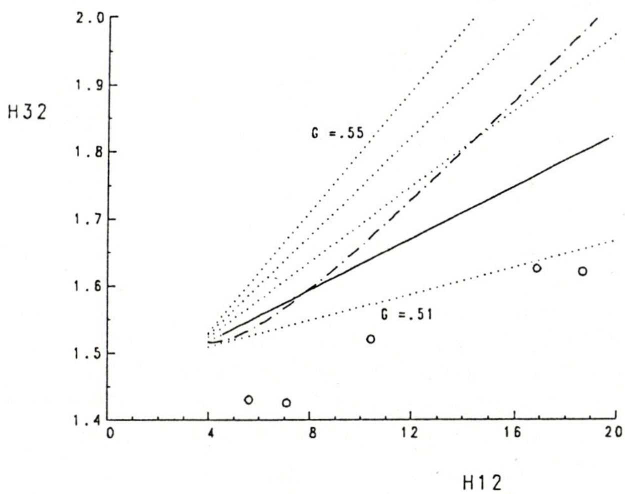

Fig. 6: Comparison of shape factor correlations. — · — : Ref. 5, from Stewartson profiles; ——— : Ref. 23, from fitted Green profiles; · · · : Eq. (9); ∘ : measured, Ref. 23.

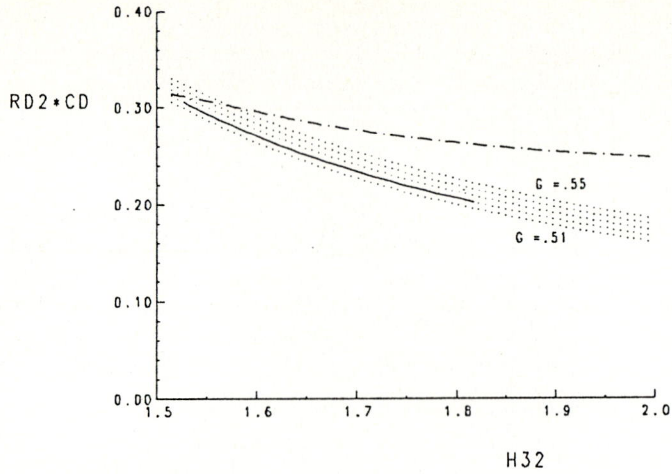

Fig. 7: Comparison of dissipation coefficient correlations. —·— : Ref. 5, from Stewartson profiles; ——— : Ref. 23, from fitted Green profiles; · · · : Eq. (10).

from the governing equations, and G, whose behavior within each bubble can be correlated to local flow conditions. Thus, the closure correlations are

$$H_{12} = \frac{3(1-G) - H_{32}}{(1-G)(1-2G)} \tag{9}$$

$$R_{\delta_2} C_D = \frac{\pi^2 G^3}{2}\left[1 - \frac{3}{2}G - \frac{(4-5G)(1-G) - (2-3G)H_{32}}{4(1-G) - 2H_{32}}\right] \tag{10}$$

The unknown quantities in the boundary-layer method for the laminar part of the bubble have been reduced to G and c_f. Physically grounded assumptions can be made about the dependence of these variables on local flow conditions. For instance, it seems reasonable to expect that, as the pressure gradient along the bubble increases, so does the strength of the recirculation and, therefore, $-c_f$. In the present version of the model, however, G is related to P such that $(\delta_2)_S$ plays an important part in determining the ratio of reverse velocity to forward boundary-layer edge velocity. The value of c_f is held constant between separation and transition.

Transition

The criterion for predicting transition in the free shear layer presently used was arrived at only after establishing that the addition of a second parameter in the laminar reversed velocity profiles still did not provide the bubble model with enough generality. The Green profiles correlations have been retained, however, as the present form of the transition criterion is believed to embody both the stability characteristics of free shear layers of varying velocity ratios (varying amounts of back-flow) as well as the perhaps more important input from the unsteady reattaching turbulent shear layer.

In the present separation bubble model, transition is predicted when a given value of the shape factor, $(H_{32})_T$, is achieved. As with other features of the bubble, $(H_{32})_T$ is taken to be dependent on both $(R_{\delta_2})_S$ and P. This relationship is

$$(H_{32})_T = \left[\frac{875 - (R_{\delta_2})_S}{35000000P^2 + 1800000P + 40000}\right]^{\frac{1}{3}} + 1.515095 \tag{11}$$

To explain the transition criterion, it is plotted together with the shear layer development on the same plot that Eppler uses to describe the boundary-layer development.[7] Since H_{32} and R_{δ_2} are calculated at each point along the boundary layer, by connecting subsequent (H_{32}, R_{δ_2})-pairs on a plot whose axes correspond to these two variables, the boundary-layer development from the stagnation point to the trailing edge can be described in a very concise way. Fig. 8 shows one such boundary-layer development together with all the transition and separation criteria.

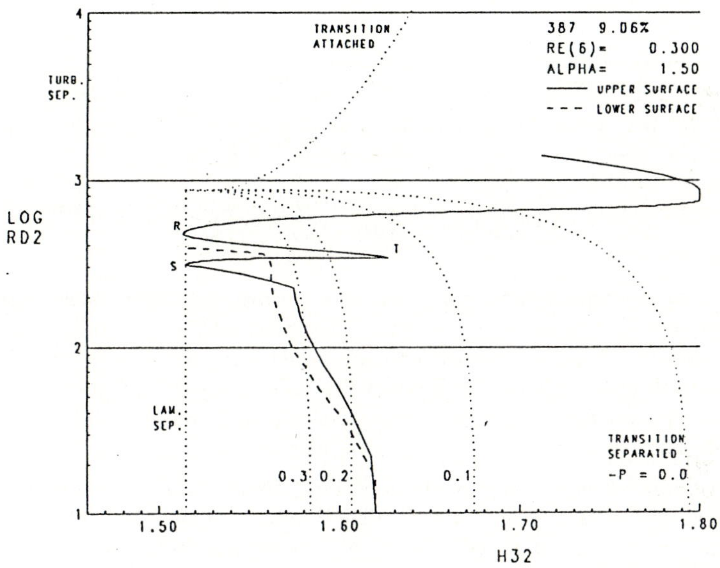

Fig. 8: Modified Eppler boundary-layer development plot showing the boundary-layer development inside the bubble.

Turbulent Part of the Bubble and Reattachment

An analysis recently completed by R. Eppler that establishes the greatest possible pressure recovery for an attached turbulent boundary layer provided inspiration for developing a new approach to the treatment of the turbulent part of the bubble. Given that the value as well as the slope of the H_{32} distribution is always known at the transition and reattachment points, a general function has been developed which allows the solution of the turbulent part of the bubble in the inverse mode. The distribution of H_{32} is specified as

$$H_{32}(\sigma) = \begin{cases} \frac{(H_{32})_{\mathcal{T}}+(H_{32})_{\mathcal{R}}}{2} + \frac{(H_{32})_{\mathcal{T}}-(H_{32})_{\mathcal{R}}}{2}\sin\left(\frac{\pi}{\frac{1.6}{6}\sigma+.4}\right) & 0 \le \sigma \le 1 \\ (H_{32})_{\mathcal{R}} + A\left[1+\sin\left(\frac{\pi}{\frac{1.6}{6}\sigma+.4}\right)\right] & \sigma > 1 \end{cases} \tag{12}$$

where

$$\sigma = \frac{s - s_{\mathcal{T}}}{\ell_2} \tag{13}$$

The corresponding distributions of $U(s)$ and δ_2 are obtained from

$$\frac{dU}{ds} = \frac{H_{32}c_f - C_D + \delta_2 H'_{32}}{\delta_2 H_{32}(H_{12}-1)}U \tag{14}$$

$$\frac{d\delta_2}{ds} = -\frac{3c_f}{H_{12}-1} + \frac{H_{12}+2}{(H_{12}-1)H_{32}}(C_D - \delta_2 H'_{32}) \tag{15}$$

where H'_{32} denotes the derivative with respect to s of Eq. (12). The closure relations for separated and attached turbulent boundary layers developed by Drela[5] are used, enforcing continuity in H_{32} at transition.

In the expression defining the reattachment process, Eq. (12), there are two free constants, ℓ_2 and A, such that two additional conditions are necessary to achieve a unique solution. One condition is that the undershoot merge smoothly with the inviscid velocity distribution. Simultaneously, the reattachment condition derived by Horton[9] must be satisfied,

$$\left[\frac{\delta_2}{U}\frac{dU}{dx}\right]_{\mathcal{R}} = -\left[\frac{C_D}{H_{32}(H_{12}-1)}\right]_{\mathcal{R}} \tag{16}$$

The accuracy of the self-consistent reattachment process described above depends on the choice of the function that represents the input H_{32} distribution and the value of $(H_{32})_{\mathcal{T}}$. As the value of H_{32} at the end of the undershoot, which depends on the amplitude A, does not vary significantly with Reynolds number or pressure gradient, the function employed is believed to provide a satisfactory approximation to the actual boundary-layer development in the reattachment region as it is affected by the local interaction. Therefore, the only parameter requiring further calibration is $(H_{32})_{\mathcal{T}}$. The transition criterion, Eq. (11), can be refined by matching predicted and measured pressure distributions. Having fixed $(H_{32})_{\mathcal{T}}$, the correct dependence of G on P and, possibly, on $(R_{\delta_2})_S$, can be determined by matching predicted and measured transition lengths.

Predictions Based on the Present Bubble Model

At the present time, all of the components of the bubble model are in place with enough flexibility to capture the physics for a wide range of conditions. It remains, however, to determine the exact forms of the correlation functions that will "tune" the model to achieve the generality and robustness that this wide range requires. With these qualifications in mind, the pressure distribution for the Eppler E387 airfoil at an angle of attack of 1.5 degrees as predicted using the Eppler and Somers program with the bubble model included is shown in Fig. 9. Also included in the figure is the distribution obtained with the Drela and Giles program,[6] with a critical amplification factor of 12, along with that of McGhee et al.[21], both at 2 degrees. For this case, the predicted pressure distributions agree reasonably well with one another and with the experimental results. As expected, because this airfoil is not highly aft-loaded, the influence

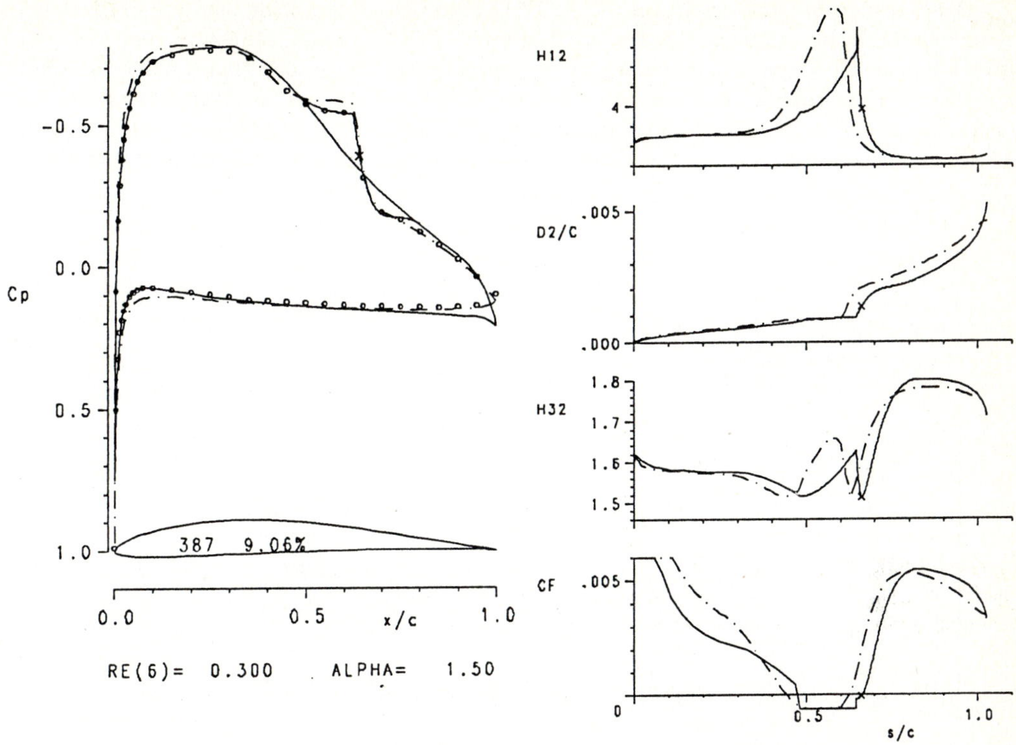

Fig. 9: Velocity distribution and upper-surface boundary-layer development predicted by present method compared to those of the Drela and Giles program.[6] Eppler E387 airfoil, $c_\ell = .55$. ——— : present method; — · — : method of Ref. 6; o : experimental, Ref. 21.

of viscous/inviscid interactions is not strong. Consequently, outside of the bubble region the inviscid prediction is comparable to that of the fully interactive method and, within the bubble region, the local interactive calculation of the present model produces results which are very close to those of the fully interactive method.

A comparison of the development of the boundary-layer properties for the E387 airfoil, H_{12}, δ_2, H_{32}, and c_f, calculated using the present method with those obtained using the Drela and Giles program are also presented in Fig. 9. As with the pressure distributions, the agreement between the predictions obtained using the present bubble model with those of the fully interactive method is good. The apparent difference in bubble length may be due to the different treatment of the transition region between the two methods. As in most transition prediction methods based on linear stability, the critical amplification factor used in the Drela and Giles program marks the beginning of the transition region while, in the present model, the transition point used corresponds to the end of this region and the beginning of the full turbulent calculations. The difference in the distribution of H_{32} can be attributed to the difference in pressure distributions in the laminar part of the bubble, shown in Fig. 9, as predicted by the two methods. The steeper pressure gradient predicted by the present model leads to a larger growth in δ_2 and, therefore, a smaller H_{32}.

The aerodynamic characteristics for the E387 airfoil, predicted using both the original and the modified Eppler and Somers program, are given in Fig. 10. Clearly, the drag predicted with the bubble model in use is much closer to the experimental values than that obtained without it. In fact, the largest difference between the two that is observed is less than four counts. In considering such plots, it should be noted that because the original program assumes transition at the laminar separation point, the width between the original transition development and that obtained with the modified program is the length of the laminar part of the bubble.

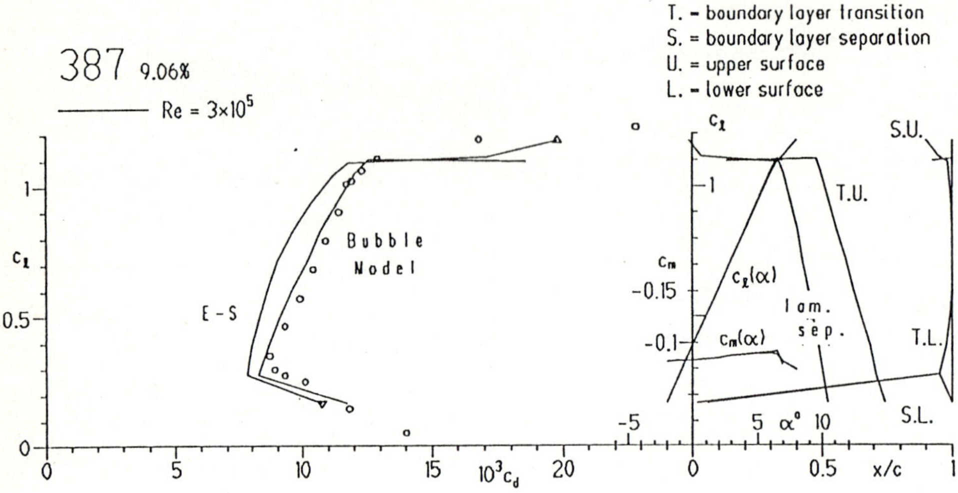

Fig. 10: Aerodynamic characteristics for the Eppler 387 airfoil obtained with the original Eppler and Somers program compared with those obtained using the program incorporating the present bubble model. Experimental data is from Ref. 21.

Because of the strong viscous interaction due to the large amount of aft-loading present, the NLF(1)-1015 airfoil represents a much more difficult case for the present method than does the E387 airfoil. The pressure distribution for this airfoil at an angle of attack relative to the chordline of -3.0 degrees is shown in Fig. 11. The greater impact of viscous interaction in this case compared to the last is apparent. While it should be emphasized again that the present model has yet to be fully calibrated, it is seen in this case that the overall method does seem to model the physical phenomenon sufficiently well to capture the two different bubble developments that occur on this airfoil at this angle of attack.

Concluding Remarks

While its development is still in progress, the laminar separation bubble model described seems to properly predict the behavior of the various parts of the bubble and possess the flexibility required to work over a wide range of cases. For some cases, such as those in which the viscous interactions greatly modify the inviscid pressure distribution, high accuracy in predicting the drag and the influence of the bubble clearly requires a fully interactive method. For many other cases, however, the present local interactive bubble model should provide a computationally efficient method which is well suited for the aerodynamic analysis which is required during the process of airfoil design.

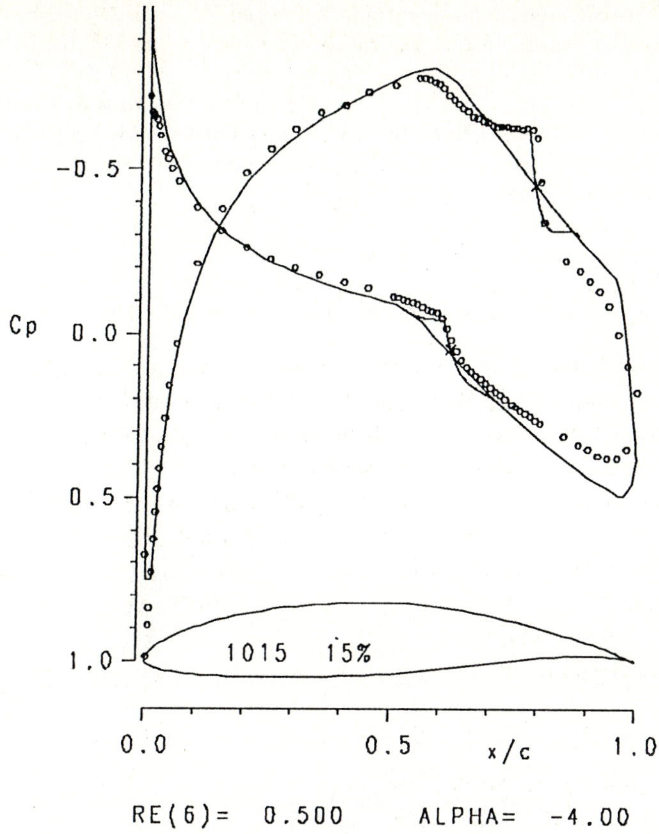

Fig. 11: Velocity distribution and boundary-layer development predicted by present method for NLF(1)-1015 airfoil, $c_\ell = .27$. o : experimental, NASA LaRC, June 1987.

Acknowledgements

This research was supported by the NASA Langley Research Center under Grant NAG 1-778. The authors also wish to acknowledge the contributions and support of the technical monitor, Mr. Dan M. Somers.

References

1. Mueller, T. J., "Low Reynolds Number Vehicles," AGARDograph No. 288, Feb. 1985.
2. Squire, H. B. and Young, A. D., "The Calculation of the Profile Drag of Aerofoils," Aeronautical Research Council R & M 1838, 1937.
3. Schmidt, G. S., "The Prediction of Transitional Separation Bubbles at Low Reynolds Numbers," Ph.D. Thesis, Department of Aerospace and Mechanical Engineering, University of Notre Dame, December 1986.
4. Davis, R. L. and Carter, J. E., "Analysis of Airfoil Transitional Separation Bubbles," NASA CR-3791, 1984.

5. Drela, M., "Two-Dimensional Transonic Aerodynamic Design and Analysis Using the Euler Equations," Massachusetts Institute of Technology, Gas Turbine Laboratory Report 187, February, 1986.
6. Drela, M. and Giles, M. B., "Viscous-Inviscid Analysis of Transonic and Low-Reynolds Number Airfoils," *AIAA Journal*, Vol. 25, No. 10, October, 1987, pp. 1347-1355.
7. Eppler, R. and Somers, D. M., "A Computer Program for the Design and Analysis of Low-Speed Airfoils," NASA TM-80210, 1980.
8. Eppler, R. "Recent Developments in Boundary Layer Computation," *International Conference on Aerodynamics at Low Reynolds Numbers* $10^4 < R < 10^6$, Vol. II, Royal Aeronautical Society, London, October 1986, pp. 12.1-12.18.
9. Horton, H. P., "A Semi-empirical Theory for the Growth and Bursting of Laminar Separation Bubbles," Aeronautical Research Council C. P. 1073, June 1967.
10. Roberts, W. B., "Calculation of Laminar Separation Bubbles and Their Effect on Airfoil Performance," *AIAA Journal*, Vol. 18, No. 1, January 1980, pp. 25-31.
11. Eppler, R. "Practical Calculation of Laminar and Turbulent Bled-Off Boundary Layers," NASA TM-75328, 1978. (Translated from *Ingenieur Archiv*, Vol. 32, 1963, pp. 221-245.)
12. Maughmer, M. D., "A Computationally Efficient Modelling of Laminar Separation Bubbles, Semi-Annual Status Report," NASA CR-182417, February 1988.
13. Dini, P. and Maughmer, M. D., "A Computationally Efficient Modelling of Laminar Separation Bubbles, Semi-Annual Status Report," February 1989.
14. Coton, F. N. and Galbraith, R. A. McD., "A Direct Aerofoil Performance Code Incorporating Laminar Separation Bubbles Effects," *Proceedings of the 16th Congress of the International Council of the Aeronautical Sciences*, ICAS 88-4.2.4, Vol. 1, 1988, pp. 329-338.
15. Gaster, M., "The Structure and Behaviour of Separation Bubbles," Aeronautical Research Council R. & M. No. 3595, March 1967.
16. Pauley, L. L., Moin, P., and Reynolds, W. C., "On the Structure of Two- and Three-Dimensional Separation," AIAA Paper No. 89-0287, January, 1989.
17. Green, J. E., "Two-Dimensional Turbulent Reattachment as a Boundary-Layer Problem," *Separated Flows*, AGARD Conference Proceedings No. 4, Part 1, May 1966.
18. Ingen, J. L. van, "On the Calculation of Laminar Separation Bubbles in Two-Dimensional Incompressible Flow," *Flow Separation*, AGARD Conference Proceedings No. 168, 1975.
19. Ingen, J. L. van and Boermans, L. M. M., "Aerodynamics at Low Reynolds Numbers: A Review of Theoretical and Experimental Research at Delft University of Technology," *International Conference on Aerodynamics at Low Reynolds Numbers* $10^4 < R < 10^6$, vol. I, Royal Aeronautical Society, London, October 1986, pp. 1.1-1.40.
20. Maughmer, M. D. and Somers, D. M., "Design and Experimental Results for a High-Altitude, Long-Endurance Airfoil," *Journal of Aircraft*, Vol. 26, No. 2, February 1989, pp. 148-153.
21. McGhee, R. J., Walker, B. S., and Millard, B. F., "Experimental Results for the Eppler 387 Airfoil at Low Reynolds Numbers in the Langley Low-Turbulence Pressure Tunnel," NASA TM-4062, October 1988.
22. Stewartson, K., "Further Solutions of the Falkner-Skan Equation," *Proceedings of the Cambridge Philosophical Society*, vol. 50, 1954, pp. 454-465.
23. Fitzgerald, E. J. and Mueller, T. J., "Laser Doppler Velocimeter Measurements of the Transitional Separation Bubble on an Airfoil at a Low Reynolds Number," to appear in *AIAA Journal*.

A Comparison Between Boundary Layer Measurements in a Laminar Separation Bubble Flow and Linear Stability Theory Calculations

P. LeBlanc, R. Blackwelder, R. Liebeck
Department of Aerospace Engineering
University of Southern California
Los Angeles, California 90089-1191 USA

Abstract

This research examines the details of the boundary layer flowfield from wind tunnel measurements of a two-dimensional Liebeck LA2573A airfoil over a range of Reynolds numbers from 235000 to 500000. In this range, a laminar separation bubble becomes significant in the boundary layer and provides a measurable contribution to the airfoil drag. Measurements include airfoil drag, mean and turbulent boundary layer velocity profiles, a calculation of integral parameters associated with these profiles, and energy spectra of the velocity signal inside the boundary layer. Evidence of the growth of boundary layer velocity fluctuations within a range of frequencies in the laminar separation and transition regions has been found in these spectral measurements. Results have shown that the peak frequencies measured in the velocity spectra for the instability region agree with the most amplified wave number and frequency scaling predicted by linear stability theory for these inflectional profiles. Additionally, the maximum measured growth rates at this peak frequency correlate with growth rates calculated from similarly shaped Falkner-Skan profiles at the corresponding frequency of maximum amplification. This agreement between experimental and theoretical peak frequencies and growth rates was confirmed for the range of Reynolds numbers and for airfoil incidence ranging from zero lift to stall.

Introduction

The primary goal of this research is a better understanding of the transitional instability mechanism which has a controlling effect on the extent of laminar separation bubbles occurring on airfoils operating at low chord Reynolds numbers, generally for $R_c < 10^6$. In this range, the local Reynolds number based on the boundary layer development is often insufficiently high for Tollmein-Schlichting (viscous type) instabilities to promote a natural transition before the laminar boundary layer reaches an adverse pressure gradient and is subject to separation. Once laminar separation has occurred, the resulting inflectional velocity profiles promote a more rapid amplification of boundary layer fluctuations which eventually reach transition levels. Laminar separation is also associated with a significant increase in the boundary layer thickness. After transition, increased mixing promotes growth of the turbulent separated shear layer which eventually reattaches to the airfoil surface in cases where the divergence of the separated laminar layer from the surface has not exceeded some critical point.

Research has shown that decreases in R_c tend to delay the transition region within the bubble. These decreases have been correlated with a thicker turbulent boundary layer and additional drag penalty due to the separation bubble. Meanwhile, the addition of external forcing conditions such as the free stream turbulence [1], acoustic disturbances [2] [3] [4] , or surface roughness [4] has been found to encourage earlier transition, resulting in a smaller bubble and reduced drag. Accordingly, the transition position within the bubble can be considered a primary factor in controlling the low Reynolds number boundary layer flowfield, and its prediction becomes critical to airfoil performance calculations.

Early attempts at transition prediction focused on correlations of local flow parameters and Reynolds numbers in the laminar bubble region [5], [6], [7]. Later work in conjunction with additional data has indicated the limited success of such correlations [8]. More recently, numerical calculation methods have been proposed for predicting low Reynolds number separation bubble flows [9], [4], [10]. These methods involve linear stability calculations to some extent in determining transition within the viscous / inviscid interaction schemes. The transition calculations generally assume an e^n transition criterion based on the stability of Falkner-Skan (β) reverse flow profile solutions which is described in detail by van Ingen and Boermans [4]. Considering the more promising results in predicting separation bubble flows and airfoil performance with these methods, an attempt is made in the present study to compare peak frequencies and the corresponding growth rates from measured boundary layer velocity fluctuations in the laminar bubble with the most amplified waves and corresponding growth rates predicted by linear stability theory. Such a comparison would attempt to verify the stability calculations which are implemented in the e^n transition method.

A correlation between experimental and theoretical results is made assuming the standard small amplitude wave-like disturbance in the linear stability formulation for theoretical boundary layer profiles. A viscous Orr-Sommerfeld calculation has been used to determine the stability of various theoretical (β) profiles. The wave number and frequency have been nondimensionalized by the local edge velocity and displacement thickness: $\bar{k} = (2\pi\delta^*)/\lambda$, $\bar{\omega} = (2\pi f\delta^*)/U_e$. For calculations assuming spatially growing waves ($\bar{\omega}_i = 0$), an example of growth rates $-\bar{k}_i$ is given by figure 1 for the Falkner-Skan reverse flow profile at $\beta = -.14$ and $R_{\delta^*} = 10^3$. The disturbance energy profile at some position $\bar{x}$ downstream of the point of neutral stability ($\bar{x}_0$) for a particular wave number may be written as

$$E_\beta(x,y) = u^2(x_0,y)\exp\left[-2\int_{\bar{x}_0}^{\bar{x}} \bar{k}_i d\bar{x}\right] \tag{1}$$

and the local dimensional energy growth rate follows as

$$\frac{d(\ln E_\beta/u^2)}{dx} = \frac{-2\bar{k}_i}{\delta^*} \tag{2}$$

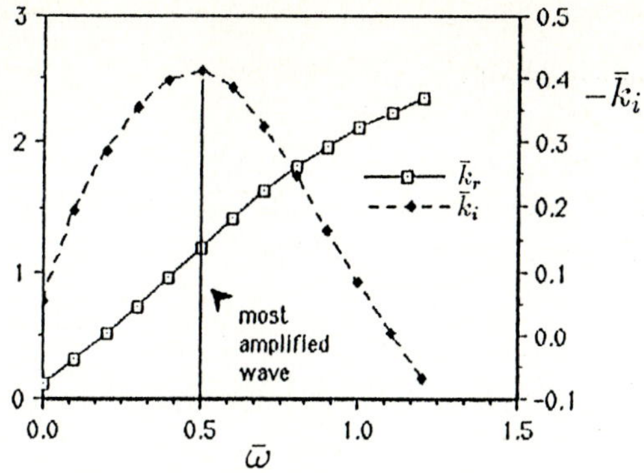

Figure 1: An example of typical growth rates for separated theoretical β profiles; $\beta = -.14$, $R_{\delta^*} = 1000$.

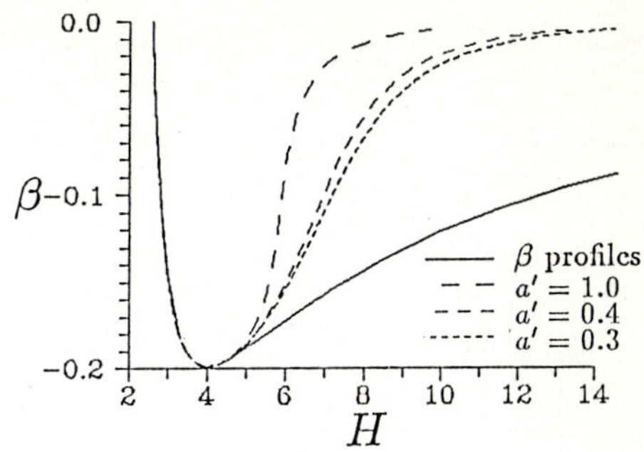

Figure 2: Falkner-Skan profile parameter β plotted against the profile shape factor H. Dashed curves represent rectified β profiles from eq. (3) for different a'.

This relation will be used in comparing calculated amplification rates with experimental data derived from boundary layer fluctuation energy spectra (E_s).

The subscript β in equation (2) refers to the pressure gradient parameter of the Falkner-Skan profiles which are used as an input to the stability calculation. These profiles are assumed for the separation bubble region in the numerical prediction schemes, and the shape factor H is essentially used as an independant variable since it uniquely describes both attached and separated profiles for decelerating flows ($-.1988 < \beta < 0$) as is indicated in figure 2. The β profiles may also be used to approximate measured mean velocity profiles upstream of transition. The shape factor H could then be used to associate experimental data with theoretical stability calculations. A problem arises, however, in the case of reverse flow profiles, where hot-wire anemometer velocity measurements cannot distinguish between velocity direction with respect to the flow sensor. The resulting experimental mean velocity profiles reflect a low magnitude rectified version of the separated profiles in the region near the wall. An example may be found in figure 3 which compares the $\beta = -.03$ reverse flow (solid line) profile with a profile measured in the separation region at $x/c = .385$, $\alpha = 4°$ and $R_c = 235000$. Thus, while a theoretical reverse flow β profile could be matched to a particular experimental profile, (excluding the region of reverse flow), a calculation of the integral thickness parameters δ^*, θ and H would not indicate a match between the two shape factors. A solution to this problem is to modify the theoretical profile to simulate the experimental data in the reverse flow region strictly for the purpose of calculating δ^*, θ and H. A rectified theoretical profile is given by

$$\bar{u}^*_\beta(y) = \begin{cases} \bar{u}_\beta(y) & \bar{u}_\beta > 0 \\ a'|\bar{u}_\beta(y)| & \bar{u}_\beta < 0 \end{cases} \tag{3}$$

where an additional amplitude factor a' has been introduced to allow a better fit of the data. Two modified theoretical profiles ($a' = 1.0$ and $a' = 0.4$) are also compared with the experimental profile in figure 3. A reasonable match is indicated for an amplification factor of $a' = 0.4$;

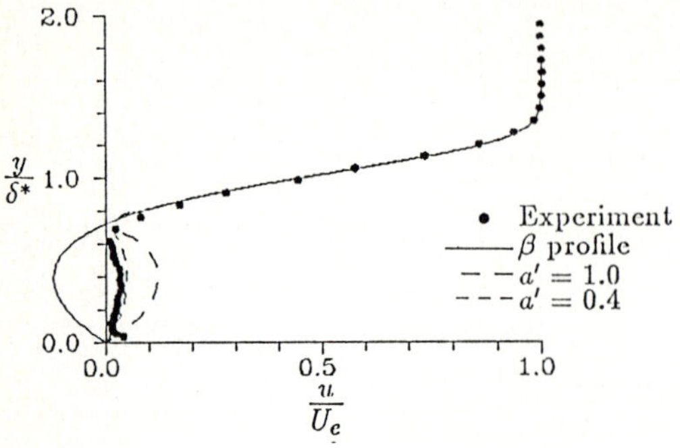

Figure 3: Comparison between typical separated experimental mean velocity profile ($R_c = 235000, \alpha = 4°, x/c = .385$) and "matching" theoretical profile ($\beta = -.03$). Dashed curves represent rectified β profiles from eq. (3) for different a'.

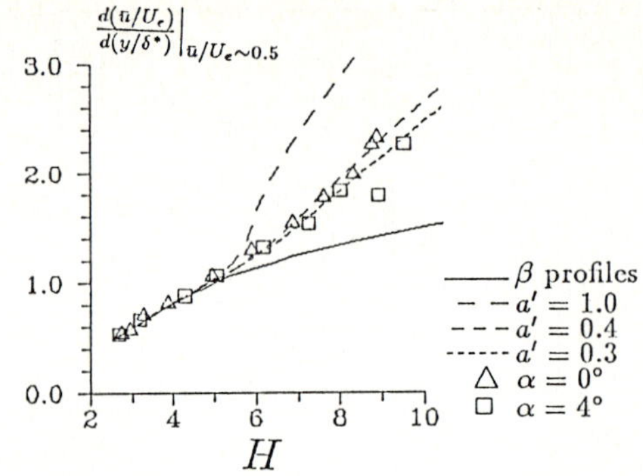

Figure 4: Nondimensional velocity shear near the profile inflection for theoretical (solid) and modified (dash) β profiles. Comparison is made with sample experimental data at $\alpha = 0°, 4°$ and $R_c = 235000$.

therefore, this value will be used in calculating a modified theoretical shape factor (H^*_β) for later comparisons with experimental results. This choice for a' is also supported by an examination of the nondimensional velocity shear ($d(\bar{u}/U_e)/d(y/\delta^*)|_{\bar{u}/U_e \sim 0.5}$) which has been approximated for the profile inflection. The theoretical curve in figure 4 for the case with $a' = 0.4$ shows a good representation of the sample data points at $\alpha = 0°$ and $4°$ ($R_c = 235000$) over a range of profile shapes.

An estimate of the most amplified or peak frequency expected from experimental results can be made from the non-dimensional frequency scaling given above, with a few additional approximations. At a fixed airfoil incidence and chord position upstream of transition, the pressure distribution is assumed approximately constant; therefore, the profile shape, most unstable wave $\bar{k}_r$ and convection speed $\bar{c}$ (non-dimensional) can be approximated as constant over a range of R_c or for different chord lengths. In addition, if δ^* is approximated according to the Blasius relation $\delta^*/x \sim 1/\sqrt{R_x}$, then the frequency ratio should scale according to equation (4) for fixed chord, or equation (5) for fixed R_c.

$$\frac{f_{p_1}}{f_{p_2}} = \left(\frac{R_{c_1}}{R_{c_2}}\right)^{1.5}\Bigg|_c \tag{4}$$

$$\frac{f_{p_1}}{f_{p_2}} = \left(\frac{c_2}{c_1}\right)^{2}\Bigg|_{R_c} \tag{5}$$

An estimate of experimental growth rates of spectral energy corresponding to peaks in the frequency spectra may be made by assuming an exponential growth of spectral energy in x. By differentiating spectral distributions with respect to the streamwise coordinate (x) at a fixed frequency f, and associating the result with equation (2), the experimental and theoretical amplification rates can be compared by with the following relation

$$\bar{k}_i(\bar{\omega}) = \frac{-\delta^*}{2c}\frac{d(\ln \bar{E}_s)}{d(x/c)} \tag{6}$$

provided that the frequency ($\bar{\omega}$), local Reynolds number (R_{δ^*}) and profile shapes (H) have been matched. All terms on the right side of equation (6) are obtained from experimental data.

Alternatively, given the experimental spectral energy at a position corresponding to the maximum measured growth rate for the peak frequency (E_{s_0}), theoretical estimates for the spectral energy upstream of that position could be predicted by integrating the relation in (6) and using theoretical amplification rates as a function of profile shape:

$$E_s(x/c) = E_{s_0} \exp\left[2c\int_{x/c}^{(x/c)_0} \frac{\bar{k}_i}{\delta^*} d(x/c)\right] \tag{7}$$

This estimate would provide a check between values expected for the spectral energy and actual experimental values which could be associated with the signal noise level at a point sufficiently far uptsream of the reference point.

Experimental Technique

Experimental measurements were conducted in the 37 by 54 inch semi-hexagonal cross section of the USC Dryden wind tunnel. This facility has a closed return with a contraction ratio of 7.1 and a maximum test section velocity of approximately 34 m/s. The free stream turbulence level in the test section has been measured at approximately .1% at airfoil test velocities for frequencies above 1 Hz.. Further details in reference [1] indicated that low frequency fluctuations on the order of 20 Hz. or less comprised a significant contribution to this turbulence.

Measurements were made on two LA2573A airfoil models designed by Liebeck [11] to operate at chord Reynolds numbers of approximately 250000. The airfoil contour is included in figures 7 and 9. Both 6 inch and 11.75 inch models were numerically milled from aluminum. In addition, the 6 inch chord model was equipped with pressure taps for lift measurements. Flowfield, pressure distribution and drag data were collected at $R_c = 235000$ (24 m/s) with the smaller model while the larger model was used to obtain flowfield and drag data at Reynolds numbers from 235000 to 500000. Drag measurements were calculated from the momentum deficit after a total pressure rake was centered in the airfoil wake. Boundary layer velocities were measured with a single hot-wire anemometer sensor, and the details of these measurements are given in [1].

Results / Discussion

Airfoil drag and boundary layer measurements were taken at Reynolds numbers of 235000, 300000, 375000 and 500000 for airfoil incidence ranging from $\alpha = -2°$ to $\alpha = 12°$ (near stall) in 2° increments. Measurements indicated that a separation bubble could still be found at $R_c = 500000$, except near stall. Boundary layer data on the airfoil lifting surface ranged in

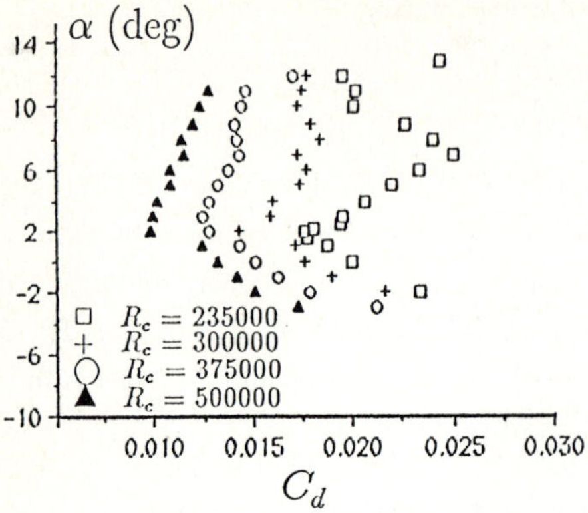

Figure 5: Experimental drag coefficients.

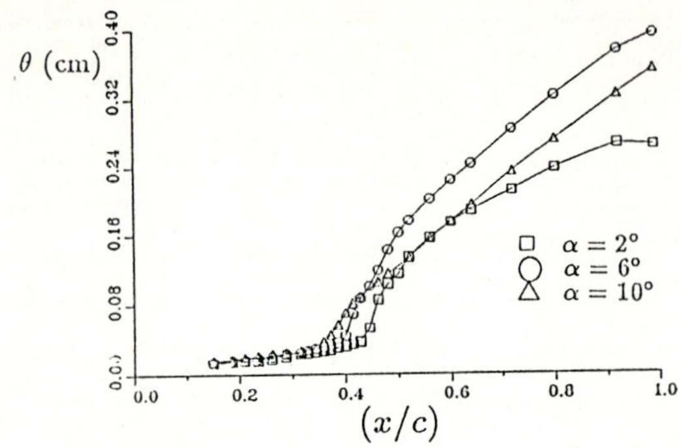

Figure 6: Boundary layer momentum thickness at $R_c = 235000$.

chord position from x/c = .150 to the trailing edge, including approximately 25 positions at each incidence with the highest concentration of data points located in the region of the laminar separation bubble. Boundary layer integral parameters δ^*, θ, H and δ_t were calculated directly from velocity profile measurements for all chord positions, including the separation region where the hot-wire measurement does not reflect the flow direction. Therefore, some error was introduced into these calculations using experimental data in the separation region. Brendel and Mueller [12] have estimated this error to be on the order of a few percent for δ^* while more significant for θ. Spectral measurements of the boundary layer velocity fluctuation were recorded at most profile data chord positions and were located at a distance normal to the surface corresponding approximately to the midpoint of the velocity shear layer (or approximately the maximum point in the velocity fluctuation ($\overline{u'}$) profile).

Airfoil Performance

Figure 5 illustrates the incremental drag penalty caused by the laminar separation bubble as Rc is reduced from 500000 to 235000 for airfoil incidence over the operating range. It can be seen that the penalty is more severe at midrange α for $R_c = 235000$ where the bubble is close to the point of bursting (failure of the turbulent separated shear layer to reattach). An examination of momentum thickness distribution in figure 6 shows that this drag penalty at lower R_c can be associated with the boundary layer growth downstream of the bubble. The case at $\alpha = 6°$ shows the highest momentum thickness approaching the trailing edge, corresponding to the maximum drag for the three incidence cases plotted. Lower trailing edge values are found for θ at $\alpha = 10°$ and $\alpha = 2°$, which corresponds to their lower respective drag values. The airfoil lift has been measured for R_c in this range in [11] and [1] and does not show any significant sensitivity to bubble effects at these Reynolds numbers.

Boundary Layer Measurements

An example of experimental mean velocity profiles and mean fluctuation profiles measured with the 11.75 inch chord model are given in figure 7, for a Reynolds number of 235000 at $\alpha = 4°$. The physical location of the measurement points are indicated on the airfoil in the upper right corner of figure 7. Profiles normalized with δ^* are plotted with increasing chord position from right to left and have been chosen to indicate the character of the boundary layer flowfield. The initial profiles at $x/c = .150$ and .235 show an attached boundary layer with a very low level in the (rms) fluctuation distribution. At $x/c = .287$, the inflectional profile nears separation. The following three normalized profiles at $x/c = .343$, .371, and .399 indicate a growing laminar separation region. (Measurements for δ^* in figure 8 illustrate that the seperated laminar boundary layer region is in fact growing.) It can be seen that the hot-wire measurements fail to indicate reverse flow in the region near the wall, but do give an indication of the extent of this region. At $x/c = .427$, a jump in the magnitude of the mean fluctuation profile can be seen to indicate transition. Note that the large increase in the fluctuations does not occur near the wall as is often found in transition but corresponds to the location of the inflection point in the mean profile. A more precise definition of transition can be obtained by defining the "integral turbulence scale," $\delta_t^n(x)$, as

$$\delta_t^n(x) = \int_0^\delta \left(\frac{\overline{u'}}{U_e}\right)^n dy \tag{8}$$

This parameter is plotted for $n = 1$ at several R_c in figure 8. Transition can be clearly defined as the location where a sharp increase in this thickness parameter occurs. Finally, a reattached turbulent boundary layer can be found in figure 7 at $x/c = .520$. Profiles downstream of this position had a characteristic turbulent shape with normal growth in $\delta^*(x)$ and an approximately constant value of H. Note that $\delta_t^1(x)$ seems to increase proportionally with $\delta^*(x)$ in the reattached region.

In general, all boundary layer data where separation bubbles were found to occur showed the same general progression as that in figure 7, and this general character of the separation bubble flowfield has been measured by several other researchers [13], [12], [9], [4], [8], and [6]. At higher Reynolds numbers, normalized profiles have been found to match lower Reynolds number profiles quite well at the same nondimensional chord positions, except near transition which moves upstream with increasing Reynolds number. The experimental laminar and separated profiles in figure 7 are also compared with Falkner-Skan profiles. They generally correspond well at different x/c positions except in the reverse flow regions. Experimental shape factors in figure 8 were used to match the data with theory (H_β^*).

Spectral Measurements

Boundary layer velocity spectra associated with the profiles given in figure 7 are plotted in figure 9 at seven different chord positions extending through the separation bubble. It is noted that the chordwise progression proceeds from bottom to top, and that each curve has been

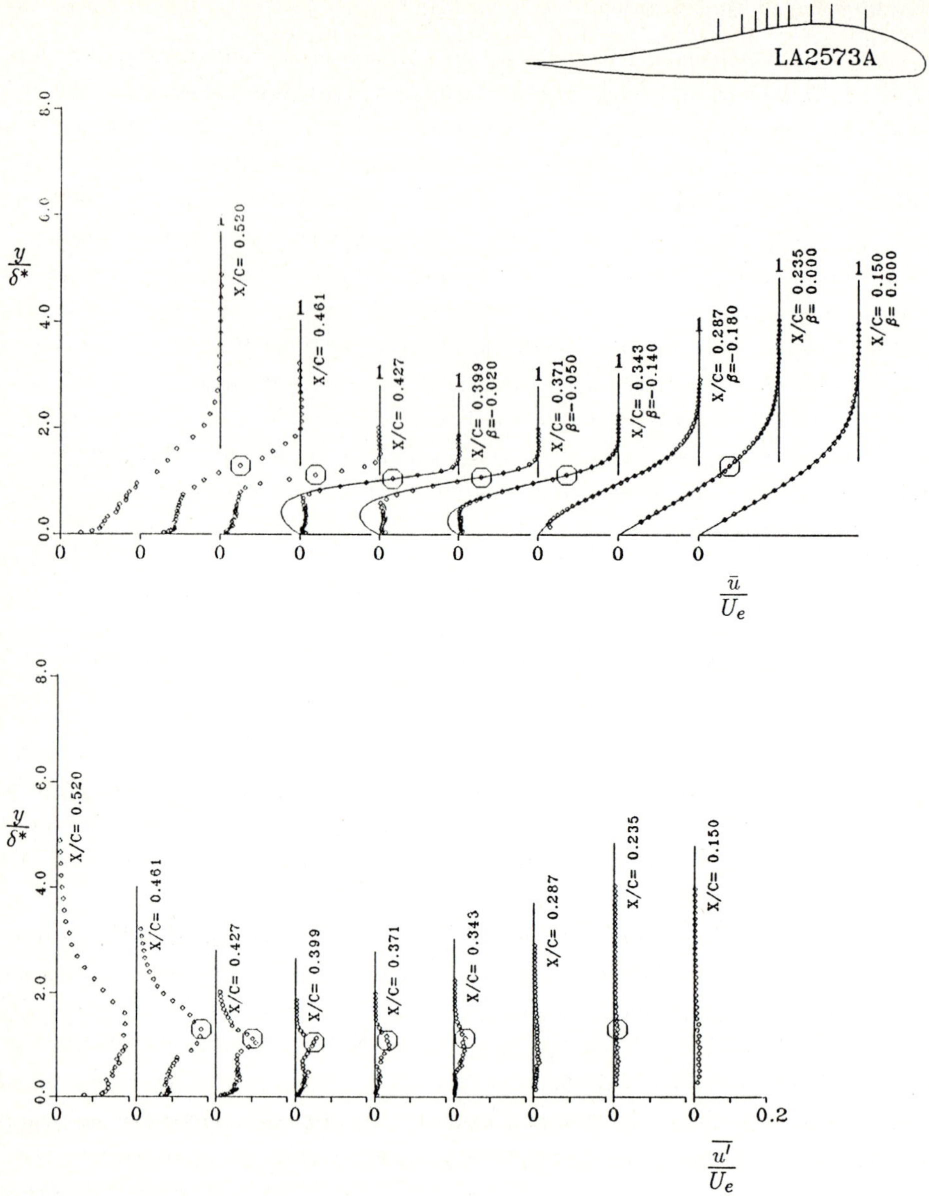

Figure 7: Mean and fluctuation velocity profiles at $\alpha = 4°$, $R_c = 235000$. Symbols ⊙ indicate position at which spectral distributions in figure 9 were recorded. Solid curves show comparison with theoretical β profiles matched using the modified shape factor H^*_β.

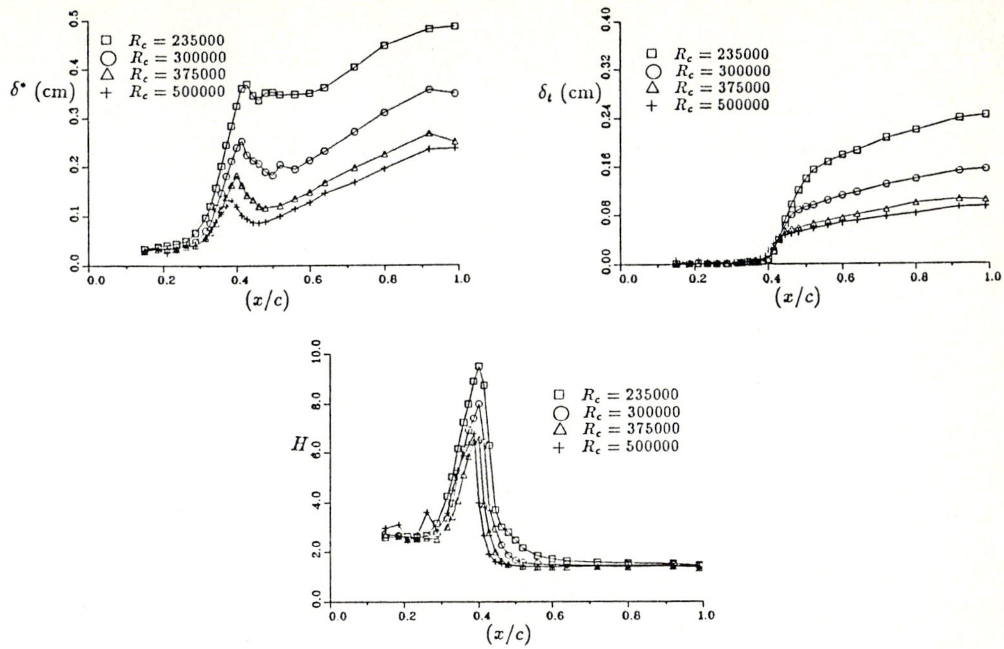

Figure 8: Integral thickness parameters δ^*, δ_t, and H at $\alpha = 4°$.

offset from the previous curve by a distance of 1 decade. A distinct peak centered at 900 Hz. can be seen to grow from $x/c = .371$, at a point well into the laminar region of the bubble, to $x/c = .427$ where the peak seems to reach a maximum. This point also corresponds to the transition region as established in figures 7 and 8. At $x/c = .461$ the spectrum of a turbulent boundary layer velocity signal is observed.

Spectral distributions at higher Reynolds numbers and other airfoil incidence cases have shown the same general character as the data given in figure 9, with the exception that the peak is found to occur at a different frequency and over a different range of chord positions. In addition, similar boundary layer spectral progressions for separation bubble flowfields have been reported in [9], [12], [14], [1], and [3].

A comparison of peak frequency scaling with chord Reynolds number for different airfoil incidence cases is given in figure 10. The log-log plot indicates that peak frequencies follow a slope of approximately 1.5 which was was anticipated from the stability theory formulation of equation (4). In addition, data at $R_c = 235000$ from the two different airfoil models confirmed the expected chord scaling for the frequency ratio in equation (5).

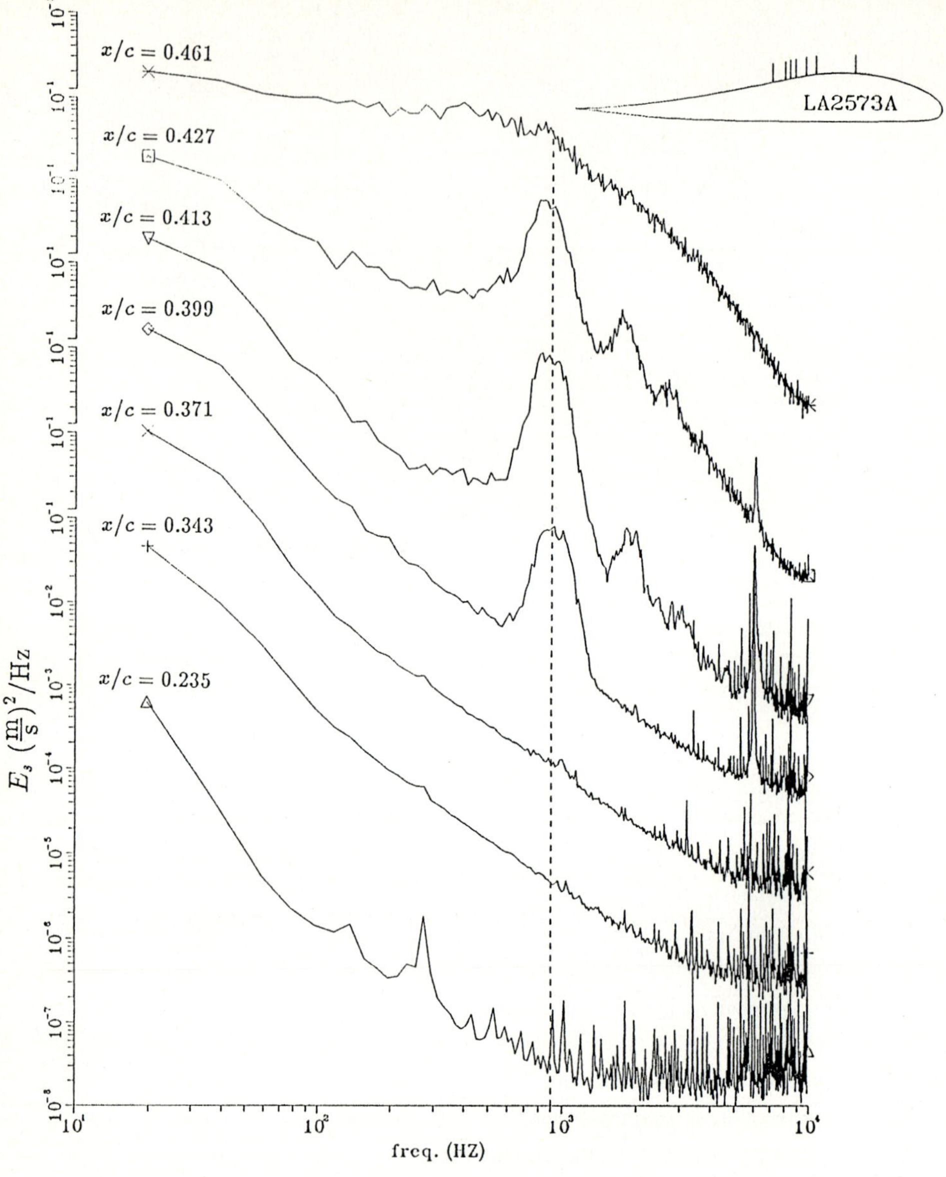

Figure 9: Boundary layer fluctuation spectral distributions for successive chord positions at $\alpha = 4°$, $R_c = 235000$. Dashed line indicates instability peak growing at $f_p = 900$ Hz. Note: curves are vertically offset by 1 decade.

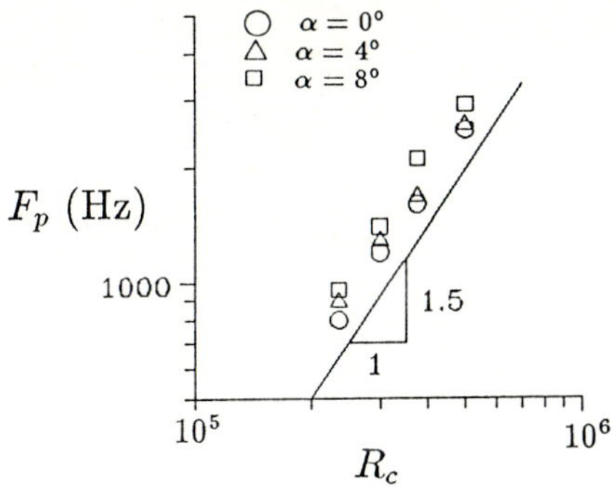

Figure 10: Scaling for peak frequency f_p with R_c.

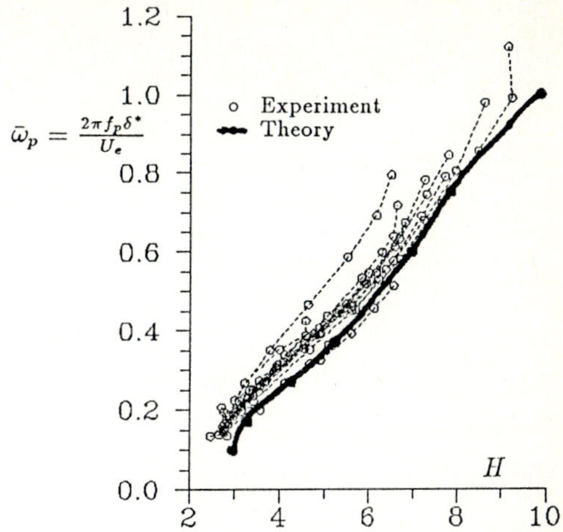

Figure 11: Correlation between $\bar{\omega}$ and profile shape H. Curves (o) include cases at $\alpha = 0°$, $4°$, $8°$, $10°$, $12°$ and $R_c = 235000$, 300000, 375000, 500000. Comparison is made with theoretical values (•) assuming H^*_β.

Comparison With Stability Theory

For a comparison between the measured peak frequency and the most amplified waves expected from linear stability theory, the experimental peak frequency was non-dimensionalized as $\bar{\omega}_p = \frac{2\pi f_p \delta^*}{U_e}$ and plotted against the experimental shape factor in figure 11. It is noted that both δ^* and H are changing with the chordwise coordinate x in the region of the separation bubble; thus, an increase in H upstream of transition reflects a parameterized increase in x. The resulting curves represented by open symbols and dashed lines are given in figure 11 for a number of different Reynolds number and airfoil incidence cases ($235000 < R_c < 500000$, $\alpha = 0°, 4°, 8°, 10°, 12°$). Curves are plotted for shape factors corresponding to x positions upstream of transition, and the rolloff at the upper end of each curve can be associated with the maximum in the H distribution found near transition (see figure 8 for example). In general, all curves tend to follow a straight line up until transition. These lines generally agree with the frequency corresponding to the most amplified wave for β profiles with varying shape factor. The theoretical peak frequencies are also given in figure 11 for $R_{\delta^*} = 2000$, and are designated by the solid symbols and line.

In order to provide a more complete comparison between experimental results and predicions estimated by linear stability theory, peak fluctuation growth rates will be examined for a variety of flowfield cases. The spectral energy at the peak frequency has been plotted against x in figure 12 for the Reynolds numbers investigated at $\alpha = 4°$. This plot includes the values at $f_p = 900$ Hz. from figure 9. The curves show a sharp amplification just prior to transition, then the spectral energy generally levels off in the turbulent boundary layer region downstream. The

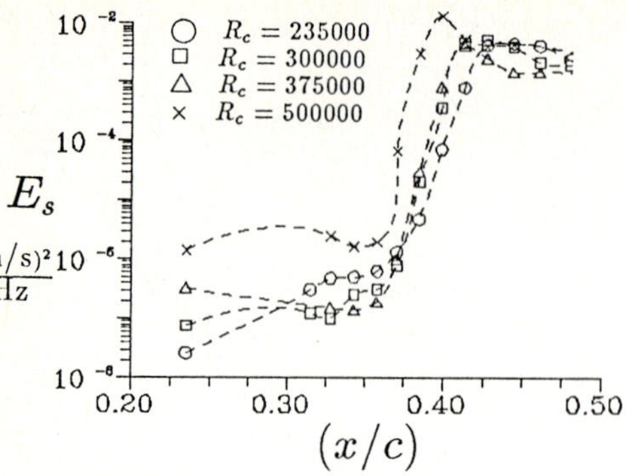

Figure 12: Spectral energy at f_p: $\alpha = 4°$.

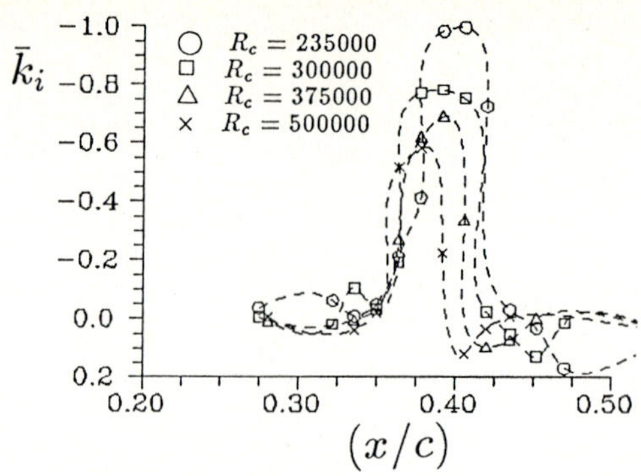

Figure 13: Experimental growth rates ($\bar{k}_i$) at f_p: $\alpha = 4°$.

maximum spectral energy occurs at a position just downstream of transition and the peak in the distribution for H in figure 8.

Using the relation from equation (6), non-dimensional growth rates can be calculated from the spectral energy curves of figure 12. These growth rates at f_p are given in figure 13 for the four R_c cases at $\alpha = 4°$. The maximum growth rate for each curve is found to decrease and move upstream as Reynolds number is increased. The chord position of this maximum can be associated with the transition positions indicated by the jump in the δ_t distributions of figure 8.

For a comparison between experimental growth rates and theoretical estimates, the profile shape factor H must be substituted for the streamwise coordinate to provide a parameter with which to associate the two sets of results. This substitution may be associated with the use of H as an independant variable in the numerical flow prediction schemes such as [10]. Thus the experimental growth rates given in figure 13 have been replotted in terms of the experimental profile shape factor. The result in figure 14 also include growth rates at f_p for $\alpha = 0°$, $8°$, $10°$, and $12°$ at the same chord Reynolds numbers as the $\alpha = 4°$ data. The curve for each airfoil incidence and Reynolds number case is plotted from a profile shape corresponding to a chord position upstream of laminar separation, to the point of the maximum measured growth rate. Presumably, the data would be expected to follow a single trend in order to confirm the result predicted by linear stability theory. However, the experimental curves do not seem to collapse in general for the variety of flowfield cases included. The measured growth rates appear to be limited approximately by the dashed line indicated in figure 14. This line represents a best fit of the maximum measured growth rate for each case. This limiting relation is more clearly defined in figure 15 where these maximum measured growth rates are represented by open symbols. These maximum experimental amplification rates show a general agreement with the growth rates predicted by stability theory for the most amplified wave. The theoretical curve is given

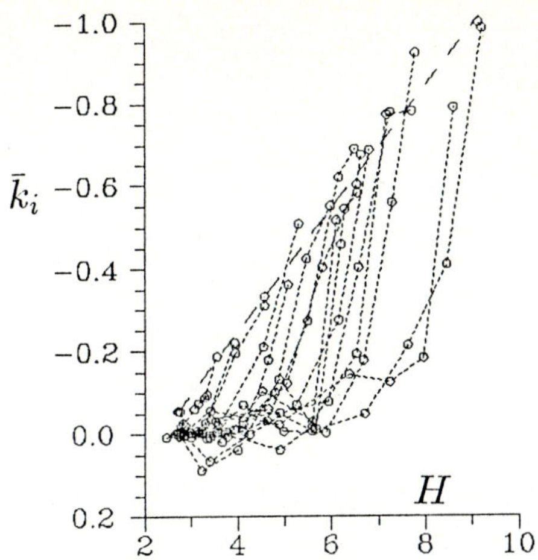

Figure 14: Correlation between experimental growth rates measured at f_p and profile shape. Curves (o) include cases at $\alpha = 0°, 4°, 8°, 10°, 12°$ and $R_c = 235000, 300000, 375000, 500000$.

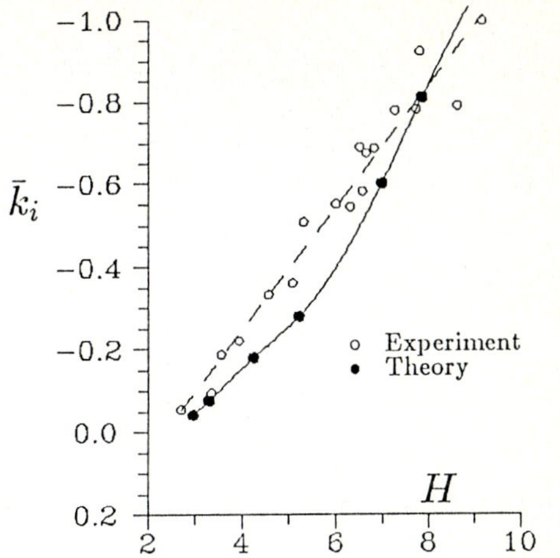

Figure 15: Correlation between maximum measured growth rates (o) in figure 14 at f_p and growth rates predicted by linear stability calculations (•) for profile shape factor H. (Theoretical values assume H^*_β.)

by the solid symbols for $R_{\delta^*} = 2000$.

One possible explanation for the disagreement between theoretical estimates and growth rates measured below the maximum for each case in figure 14 is the effect of the hot-wire velocity signal noise on the resolution of fluctuation amplitudes well below the maximum amplified values. A check of the expected spectral energy against the noise level may be made by extrapolating spectral energy levels at f_p upstream of the amplitude corresponding to the maximum measured growth rate. The extrapolation would use the relation given in equation (7) and assume that growth rates followed the theoretical curve in figure 15 in order to compare the amplitude of the spectral energy which would be expected upstream with the amplitudes actually measured. An example of this calculation is given in figure 16 in a format similar to that of figure 12. Two curves are plotted for the case $\alpha = 4°$, $R_c = 235000$, using the spectral energy at $x/c = .406$ as a reference for the theoretical extrapolation. The experimental and predicted spectral energies agree for a short distance upstream of the reference point and then diverge as the experimental curve levels out. Thus, if this upstream level of the measured spectral energy were indicative of the signal noise level at the peak frequency, then the amplitude of fluctuations which would be expected below this noise level would be lost in the signal noise. The figure indicates that the actual spectral energy of fluctuations being amplified might only be resolved in measurements over a short region in x (or alternatively for a short variation in H.) Consequently, the comparison in figure 16 would infer that the disagreement found in figure 14 might be due to an experimental difficulty in the resolution of very low amplitude velocity fluctuations.

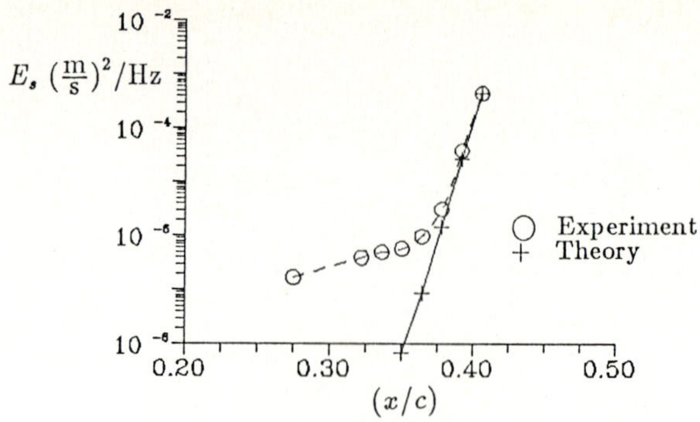

Figure 16: Comparison between measured and predicted spectral energy E_s at $\alpha = 4°$, $R_c = 235000$. Data at $x/c = .406$ is used for theoretical extrapolation.

Conclusions

Comparisons between experimental boundary layer spectral fluctuation data for low Reynolds number separation bubble flows and linear stability theory calculations have shown that the peak frequency observed in the disturbance spectra corresponds to the most amplified waves predicted by theory for a large range of experimental conditions (R_c and α). These comparisons were made primarily for inflectional profile shapes in the separation region, leading to inflectional inviscid type instabilities, calculated with the Falkner-Skan profiles. In particular, measurements of f_p followed the predicted scaling for chord Reynolds number or chord variations. Comparisons between growth rates showed an agreement between the maximum experimentally measures growth rates at peak frequencies and predicted growth rates for the most amplified waves of similarly shaped profiles at a given airfoil incidence and Reynolds number. Discrepancies between measurements and theory at growth rates lower than the maximum were consistent with limitations in the resolution of the hot-wire velocity signal.

These results generally seem to show a favorable comparison between measurements upstream of the transition region and estimates predicted by linear stability theory associated with an inviscid inflectional profile. This agreement lends support to the use of linear stability theory calculations in determining the transition region for prediction methods designed to provide low Reynolds number laminar separation bubble flowfield calculations.

Acknowledgements

This work was supported by the Office of Naval Research under their University Research Initiative Contract N00014-86-K-0679 monitored by Mike Reischman. The authors would also like to thank P. Monkewitz for supplying the Orr-Sommerfeld code and S.-I. Chang for adapting it to our use.

Notation

a'	amplitude factor in eq. (3)	
c	airfoil chord	
$\bar{c}$	non-dimensional convection velocity	$\bar{\omega}/\bar{k}_r$
C_d	airfoil drag coefficient per unit span	$\frac{D}{Qc}$
D	airfoil drag	
E_β	theoretical disturbance energy in eq. (1)	
E_s	experimental spectral fluctuation energy ($\bar{E}_s$ - nondimensional)	
f	frequency	
f_p	f corresponding to instability peak in E_s distribution	
H	shape factor	$\frac{\delta^*}{\theta}$
H^*_β	modified β profile shape factor (calculated with $\bar{u}^*(y)$)	
$\bar{k}$	non-dimensional complex wave number	$\bar{k}_r + i\bar{k}_i$
$\bar{k}_r$	non-dimensional wave number	$\frac{2\pi\delta^*}{\lambda}$
$\bar{k}_i$	non-dimensional growth rate	
L	airfoil lift	
p	static pressure	
Q	dynamic pressure	$\frac{1}{2}\rho(U_\infty)^2$
R_c	Reynolds number, chord	$(U_\infty c)/\nu$
R_{δ^*}	Reynolds number, displacement thickness	$(U_e\delta^*)/\nu$
R_x	Reynolds number, local	$(U_e x)/\nu$
u	time dependant streamwise velocity	
$\bar{u}, U$	mean velocity	$\frac{1}{N}\left[\sum_{j=0}^{N-1} u_j\right]$
u'	fluctuation velocity	$u - \bar{u}$
$\overline{u'}$	mean fluctuation velocity	$\left[\frac{1}{N}\sum_{j=0}^{N-1}(u_j - \bar{u})^2\right]^{1/2}$
$\bar{u}^*_\beta$	modified β profile mean velocity from eq. (3)	
U_e	boundary layer edge velocity	
U_∞	free stream velocity	
x	chordwise position downstream of the airfoil leading edge	
$\bar{x}$	non-dimensional distance	x/δ^*
y	coordinate normal to the free stream or airfoil surface	

Greek Symbols

α	airfoil angle of attack

β	Falkner - Skan pressure gradient parameter	
δ	boundary layer thickness	
δ^*	displacement thickness	$\int_0^\delta (1 - \frac{\bar{u}}{U_e}) dy$
δ_t	"turbulence thickness scale"	$\int_0^\delta (\frac{\overline{u'}}{U_e}) dy$
θ	momentum thickness	$\int_0^\delta \frac{\bar{u}}{U_e}(1 - \frac{\bar{u}}{U_e}) dy$
λ	disturbance wavelength	
ν	kinematic viscosity	
ρ	density	
$\bar{\omega}$	non-dimensional frequency	$(2\pi f \delta^*)/U_e$

References

[1] P. J. LeBlanc, R. H. Liebeck, and R. Blackwelder. Boundary layer and performance characteristics from wind tunnel tests of a low Reynolds number Liebeck airfoil. In *Aerodynamics at Low Reynolds Numbers*, pages 8.1–8.19, London, Oct. 1986. The Royal Aeronautical Society.

[2] K. B. Zaman, A. Bar-Sever, and S. M. Mangalam. Effect of acoustic excitation on the flow over a low-Re airfoil. *Journal of Fluid Mechanics*, 182:127–148, 1987.

[3] C. F. Liu. *Control of Airfoil Performance by Local Excitation.* PhD thesis, National Cheng Kung University, Tainan, Taiwan, October 1988.

[4] J. L. van Ingen and L. M. M. Boermans. Aerodynamics at low Reynolds numbers: A review of theoretical and experimental research at Delft University of Technology. In *Aerodynamics at Low Reynolds Numbers*, pages 1.1–1.40, London, Oct. 1986. The Royal Aeronautical Society.

[5] D. E. Gault. An experimental investigation of regions of separated laminar flow. Technical report, NACA TN 3505, Sept. 1955.

[6] M. Gaster. The structure and behavior of separation bubbles. Technical report, ARC R&M No. 3595, March 1967.

[7] H. P. Horton. A semi-empirical theory for the growth and bursting of laminar separation bubbles. Technical report, ARC CP 1073, June 1967. (replaces ARC 29 185).

[8] M. M. O'Meara and T. J. Mueller. Laminar separation bubble characteristics on an airfoil at low Reynolds numbers. *AIAA Journal*, 25(8):1033–1041, Aug. 1987.

[9] C. Gleyzes, J. Cousteix, and J. L. Bonnet. Theoretical and experimental study of low Reynolds number transitional separation bubbles. In T. J. Mueller, editor, *Proceedings of*

the Conference on Low Reynolds Number Airfoil Aerodynamics, pages 137–152, University of Notre Dame, June 1985.

[10] M. Drela and M. B. Giles. Viscous-inviscid analysis of transonic and low Reynolds number airfoils. Technical report, AIAA-86-1786, 1986.

[11] R. H. Liebeck. Design and testing of an airfoil with positive pitching moment at low Reynolds numbers. Technical report, Douglas Aircraft Co., MDC 18893, Nov. 1982.

[12] M. Brendel and T. J. Mueller. Boundary layer measurements on an airfoil at low Reynolds numbers. Technical report, AIAA-87-0495, 1987b.

[13] M. Brendel and T. J. Mueller. Boundary layer measurements on an airfoil at a low Reynolds number in an oscillating free stream. Technical report, AIAA-87-1332, 1987a.

[14] F. Hsiao, C. Liu, and Z. Tang. Experimental studies of airfoil performance and flow structures on a low Reynolds number airfoil. Technical report, AIAA-87-1267, 1987.

UNSTEADY AERODYNAMICS OF WORTMANN FX63-137 AIRFOIL AT LOW REYNOLDS NUMBERS [1]

Andrew M. Wo[2] and Eugene E. Covert[3]
Department of Aeronautics and Astronautics
Massachusetts Institute of Technology
Cambridge, MA 02139

Abstract

Unsteady pressure data are discussed for a 2-D Wortmann FX63-137 airfoil at Reynolds numbers 125,000 and 400,000, $\alpha = 0^o$, and over a range of reduced frequencies, $k = \omega c/2U_\infty$, from 0.15 to 6.4. Unsteady boundary layer data near the region of laminar separation are also presented for Re = 125,000, k = 0.15 and 2.0. Surface pressure data show that the unsteady excitation imposed interacts most actively in the transition region with large amplitude and higher harmonics. Unsteady boundary layer results in the region of laminar separation reveal that the time mean unsteady velocity profile for both reduced frequencies tested are quite similar. The ensemble averaged profiles for $k = 0.15$ suggest that the unsteady boundary layer tends to be more attached when the free stream velocity is minimum. Data show that this effect is less pronounced at $k = 2.0$.

1 Introduction

Real life aerodynamics is time dependent by nature. For example, helicopter and gas turbine blades operate in the wakes of the previous blades, vortices shedding off the leading edge of a delta wing inducing velocities on the entire wing, and aerodynamic bodies may flutter causing very unsteady loading and unsteady shed vortex fields downstream.

Unsteadiness is expected at all Reynolds numbers. For example, Reynolds number associated with an aircraft subjected to unsteady motion may be high as 40

[1]This research was conducted under ONR grant N00014-85-K-0513.
[2]Graduate Research Assistant
[3]Professor and Chairman

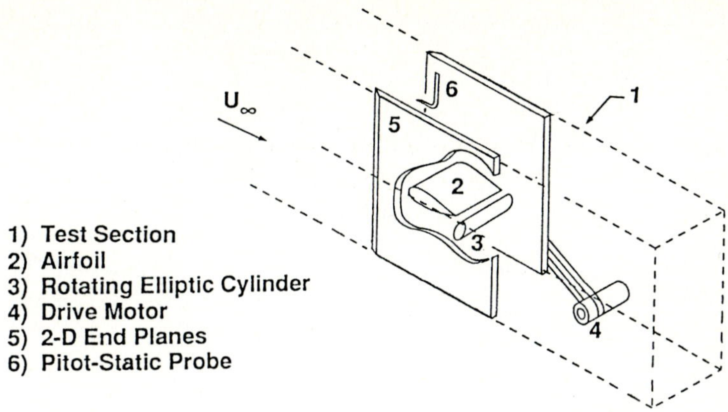

Figure 1: Test Configuration in Low Turbulence Wind Tunnel

million while Reynolds number of the flow associated with rotor-stator interaction, remotely piloted vehicles (RPV) and wind turbines may be as low as 100,000.

Some recent works has been done on the unsteady aerodynamics of low Reynolds number flow [1][2][3]. However, the unsteady behavior of the separation bubble is still considered not well understood especially related to the region of transition and flow reattaching as turbulent boundary layer. In an attempt to further study the complicated unsteady effects, in this paper, we will present steady and unsteady pressure data on the Wortmann FX63-137 airfoil at $\alpha = 0^o$, $Re = 125{,}000$ and 400,000, and over a range of reduced frequencies, $k = 0.15$ to 6.4. Boundary layer results, for $k = 0.15$ and 2.0, near the region of laminar separation are also discussed. We will draw basic observations on phenomena associated with the range of unsteady frequencies studied. Results presented are somewhat preliminary and further analyses are underway.

2 Experimental Apparatus and Unsteady Flow Field

The experimental setup is shown in figure 1. Airfoil is mounted between two end planes to provide two-dimensionality in the external flow. Tests were conducted in the M.I.T. Low Turbulence Wind Tunnel which is a modern tunnel having a maximum turbulence level of 0.05% in the free stream velocity range of 15 to 150 ft/s [4]. This level of turbulence has almost no effect on the critical Reynolds number of a sphere as measured by Dryden et al. [5] so we assumed that free stream turbulence had a negligible effect on our data. Effects of free stream turbulence in low Reynolds number unsteady had been studied by Wo and Covert [6]. Other

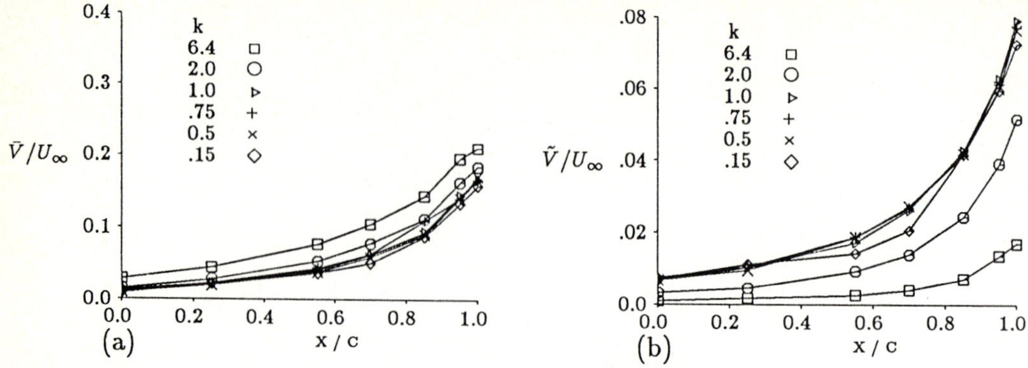

Figure 2: Induced Upwash: a)Time Mean and b)Fundamental Amplitude (Re = 125,000)

features of the apparatus have also been discussed in detail [7].

Airfoil tested was a modified Wortmann FX63-137. The Wortmann was modified in two ways. First, the last 1.1% of the airfoil was too thin to manufacture so it was truncated. The actual chord is 19.78 inches rather than the designed 20.00 inches. Second, near the trailing edge of the upper surface a bump was found in the coordinates. This could not be machined accurately, so the bump was smoothed.

The airfoil is instrumented with 38 centerline pressure taps. Two Setra model 237 (+/- 0.1 *psid*) capacitance type pressure transducers were mounted inside the airfoil to minimize the length of tubing needed hence maximizing the frequency response. Tygon tubings used were 0.05 inches I.D. with length of 10 and 12 inches depending on the distance of the pressure tap from the scannivalve. At the excitation frequency tested of 45 *Hz.*, corresponds to Re = 400000 and k = 6.4, the amplitude attenuation is only about 2%.

TSI 9100-7 two color LDV system was used to measure the unsteady boundary layer. The beam width at the measurement point is calculated to be 0.2mm and the half angle of beam crossing is 5.1 degrees. Frequency shifting using a Bragg cell allows measurement of reverse flow in the region of laminar separation bubble. Vaporized ethelene glycol is used to seed the flow field. Due to the proximity to the wall, seeding was insufficient for the v-component velocity hence only streamwise velocity is processed.

Our unsteady flow field is periodic external flow generated by rotating an ellipse downstream and below the trailing edge (see fig. 1). The induced upwash along the chordline without the airfoil was measured with a two component LDV system and the results are shown in figures 2a and 2b. The range of reduced frequencies is $0.15 \leq k \leq 6.4$ and Reynolds number are Re= 125,000 and 400,000. The time mean vertical velocity at Re= 125,000, figure 2a, shows $1/r^n$ type decay as would be expected since the time mean flow field can be modeled by flow around a

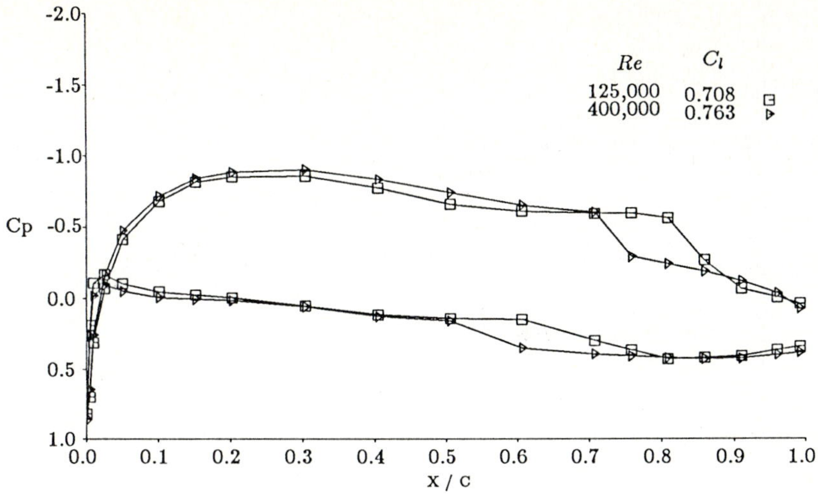

Figure 3: Wortmann Steady Pressure, $\alpha = 0^o$

rotating circular cyclinder. Data show that the higher the value for k, the higher the mean component of induced velocity. Figures 2b shows the periodic amplitude of vertical velocity as distributed along the chord at $Re=$ 125,000 and different values of k. The two important aspects in our unsteady flow field are that the unsteady excitation is greatest near the trailing edge position and the excitation is essentially constant phase since the flow field is incompressible; the free stream Mach number ranges from 0.01 to 0.03.

3 Results and Discussions

3.1 Steady and Unsteady Surface Pressure

Steady data tested at $Re = 125{,}000$ and 400,000, and $\alpha = 0^o$ are shown in figure 3. The most drastic difference is the location of the region of flow transition and subsequent reattachment for the two Reynolds numbers tested. Transition, being a strong function of Reynolds number, is delayed for $Re = 125{,}000$ compared to 400,000. On the upper surface the turbulent reattachment region is located near 90% chord and 75% chord for $Re = 125{,}000$ and 400,000 respectively. This suggests mixing of the free-stream flow with the transitional bubble is more vigorous at higher Reynolds number. Hence, the length of the transitional separation bubble is longer for $Re = 125{,}000$ than 400,000; the overall bubble lengths differ by 15% chord. Similar Reynolds number effect is also shown on the lower surface.

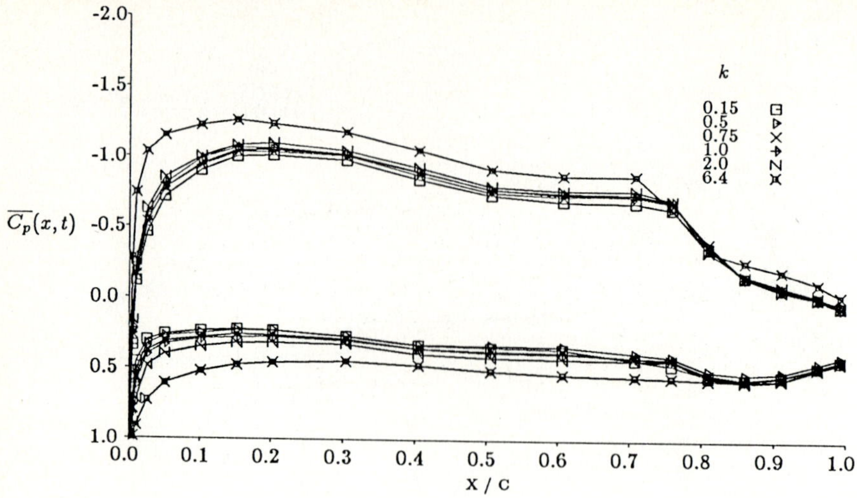

Figure 4: Wortmann Time Mean Pressure ($Re = 125{,}000$, $\alpha = 0^o$)

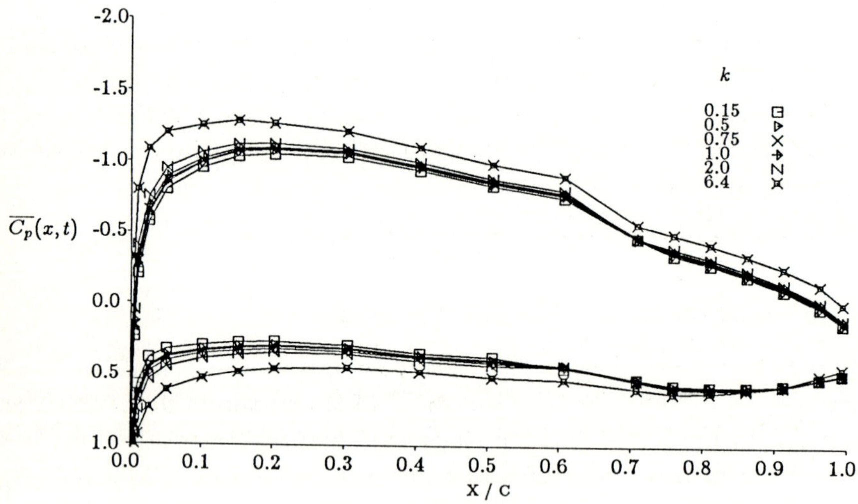

Figure 5: Wortmann Time Mean Pressure ($Re = 400{,}000$, $\alpha = 0^o$)

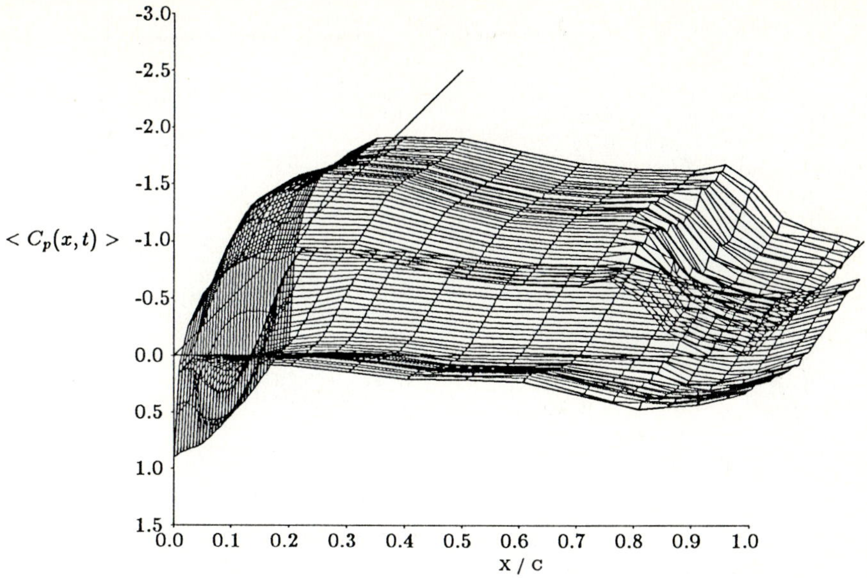

Figure 6: Wortmann Ensemble Averaged Pressure ($Re = 125{,}000$, $\alpha = 0^o$, $k = 0.15$)

Figures 4 and 5 show the chordwise time mean pressure for values of reduced frequencies tested. The most prominent effect is that the overall lift increases with reduced frequency. The mean upwash, figure 2a, that the airfoil sees increases with reduced frequency hence greater overall lift. Also note that the region of transition corresponds to the corner of the pressure distribution is more upstream than that in steady flow at the same Reynolds number. This is probably due to higher adverse pressure gradient which is the sum of the steady pressure gradient and the unsteady rectified pressure gradient. That is,

$$-\frac{1}{\rho}\overline{\frac{\partial p}{\partial x}} = U\frac{dU}{dx} + \overline{\tilde{u}\frac{\partial \tilde{u}}{\partial x}}$$

where $u(x,t) = U(x) + \tilde{u}(x,t)$.

Figures 6 an 7 plots the ensemble averaged surface pressure for $k = 0.15$ and 2.0. Each plotting surface corresponds to the spatial and temporal pressure distributions. Zero phase or initial time is defined when the ellipse is at horizontal position. The temporal coordinate plotted extends over a completed unsteady period.

At low reduced frequency, $k = 0.15$, upstream of the transition region on both surfaces, figure 6 shows a dominant frequency which is the excitation frequency. However, near the region of transition, higher harmonics can be found. This effect is especially pronounced when the boundary layer is undergoing reattachment as turbulent boundary layer. For $k = 2.0$, figure 7, the upper surface is almost constant in time since the unsteady excitation decreases with reduced frequency

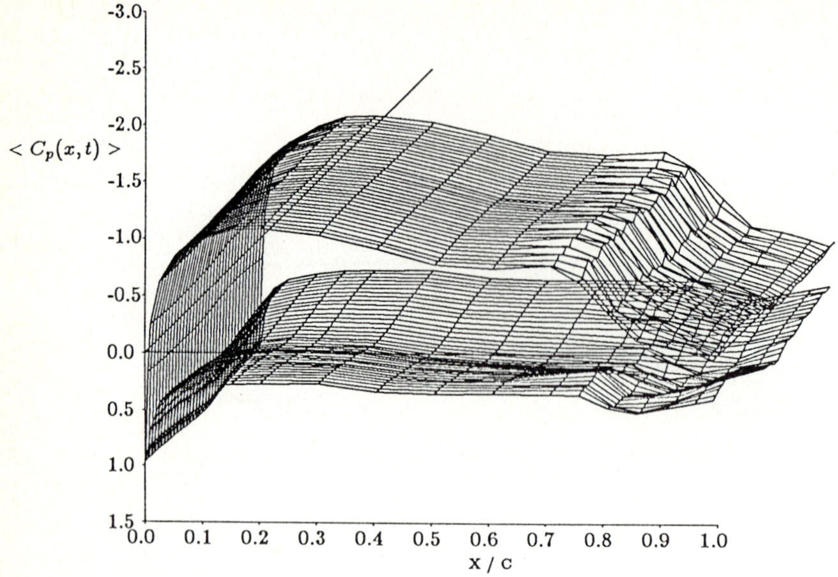

Figure 7: Wortmann Ensemble Averaged Pressure ($Re = 125{,}000$, $\alpha = 0^o$, $k = 2.0$)

(see figure 2b).

Although data suggest the pressure jump near the trailing edge is tending towards zero yet at 99% physical chord position, mean $\triangle C_p$ of the data presented is about 0.25. Comparison with steady airfoil viscous-inviscid calculation by Drela [8] show that at 0.99% chord $\triangle C_p$ is about 0.20. Since the Wortmann is an aft-loaded airfoil, the main contribution to the discontinuity in $\triangle C_p$ observed is due to lack of data beyond the 0.99% chord.

In order to better understand the flow behavior near the aft portion of the upper surface, figures 8a and 8b show the evolution of the pressure amplitude, without time mean, for several chordwise positions and reduced frequencies. At $k = 0.15$, figure 8a, several trends are clearly shown. First, at 40% to 70% chord, the pressure consists essentially the fundamental unsteady frequency. In the region of transition, 75% to 80% chord, not only the amplitude of the fundamental harmonic is higher but also a much higher frequency scale appears. This trend is consistant at all reduced frequencies tested with the effect being less at higher reduced frequencies.

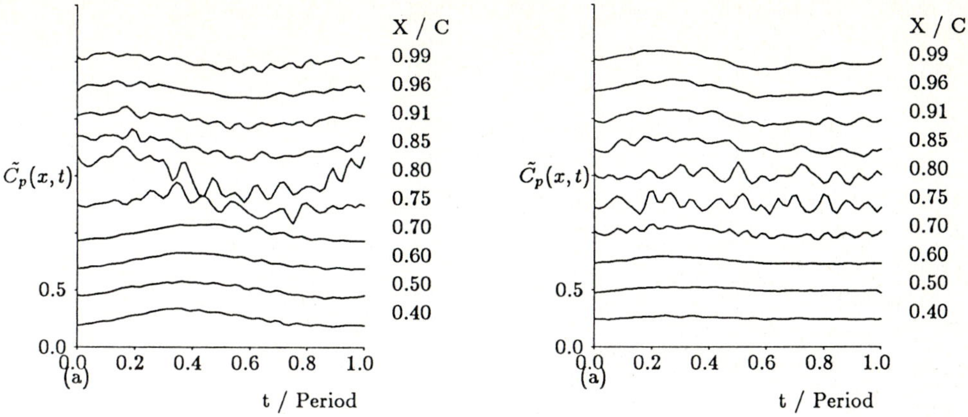

Figure 8: Wortmann C_P Amplitude, $\alpha = 0^o$, $Re = 125{,}000$, a)$k = 0.15$ and b)$k = 2.0$

3.2 Steady and Unsteady Boundary Layer

Boundary layer flow on the upper airfoil surface at zero degree angle of attack was measured using a two component LDV system (see section 2). The external flow condition is $Re = 125{,}000$, $k = 0.15$ ($0.33\ Hz.$) and $k = 2.0$ ($4.4\ Hz.$). The chordwise measurement locations correspond to the region of laminar separation as is suggested by the pressure plateau of figure 3 for steady flow and figure 4 for the time mean unsteady flow. The emphasis of this particular study is on the unsteady boundary layer behavior in the region of laminar separation.

Figure 9 presents the steady and time mean unsteady streamwise boundary layer flow. For steady flow, vanishing of wall shear occurs closest to 52% chord. This agrees with a Falkner-Skan calculation by the first author. Flow visualization by Brendel [9] also confirmed this fact. In general, the steady flow value is larger than that of the mean unsteady flow. It should be noted that the steady flow data were taken without the ellipse mounted. The mean unsteady profiles are almost identical for the two cases of reduced frequencies tested. Figure 10 shows the time mean displacement and momentum thicknesses. Hence, these two figures suggest that the time mean boundary layer properties at these two reduced frequencies are quite similar.

The main difference in the boundary layer flow is clearly shown by the ensemble averaged, or phase-locked, data in figures 11 and 12. Since the free stream unsteady velocity amplitude is different for the two reduced frequencies tested, each velocity profile plotted is normalized by the edge velocity at that particular ensemble. Figure 11 shows the velocity profiles at $k = 0.15$ when the free stream is maximum and minimum. In the region near wall, viscous stresses cause the unsteady amplitude to vanish as expected. Further away from the wall, velocity

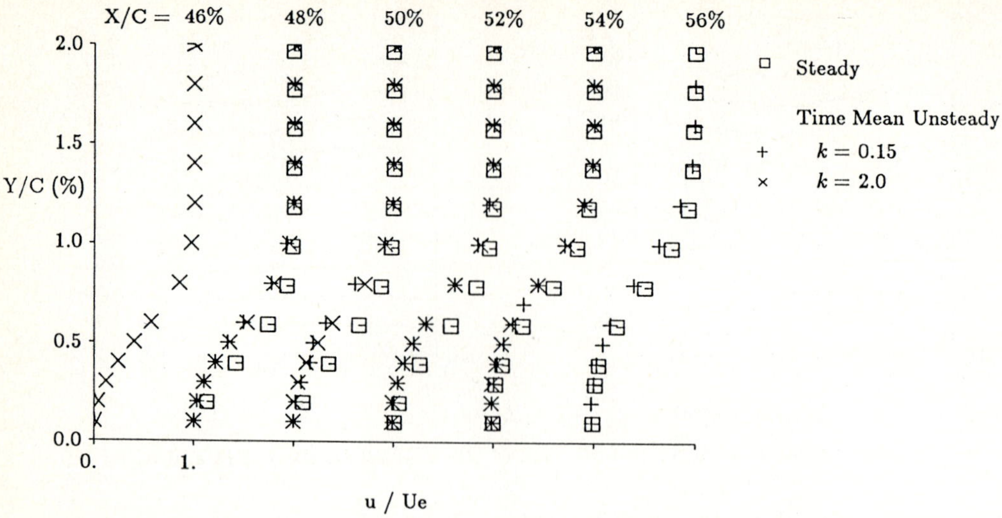

Figure 9: Steady and Time Mean Unsteady Velocity Profile ($Re = 125{,}000$, $\alpha = 0^o$)

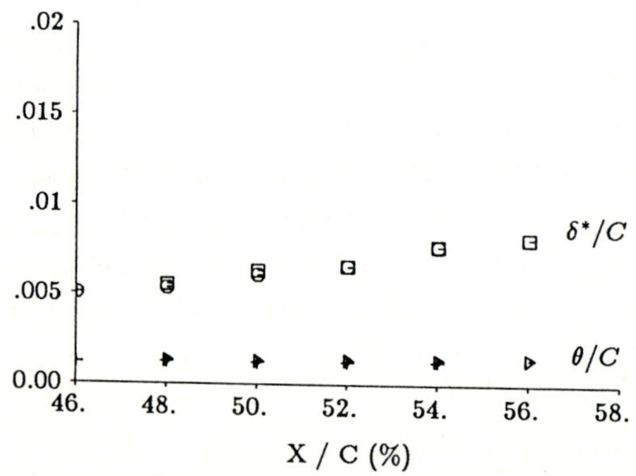

Figure 10: Time Mean Unsteady Displacement and Momentum Thicknesses ($Re = 125{,}000$, $\alpha = 0^o$, $k = 0.15$ and 2.0)

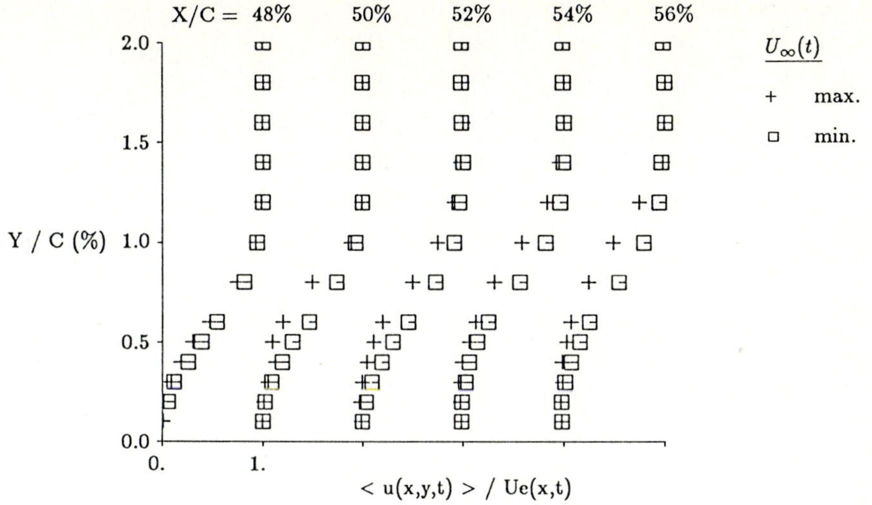

Figure 11: Ensemble Averaged Velocity Profile ($Re = 125{,}000$, $\alpha = 0^o$, $k = 0.15$)

variation increases. At the outer portion of the boundary layer, the velocities are normalized to be unity. Note that the largest variation in the unsteady amplitude, at each chordwise location, occurs near the inflection region of the time mean velocity profile.

The boundary layer is more attached when the free stream flow is minimum. Perhaps this is not too surprising since at $k = 0.15$ the imaginary part of the Theodorsen's function [10] almost reaches a minimum hence the flow is in the mid-frequency range and some phase shift is expected. This effect is also found by Brendel and Mueller [1] for Wortmann FX63-137 at $Re = 100{,}000$, $\alpha = 7^o$, $k = 0.35$. At $k = 2.0$, figure 12 suggests that the ensemble averaged profiles approach that of the time mean profile in figure 9. This means that the unsteady boundary layer flow at $k = 2.0$ can be modeled by the high frequency limiting case which approaches steady flow.

4 Concluding Remarks

Steady and unsteady surface pressure data are presented for the Wortmann FX63-137 airfoil at $Re = 125{,}000$ and $400{,}000$, $\alpha = 0^o$, and reduced frequencies of 0.15 to 6.4. The focus of the discussion is placed on the data obtained at $Re = 125{,}000$ since most features of unsteady flow measured can be illustrated. Results of the upper surface steady and unsteady boundary layer near the laminar separation region are also presented.

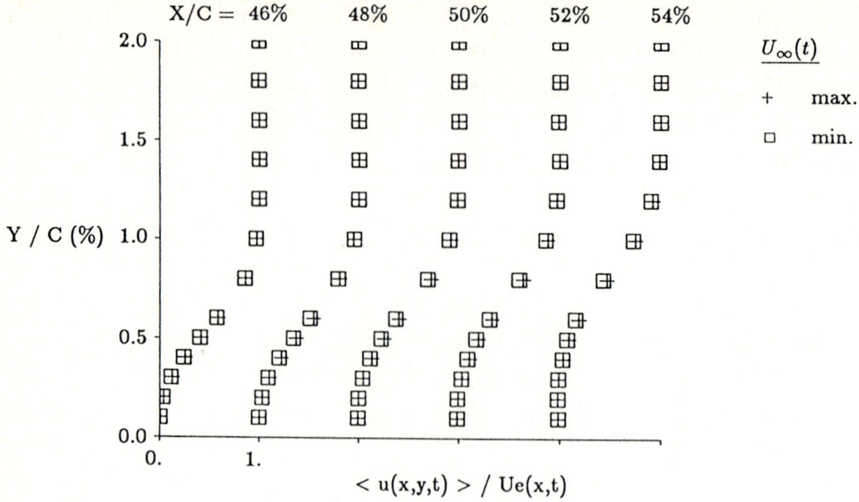

Figure 12: Ensemble Averaged Velocity Profile (Re = 125,000, $\alpha = 0^o$, $k = 2.0$)

Ensemble averaged unsteady surface pressure reveal that the laminar portion of the airfoil basically response to the periodic excitation imposed. However, the transition region exhibit large amplitude and higher harmonics. Unsteady boundary layer results in the region of laminar separation show that the time mean profile, normalized by the edge velocity, seems to be independent of reduced frequency, at least for $k = 0.15$ and 2.0. At reduced frequency of 0.15, figure 11, the maximum variation of ensemble averaged velocity is located near the inflection point. It is believed that this fact contributes to both the spatial and temporal growth of instability waves which eventually leads to transition to turbulent flow. The ensemble averaged profiles suggest that the unsteady boundary layer tends to be more attached when the free stream velocity is a minimum hence a phase shift across the boundary layer is present. For $k = 2.0$, figure 12, the variation in phase for the ensemble averaged profile is less than that at $k = 0.15$. Analyses are underway to correlate the boundary layer data with the pressure data.

5 References

[1] Brendel, M. and Mueller, T.J., "Boundary Layer Measurements on an Airfoil at a Low Reynolds Number in an Oscillating Freestream," *AIAA Journal*, Vol. 26, No. 3, Mar. 1988, pp. 257-263.

[2] Ratelle, J.P., "Unsteady Boundary Layer Flow Reversal in a Longitudinally Oscillating Flow," USAF-FJSRL, SRL-TR-78-0006, 1978.

[3] Saxena, L.S., "An Experimental Investigation of Oscillating Flows Over an Airfoil," Ph.D Thesis, Illinois Institute of Technology, 1977.

[4] Mangus, John, "Measurement of Drag and Bursting Frequency Downstream of Tandem Spanwise Ribbons in a Turbulent Boundary Layer," M.I.T SM Thesis in Aeronautics and Astronautics, 1984.

[5] Dryden, H.L., Schubauer, G.B., Mock, W.C. Jr., and Skramstad, H.K., "Measurements of Intensity and Scale of Wind-Tunnel Turbulence and their Relation to the Critical Reynolds Number of Spheres," *NACATN*581, 1937.

[6] Wo, A.M. and Covert, E.E., "On the Effects of Wind Tunnel Turbulence on Steady and Unsteady Airfoil Characteristics," *4th* Symposium on Numerical and Physical Aspects of Aerodynamic Flows, Jan 16-19, 1989.

[7] Lober, P.F. and Covert, E.E., "Unsteady Airfoil Pressures Produced by Periodic Aerodynamic Interference," *AIAAJournal*, Vol.20, No.9, Sept. 1982, pp. 1153-59.

[8] Drela, M., "Two-Dimensional Transonic Aerodynamic Design and Analysis Using the Euler Equations," Ph.D Thesis, M.I.T., 1985.

[9] Brendel, M., "Experimental Study of the Boundary Layer on a Low Reynolds Number Airfoil in Steady and Unsteady Flow," Ph.D. Thesis, University of Notre Dame, 1986, p. 267

[10] Bisplinghoff, R.L., Ashley, H., and Halfman, R.L., *Aeroelasticity*, Addison Wesley Publishing Co., Reading, Mass., 1955, p. 277

A METHOD TO DETERMINE THE PERFORMANCE OF LOW-REYNOLDS-NUMBER AIRFOILS UNDER OFF-DESIGN UNSTEADY FREESTREAM CONDITIONS

Helen L. Reed and Beth A. Toppel
Mechanical and Aerospace Engineering
Arizona State University
Tempe, Arizona 85287-6106

Abstract

A procedure is described to analyze the boundary layer on airfoils experiencing unsteady flight conditions and to predict the changes in the performance characteristics during off-design. The method predicts the flow in the boundary-layer region near the separation bubble using the incompressible Navier-Stokes equations with boundary conditions from inviscid and laminar boundary-layer solutions. The rate at which the separation bubble develops and decays is of primary interest in this study. Unsteady surface-pressure-coefficient distributions and velocity profiles are presented. Immediate-future work will include application of this technique to two state-of-the art airfoils, the ASM-LRN-010 (see Reference [1]) and Wortmann FX63-137 airfoil sections with chord Reynolds numbers of 250,000 and 500,000 and operating under light dynamic stall conditions.

1. Introduction

Separation is a critical consideration in a variety of technical situations, including the design of low-speed aircraft. Separation depends on many factors including unsteadiness, adverse pressure gradient, freestream turbulence intensities and spectra, compressibility, and three-dimensionality of the flow. In this effort the effects of adverse pressure gradient and unsteadiness are studied by computation. In order to isolate these effects, the test flow consists of a laminar, incompressible boundary-layer flow with a predictable, well-described unsteady freestream pressure gradient applied. Cases for impulsively (gusts and lolls) and sinusoidally fluctuating pressure distributions are possible in the computations. The response of the boundary layer to these fluctuations is observed, in particular, separation location and structure and the rate at which separated regions develop and decay. If these zones develop slowly, there may not be serious problems; but if the off-design unsteady conditions persist sufficiently long, the buildup of these separated regions could be fatal to a low-Reynolds-number flight vehicle (Reynolds number 50,000-500,000).

2. Computational Approach

Direct numerical simulations are playing an increasingly important role in the investigation of separation; the literature is growing, especially recently. This trend is likely to continue as considerable progress is expected towards the development of new, extremely powerful supercomputers. In such simulations, the full Navier-Stokes equations are solved directly by employing numerical methods, such as finite-difference or spectral methods. The direct simulation approach is widely applicable since it avoids many of the restrictions that usually have to be imposed in theoretical models.

The principal goal of this work is the development of a computational method for the simulation of the process of unsteady separation on the upper surface of a wing. By varying the freestream conditions (oscillations, ramps), the intermittent off-design conditions low-Reynolds-number flight vehicles are likely to encounter, such as variations in angle-of-attack or the passage of a gust, can be simulated and the rate at which separated regions develop and decay can be studied. In this way it is also possible to verify the length and time scales predicted by triple-deck theory. The existence of such a method will provide a tool which will enable computation to complement theoretical and experimental contributions to further the understanding of the physics of these flows and, ultimately, will provide a tool for the prediction and modeling of these flows.

Basically, this model involves the use of the unsteady inviscid solution for the airfoil shape as an edge boundary condition for a Navier-Stokes solution near the wall. The inviscid solution is described in Section 2.1 and the viscous solution is described in Section 2.2. Once the viscous solution is obtained, the pressure distribution is determined at the wall and the "actual" lift and drag histories for the airfoil under unsteady, off-design conditions can be obtained.

2.1 Unsteady Inviscid Method

A vortex panel method is used to compute the inviscid unsteady flow around a given airfoil shape. The unsteady Kutta condition is used as well as conservation of vorticity and the fact that each point of the wake must be aligned in the force-free position. Changes in angle of attack or speed cause vorticity to be shed from the trailing edge of the airfoil. This vorticity is then convected along aft of the airfoil according to the local velocity vector.

The resulting velocity field is then applied as the edge condition for the Navier-Stokes calculation presented in the next section.

2.2 Unsteady Viscous Method

The governing dimensionless, two-dimensional Navier-Stokes equations in terms of vorticity and stream function are:

Vorticity Transport Equation

$$\frac{\partial \omega}{\partial t} + u\frac{\partial \omega}{\partial x} + v\frac{\partial \omega}{\partial y} = \frac{1}{Re}\left(\frac{\partial^2 \omega}{\partial x^2} + \frac{\partial^2 \omega}{\partial y^2}\right) \qquad (1)$$

Stream Function Equation

$$\omega = \frac{1}{Re}\left(\frac{\partial^2\psi}{\partial x^2} + \frac{\partial^2\psi}{\partial y^2}\right) \tag{2}$$

u, v Component of Velocity

$$u = \frac{1}{\sqrt{Re}}\frac{\partial\psi}{\partial y}, \quad v = \frac{-1}{\sqrt{Re}}\frac{\partial\psi}{\partial x} \tag{3}$$

where Re is the chord Reynolds number defined by

$$Re = \frac{\overline{U}_o\,\overline{c}}{\overline{\upsilon}} \tag{4}$$

The overbars represent dimensional quantities. $\overline{U}_o$ is the undisturbed velocity, $\overline{c}$ is the airfoil chord and $\overline{\upsilon}$ is the kinematic viscosity. In equations (1) through (3), ω is the vorticity, ψ is the stream function, x is the streamwise coordinate, and y is the transverse coordinate. The present study has assumed that the body surface in the region near the separation bubble can be represented by a flat plate. The domain of solution is shown in Figure 1.

2.2.1 Boundary Conditions

Upstream Boundary

It was pointed out in Reference [2] that for oscillating airfoils the front stagnation point is moving in time, but the boundary layer calculations need to be started at the instantaneous positions of the stagnation points which are known from the inviscid calculations of Section 2.1. Since this paper only considered small amplitude oscillations, it was assumed that the unsteady effects in the boundary layer in the region near the stagnation point are negligible. Therefore the upstream boundary conditions were obtained by the solution of the two-dimensional laminar boundary layer equations on a quasi-steady basis. The modification to include stagnation point motion as in Reference [2] is straightforward.

The two dimensional boundary layer equations are:

$$\frac{\partial u}{\partial s} + \frac{\partial v}{\partial y} = 0 \tag{5}$$

$$u\frac{\partial u}{\partial s} + v\frac{\partial u}{\partial y} = U_e\frac{dU_e}{ds} + \upsilon\frac{\partial^2 u}{\partial y^2} \tag{6}$$

with the boundary conditions

$$u(s, y = 0) = v(s, y = 0) = 0 \tag{7}$$

$$u(s, y \rightarrow \infty) = U_e(s) \tag{8}$$

$$u(s = 0, y) = u_i \tag{9}$$

Introducing the Falkner-Skan-like variables

$$\xi = s, \eta = y \left[\frac{\xi}{ReU_e(\xi)} \right]^{-1/2} \tag{10}$$

$$u = U_e(\xi) f'(\xi,\eta) \tag{11}$$

$$V(\xi,\eta) = v \left[\frac{Re\xi}{U_e(\xi)} \right]^{1/2} \tag{12}$$

where s is the dimensionless distance measure along the surface of the airfoil, y is the dimensionless distance normal to the surface and the prime indicates differentiation with respect to η.

We rewrite equations (5) and (6) as

$$\xi \frac{\partial f'}{\partial \xi} + \beta f' + \frac{\eta}{2}(\beta-1)\frac{\partial f'}{\partial \eta} \frac{\partial V}{\partial \eta} = 0 \tag{13}$$

$$f'\xi \frac{\partial f'}{\partial \xi} + \bar{V}\frac{\partial f'}{\partial \eta} = [1-f'^2]\beta + \frac{\partial^2 f'}{\partial \eta^2} \tag{14}$$

where

$$\beta(\xi) = \frac{\xi}{U_e} \frac{dU_e}{d\xi} \tag{15}$$

and

$$\bar{V} = V + \frac{1}{2}\eta f' (\beta-1) \tag{16}$$

The boundary conditions transform to

$$f'(\xi,0) = 0, \; V(\xi,0) = 0 \tag{17}$$
$$f'(\xi,\eta \to \infty) = 1 \tag{18}$$
$$f'(0,\eta) = f_i'(\eta) \tag{19}$$

Equations (13) through (19) are solved using the Crank-Nicolson method described in Reference [3] with initial conditions supplied by the Falkner-Skan profile of $\beta_i = 1$, the plane stagnation-point solution.

Solid Wall

The no slip conditions are used at the wall. The vorticity at the wall is calculated by a second-order accuracy formula used by Briley [4]. Briley fit the values of ψ near the wall to a third-order Lagrange interpolating polynomial, with the no slip condition

$$\frac{1}{\sqrt{Re}} \frac{\partial \psi}{\partial y} \Big|_{wall} = u_{wall} = 0 \tag{20}$$

The result is

$$\omega_{wall} = \frac{-7\psi_{i,2} + 8\psi_{i,2} - \psi_{i,3}}{2\Delta y^2} \tag{21}$$

Briley also pointed out that the use of Equation (21) differencing requires a special formula for the u-component of velocity at the row of points adjacent to the wall which differs from the standard three-point central-difference form. The result is

$$u_{i,2} = \frac{1}{\sqrt{Re}} \frac{\partial \psi}{\partial y} \Big|_{i,2} = \frac{1}{\sqrt{Re}} \left[\frac{-5\psi_{i,1} + 4\psi_{i,2} + \psi_{i,3}}{4\Delta y} \right] \tag{22}$$

Outer-Flow Boundary

At this boundary the u-component of velocity is prescribed by the unsteady inviscid calculations of Section 2.1. The flow is required to be irrotational by setting the vorticity equal to zero. The boundary is located at a distance approximately equal to twice the boundary layer thickness.

Downstream Boundary

At this boundary the flow is assumed to return to a laminar boundary layer. Therefore, the boundary conditions are given by the boundary layer approximations,

$$\frac{\partial^2 \psi}{\partial x^2} = \frac{\partial^2 \omega}{\partial x^2} = 0 \tag{23}$$

2.2.2 Computational Scheme

The computational scheme used to solve the vorticity transport equation is an alternating-direct-implicit (ADI) method for parabolic equations based on that of Peaceman and Rachford [5]. The advancement of Equation (1) over a time step Δt is accomplished by splitting a finite-difference algorithm into a sequence of one-dimensional operations. As a result, only triadiagonal systems of linear algebraic equations need to be solved by the Thomas algorithm. The advancement of Equation (1) over Δt is accomplished in the following two steps:

$$\frac{\omega_{i,j}^{n+1/2} - \omega_{i,j}^{n}}{\Delta t/2} = u_{i,j} \frac{\delta \omega}{\delta x}^{n+1/2} - v_{i,j} \frac{\delta \omega}{\delta y}^{n} + \frac{1}{Re} \left(\frac{\delta^2 \omega}{\delta x^2}^{n+1/2} + \frac{\delta^2 \omega}{\delta y^2}^{n} \right) \tag{24a}$$

$$\frac{\omega_{i,j}^{n+1} - \omega_{i,j}^{n+1/2}}{\Delta t/2} = u_{i,j} \frac{\delta \omega}{\delta x}^{n+1/2} - v_{i,j} \frac{\delta \omega}{\delta y}^{n+1} + \frac{1}{Re} \left(\frac{\delta^2 \omega}{\delta x^2}^{n+1/2} + \frac{\delta^2 \omega}{\delta y^2}^{n+1} \right) \tag{24b}$$

where the central-difference operators δ and δ^2 are defined by

$$\frac{\delta f}{\delta x} \Big|_{i,j} = \frac{f_{i+1,j} - f_{i-1,j}}{2\Delta x} + 0(\Delta x^2) \tag{25}$$

$$\frac{\delta^2 f}{\delta x^2}\Big|_{i,j} = \frac{f_{i+1,j} - 2f_{i,j} + f_{i-1,j}}{\Delta x^2} + 0(\Delta x^2) \tag{26}$$

This two-dimensional method is unconditionally stable and has a truncation error of $0(\Delta t^2, \Delta x^2, \Delta y^2)$.

The solution of steady elliptic equations by iteration is analogous to solving a time-dependent problem to an asymptotic steady state [6]. Therefore, the stream function equation was solved by the ADI method for elliptic equations based on that of Peaceman and Rachford [5]. The iteration is first performed implicitly by row and then by columns using the following difference equations:

$$\frac{\delta^2 \psi^{q+1/2}}{\delta x^2} - \rho\psi_{i,j}^{q+1/2} = Re\cdot\omega_{i,j} + \rho\psi_{i,j}^{q} - \frac{\delta^2 \psi}{\delta y^2}^{q+1/2} \tag{27a}$$

$$\frac{\delta^2 \psi^{q+1}}{\delta y^2} - \rho\psi_{i,j}^{q+1} = Re\cdot\omega_{i,j} + \rho\psi_{i,j}^{q+1/2} - \frac{\delta^2 \psi^{q+1/2}}{\delta x^2} \tag{27b}$$

where q is the iteration index and ρ is a positive parameter chosen to accelerate the convergence of the method.

A summary of the algorithm which advances Equations (1) through (3) from n to the n+1 time level is as follows:

(i) Obtain boundary values form ψ^{n+1}, ω^{n+1} or their derivatives. These values are known except for ω^{n+1} at the wall.

(ii) Solve (24a) and (24b) to obtain ω^{n+1}.

(iii) Solve for ψ^{n+1} using the iterative scheme given by (27a) and (27b).

(iv) Compute u^{n+1} and v^{n+1} from ψ^{n+1} using three-point central differencing.

(v) Calculate new values for ω^{n+1} at the wall using (21).

(vi) Return to Step (ii) and repeat this procedure until the values of ω^{n+1} at the wall computed in Step (v) converge.

The maximum normalized change in the wall velocity was used as convergence criteria in Steps (iii) and (vi).

3. Results

Calculations are presented for two freestream velocity distributions representing two different Howarth retarded flows; that is, two different constant downstream velocities are specified for the problem

$$u_e(x)=1-x \quad x_1 \leq x \leq x_{end} \tag{28a}$$

$$u_e(x)=1-x_{end} \quad x_{end}<x\leq x_2 \qquad (28b)$$

For these test cases, the upstream boundary conditions are given by Howarth boundary-layer solutions [7] and initial conditions are given by Blasius solutions. A solution is accepted as the steady-state solution when the maximum normalized change in vorticity during a time step fell within a value of 1×10^{-4} in accordance with [4]. For comparison in Figures 2 and 3, distributions of shearing stress along the wall are given for values of x_{end} equal to 0.16296 and 0.20058, respectively. In Figure 2, the flow is near separation, whereas in Figure 3, the freestreaam velocity produces separation and reattachment.

Figure 4 shows a vorticity profile downstream of the separation bubble at x=0.40122. Briley points out that the presence of this additional inflection point in the vorticity profile indicates that diffusion has not had a sufficient distance over which to smooth the new vorticity being generated at the wall with the vorticity being convected and diffused from further upstream.

To validate the downstream boundary conditions, the u-component of velocity at the downstream boundary was compared to the Blasius profile. This comparison is shown in Figure 5 for the solution near separation and in Figure 6 for the solution which contains the separation bubble. Moreover, when the domain is lengthened in the downstream direction, the solution within the domain is unaffected. All the above results are in good agreement with Briley's.

To test the proposed method, we consider the case of oscillations about an established steady Howarth retarded flow. Wall shear stress distribution is shown in Figure 7 where the freestream velocity is given by

$$u_e(x)=1-(1-.1\cos(2\pi t))x \quad x_1+1\leq x\leq x_{end} \qquad (29a)$$

$$u_e(x)=1-(1-.1\cos(2\pi t))x_{end} \quad x_{end}<x<x_2 \qquad (29b)$$

and is denoted by a solid line. The initial conditions are obtained by the steady-state solution, t=0, where the freestream velocity distribution is denoted by a line of asterisks. Five complete cycles were required for convergence. Figure 8 displays the velocity profiles, the "separation" line, the edge of the region of reversed flow and the boundary-layer thickness for a Reynolds number of 20,833. Surface pressure coefficients are shown in Table 1 for inviscid and viscous flows. The viscous pressure coefficients were calculated by evaluating the x-momentum equation at the wall. The result is

$$\frac{\partial C_p}{\partial x} = \frac{2}{\sqrt{Re}} \frac{\partial \omega}{\partial y}\bigg|_{wall} \qquad (30)$$

Future work will include application of this technique to the ASM-LRN-010 and Wortmann FX 63-137 airfoil sections. Inviscid surface pressure distributions for the ASM-LRN-010 (Figure 9) at various angles-of-attack are shown in Figures 10-13. These solutions and their unsteady counterparts will be applied as outer-flow boundary conditions in unsteady calculations as above.

4. Acknowledgments

The authors thank the Office of Naval Research for support under N000014-86-K-0066.

5. References

1. Pfenninger, W., Vemuru, C.S., Mangalam, S.M. and Evangelista, R., "Design of Low Reynolds Number Airfoils--11," AIAA-88-3766, pp. 1-15, 1988.

2. Geibler, W., Carr, L.W. and Cebeci, T., "Viscous/Inviscid Interaction Procedure for High-Amplitude Oscillating Airfoils," ICAS-88-5.5.1, pp. 766-778, 1988.

3. Panton, R.L., Incompressible Flow, John Wiley and Sons, Inc., 1984.

4. Briley, W.R., "A Numerical Study of Laminar Separation Bubbles Using the Navier-Stokes Equations," J. Fluid Mechanics, Vol. 47, Part 4, pp. 713-736, 1970.

5. Peaceman, D.W. and Rachford, H.H., Jr., "The Numerical Solution of Parabolic and Elliptic Differential Equations," J. Soc. Indust. Applied Mathematics, Vol. 3, No. 1, pp. 28-41, 1985.

6. Frankel, S.P., "Convergence Rates of Iterative Treatments of Partial Differential Equations," Math Tables and Other Aids to Computation, Vol. 4, pp. 65-75, 1950.

7. Howarth, L., "On the Solution of the Laminar Boundary-Layer Equations," Proc. Roy. Soc. A, Vol. 164, 547, 1938.

TABLE 1

x	C_p (inviscid)	C_p (viscous)
.05	.098	.098
.075	.16	.17
.11	.22	.23
.14	.28	.29
.17	.34	.32
.20	.40	.35
.23	.39	.37
.26	.39	.40
.30	.39	.40
.33	.39	.40

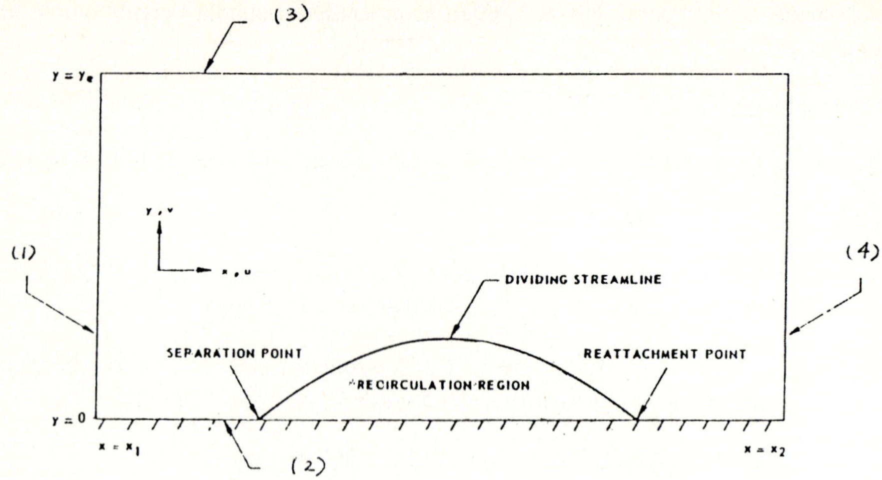

Figure 1. Schematic of solution domain for separation bubble (Briley [4])

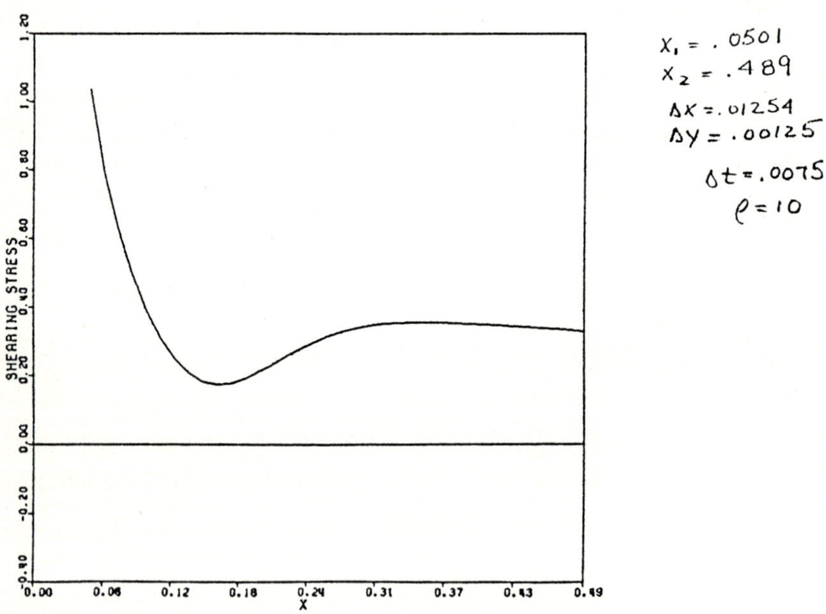

Figure 2. Wall shearing stress for $x_{end} = 0.16296$

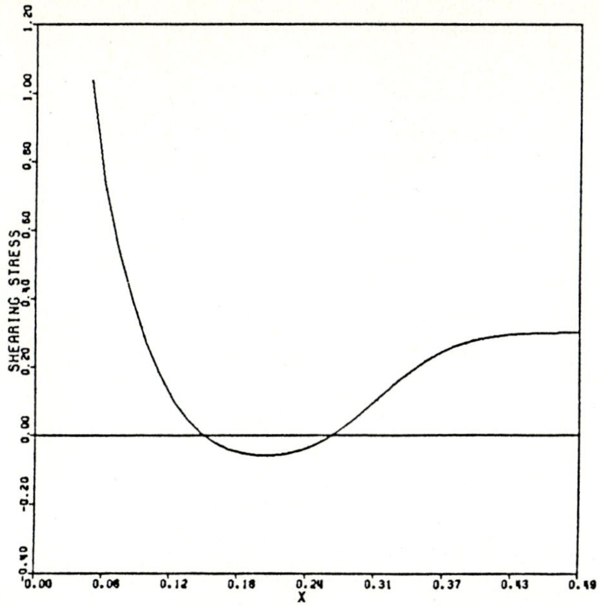

$X_1 = .0501$
$X_2 = .489$
$\Delta X = .01254$
$\Delta Y = .00125$
$\Delta t = .0075$
$e = 10$

Figure 3. Wall shearing stress for $x_{end} = 0.20058$

$ETA = \frac{Y}{2}\sqrt{\frac{Re}{X}}$, $RE = 20833$

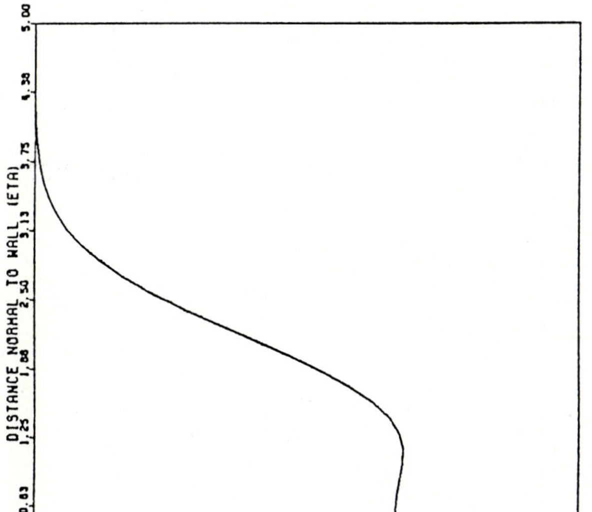

Figure 4. Vorticity profile at x = 0.40122

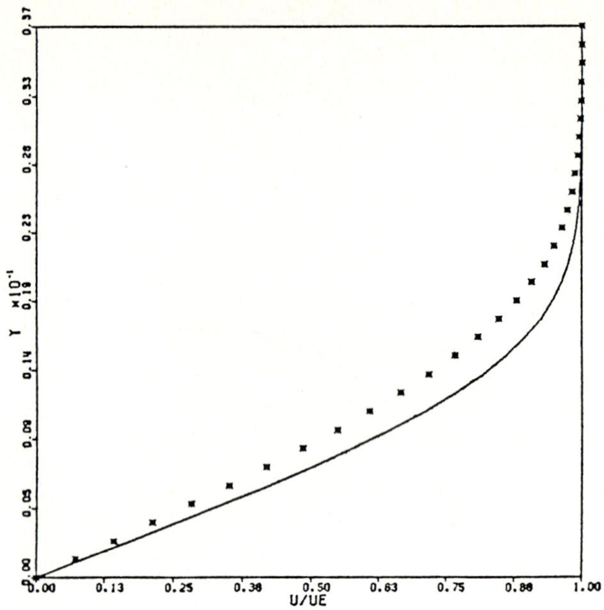

Figure 5. Comparison of velocity profile at downstream boundary with Blasius profile for solution near separation

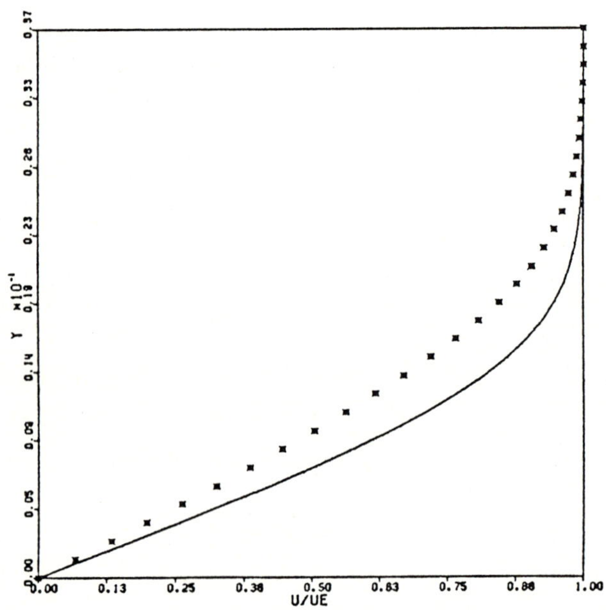

Figure 6. Comparison of velocity profile at downstream boundary with Blasius profile for solution with separation bubble

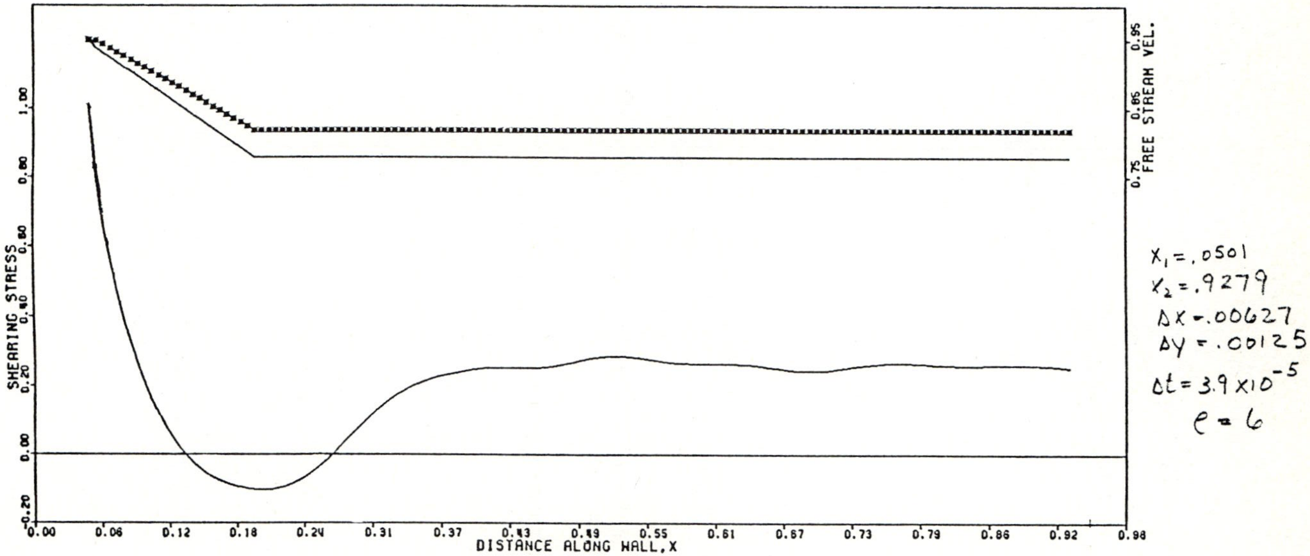

Figure 7. Wall shearing stress for t=4.5

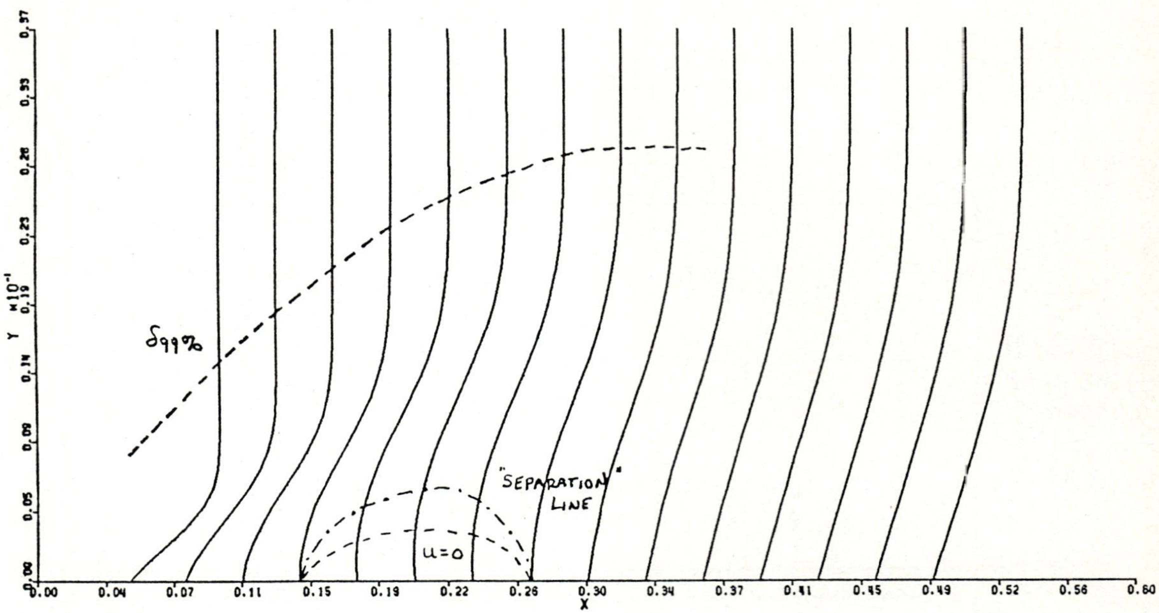

Figure 8. Velocity profiles with "separation" line, edge of reversed flow region, and boundary layer thickness indicated

AN UNSTEADY MODEL OF ANIMAL HOVERING.

P. Freymuth, F.J. Seiler Research Laboratory, USAF Academy, Colorado Springs (on sabbatical leave from the University of Colorado, Boulder, Colorado 80309-0429)

Abstract

An unsteady, two-dimensional model of animal hovering is experimentally investigated. The model consists of an airfoil which executes combined plunging and pitching sinusoidal motions in still air. While the kinematics is rather simple, the model has sufficient flexibility to qualitatively capture various hover modes of insects, birds and fishes. By constructive use of dynamic stall vortices the model creates a hover-jet indicative of the high thrust known for many insects.

Introduction

Unsteady aerodynamics principles play an important role in animal hovering, where the weight of the animal is suspended in still air by the actions of the wings. In contrast, steady state principles seem sufficient to model the mean thrust during sustained forward flight, where according to Freymuth (1988) trailing edge vortex shedding during wing flapping is the dominant flow feature. A contribution to the hovering problem by Maxworthy (1981) showed on the other hand that leading edge separation is dominant during the complex clap and fling of chalcid wasps. Similarly, Somps and Luttges (1985) and Luttges (1989) showed the occurrence of leading edge separation for hovering dragonflies. According to the sketches by Weis-Fogh (1973-1975) and Ellington (1984) high angles of attack seem to occur for other hovering insects as well which makes leading edge separation likely. Unfortunately, leading edge separation is incompatible with steady state principles and this and the need for simplicity provides the overriding motivation to propose a basic unsteady model. Many hovering insects attain lift coefficients too high for steady state principles which further strengthens the need for a basic unsteady model.

The Basic Model

Consider a thin, flat-plate airfoil in still air which executes combined plunging and pitching motions described by

$$h = h_a \sin 2\Pi ft \qquad (1)$$

$$\alpha = \alpha_o + \alpha_a \sin (2\Pi ft + \phi) \qquad (2)$$

where h is a linear translation in horizontal direction and with amplitude h_a, f is frequency, t is time, α is the pitch angle with respect to the horizontal, α_o is the mean pitch angle, α_a is the pitch amplitude and ϕ is the phase difference between pitching and plunging. Pitching is around the half chord axis although this requirement can be relaxed.

Dimensionless parameters of this basic, two-dimensional, unsteady model are: α_o, α_a, ϕ, the plunge amplitude normalized with the chord length c of the airfoil h_a/c and a Reynolds number $R_f = 2\Pi \, fh_a \, c/v$, where v is the kinematic viscosity. Allowing for adjustment of these parameters keeps the model quite flexible. The basic model could be extended in many directions at the expense of added complexity.

Modes of Hovering

A strong air jet is generated by vigorous hovering which exerts a thrust on the airfoil in opposite direction. Not all parameter combinations produce strong hover-jets but two vigorous hover-modes were initially identified based on intuition and animal observation:

Mode 1 or the "water-treading mode" characterized by $\alpha_o = 0°$ and $\phi = 90°$ generates a hover-jet directed upward as sketched in Fig. 1. Leading and trailing edges switch their roles during one oscillation cycle. This mode resembles the action of the ventral fins of fishes for stable positioning in water. Since the experimental apparatus obstructs the view in downward direction the jet is intentionally directed upward. Switching the jet downward requires a sign switch for ϕ.

Mode 2 or the "normal hovering mode" sketched in Fig. 2 is characterized by $\alpha_o = 90°$ and $\phi = -90°$. In this mode leading and trailing edges do not switch their roles. This mode resembles the hovering of humming birds and most flying insects.

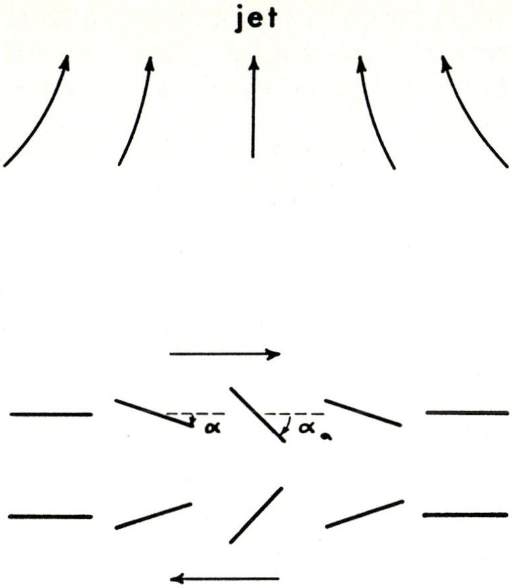

Fig. 1 Sketch of combined translating-pitching motions of the airfoil for one cycle of mode 1 hovering.

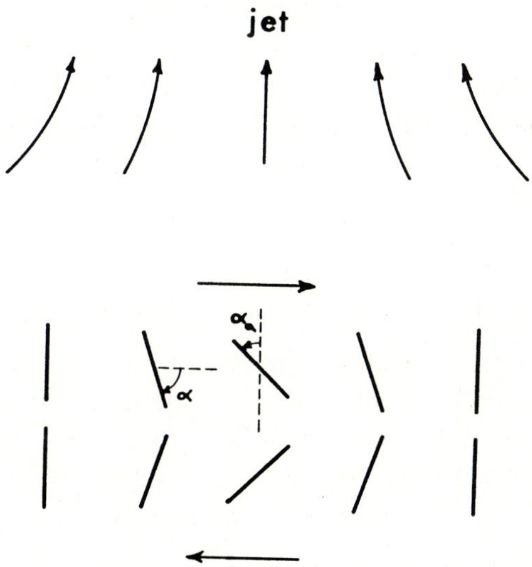

Fig. 2 Sketch of combined translating-pitching motions of the airfoil for one cycle of mode 2 hovering.

Subsequently another class of hovering was discovered and termed mode 3 or "dragonfly mode". It is characterized by an oblique mean pitch angle α_o while orthogonality of ϕ is still optimal. In this mode the hover-jet is directed oblique to the plunge or stroke plane as has also been observed for hovering dragonflies by Norberg (1975) and for hover and fruit flies by Weis-Fogh (1973).

If the airfoil pitches around a pivot point eccentric to the half chord point, for instance around the leading edge another hover-mode should be possible without the need for plunging. This mode 4 or "oriental fan mode" needs future investigation.

Bennet et al, (1975) observed some thrust generation in still air for an airfoil in pure plunging motion, due to asymmetry between round leading and sharp trailing edges of their profile which may be termed mode 5.

Experimental Apparatus

A flat airfoil with chord $c = 2.54$cm, span $l = 30$cm and a thickness of 1.6mm is mounted on a sliding frame, where it executes combined plunging and pitching motions by means of an assembly of wheels, rods and a pitch lever as shown in Fig. 3. By rotation of the pitch wheel against the slave wheel ϕ can be adjusted. By a rotation of the airfoil against the pitch lever α_o can be adjusted. Plunge and pitch amplitudes can be adjusted by inserting the drive and pitch rods into appropriate holes in the wheels. The rotation frequency f of the 40W dc-motor can be regulated between 1 Hz and 6 Hz as measured by means of a reed switch in conjunction with an electronic counter.

Plexiglas plates were vertically and stationary mounted slightly above and one chord length inboard of the tips of the airfoil to avoid three-dimensional effects which unknowingly plagued some of the initial and preliminary results by Freymuth (1989). Flow visualization by means of the titanium tetrachloride method described by Freymuth (1985) was an important tool of flow investigation. Airfoil thrust was determined from the mean dynamic pressure profile of the hover-jet which was measured with a Pitot tube in conjunction with a pressure transducer. The apparatus allowed the investigation of modes 1 to 3.

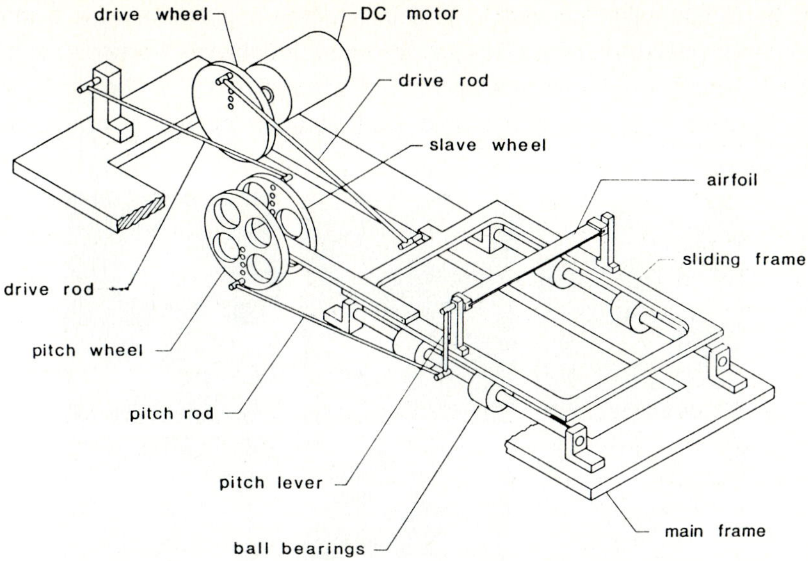

Fig. 3 Sketch of hover apparatus.

Results of Visualization

Vortex visualizations of vigorous hover-jets in side view will be shown for all three hover modes. Each figure is composed of a sequence of movie frames representing one oscillation cycle. Frames are ordered into columns from top to bottom and columns are ordered from left to right, with a time Δt between consecutive frames. Parameter details will be listed in the figure captions.

An example of a mode 1 hover-jet photographed slightly from the left is presented in Fig. 4. In the first frame the airfoil passes through the $h = 0$ position from left to right where it reaches its maximum pitch angle. The first column shows the shedding of a counterclockwise rotating trailing edge vortex as well as the development of a clockwise rotating leading edge stall vortex. In column 2 this stall vortex rolls partly over the left edge of the airfoil and then gets severed during the leftward motion of the airfoil. The severed parts of the stall vortex then amalgamate with the vorticity produced and shed from the right edge of the airfoil forming a single clockwise vortex by the end of column 2. An analogous process creates the counter-clockwise vortex but would need observation slightly from the right. Vortex severing has been previously discovered by Ziada and Rockwell (1982) in a different context. On a larger scale, counterclockwise vortices to the left and clockwise vortices to the right of the frame center form a vortex street representative of an upward moving hover-jet. Constructive use of dynamic stall vortices played a prominent role in creating the jet.

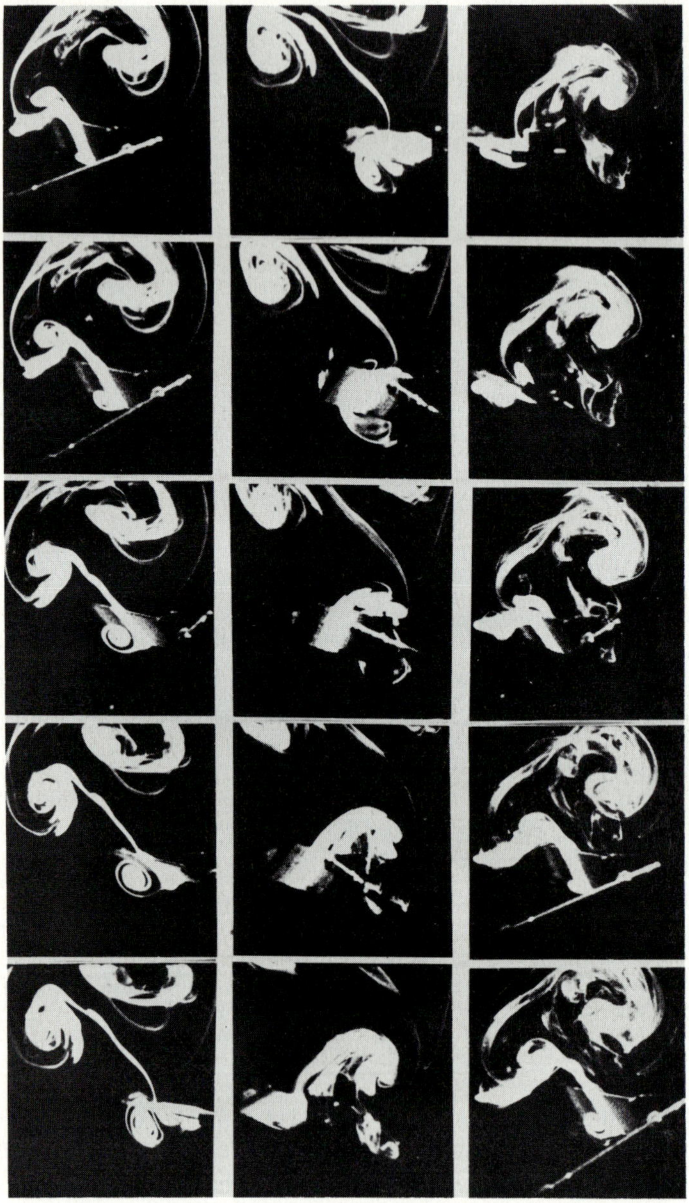

Fig. 4 Closeup sequence of mode 1 hover-jet. $\alpha_o = 0°$, $\alpha_a = 66°$, $\phi = 90°$, $h_a/c = 1.5$, $R_f = 400$, $\Delta t = 1/16$ sec.

As Fig. 5 shows, amalgamation of stall vortices with trailing edge vortices have also been observed in mode 2 where a vigorous hover-jet develops. In this mode vortex severing has not been observed.

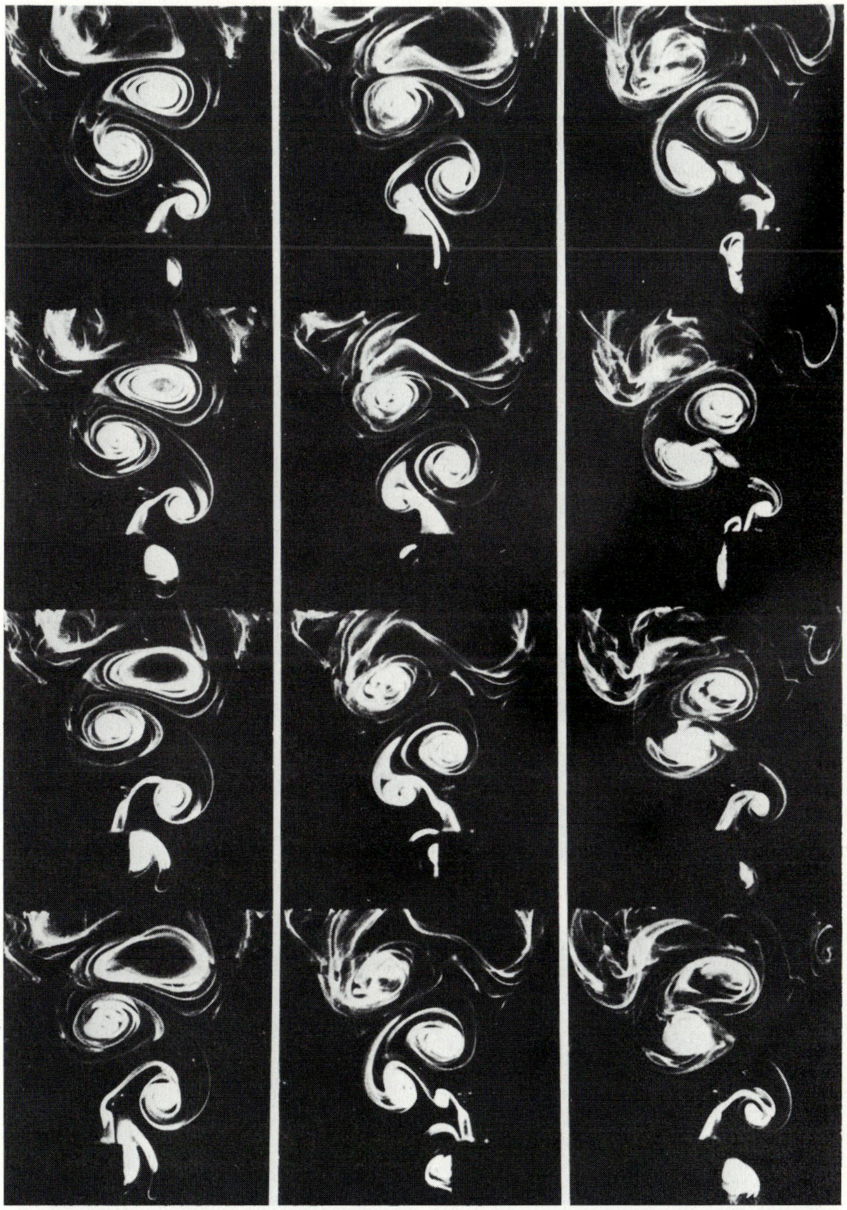

Fig. 5 Sequence of mode 2 hover-jet. α_o = 90°, α_a = 25°, ϕ = -90°, h_a/c = 0.5, R_f = 340, Δt = 1/32 sec.

Fig. 6 shows the development of an oblique hover-jet by operating the airfoil in mode 3. Again the plunge or stroke oscillation is horizontal. Stall vortices play a constructive role also in the formation of this jet. By changing the mean pitch

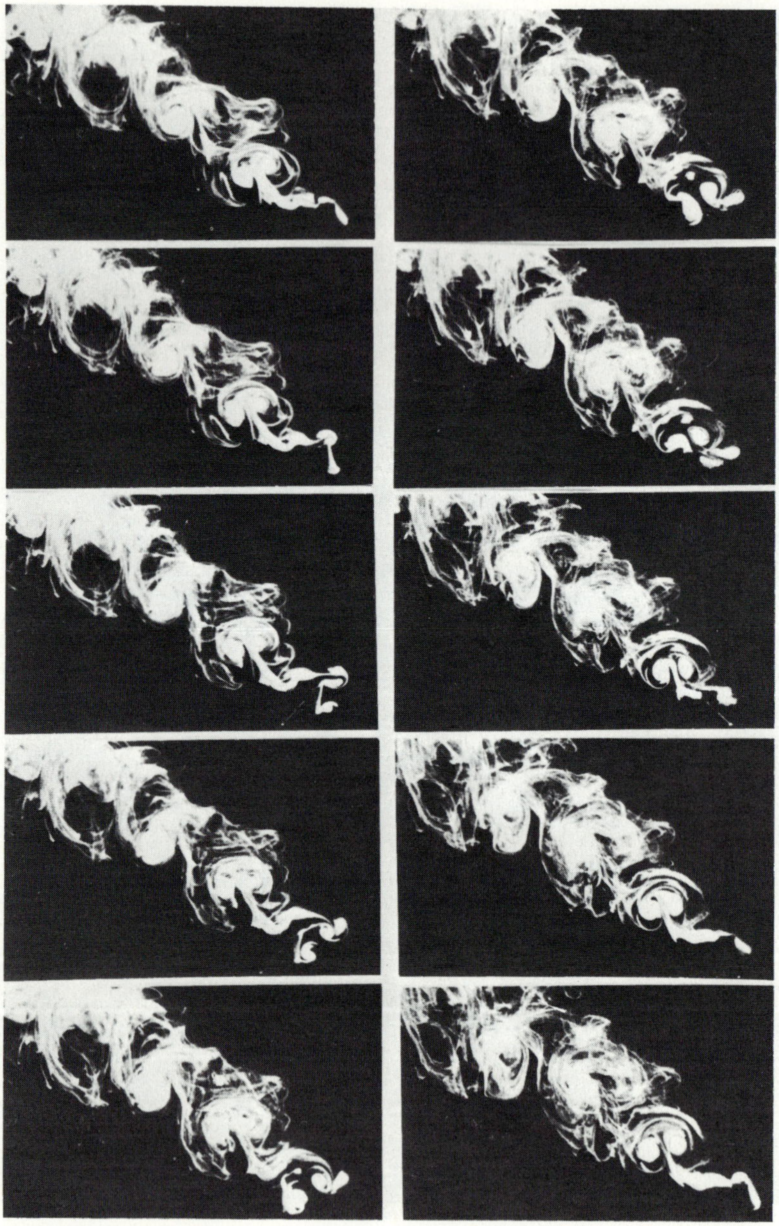

Fig. 6 Sequence of mode 3 hover-jet. $\alpha_o = 60°$, $\alpha_a = 45°$, $\phi = -90°$, $h_a/c = 1$, $R_f = 400$, $\Delta t = 1/16$ sec.

angle α_o the jet can be vectored in any desired direction. Crudely speaking, for every degree of change in angle of attack α_o there is a 2° change in the direction of the hover jet.

Measurement of the Thrust Coefficient

To characterize the time averaged thrust T of the airfoil a thrust coefficient C_T is defined by:

$$C_T = \frac{T}{\frac{\varrho}{2}\overline{V_t^2}\, c\, \ell} \qquad (3)$$

where

$$\overline{V_t^2} = 0.5\,(2\pi f h_a)^2 \qquad (4)$$

is the mean square speed of linear airfoil oscillation, ϱ is the air density and $l \gg c$ is the airfoil span. From the momentum theorem

$$T = \varrho \ell \int_{-\infty}^{\infty} \overline{V^2}\, dx \qquad (5)$$

where $\overline{V^2}$ is the mean square velocity in the hover-jet measured with a Pitot tube along the coordinate x across the jet. Thus

$$C_T = \frac{\int_{-\infty}^{\infty} \overline{V^2}\, dx}{(\pi f h_a)^2 c} \qquad (6)$$

In actual animal hovering thrust is directed upward and thus C_T may also be considered a lift coefficient of the hovering airfoil. For the determination of C_T according to eq. 6 the mean square velocity profiles where measured four chord lengths above the airfoil and graphically integrated. C_T was determined for some parameter range of h_a/c and α_a for modes 1 and 2 at a Reynolds number $R_f = 1700$. Figs. 7 and 8 sum up the results. Most prominent are the high thrust coefficients which reach into the range 5 to 7 when the airfoil is properly tuned to rather low values of h_a/c and which reflect constructive use of the dynamic stall vortices during jet formation. Thrust coefficients in the oblique mode 3 were not investigated in detail but a few checks indicated lesser values in the range 2 to 3.

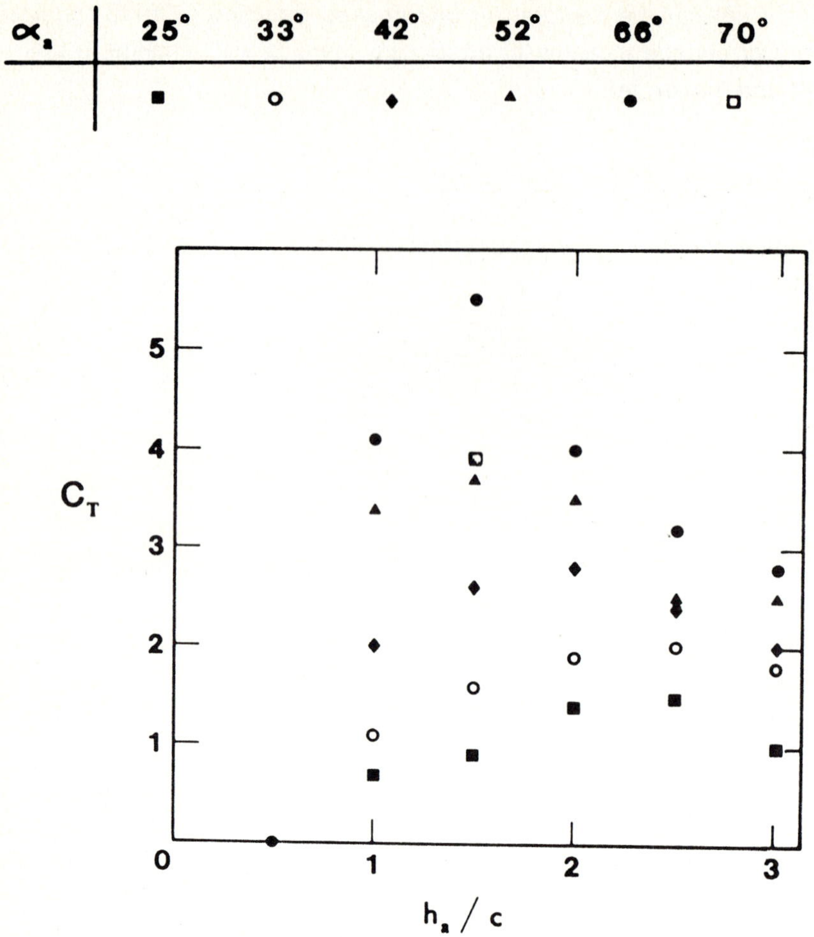

Fig. 7 Thrust coefficient C_T versus h_a/c and α_a, mode 1.

Concluding Remarks

It has been shown that a relatively simple unsteady model of animal hovering has sufficient flexibility to qualitatively cover most hovering techniques available to animals. The high thrust coefficients achieved by the constructive use of dynamic stall vortices strengthens the value of the model. This model seems to be the first device which deliberately creates and constructively uses dynamic stall vortices.

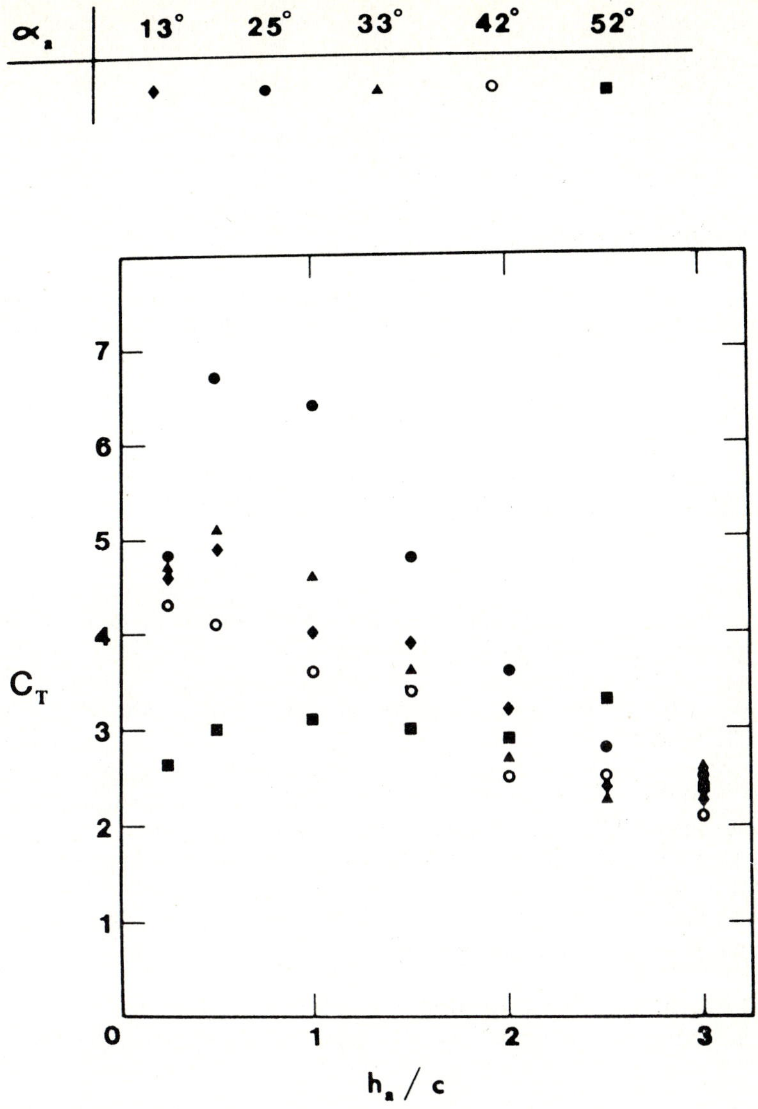

Fig. 8 Thrust coefficient C_T versus h_a/c and α_a, mode 2. (For the smallest value of h_a/c the standard airfoil was substituted by one with twice the chord length.

References

Ellington, C.P. (1984). The aerodynamics of hovering insect flight. III. Kinematics. Phil. Trans. Roy. Soc. London, Vol. B 305, pp. 41-78.

Freymuth, P., Bank, W., Palmer, M. (1985). Use of titanium tetrachloride for visualization of accelerating flow around airfoils. Flow Visualization III, pp. 99-105, Hemisphere Publishing Corp.

Freymuth, P. (1988). Propulsive vortical signature of plunging and pitching airfoils. AIAA J. Vol. 26, pp. 881-883. For additional details see Paper AIAA-88-0323.

Freymuth, P. (1989). Thrust generation by an airfoil in hovering mode. F.J. Seiler, Research Laboratory Report FJSLR-TR-89-0002.

Luttges, M.W. (1989). Accomplished insect fliers. Frontiers in Fluid Mechanics. M. Gad-el-Hak (ed), Springer-Verlag, in press.

Maxworthy, T. (1981). The fluid dynamics of insect flight. Ann Rev. Fluid Mech., Vol. 13, pp. 329-350.

Norbert, R.A. (1975). Hovering flight of the dragonfly. Aeschna Juncea L., kinematics and aerodynamics. Swimming and Flying in Nature. Wu, Y., Brokaw, C., Brennen, C. (eds.) Vol. II, pp. 7863-781, Plenum Press.

Somps, C., Luttges, M. (1985). Dragon fight—Novel uses of unsteady separated flows. Science, Vol. 228, pp. 1326-1329.

Weis-Fogh, T. (1973). Quick estimates of flight fitness in hovering animals, including novel mechanisms for list repoduction. J. Exp. Biol., Vol. 59, pp. 169-230.

Weis-Fogh, T. (1975). Flapping flight and power in birds and insects, conventional and novel mechanisms. Swimming and Flying in Nature, Wu, Y., Brokaw, C., Brennen, C. (eds.), Vol. 11, pp. 729-762, Plenum Press.

Ziada, S., Rockwell, D. (1982). Vortex-leading edge interaction. J. Fluid Mech., Vol. 118, pp. 79-107.

ADDENDUM

Hovering in Modes 4 and 5.

To round out our view of hovering in still air, modes 4 and 5 were explored in passing.

Fig. 9 visualizes an upward moving mode 4 hover-jet. The airfoil chord was c=10cm, its span l=86 cm and pitching was around the leading edge with pitch amplitude of $\alpha_a = 30^{\circ}$ at a frequency of 2 Hz. Clearly, a vigorous hover-jet develops. Fig. 10 shows limited results obtained for the thrust coefficient

$$C_T = \frac{\int_{-\infty}^{\infty} \overline{V^2}\, dx}{\alpha_a^2 (\pi f c)^2 c} \qquad (7)$$

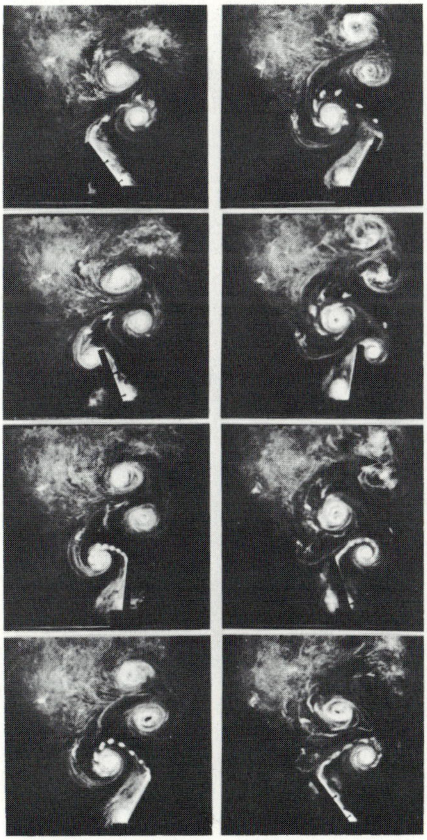

Fig. 9 Sequence of mode 4 hover-jet. $\alpha_a = 30^{\circ}$, $\Delta t = 1/16$ sec., $f = 2$ Hz.

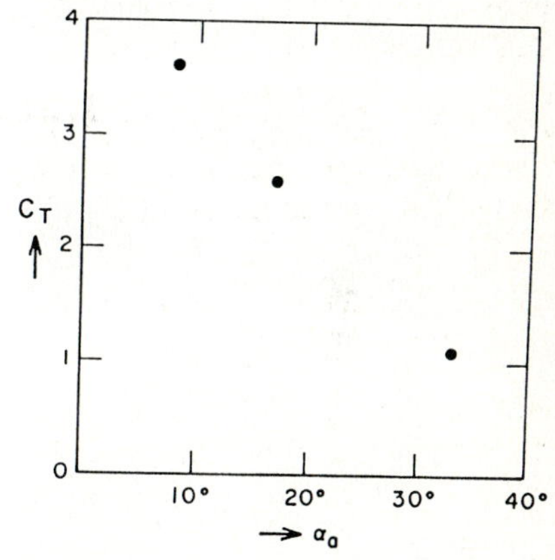

Fig. 10 Thrust coefficient C_T versus α_a, mode 4.

as function of α_a, suggesting usefulness of the device as a fan.

The mode 5 hovering discovered by Bennet et al. (1975) generates thrust by an airfoil in pure plunge provided the airfoil has a sharp trailing edge and a well rounded leading edge. In Fig. 11, vortex development near the upward pointing sharp trailing edge of an NACA 0015 airfoil which executes horizontal plunge motions is depicted. The airfoil chord is c=15 cm and the span is l=91 cm.

The dimensionless plunge amplitude was ha/c=0.2. An upward moving vortex street develops which quickly becomes turbulent. The airfoil then was remounted such that the round leading edge showed upward and the airfoil was plunged again. Visualizations near the leading edge are shown in Fig. 12. The flow is rather stagnant in this case. At considerable larger plunge amplitudes both leading and trailing edges develop vortex streets in opposite directions, rendering the device ineffective for thrust generation.

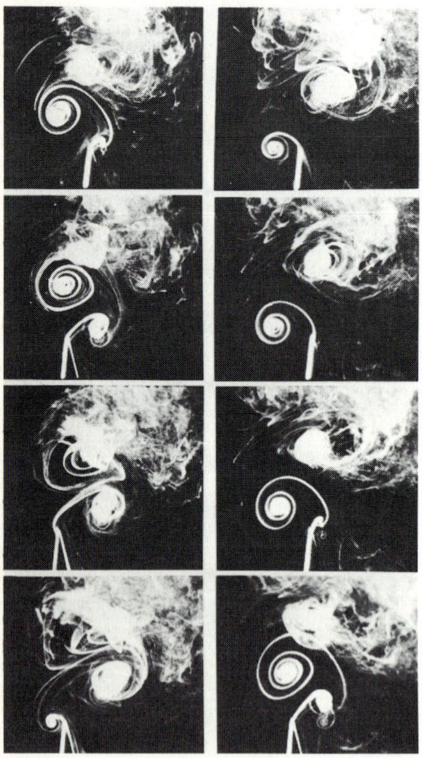

Fig. 11 Mode 5, hover-jet above the sharp trailing edge, $h_a/c = 0.2$, $\Delta t = 3/32$ sec., $f = 1.8$ Hz.

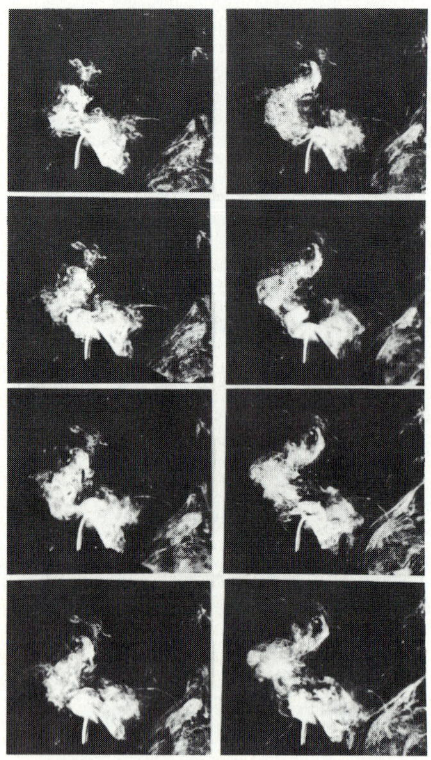

Fig. 12 Mode 5, view at the round leading edge, $h_a/c = 0.2$, $\Delta t = 3/32$ sec., $f = 1.8$ Hz.

Acknowledgements

The visiting professorship at the F. J. Seiler Research Laboratory was sponsored by the Air Force Systems Command. An AFOSR Grant F 49620-84-0065, H. Helin program manager, provided additional assistance. Photocredit goes to W. Bank, the experimental apparatus were built by R. Hatfield and R. Tarasewicz provided the illustrations. Discussions with R. Reilman have been helpful.

Bennett, A. G., Obye, R. C., Jeglum, P. A. (1975). Ornithopter aerodynamic experiments. Swimming and Flying in Nature, Wu, Y., Brokaw, C., Brennen, C. (eds.), Vol. II, pp. 986-1000, Plenum Press.

CONTROL OF LOW-REYNOLDS-NUMBER AIRFOILS: A REVIEW

MOHAMED GAD-EL-HAK
Department of Aerospace and Mechanical Engineering
University of Notre Dame
Notre Dame, Indiana 46556
U.S.A.

ABSTRACT

Among the modern goals of external flow control are separation postponement, lift enhancement, transition delay or advancement, and drag reduction. These objectives are not necessarily mutually exclusive. For low-Reynolds-number lifting surfaces, where the formation of a laminar separation bubble may have a dominant effect on the flow field, the interrelation between the above goals is particularly salient, presenting an additional degree of complexity. The present article is an overview of passive and active techniques used to control a low-Reynolds-number boundary layer to achieve an improved performance. A unified view based on a vorticity framework will be presented to explain many of the available or contemplated control methods. Among the control techniques to be considered are wall suction/injection, shaping, heat transfer through the surface, introduction of a foreign substance into the boundary layer, fixed or moving geometric modifications, and turbulence manipulation. Among the practical considerations that will be reviewed for these devices are their cost of construction and operation, complexity, and potential trade-off 's or penalties associated with their use.

1. INTRODUCTION

Insects, birds and bats have perfected the art of flight through millions of years of evolution. Man's dream of flying dates back to the early Greek myth of Daedalus and his son Icarus, but the first successful heavier-than-air flight took place a mere 86 years ago. Today, the Reynolds numbers for natural and man-made fliers span the amazing range from 10^2 to 10^9; insects being at the low end of this spectrum and huge airships occupying the high end (Carmichael, 1981).

The function of the airfoil section on those fliers is to produce lift. Inevitably, viscous effects, compressibility effects and the finite span of the lifting surface all ensure that drag is also produced. A thrust must be generated by some sort of a power plant to overcome this streamwise resistance to the motion. The lift-to-drag ratio is a measure of the effectiveness of the airfoil. In general, this ratio is very low at low Reynolds numbers and improves with increases in this parameter. As shown in Figure 1, taken from McMasters and Henderson (1980), the maximum $[C_L/C_D]$ improves dramatically in the range of Reynolds numbers of 10^4–10^6. Below 10^4, typical of insects and small model airplanes, the boundary layer around the lifting surface is laminar. Stalling in this case is caused by an abrupt separation of the laminar flow near the leading edge as the angle of attack is increased to modest values. The maximum lift is limited and the drag increases significantly when the lifting surface stalls.

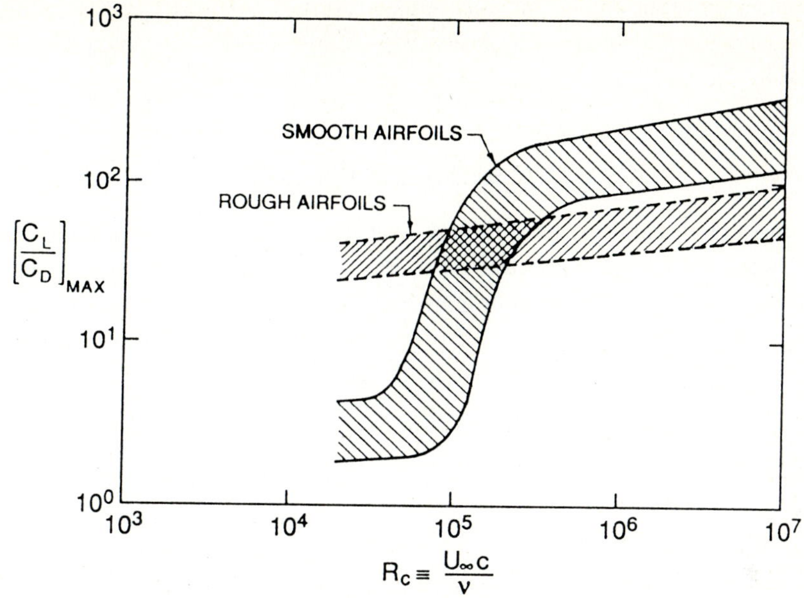

Figure 1. Airfoil Performance as a Function of Chord Reynolds Number. (from McMasters and Henderson, 1980).

For Re > 10^6, typical of large aircraft, boundary layer transition to turbulence usually takes place ahead of the theoretical laminar separation point. A turbulent boundary layer can negotiate quite severe adverse pressure gradients without separation, and this kind of lifting surface often experiences a trailing-edge stall at relatively high angles of attack. The stall is preceded by a movement of the separation point forward from the trailing edge with increasing incidence (McCullough and Gault, 1951).

In the range of Reynolds numbers of 10^4–10^6, termed low Reynolds number for the purpose of this article (Lissaman, 1983), many complicated phenomena take place within the boundary layer. Separation, transition and reattachment could all occur within a short distance and dramatically affect the performance of the lifting surface. The laminar separation bubble that commonly forms in this range of Reynolds numbers plays an important role in determining the boundary layer behavior and the stalling characteristics of the airfoil (Tani, 1964). As indicated in Figure 1, the maximum lift-to-drag ratio for a smooth airfoil increases by two orders of magnitude in this Reynolds number regime.

The skilled designer has available to him a variety of passive and active techniques to effect a beneficial change in the complex flow field that characterize this intermediate range of Reynolds numbers. Roughness and shaping are among the simplest passive methods to ensure flow attachment beyond a critical angle of attack and, thus, an improved performance. Wall transpiration or heat transfer provides examples of active control methods to improve the lift-to-drag ratio.

The object of this paper is to survey available and contemplated flow control methods particularly suited for low-Reynolds-number airfoils. Section 2 gives a brief overview of the fluid dynamics of these lifting surfaces. The different control goals and their interrelations are summarized in Section 3. The governing equations are recalled in the following section. Sections 5 through 7 details the methods for separation/reattachment, transition, and drag control, respectively. Finally, brief concluding remarks are given in Section 8.

2. LOW-REYNOLDS-NUMBER AIRFOILS

In the range of Reynolds numbers of 10^4–10^6, a substantial improvement in the lift-to-drag ratio of an airfoil takes place. According to Carmichael (1981), this is the Reynolds number regime where we find man and nature together in flight; large soaring birds, large radio-controlled model aircraft, foot-launched ultralight, man-carrying hangliders, human-powered aircraft, and the more recently developed remotely-piloted-vehicles (RPVs) used for military and scientific sampling, monitoring and surveillance. Three review articles of low-Reynolds-number aerodynamics are cited in here; Tani (1964), Lissaman (1983), and Mueller (1985).

In this range of Reynolds numbers, very complex flow phenomena take place within a short distance on the upper surface of an airfoil at incidence. Unless artificially tripped, the boundary layer remains laminar at the onset of pressure recovery and the airfoil's performance is then entirely dictated by the laminar flow's poor resistance to separation. The separated flow forms a free-shear layer which is highly unstable and transition to turbulence is readily realized. Subsequent reattachment of the separated region may take place because of the increased entrainment associated with the turbulent flow. Provided that the high-speed fluid entrained into the wall region supplies sufficient energy to maintain the circulating motion against dissipation, a separation bubble forms.

The precise conditions for the occurrence of separation, transition and reattachment, in other words for the formation of a laminar separation bubble, depend on the Reynolds number, the pressure distribution, the surface curvature, the surface roughness, and the freestream turbulence as well as other environmental factors. If the Reynolds number is sufficiently high, transition takes place near the minimum pressure point ahead of the location at which separation would have occurred if the boundary layer had remained laminar. For moderate Reynolds numbers, separation takes place prior to transition. The laminar boundary layer can only support very small adverse pressure gradient without separation. In fact, if the ambient incompressible fluid decelerates in the streamwise direction faster than $U_\infty \sim x^{-0.09}$, the flow separates (Schlichting, 1979). The separated flow will not reattach to the surface and no bubble will be formed if the Reynolds number is sufficiently low. However, for the intermediate Reynolds number range (typically 10^4–10^6), the separated flow proceeds along the direction of the tangent to the surface at the separation point (von Doenhoff, 1938) and transition to turbulence takes place in the free-shear layer due to its increased transition susceptibility. Subsequent turbulent entrainment of high-speed fluid causes the flow to return to the surface, thus forming what is known as a laminar separation bubble, as sketched in Figure 2. Downstream of the point of reattachment, the newly formed turbulent boundary layer is capable of negotiating quite severe adverse pressure gradients without separation. The ability of a turbulent boundary layer to resist separation improves as the Reynolds number increases (Lissaman, 1983).

It is clear from the above arguments that bubble formation is confined to a certain range of Reynolds numbers and that this range changes from one airfoil to another as well as from one environment to another. A rough rule according to Carmichael (1981) is that the Reynolds number based on freestream velocity and the distance from separation to reattachment is approximately $5x10^4$. In general, then, an airfoil with chord Reynolds number less than $5x10^4$ will experience laminar separation with no subsequent reattachment. For chord Reynolds numbers slightly higher than $5x10^4$, a long bubble is expected. Shorter bubbles are formed at higher Reynolds numbers. Tani (1964) asserts that a Reynolds number typical of local conditions in the boundary layer is more appropriate to characterize a separation bubble than the chord Reynolds number. Typically, the Reynolds number based on the boundary-layer's displacement thickness and the velocity just outside the rotational flow region at the point of separation is more than 500 for a short bubble and less than 500 for a long one. The corresponding bubble's streamwise extent, normalized with the displacement thickness at the point of separation, is 10^2 and 10^4, respectively (Tani, 1964).

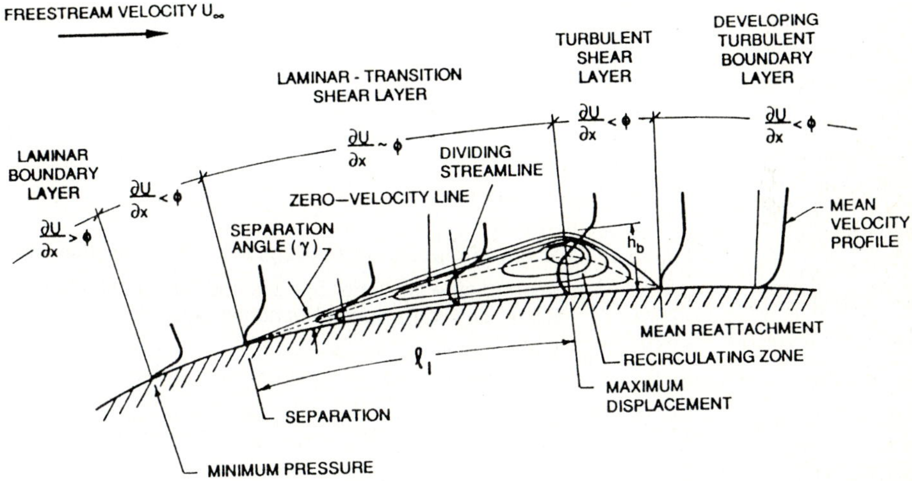

Figure 2. Sketch of a Laminar Separation Bubble.

The short separation bubble generally has a length of the order of a few percent of the chord. It merely represents a transition-forcing mechanism and does not greatly affect the peak suction as determined from the potential flow solution around the airfoil. Except for the appearance of a minute bump in the lift curve (C_L versus α curve), the presence of a short bubble has no significant effect on the pressure distribution around the lifting surface, as depicted in Figure 3a. On the other hand, a long bubble significantly changes the pressure distribution by effectively altering the shape over which the outer potential flow is developed. In this case, the sharp suction peak near the leading edge is generally not realized and a suction plateau of a reduced level extends over the region occupied by the bubble (Figure 3b).

In general, the lift-to-drag ratio is higher for an airfoil having shorter bubble. Depending on many factors, a short bubble forming at low incidence may move forward and contract in streamwise extent as the angle of attack is increased (Tani, 1964). Within the bubble, a small region of constant pressure exists followed by pressure recovery. At higher incidence, the bubble bursts and no longer reattaches thus ensuing a leading edge stall (Jones, 1934). This process is often irreversible, meaning that reducing the angle of attack will not immediately "unburst" the bubble. Strong hysteresis effects are thus observed as the attack angle is recycled (Carmichael, 1981).

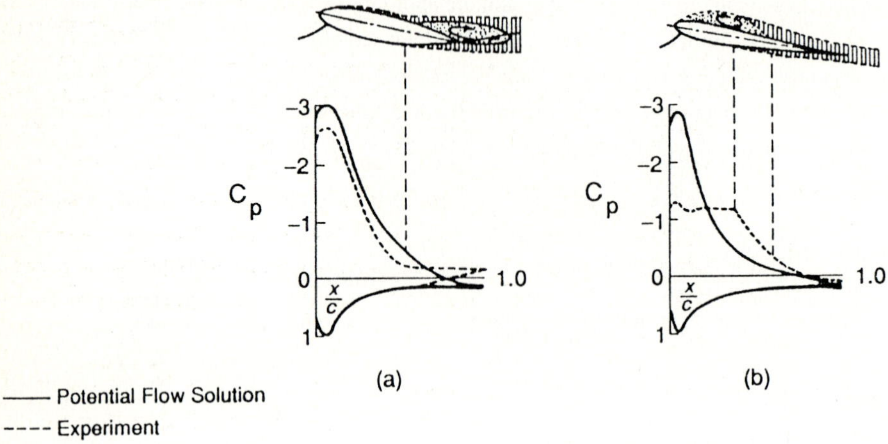

Figure 3. Qualitative Pressure Distributions for Two Airfoils at Incidence.

a. Short Bubble on Upper Surface and Subsequent Rear Separation of Turbulent Boundary Layer.

b. Long Bubble.

For thin airfoils of small nose radius, pressure recovery commences very near the leading edge and the adverse pressure gradients are severe at high angles of attack. Separation bubble may occur on these airfoils even at chord Reynolds numbers exceeding 10^6. At large incidence, the short bubble breaks down into a long bubble. With increasing angle of attack, the reattachment point moves progressively backward until it reaches the trailing edge, at which stage its maximum thickness is typically 3% of the chord. A further increase in incidence leads to completely detached flow and the so-called thin airfoil stall. A comprehensive review of the different kinds of stall on thin airfoils is given by Crabtree (1957).

Available experimental data on bubble's formation and bursting indicates transition to turbulence in the separated shear-layer and subsequent reattachment will occur if the Reynolds number based on displacement thickness at the point of laminar separation exceeds a critical value that is not necessarily universal (Tani-Owen-Klanfer criterion). A lower limit for this Reynolds number seems to be $R_{\delta *} \simeq 350$. Bursting occurs if the pressure recovered in the reattachment process in terms of the dynamic pressure at separation (pressure recovery coefficient) exceeds a certain critical value (Crabtree criterion). Again, this critical value changes from one airfoil to another but an upper limit of 0.35 appears to be valid for many shapes. Crabtree (1954) assumes that bubble's breakdown occurs because there exists a maximum possible value of pressure that can be recovered in the turbulent entrainment process that causes the flow reattachment. This implies the existence of a maximum possible value of the shear stress set up in the turbulent entrainment region so as to counteract the pressure gradient. At breakdown, caused by either an increase in incidence or a decrease in Reynolds number, the Tani-Owen-Klanfer criterion is satisfied but the Crabtree criterion is about to be violated.

The question of primary concern to us in this article is how to control the flow around a low-Reynolds-number airfoil to achieve an improved performance. The interrelation between the different control goals is particularly salient when a separation bubble exists, and this issue will be tackled in the next section.

3. CONTROL GOALS AND THEIR INTERRELATIONS

According to Tani (1964), all three kinds of stall, trailing-edge stall, leading-edge stall and thin-airfoil stall, may occur for a given airfoil at different Reynolds numbers or for different airfoils at a given Reynolds number. A particular lifting surface produces higher lift at higher incidence, limited by the angle at which the airfoil stalls. At that point, drag increases dramatically and the lifting surface performance deteriorates rapidly. Flow control is aimed at improving this performance. Among the practical considerations that must be considered for both active and passive control devices are their cost of construction and operation, complexity, and potential trade-off 's or penalties associated with their use. It is this latter point in particular that presents an additional degree of complexity for controlling low-Reynolds-number lifting surfaces. Achieving a beneficial effect for one control goal may very well adversely affect another goal and design compromises must often be made.

Among the desired goals of external flow modification are separation/reattachment control, lift enhancement, transition delay/advancement, and drag reduction. These objectives are not necessarily mutually exclusive, and for low-Reynolds-number lifting surfaces the interrelations between these goals are particularly conspicuous, presenting an additional degree of complexity. As mentioned before, in the range of Reynolds numbers of 10^4–10^6, a laminar separation bubble may form and may have a dominant effect on the flow field and the airfoil's performance. Figure 4 is a schematic representation of the interrelation between one control goal and another. If the boundary layer becomes turbulent, its resistance to separation is enhanced and more lift could be obtained at increased incidence. On the other hand, the skin-friction drag for a laminar boundary layer can be as much as an order of magnitude less than that for a turbulent one. If transition is delayed, lower skin friction is achieved. However, the laminar boundary layer can only support very small adverse pressure gradient without separation and subsequent loss of lift and increase in form drag occur. Once the laminar boundary layer separates, a free-shear layer forms and for moderate Reynolds number transition to turbulence takes place. Increased entrainment of high-speed fluid due to the turbulent mixing may result in reattachment of the separated region and formation of a laminar separation bubble. At higher incidence, the bubble breaks down either separating completely or forming a longer bubble. In either case, the form drag increases and the lift-curve's slope decreases. The ultimate goal of all this is to improve the airfoil's performance by increasing the lift-to-drag ratio. However, induced drag is caused by the lift generated on a lifting surface with a finite span. Moreover, more lift is generated at higher incidence but form drag also increases at these angles.

All of the above points to potential conflicts as one tries to achieve a particular control goal only to adversely affect another goal. An ideal method of control that is simple, inexpensive to build and operate, and does not have any trade-off 's does not exist, and the skilled engineer has to make continuous compromises to achieve a particular design goal. Keeping this in mind, we now proceed to review the control methods available to the designer of a low-speed lifting surface. A brief review of the governing equations will be useful in presenting a unified view, based on vorticity considerations, of the different control methods to achieve a variety of end results.

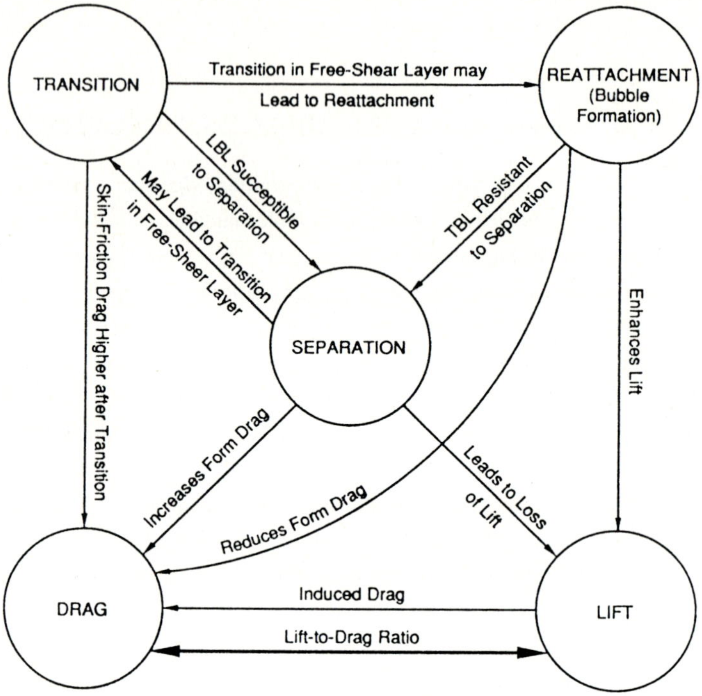

Figure 4. Interrelation Between Flow Control Goals.

4. ANALYTICAL CONSIDERATIONS

The principles of conservation of mass, momentum and energy govern all fluid motions. In general, a set of partial, nonlinear differential equations expresses these principles, and together with appropriate boundary and initial conditions constitute a well-posed problem. The equations are valid at each point in space and time for nonturbulent as well as turbulent flows. However, in the latter case the dependent variables are in general random functions of space and time. No straightforward method exists for solving stochastic, nonlinear partial differential equations. A statistical approach, where a temporal, spatial or ensemble mean is defined and the equations of motion are written for the various moments of the fluctuations about this mean, is the only route available to get meaningful engineering results. Unfortunately, the nonlinearity of the Navier-Stokes equations guarantees that the process of averaging to obtain moments results in an open system of equations, where the number of unknowns is always more than the number of equations.

For external flows at high Reynolds number, viscous forces are confined to a relatively thin layer along the surface of a body, although this layer's thickness increases in the downstream direction. Outside the boundary layer, the flow could be computed using the potential flow theory. Within the viscous region, the classical boundary-layer approximations apply. Ignoring body forces, the continuity equation and the streamwise and normal momentum equations for a steady, two-dimensional turbulent flow of a Newtonian, incompressible fluid read:

$$\frac{\partial \overline{U_1}}{\partial x_1} + \frac{\partial \overline{U_2}}{\partial x_2} = 0\,, \tag{4.1}$$

$$\rho\left(\overline{U_1}\,\frac{\partial \overline{U_1}}{\partial x_1} + \overline{U_2}\,\frac{\partial \overline{U_1}}{\partial x_2}\right) = -\frac{\partial \overline{P}}{\partial x_1} + \frac{\partial}{\partial x_2}\left(\mu\,\frac{\partial \overline{U_1}}{\partial x_2} - \rho\,\overline{u_1 u_2}\right), \tag{4.2}$$

$$0 = -\frac{\partial \overline{P}}{\partial x_2} + \frac{\partial}{\partial x_2}\left(-\rho\,\overline{u_2^2}\right), \tag{4.3}$$

where $\overline{U_1}$ and $\overline{U_2}$ are the time-averaged velocity in the streamwise and normal directions, respectively, $\overline{P}$ is the mean pressure, ρ is the constant density, μ is the variable viscosity, $-\rho\,\overline{u_1 u_2}$ is the tangential Reynolds stress, and $-\rho\,\overline{u_2^2}$ is the normal Reynolds stress.

Equation (4.3) can be integrated with respect to x_2 then differentiated with respect to x_1 to yield:

$$\frac{\partial \overline{P}}{\partial x_1} = \frac{d P_0}{d x_1}\,, \tag{4.4}$$

leaving (4.1) and (4.2) as two equations for the three unknowns $\overline{U_1}$, $\overline{U_2}$ and $\overline{u_1 u_2}$. Obviously, no solution can be obtained from first principles, and we must rely on more or less heuristic models to close the equations. Nevertheless, both equations can be integrated in the normal direction to yield the von Karman integral momentum-balance equation. For a steady, incompressible turbulent flow around a two-dimensional or axisymmetric surface of small curvature, this equation reads:

$$C_f = 2\,\frac{d\,\delta_\theta}{d\,x_1} + 2\,\delta_\theta\left[\left(2 + \frac{\delta^*}{\delta_\theta}\right)\frac{1}{U_\infty}\,\frac{dU_\infty}{d\,x_1} + \frac{1}{R}\,\frac{dR}{d\,x_1}\right] - 2\,\frac{v_0}{U_\infty}\,, \tag{4.5}$$

where C_f is the local skin-friction coefficient, δ^* and δ_θ are the displacement and momentum thicknesses, respectively, U_∞ is the freestream velocity, R is the radius of curvature of the wall, and v_0 is the normal velocity of fluid injected through the surface (positive for injection and negative for suction). Equation (4.5) is valid for both laminar and turbulent boundary layers. In the former case, $\overline{u_1 u_2} = 0$. In the latter case, the mean streamwise velocity is used in the definition of δ^* and δ_θ.

A second useful equation is obtained from (4.2) by taking the limit $x_2 \rightarrow 0$. At a fixed wall, the equation becomes after some rearranging:

$$\rho v_0 \left[\frac{\partial \overline{U_1}}{\partial x_2} \right]_0 + \frac{d P_0}{d x_1} - \frac{d \mu}{d T} \left[\frac{\partial \overline{T}}{\partial x_2} \frac{\partial \overline{U_1}}{\partial x_2} \right]_0 + \rho \left[\frac{\partial \overline{u_1 u_2}}{\partial x_2} \right]_0 = \left[\mu \frac{\partial^2 \overline{U_1}}{\partial x_2^2} \right]_0 , \qquad (4.6)$$

where $\overline{T}$ is the mean temperature field, and the subscript $[\ \ ]_0$ indicates flow quantities computed at the wall. The right-hand side of (4.6) is the flux of mean spanwise vorticity, $\overline{\Omega_3} = -\partial \overline{U_1} / \partial x_2$, at the surface. In the absence of suction/injection, pressure gradient, and surface heating/cooling, the first three terms on the left-hand side of (4.6) vanish.

For a laminar boundary layer, the Reynolds stress term vanishes. For the turbulent case, information regarding $\overline{u_1 u_2}$ has to come from experiment. Accurate measurement of the Reynolds shear stress near the wall is, however, extremely difficult. Returning to equation (4.6), the fourth term on the left-hand side is the slope of the normal profile of $\overline{u_1 u_2}$ at $x_2 = 0$. This term could be asymptotically estimated as the wall is approached. Consider a Taylor's series expansion in powers of x_2 in the neighborhood of the point $x_2 = 0$. As a result of the no-slip condition, the streamwise velocity fluctuations u_1 varies at least as x_2. To conserve mass, the normal velocity fluctuations must vary as x_2^2. It follows then that very near the wall (within the viscous sublayer), the tangential Reynolds stress $\overline{u_1 u_2}$ varies at least as x_2^3 and that $\partial \overline{u_1 u_2} / \partial x_2$ varies as x_2^2. At the wall itself, $x_2 = 0$ and $[\partial \overline{u_1 u_2} / \partial x_2]_0 = 0$, although close to the wall the slope of the tangential Reynolds stress profile is quite large.

The above arguments together with equation (4.6) indicate that the streamwise mean velocity profile for the canonical turbulent boundary layer (two-dimensional, isothermal, zero pressure gradient, over an impervious, rigid surface) will have a zero curvature at the wall. Notwithstanding this common characteristic with the Blasius boundary layer, the turbulent boundary layer is quite different from the laminar one. As pointed out by Lighthill (1963), the turbulent mixing concentrates most of the mean vorticity much closer to the wall as compared to the laminar case. The mean vorticity at the wall, $[\partial \overline{U}_1 / \partial x_2]_0$, is typically an order of magnitude larger than that in the laminar case. This explains the higher skin-friction drag associated with a turbulent flow. The turbulent mixing also causes the mean vorticity to migrate away from the wall and about 5% of the total is found much farther from the surface. The flux of mean spanwise vorticity is zero at the wall itself but very large close to it reaching a maximum at about the same location where the root-mean-square vorticity fluctuations peaks (near the edge of the viscous sublayer). This trait is responsible for the turbulent boundary layer's resistance to separation.

In the next three sections, available and contemplated flow control methods for low-Reynolds-number lifting surfaces will be discussed. The equations developed in this section for laminar and turbulent boundary layers will help in presenting a unified view of the different control techniques. A recent comprehensive review of boundary layer control is available (Gad-el-Hak, 1989).

5. SEPARATION/REATTACHMENT CONTROL

Fluid particles in a boundary layer are slowed down by wall friction. If the external potential flow is sufficiently retarded, for example due to the presence of an adverse pressure gradient, the momentum of those particles will be consumed by both the wall shear and the pressure gradient. At some point (or line), the viscous layer departs or breaks away from the bounding surface. The surface streamline nearest to the wall leaves the body at this point and the boundary layer is said to separate (Maskell, 1955). At separation, the rotational flow region next to the wall abruptly thickens, the normal velocity component increases, and the boundary-layer approximations are no longer valid. Due to the large energy losses associated with boundary-layer separation, the performance of a lifting surface is often controlled by the separation location. If separation is postponed, the pressure drag is decreased and the circulation and hence the lift at high angles of attack is enhanced.

Prandtl (1904) was the first to explain the mechanics of separation. He provided a precise criterion for its onset for the case of a steady, two-dimensional boundary layer developing over a fixed wall. In case such a flow is retarded, the near-wall fluid may have insufficient momentum to continue its motion and is brought to rest at the point of separation. Fluid particles behind this point move in a direction opposite to the external stream and the original boundary-layer fluid passes over a region of recirculating flow. Since the velocity at the wall is always zero, the gradient $[\partial U_1 / \partial x_2]_0$ will be positive upstream of separation, zero at the point of separation, and negative in the reverse flow region. The velocity profile at separation must then have a positive curvature at the wall. However, $[\partial^2 U_1 / \partial x_2^2]$ is negative at a large distance from the wall, which means the velocity profile at separation must have a point of inflection somewhere above the wall as shown in Figure 5d. Since $[\partial^2 U_1 / \partial x_2^2]_0 > 0$ is a necessary condition for a steady, two-dimensional boundary layer to separate, the opposite, i.e., a negative curvature of the velocity profile at the wall (Figure 5a), is a sufficient condition for the boundary-layer flow to remain attached.

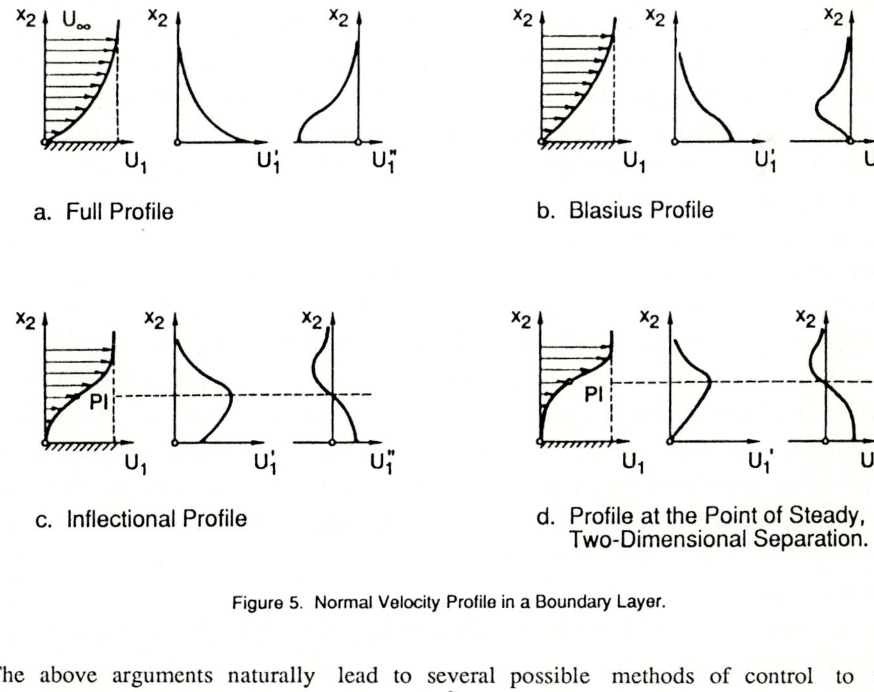

Figure 5. Normal Velocity Profile in a Boundary Layer.

The above arguments naturally lead to several possible methods of control to delay separation. Namely, the object is to keep $[\partial^2 U_1 / \partial x_2^2]_0$ as negative as possible, or in other words to make the velocity profile as full as possible. In this case, the spanwise vorticity decreases monotonically away from the wall and the surface vorticity flux is in the positive x_2 direction. Considering the effects of the terms in the left-hand side of (4.6) on the curvature of the velocity profile (or the vorticity flux) at the wall, separation control methods include the use of wall suction ($v_0 < 0$), favorable pressure gradient ($dP_0 / dx_1 < 0$), surface cooling in gases ($d\mu / d\overline{T} > 0$; $[\partial \overline{T} / \partial x_2]_0 > 0$), or surface heating in liquids ($d\mu / d\overline{T} < 0$; $[\partial \overline{T} / \partial x_2]_0 < 0$). Obviously any one or a combination of these methods could be used in a particular situation. For example, beyond the point of minimum pressure on a streamlined body the pressure gradient is adverse and the boundary layer may separate if the pressure rise is sufficiently steep; however, enough suction could be applied there to overcome the retarding effects of the adverse pressure and to prevent separation. Each of these control methods is covered in more details in the article by Gad-el-Hak (1989).

As an example of the effects of the airfoil's shape on its performance, consider the lift curves for the three sections NACA 63_3-018, NACA 63-009, and NACA 64A006. These airfoils have maximum thicknesses of 18%c, 9%c and 6%c, respectively, where c is the chord. The respective leading edge radii are 2.12%c, 0.631%c and 0.256%c. Figure 6, adapted from the measurements by McCullough and Gault (1951), depicts C_L vs. α curves for the three sections at chord Reynolds number of 5.8×10^6. For the thick section, NACA 63_3-018, transition takes place near the minimum pressure point. Stalling in this case is of the trailing edge type and is preceded by a gradual movement of the separation point of the turbulent boundary layer forward from the trailing edge as α increases. A laminar separation bubble is formed on the other two sections at small incidence. However, the NACA 63-009 section experiences a sudden leading edge stall when the bubble bursts with no subsequent reattachment, while the NACA 64A006 section experiences a more gradual thin-airfoil stall. In the latter case, the short bubble breaks down into a longer bubble at an angle of attack of 5° causing a slight discontinuity in the lift curve. Subsequent increase in α leads to a movement of the reattachment point towards the trailing edge. The maximum lift in this case is about 40% lower than that for the thick airfoil. The stall angle is also lower.

In his 1976 monograph, Chang reviews several other passive and active methods to postpone separation for low- and high-speed flows. Common to all these control methods is an attempt to supplying additional energy to the near-wall fluid particles which are being retarded in the boundary layer. Passive techniques do not require auxiliary power, but do have an associated drag penalty, and include intentional tripping of transition from laminar to turbulent flow upstream of what would be a laminar separation point, boundary-layer fences to prevent separation at the tips of swept-back wings, placing an array of vortex generators on the body to raise the turbulence level and enhance the momentum and energy in the neighborhood of the wall (Mehta, 1985; Rao and Kariya, 1988), geometric design to avoid shock-induced separation, machining a series of lateral grooves on the surface of the body upstream of separation, or using a screen to divert the flow and increase the velocity gradient at the wall. Howard and Goodman (1985; 1987) recently investigated the effectiveness of two passive techniques to reduce the flow separation, transverse rectangular grooves and longitudinal V-grooves placed in the aft shoulder region of a bluff body. Both types of grooves were beneficial in reducing the form drag on a body at zero and moderate angles of yaw.

Active methods to postpone separation require energy expenditure. Obviously, the energy gained by the effective control of separation must exceed that required by the device. In addition to suction or heat transfer mentioned earlier, fluid may be injected parallel to the wall to augment the shear-layer momentum or normal to the wall to enhance the mixing rate. Either a blower is used or the pressure difference that exists on the aerodynamic body itself is utilized to discharge the fluid into the retarded region of the boundary layer. The latter method is found in nature in the thumb pinion of a pheasant, the split-tail of a falcon, or the layered wing feathers of some birds. In man-made devices, passive blowing through leading-edge slots and trailing-edge flaps is commonly used on aircraft wings. Although in this case direct energy expenditure is not required, the blowing intensity is limited by the pressure differentials obtainable on the body itself.

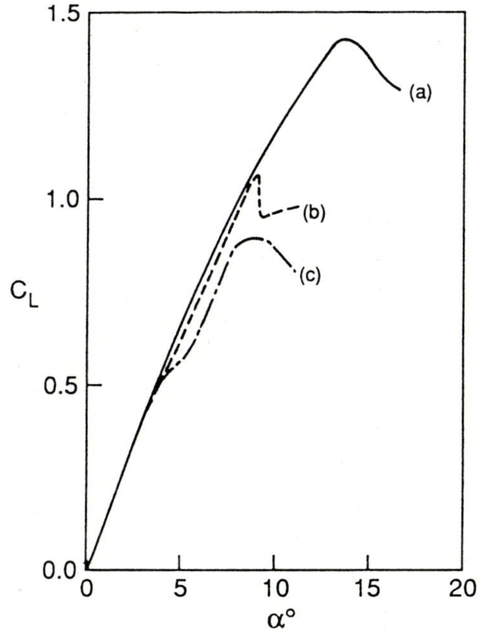

Figure 6. Lift Curves for Three Airfoils at $R_c = 5.8 \times 10^6$.
(Adapted from McCullough and Gault, 1951).

a. NACA 63_3–018.
b. NACA 63–009.
c. NACA 64A006.

More recently developed active methods for controlling boundary layer separation and reattachment include the use of acoustic excitations (Ahuja et al., 1983), oscillating surface flaps (Koga et al., 1984) and oscillatory surface heating (Maestrello et al., 1988). Ahuja and his colleagues successfully demonstrated that sound at a preferential frequency can postpone the turbulent separation on an airfoil in both pre- and post-stall regimes. The optimum frequency was found to be 4 U_∞ / c (Strouhal number = 4), where U_∞ is the freestream velocity and c is the airfoil chord. Goldstein (1984) speculates that the delay in separation in Ahuja et al.'s (1983) experiment resulted from enhanced entrainment promoted by instability waves that were triggered on the separated shear layers by the acoustic excitation. Koga et al. (1984) used a computer-controlled spoiler-like flap in a flat-plate turbulent boundary layer with and without modelled upstream separation. They were able to manipulate the separated flow region and its reattachment length characteristics by varying the frequency, amplitude and waveform of the oscillating flap. Reynolds and Carr (1985) offer a plausible explanation, from the viewpoint of a vorticity framework, for the experimental observations of Koga et al. (1984).

6. TRANSITION CONTROL

Delaying laminar-to-turbulent transition of a boundary layer has many obvious advantages. Depending on the Reynolds number, the skin-friction drag in the laminar state can be as much as an order of magnitude less than that in the turbulent condition. On the other hand, turbulence is an efficient mixer and rates of mass, momentum and heat transfer are much lower in the laminar state, so early transition may be sought in some applications as for example when rapid mixing or separation delay is desired.

Reshotko (1987) asserts that transition is a consequence of the nonlinear response of the laminar shear layer (a very complicated oscillator) to random forcing disturbances that result from freestream turbulence, radiated sound, surface roughness, surface vibrations, or combination of these environmental factors. If the initial disturbances are small, transition Reynolds number depends upon the nature and spectrum of these disturbances, their signature and excitation of the normal modes in the boundary layer (receptivity; see Morkovin, 1969), and the linear amplification of the growing normal modes. Once wave interaction and nonlinear processes set in, transition is quickly completed. If the initial disturbance levels are large enough, the relatively slow linear amplification step of Tollmien-Schlichting waves is bypassed (Morkovin, 1988) and transition can occur at much lower Reynolds numbers. In fact, a sufficiently violent disturbance, $u_{rms} / U_\infty \sim 10\%$, can cause transition of a laminar boundary layer to advance to the position upstream of which perturbations of all wave numbers decay (Klebanoff et al., 1955).

To delay transition to as far downstream position as possible, the following steps may be taken. First, since factors that affect the linear amplification of Tollmien-Schlichting waves determine the magnitude of the transition Reynolds number, these waves may be either inhibited or cancelled. In the former method of control, the growth of the linear disturbance is minimized using any or a combination of the so-called stability modifiers which alter the shape of the velocity profile. Wave cancellation of the growing perturbation is accomplished through exploiting but not altering the stability characteristics of the flow. Secondly, the forcing disturbances in the environment in which the laminar shear layer develops may be reduced. This is accomplished by using smooth surfaces, reducing the freestream turbulence and the radiated sound, minimizing body vibration, and ensuring a particulate-free incoming flow or, in the case of a contaminated environment, using a particle-defense mechanism. Practically achieved surface smoothness and levels of radiated noise place an upper limit for unit Reynolds number required for a successful laminar flow control system. For conventional aircraft, this typically translates into a requirement for high altitude operation (above 10 Km). For low-Reynolds-number aircraft, the unit Reynolds number criterion is easily met even at low altitudes. Thirdly, one may provide a flow where other kind of instabilities, e.g., Taylor-Görtler vortices or cross-flow instabilities, will not occur or at least will not grow at a rapid rate. This is done by avoiding as much as possible concave surfaces or concave streamlines, minimizing the sweep on lifting surfaces, etc.

The Reynolds number below which perturbations of all wave numbers decay is termed the critical Reynolds number or the limit of stability. For a given velocity profile, $U_1(x_2)$, the critical Reynolds number and the rate of growth of perturbations depends strongly on the shape of the velocity profile. A profile with an inflectional point $(\partial^2 U_1 / \partial x_2^2 = 0)$ above the wall provides a necessary and sufficient condition for inviscid instability. Such profiles must have a positive curvature at $x_2 = 0$, since $\partial^2 U_1 / \partial x_2^2$ is negative at a large distance from the wall (Figure 5c). Even when viscous effects are included, a velocity profile becomes more stable as its second derivative near the wall becomes negative, $[\partial^2 U_1 / \partial x_2^2]_0 < 0$. The profile is then said to be more full (Figure 5a), having a smaller ratio of displacement thickness to momentum thickness than, for example, an inflectional velocity profile. In the former case, the critical Reynolds number is increased, the range of amplified frequencies is diminished and the amplification rate of unstable waves is reduced.

Stability modifiers are those methods of laminar flow control which alter the shape of the velocity profile to minimize the linear growth of unstable waves. For a two-dimensional laminar boundary layer, vorticity is only in the spanwise direction and is given by $\Omega_3 = - \partial U_1 / \partial x_2$. Any of the terms on the left-hand side of (4.6) can affect the sign of the second derivative of the velocity profile (or the direction of the vorticity flux) at the wall and, hence, the flow stability. Stability modifiers do just that and include wall suction, favorable pressure gradient, surface cooling in gases, or surface heating in liquids. Boundary layers which are stabilized by extending the region of favorable pressure gradient are known as natural laminar flow (NLF), while the other methods to modify the stability of the shear flow are termed laminar flow control (LFC). It is clear from (4.6) that the effects of these methods are additive. The term hybrid laminar flow control normally refers to the combination of NLF and one of the LFC techniques.

We first consider the active control of transition using wall suction. As seen from (4.6), small amounts of fluid withdrawn from the near-wall region of the boundary layer change the curvature of the velocity profile at the wall and can dramatically alter the stability characteristics of the boundary layer. Additionally, suction inhibits the growth of the boundary layer, so that the critical Reynolds number based on thickness may never be reached.

Although laminar flow can be maintained to extremely high Reynolds number provided that enough fluid is sucked away, the goal is to accomplish transition delay with the minimum suction flow rate. Not only this will reduce the power necessary to drive the suction pump but also the momentum loss due to suction, and hence the skin friction, is minimized.

The case of a uniform suction from a flat plate at zero incidence is an exact solution of the Navier-Stokes equation. Assuming weak enough suction that the potential flow outside the boundary layer is unaffected by the loss of mass at the wall (sink effects), the asymptotic velocity profile in the viscous region is exponential and has a negative curvature at the wall:

$$U_1 (x_2) = U_\infty [1 - \exp (- |v_0| x_2 / \upsilon)] \ . \qquad (6.1)$$

The displacement thickness has the constant value $\delta^* = \upsilon / |v_0|$, where υ is the kinematic viscosity and $|v_0|$ is the absolute value of the normal velocity at the wall. In this case, (4.5) reads:

$$C_f = 2 C_q \ . \qquad (6.2)$$

Bussmann and Münz (1942) computed the critical Reynolds number for the above asymptotic velocity profile to be $R_{\delta^*} \equiv U_\infty \delta^* / \upsilon = 70{,}000$. From the value of δ^* given above, the flow is stable to all small disturbances if $C_q \equiv |v_0| / U_\infty > 1.4 \times 10^{-5}$. The amplification rate of unstable disturbances for the asymptotic profile is an order of magnitude less than that for the Blasius boundary layer (Pretsch, 1942). This treatment ignores the development distance from the leading edge needed to reach the asymptotic state. When this is included into the computation, a higher C_q (1.18×10^{-4}) is required to ensure stability (Iglisch, 1944; Ulrich, 1944). Wuest (1961) presented a summary of transpiration boundary layer computations up to the early 1960's.

A second method of control to delay laminar-to-turbulent transition is a passive one and involves the use of suitably shaped bodies to manipulate the pressure distribution. In (4.6), the pressure gradient term can affect the sign of the curvature of the velocity profile at the wall and, hence, change the stability characteristics of the boundary layer. The critical Reynolds number based on displacement thickness and freestream velocity changes from about 100 to 10,000 as a suitably nondimensionalized pressure gradient (the shape factor, Λ), varies from $\Lambda = -6$ (adverse) to $\Lambda = +6$ (favorable). Moreover, for the case of a favorable pressure gradient, no unstable waves exist at infinite Reynolds number. In contrast, the upper branch of the neutral stability curve in the case of an adverse pressure distribution tends to a non-zero asymptote so that a finite region of wavelengths at which disturbances are always amplified remains even as $Re \rightarrow \infty$.

Streamlining a body to prevent separation and reduce form drag is quite an old art, but the stabilization of a boundary layer by pushing the longitudinal location of the pressure minimum to as far back as possible dates back to the 1930's and led to the successful development of the NACA 6-Series NLF airfoils. Newer, low-Reynolds-number lifting surfaces used in sailplanes, low-speed drones and executive business jets have their maximum thickness point far aft of the leading edge. The recent success of the Voyager's nine-day, unrefueled flight around the world was due in part to a wing design employing natural laminar flow to approximately 50% chord.

The favorable pressure gradient extends to the longitudinal location of the pressure minimum. Beyond this point, the adverse pressure gradient becomes steeper and steeper as the peak suction is moved further aft. For an airfoil, the desired shift in the point of minimum pressure can only be attained in a certain narrow range of angles of incidence. Depending on the shape, angle of attack, Reynolds number, surface roughness and other factors, the boundary layer either becomes turbulent shortly after the point of minimum pressure or separates first and then undergoes transition. One of the design goals of NLF is to maintain attached flow in the adverse pressure gradient region and some method of separation control (Section 5) may have to be used there.

Subtle changes in the magnitude and extent of favorable pressure gradient, leading edge radius, and other shape variables can have pronounced effects on the airfoil's performance. As an example, consider the lift and drag characteristics of a conventional section (NACA 23015) and a laminar-flow section (NACA 66_2-215). Both airfoils are cambered and both have maximum thickness of 15% chord. However, the maximum thickness point is located at 30%c for the conventional section and at 45%c for the laminar-flow one. The leading edge radius is 2.48%c and 1.44%c for the two respective airfoils. The point of minimum pressure, as computed for the basic symmetric section at zero incidence, is located at 15%c for the conventional airfoil while this point is pushed back to 60%c for the laminar-flow airfoil. As seen from the lift curves depicted in Figure 7 for chord Reynolds number of 9×10^6, the laminar-flow section has slightly higher lift at small angles of attack than the conventional section, but stalling occurs at lower incidence and the maximum lift is smaller for the laminar-flow airfoil. The lift-drag polars for the same two airfoils are shown in Figure 8. The sudden increase in drag for the laminar-flow section occurs at an angle of attack of about 1.5°. This is caused by a forward movement of the minimum pressure point and corresponding accentuation of the adverse pressure gradient as the flow on the upper surface accelerates sharply to round the airfoil's sharp nose. The increased adverse pressure gradient leads to early transition and corresponding drag increase. The maximum lift-to-drag ratios for the conventional and laminar-flow airfoils are 125 and 102, respectively. However, the laminar-flow section is designed to cruise in the low-drag region ($\alpha < 1.5°$). At the design lift coefficients of 0.3 and 0.2 for the conventional and laminar-flow sections, the lift-to-drag ratios are 47.6 and 59.5, respectively.

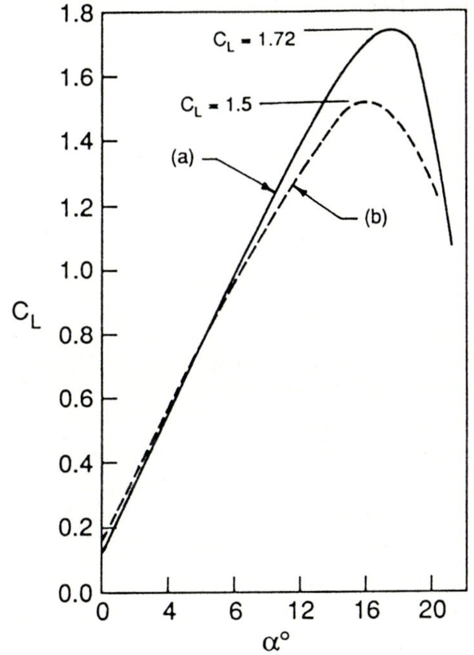

Figure 7. Lift Curves for a Conventional Section and a Laminar - Flow Section at $R_c = 9 \times 10^6$.
(Data from Abbott and Doenhoff, 1959).

a. Conventional Airfoil (NACA 23015).
b. Laminar - Flow Airfoil (NACA 66_2–215).

Factors that limit the utility of NLF include crossflow instabilities and leading edge contamination on swept wings, insect and other particulate debris, high unit Reynolds numbers of conventional aircraft at lower cruise altitudes (not a limiting factor for low-Reynolds-number aircraft), and performance degradation at higher angles of attack due to the necessarily small leading edge radius of NLF airfoils. Reductions of surface waviness and smoothness of modern production wings, special leading edge systems to prevent insect impacts and ice formation, higher cruise altitudes of newer airplanes, and higher Mach numbers all favor the application of NLF (Runyan and Steers, 1980).

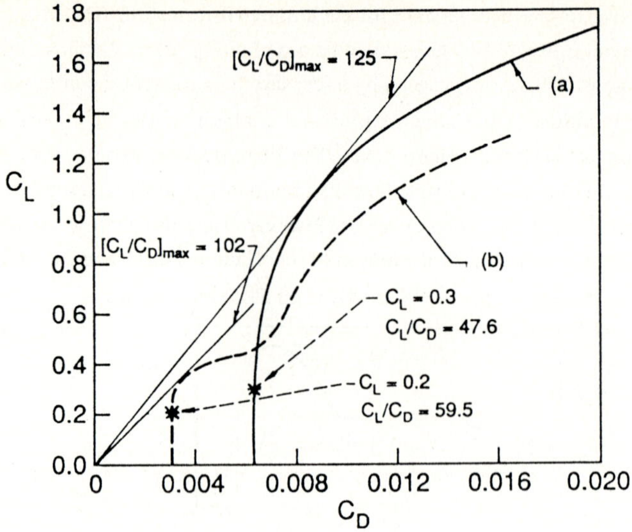

Figure 8. Lift - Drag Polars for the Two Airfoils Depicted in Figure 7. (Data from Abbott and von Doenhoff, 1959).

a. Conventional Airfoil.
b. Laminar - Flow Airfoil.

The third stability modifier is an active one and involves the addition or removal of heat from a surface, which causes the viscosity to vary with distance from the wall. In general, viscosity increases with temperature for gases, while the opposite is true for liquids. Thus, if heat is removed from the surface of a body moving in air, the third term on the left-hand side of (4.6) is negative. In that case, the velocity gradient near the wall increases and the velocity profile becomes fuller and more stable. The term containing the viscosity derivative will also be negative if the surface of a body moving in water is heated. With heating in water or cooling in air, the critical Reynolds number is increased, the range of amplified frequencies is diminished and the amplification rate of unstable waves is reduced. Substantial delay of transition is feasible with a surface that is only a few degrees hotter (in water) or colder (in air) than the freestream. For aircraft, this method of transition delay is feasible only for a vehicle which uses a cryo-fuel such as liquid hydrogen or liquid methane. In this case, a sizeable heat sink is readily available. The idea being that the fuel is used to cool the major aerodynamic surfaces of the aircraft as it flows from the fuel tanks to the engines.

An alternative approach to increase the transition Reynolds number of a laminar boundary layer is wave cancellation. If the frequency, orientation and phase angle of the dominant element of the spectrum of growing linear disturbances in the boundary layer is detected, a control system and appropriately located disturbance generators may then be used to effect a desired cancellation or suppression of the detected disturbances. In this case, the stability characteristics of the boundary layer are exploited but not altered (Reshotko, 1985). Wave cancellation is feasible only when the disturbances are still relatively small, their growth is governed by a linear equation, and the principle of superposition is still valid.

A turbulent boundary layer is more resistant to separation than a laminar one and transition advancement may be desired in some situations. In low-Reynolds-number terminology, the transition promoting devices are called *turbulators*. For a zero-pressure-gradient boundary layer, transition typically occurs at a Reynolds number based on distance from leading edge of the order of 10^6. The critical Re below which perturbations of all wave numbers decay is about 6×10^4. To advance the transition Reynolds number, one may attempt to lower the critical Re, increase the growth rate of Tollmien-Schlichting waves, or introduce large disturbances that can cause bypass transition. The first two routes involve altering the shape of the velocity profile using wall motion, injection, adverse pressure gradient, or surface heating in gases or cooling in liquids. The third route is much simpler to implement though more difficult to analyze. Morkovin (1984) broadly classify the large disturbances that can cause bypass transition into steady or unsteady ones originating into the freestream or at the body surface. The most common example is single, multiple or distributed roughness elements placed on the wall. The mechanical roughness elements, in the form of serrations, strips, bumps or ridges, are typically placed near the airfoil's leading edge. If the roughness characteristic-length is large enough, the disturbance introduced is nonlinear and bypass transition takes place. For a three-dimensional roughness element of height-to-width ratio of one, a transition Reynolds number $R_{\delta^*} \simeq 300$ (below the critical $R_{\delta^*} = 420$ predicted from the linear stability theory) is observed for a roughness Reynolds number, based on its height and the velocity in the undisturbed boundary layer at the height of the element, of about 10^3 (Tani, 1969). Transition occurs at $R_{\delta^*} \simeq 10^3$ for a roughness Reynolds number of about 600. For a smooth surface, transition typically takes place at $R_{\delta^*} \simeq 2.6 \times 10^3$. An important consideration when designing a turbulator is to produce turbulence and suppress laminar separation without causing the boundary layer to become unnecessarily thick. A thick turbulent wall-bounded flow suffers more drag and is more susceptible to separation than a thin one. Consistent with this observation, available data (Figure 1) indicates that a rough airfoil has higher lift-to-drag ratio than a smooth one for $R_c < 10^5$, but that this trait is reversed at higher Reynolds numbers.

Other large disturbances that could lead to early transition include high turbulence levels in the freestream, external acoustic excitations, particulate contamination, and surface vibration. These are often termed environmental tripping. Transition could also be effected by detecting naturally occurring T-S waves and artificially introducing in-phase waves. Transition could be considerably advanced, on demand, using this wave superposition principle.

Early transition could also be achieved by exploiting other routes to turbulence such as Taylor-Görtler or cross-flow vortices. For example, a very mild negative curvature of $(0.003/\delta^*)$ results in the generation of strong streamwise vortices. In this case, transition Reynolds number is lowered from $R_{\delta^*} \simeq 2600$ for the flat-plate case to $R_{\delta^*} \simeq 700$ for the curved surface (Tani, 1969). For high Mach number flows, the general decay in spatial amplification rate of T-S waves makes conventional tripping more difficult as the Mach number increases (Reshotko, 1976). For these flows, trips that generate oblique vorticity waves of appropriate wavelength may be most effective to advance the transition location.

The last issue to be considered in this section is how to augment the turbulence for a shear flow that has already undergone transition. For moderate Reynolds numbers, a laminar boundary layer may undergo transition directly or may separate first, undergo transition, and subsequently reattach as a turbulent boundary layer. In either case, the newly developed turbulent flow is less capable of resisting separation than a corresponding flow at higher Reynolds numbers. Turbulence augmentation in the low-Reynolds-number case is then a useful control goal to energize the flow and to enhance its ability to resist separation at higher angles of attack. Roughness will enhance the turbulence, but its associated drag must be carefully considered. Other devices to enhance the turbulent mixing include vane-type vortex generators, which draw energy from the external flow, or wheeler-type or Keuthe-type generators, which are fully submerged within the boundary layer and presumably have less associated drag penalty (Rao and Kariya, 1988).

7. DRAG REDUCTION

The total drag experienced by a lifting surface consists of skin friction and pressure drag. Flow Separation is the major source of pressure drag with additional contributions due to displacement effects of the boundary layer, wave resistance in a supersonic flow or at an air/water interface, and drag induced by lift on a finite body. For a vehicle, the reduced drag means longer range, reduced fuel cost/volume, higher payload, or increased speed; all potentially important considerations when designing an RPV for military or scientific applications.

Streamlining and other control methods summarized in Section 5 can eliminate most of the pressure drag due to flow separation. Some form drag remains, however, even when the flow remains attached to the trailing edge. Due to the displacement effects of the boundary layer, the pressure distribution around the body differs from the symmetric distribution predicted by potential flow theory. This remanent drag can be reduced by keeping the boundary layer as thin as possible. Wave resistance and induced drag can also be reduced by geometric design. By sweeping the wings of a subsonic aircraft, drag divergence is delayed to higher Mach numbers, thus allowing the aircraft to fly at higher speeds without experiencing a sudden increase in drag. Additionally, the so-called area rule or coke-bottle effect typically leads to a factor of 2 reduction in wave drag at Mach number of 1 (Whitcomb, 1956). The induced drag of an aircraft's wing is inversely proportional to its aspect ratio and, hence, a lifting surface is typically designed with as large an aspect ratio as permissible by structural considerations and desired degree of maneuverability. End plates or other vortex diffusers can also be used to further reduce the induced drag.

In the absence of transition promoters, such as cross flow, concave surfaces, adverse pressure gradient, roughness, or freestream disturbances, the boundary layer is laminar to $Re = 0[10^6]$. In this case, methods to reduce the laminar skin-friction are sought. This may be useful for some land vehicles, airborne vehicles at very high altitude, small RPVs, or small hang-gliders and the like, but obviously large underwater or air vehicles cannot benefit from these techniques.

From equation (4.6), it is clear that any or a combination of the following techniques can be used to make the curvature of the velocity profile at the wall more positive and, thus, lower C_f; injection of fluid normal to the wall $(v_0 > 0)$; adverse pressure gradient $(dP_0/dx_1 > 0)$; wall heating in air $(d\mu/dT > 0;\ [\partial T/\partial x_2]_0 < 0)$; or wall cooling in water $(d\mu/dT < 0;\ [\partial T/\partial x_2]_0 > 0)$. Note that any of these methods will promote flow instability (Section 6) and separation (Section 5). These tendencies have to be carefully considered when deciding how far to go with the attempt to lower C_f. Moreover, energy expenditure for the active techniques must be less than the energy saved due to the lower skin friction.

At large Reynolds numbers, transition can no longer be postponed and the boundary layer becomes turbulent and, thus, an excellent momentum conductor. Such a flow is less prone to separation but is characterized by large skin-friction. For low-Reynolds-number airfoils, a laminar separation bubble is commonly present and a turbulent boundary layer develops downstream of the reattachment point. A design goal is to reduce the skin friction in the turbulent region while maintaining attached flow as far aft as possible. Several techniques are available to reduce the turbulent skin-friction coefficient, but only very few are in actual use, the basic reason being that the majority of these methods are relatively new and, thus, are still in the research and development stage. These techniques are classified in four categories: reduction of near-wall momentum; introduction of foreign substances; geometrical modifications; and synergism.

Methods of skin-friction drag reduction in turbulent boundary layers that rely on reducing the near-wall momentum are similar to those used in the laminar case. The influence of wall transpiration, shaping or heat transfer on the mean velocity profile is complicated by the additional effects of these modulations on the Reynolds stress term. However, these influences are qualitatively in the same direction as in the simpler laminar case. Thus, lower skin friction is achieved by driving the turbulent boundary layer towards separation. This is accomplished by injecting fluid normal to the wall, shaping to produce adverse pressure gradient, surface heating in air, or surface cooling in water. These methods of control in general result in an increase in turbulence intensity (Wooldridge and Muzzy, 1966).

Although in the reverse flow region downstream of the separation line the skin friction is negative, the increase in pressure drag is far more than the saving in skin-friction drag. The goal of the above methods of control is to avoid actual separation, i.e., lower C_f but not any lower than zero (the criterion for steady, two-dimensional separation). The papers by Stratford (1959a; 1959b) provide useful discussion on the prediction of turbulent-boundary-layer separation and the concept of flow with continuously zero skin-friction throughout its region of pressure rise. By specifying that the turbulent boundary layer be just at the condition of separation, without actually separating, at all positions in the pressure rise region, Stratford (1959b) experimentally verified that such a flow achieve a specified pressure rise in the shortest possible distance and with the least possible dissipation of energy. An airfoil which could utilize the Stratford's distribution immediately after transition from laminar to turbulent flow would be expected to have a very low drag. Liebeck (1978) successfully followed this strategy to achieve the best lift-to-drag ratio (over 200) of any airfoil tested in the range of Reynolds numbers of $5\times10^5 - 2\times10^6$. He argued that the entire pressure-recovery region of an airfoil's upper surface would be operating at its maximum capacity if the adverse pressure distribution was uniformly critically close to separation. By assuming an incipient-separation turbulent profile, Liebeck calculated the pressure field required then used an inverse calculation procedure to derive the airfoil shape from the given critical-velocity distribution.

When attempting to reduce drag by driving the boundary layer towards separation, a major concern is the flow behavior at off-design conditions. A slight increase in angle of attack for example can lead to separation and consequent large drag increase as well as loss of lift. High performance airfoils with lift-to-drag ratio of over 100 utilize carefully controlled adverse pressure gradient to retard the near-wall fluid, but their performance deteriorates rapidly outside a narrow envelope (Carmichael, 1974).

Turbulent skin-friction drag can be also reduced by the addition of several foreign substances. Examples include long-chain molecules, surface-active agents and microbubbles in liquid flows, and small solid particles or fibers in either gases or liquids. In general, the addition of these substances leads to a suppression of the Reynolds stress production in the buffer zone and, thus, to an inhibition of the turbulent mixing and a consequent reduction in the viscous shear stress at the wall. Numerous review and technical articles of drag reduction methods involving the introduction of a foreign substance are available (see, for example, Lumley, 1969; 1977; 1978; McComb and Chan, 1981; and Gad-el-Hak, 1989).

The most recently researched techniques to reduce the turbulent skin-friction drag involve geometric modifications. These include large eddy breakup devices, riblets, compliant surfaces, wavy walls, and other surface modifications. The large eddy breakup devices (LEBUs) are designed to sever, alter or break up the large vortices that form the convoluted outer edge of a turbulent boundary layer. A typical arrangement consists of one or more splitter plates placed in tandem in the outer part of a turbulent boundary layer. It is of course very easy to reduce substantially the skin friction in a flat-plate boundary layer by placing an obstacle above the surface. What is difficult is to ensure that the device's own skin-friction and pressure drags do not exceed the saving. The original screen fence device of Yajnik and Acharya (1978) and the various sized honeycombs used by Hefner et al. (1980) did not yield a net drag reduction. In low-Reynolds-number experiments, very thin elements placed parallel to a flat plate have a device's total drag that is nearly equal to laminar skin friction. A net drag reduction of the order of 20% is feasible with two elements placed in tandem with a spacing of $0[10\delta]$ (Corke et al. 1980; 1981). These ribbons have typically a thickness and a chord of the order of 0.01δ and δ, respectively, and are placed at a distance from the wall of 0.8δ. Several experiments report a more modest drag reduction (Hefner et al., 1983) or even a drag increase, but it is believed that a slight angle of attack of the thin element can result in a laminar separation bubble and a consequent increase in the device pressure drag. Flat ribbons at small, positive angle of attack produce larger skin-friction reductions. This is consistent with the analytical result that a device producing positive lift away from the wall is more effective (Gebert, 1988). In any case, a net drag reduction should be achievable, at least for devices having chord Reynolds numbers $< 10^6$, if extra care is taken to polish and install the LEBU. This device's Reynolds number requirement is, of course, easily met for low-Reynolds-number aircraft, providing a strong incentive for using LEBU devices on RPVs and the like.

The second geometrical modification is the riblets. Small longitudinal striations in the surface, interacting favorably with the near-wall structures in a turbulent boundary layer, can produce a modest drag reduction in spite of the increase in surface area. The early work employed rectangular fins with height and spacing of $0[100\upsilon/u^*]$. The turbulent bursting rate was reduced by about 20% and a modest 4% net drag reduction was observed (Liu et al., 1966). In a later refinement of this technique, Walsh and his colleagues at NASA-Langley examined the drag characteristics of longitudinally ribbed surfaces having a wide variety of fin shapes that included rectangular grooves, V-grooves, razor blade grooves, semi-circular grooves, and alternating transverse curvature (Walsh and Weinstein, 1978; Walsh, 1980; 1982; 1983; Walsh and Lindemann, 1984). A net drag reduction of 8% is obtained with V-groove geometry with sharp peak and either sharp or rounded valley. Optimum height and spacing of the symmetric grooves are about $15\upsilon/u^*$. Although these dimensions would be extremely small for the typical Reynolds numbers encountered on a conventional airplane or a submarine (peak-to-valley height $\sim 35\ \mu m$), such riblets need not be machined on the surface. Thin, low-specific-gravity plastic films with the correct geometry on one side and an adhesive on the other side are presently available commercially and existing vehicles could be readily retrofitted. In fact, these tapes were successfully tested at $M = 0.7$ on a

T-33 airplane and on a Lear jet. The performance of the riblets in flight was similar to that observed in the laboratory (Anders et al., 1988). For the typical low-Reynolds-number aircraft, the required height and spacing of the grooves exceeds 350 μm. In water, riblets were employed on the rowing shell during the 1984 Summer Olympic by the United States rowing team. Similar riblets were also used on the submerged hull of the winner of the 1987 America's Cup yacht race, the Stars and Stripes, with apparent success.

In addition to the drag reduction techniques surveyed in here, many others are available. The interested reader is referred to the review articles by Bushnell (1983), Bandyopadhyay (1986), Wilkinson et al. (1988), Bushnell and McGinley (1989), and Gad-el-Hak (1989). In particular, the last article on this list presents an extensive discussion of the value of combining more than one drag reduction technique to achieve a synergetic effect.

8. CONCLUDING REMARKS

Available and contemplated flow control methods particularly suited for low-Reynolds-number lifting surfaces have been surveyed. The flow around these airfoils is dominantly effected by the formation of a separation bubble. The laminar separation makes the interrelation between transition, separation, lift and drag controls particularly salient, presenting an additional degree of complexity.

Low-Reynolds-numbers cover the range of 10^4–10^6. In this regime, laminar separation, transition and reattachment may all occur within a short distance on the upper surface of an airfoil at incidence. The precise conditions for the occurrence of a laminar separation bubble depend on the local Reynolds number, pressure distribution, surface curvature, surface roughness, and freestream turbulence as well as other environmental factors.

To control the flow around a low-Reynolds-number airfoil to achieve improved performance, one must carefully consider potential conflicts in trying to achieve a particular control goal while inadvertently causing an adverse effect on another goal. A laminar boundary layer is less able to resist separation but is characterized by a very low skin-friction drag. However, if the flow separates, lift decreases and form drag increases substantially. In the low-Reynolds-number regime, laminar separation is often proceeded by transition to turbulence in the separated free-shear layer and subsequent reattachment to form a closed bubble. Bubble bursting at higher incidence leads to loss of lift and increased drag.

The effect of many of the control methods reviewed in this article is explained in terms of the behavior of the spanwise vorticity flux at the wall. The fullness of the normal velocity profile is related to the direction of this vorticity flux, which in turn has a direct influence on the stability, separation, skin friction, and turbulence levels.

Stability modifiers inhibit the linear amplification of Tollmien-Schlichting waves and, therefore, determine the magnitude of the transition Reynolds number. Shaping to provide extended regions of favorable pressure gradient is the simplest method of control and is well suited for the wings of low- or moderate-speed aircraft. Flight tests have demonstrated the feasibility of using suction to maintain a laminar flow on a swept wing to $Re \simeq 4.7 \times 10^7$. The required suction rate is very modest and 20% net drag reduction is possible. Moderate surface cooling in air or heating in water also increases the transition Re by an order of magnitude, but a sink/source of heat must be available to achieve net drag reduction. For futuristic aircraft using cryo-fuel, surface cooling may be a feasible method to delay transition.

Therefore, two-dimensional, steady separation can be postponed using the same techniques. Other separation control methods include passive ones, such as intentional tripping, fences or vortex generators, and active devices, such as tangential injection, acoustic excitations or oscillating surface flaps. A drag penalty is associated with both passive and active devices.

Techniques to reduce the pressure drag are more well established than turbulent skin-friction reduction techniques. Streamlining and other methods to postpone separation can eliminate most of the form drag. The wave and induced drag contributions to the pressure drag can also be reduced by geometric design. For the purpose of reducing the skin friction, three flow regimes are identified. First, for $Re < 10^6$, the flow is generally laminar and skin friction may be lowered by reducing the near-wall momentum. Adverse pressure gradient, blowing and surface heating/cooling could lower the skin friction, but increase the risk of transition and separation. Secondly, for $10^6 < Re < 4 \times 10^7$, active and passive methods to delay transition could be used, thus avoiding the much higher turbulent skin-friction. Thirdly, for high-Reynolds-number flows, the boundary layer is turbulent and methods to reduce the large mean vorticity at the wall, $[\overline{\Omega_3}]_0$, are sought. These methods are classified in the following categories: Reduction of near-wall momentum; introduction of foreign substances; geometrical modifications; and synergism. Many of these techniques could conceivably be applied to reduce the skin-friction in the turbulent boundary layer that develops downstream of a laminar separation bubble.

Recently introduced techniques mostly fall under the third category above and seem to offer more modest net drag reduction. These methods are, however, still in the research stage and include riblets (~ 8%), large eddy breakup devices (~ 20%), and convex surfaces (~ 20%). Potential improvement in these and other methods will perhaps involve combining more than one technique aiming at achieving a favorable effect that is greater than the sum.

REFERENCES

Abbott, I.H., and von Doenhoff, A.E. (1959) Theory of Wing Sections, Dover, New York.

Ahuja, K.K., Whipkey, R.R., and Jones, G.S. (1983) "Control of Turbulent Boundary Layer Flows by Sound," AIAA Paper No. 83-0726.

Anders, J.B., Walsh, M.J., and Bushnell, D.M. (1988) "The Fix for Tough Spots," Aerospace America 26, pp. 24-27.

Bandyopadhyay, P.R. (1986) "Review-Mean Flow in Turbulent Boundary Layers Disturbed to Alter Skin Friction," J. Fluids Eng. 108, pp. 127-140.

Bushnell, D.M. (1983) "Turbulent Drag Reduction for External Flows," AIAA Paper No. 83-0227.

Bushnell, D.M., and McGinley, C.B. (1989) "Turbulence Control in Wall Flows," Ann. Rev. Fluid Mech. 21, pp. 1-20.

Bussmann, K., and Münz, H. (1942) "Die Stabilität der laminaren Reibungsschicht mit Absaugung," Jahrb. Dtsch. Luftfahrtforschung 1, pp. 36-39.

Carmichael, B.H. (1974) "Application of Sailplane and Low-Drag Underwater Vehicle Technology to the Long-Endurance Drone Problem," AIAA Paper No. 74-1036.

Carmichael, B.H. (1981) "Low Reynolds Number Airfoil Survey," NASA CR 165803, Washington, D.C.

Chang, P.K. (1976) Control of Flow Separation, Hemisphere, Washington, D.C.

Corke, T.C., Guezennec, Y., and Nagib, H.M. (1980) "Modification in Drag of Turbulent Boundary Layers Resulting from Manipulation of Large-Scale Structures," in Viscous Flow Drag Reduction, ed. G.R. Hough, pp. 128-143, AIAA Prog. in Astro. & Aero. 72, New York.

Corke, T.C., Nagib, H.M., and Guezennec, Y. (1981) "A New View on Origin, Role and Manipulation of Large Scales in Turbulent Boundary Layers," NASA Contractor Report No. 165861.

Crabtree, L.F. (1954) "The Formation of Regions of Separated Flow on Wing Surfaces. Part I: Low-Speed Tests on a Two Dimensional Unswept Wing with a 10% Thick RAE-101 Section," Royal Aircraft Establishment Report Aero. 2528, Farnborough, England.

Crabtree, L.F. (1957) "Effects of Leading-Edge Separation on Thin Wings in Two-Dimensional Incompressible Flow," J. Aeronaut. Sci. 24, pp. 597-604.

Gad-el-Hak, M. (1989) "Flow Control," Appl. Mech. Rev., in press.

Gebert, G.A. (1988) "Turbulent Boundary Layer Modification by Streamlined Devices," Ph.D. Thesis, University of Notre Dame, IN.

Goldstein, M.E. (1984) "Generation of Instability Waves in Flows Separating from Smooth Surfaces," J. Fluid Mech. 145, pp. 71-94.

Hefner, J.N., Anders, J.B., and Bushnell, D.M. (1983) "Alteration of Outer Flow Structures for Turbulent Drag Reduction," AIAA Paper No. 83-0293.

Hefner, J.N., Weinstein, L.M., and Bushnell, D.M. (1980) "Large-Eddy Breakup Scheme for Turbulent Viscous Drag Reduction," in Viscous Flow Drag Reduction, ed. G.R. Hough, pp. 110-127, AIAA Prog. in Astro. & Aero. 72, New York.

Howard, F.G., and Goodman, W.L. (1985) "Axisymmetric Bluff-Body Drag Reduction Through Geometrical Modification," J. Aircraft 22, pp. 516-522.

Howard, F.G., and Goodman, W.L. (1987) "Drag Reduction on a Bluff Body at Yaw Angles to 30 Degrees," J. Spacecraft & Rockets 24, pp. 179-181.

Iglisch, R. (1944) "Exakte Berechnung der laminaren Reibungsschicht an der längsangeströmten ebenen Platte mit homogener Absaugung," Schr. Dtsh. Akad. Luftfahrtforschung 8 B, pp. 1-51.

Jones, B.M. (1934) "Stalling," J. R. Aero. Soc. 38, pp. 753-770.

Klebanoff, P.S., Schubauer, G.B., and Tidstrom, K.D. (1955) "Measurements of the Effect of Two-Dimensional and Three-Dimensional Roughness Elements on Boundary-Layer Transition," J. Aero. Sci. 22, pp. 803-804.

Koga, D.J., Reisenthel, P., and Nagib, H.M. (1984) "Control of Separated Flowfields Using Forced Unsteadiness," Illinois Institute of Technology, Fluids & Heat Transfer Report No. R84-1, Chicago, IL.

Liebeck, R.H. (1978) "Design of Subsonic Airfoils for High Lift," J. Aircraft 15, pp. 547-561.

Lighthill, M.J. (1963) "Introduction – Boundary Layer Theory," in Laminar Boundary Layers, ed. L. Rosenhead, pp. 46-113, Clarendon Press, Oxford.

Lissaman, P.B.S. (1983) "Low-Reynolds-Number Airfoils," Ann. Rev. Fluid Mech. 15, pp. 223-239.

Liu, C.K., Kline, S.J., and Johnston, J.P. (1966) "Experimental Study of Turbulent Boundary Layer on Rough Walls," Dept. of Mechanical Engineering, Report No. MD-15, Stanford University, CA.

Lumley, J.L. (1969) "Drag Reduction by Additives," Ann. Rev. Fluid Mech. 1, pp. 367-384.

Lumley, J.L. (1977) "Drag Reduction in Two Phase and Polymer Flows," Phys. Fluids 20, pp. S64-S71.

Lumley, J.L. (1978) "Two-Phase and Non-Newtonian Flows," in Topics in Applied Physics 12, ed P. Bradshaw, Second Edition, pp. 289-324, Springer-Verlag, Berlin.

Maestrello, L., Badavi, F.F., and Noonan, K.W. (1988) "Control of the Boundary Layer Separation about an Airfoil by Active Surface Heating," AIAA Paper No. 88-3545-CP.

Maskell, E.C. (1955) "Flow Separation in Three Dimensions," RAE Report Aero. 2565, Royal Aircraft Establishment, Farnborough, England.

McComb, W.D., and Chan, K.T.J. (1981) "Drag Reduction in Fibre Suspension," Nature 292, pp. 520-522.

McCullough, G.B., and Gault, D.E. (1951) "Examples of Three Representative Types of Airfoil-Section Stall at Low Speed," NACA TN 2502, Washington, D.C.

McMasters, J.H., and Henderson, M.L. (1980) "Low Speed Single Element Airfoil Synthesis," Tech. Soaring 6, pp. 1-21.

Mehta, R.D. (1985) "Effect of a Longitudinal Vortex on a Separated Turbulent Boundary Layer," AIAA Paper No. 85-0530.

Morkovin, M.V. (1969) "Critical Evaluation of Transition from Laminar to Turbulent Shear Layers with Emphasis on Hypersonically Traveling Bodies," Air Force Flight Dynamics Laboratory Report No. AFFDL-TR-68-149, Wright-Patterson AFB, OH.

Morkovin, M.V. (1984) "Bypass Transition to Turbulence and Research Desiderata," in Transition in Turbines Symposium, NASA CP-2386.

Morkovin, M.V. (1988) "Recent Insights into Instability and Transition to Turbulence in Open-Flow Systems," AIAA Paper No. 88-3675.

Mueller, T.J. (1985) "Low Reynolds Number Vehicles," AGARDograph No. 288, Neuilly sur Seine, France.

Prandtl, L. (1904) "Über Flüssigkeitsbewegung bei sehr kleiner Reibung," Proc. Third Int. Math. Congr, pp. 484-491, Heidelberg, Germany.

Pretsch, J. (1942) "Umschlagbeginn und Absaugung," Jahrb. Dtsch. Luftfahrtforschung 1, pp. 54-71.

Rao, D.M. and Kariya, T.T. (1988) "Boundary-Layer Submerged Vortex Generators for Separation Control - An Exploratory Study," AIAA Paper No. 88-3546-CP.

Reshotko, E. (1976) "Boundary-Layer Stability and Transition," Ann. Rev. Fluid Mech. 8, pp. 311-349.

Reshotko, E. (1985) "Control of Boundary Layer Transition," AIAA Paper No. 85-0562.

Reshotko, E. (1987) "Stability and Transition, How Much Do We Know? "Proc. 10th U.S. National Cong. of App. Mech., ed. J.P. Lamb, pp. 421-434, ASME, New York.

Reynolds, W.C., and Carr, L.W. (1985) "Review of Unsteady, Driven, Separated Flows," AIAA Paper No. 85-0527.

Runyan, L.J., and Steers, L. L. (1980) "Boundary Layer Stability Analysis of a Natural Laminar Flow Glove on the F-111 TACT Airplane," in Viscous Flow Drag Reduction, ed. G.R. Hough, AIAA Progress in Astronautics & Aeronautics 72, pp. 17-32.

Schlichting, H. (1979) Boundary-Layer Theory, Seventh Edition, McGraw-Hill, New York.

Stratford, B.S. (1959 a) "The Prediction of Separation of the Turbulent Boundary Layer," J. Fluid Mech. 5, pp. 1-16.

Stratford, B.S. (1959 b) "An Experimental Flow with Zero Skin Friction Throughout its Region of Pressure Rise," J. Fluid Mech. 5, pp. 17-35.

Tani, I. (1964) "Low-Speed Flows Involving Bubble Separations," Progress in Aeronautical Sciences, Vol. 5, eds. D. Küchemann and L.H.G. Sterne, pp. 70-103, Pergamon Press, New York.

Tani, I. (1969) "Boundary-Layer Transition," Ann. Rev. Fluid Mech. 1, pp. 169-196.

Ulrich, A. (1944) "Theoretische Untersuchungen über die Widerstandsersparnis durch Laminarhaltung mit Absaugung," Schriften Dtsch. Akad.Luftfahrtforschung 8 B, p. 53.

Van Doenhoff, A.E. (1938) "A Preliminary Investigation of Boundary-Layer Transition Along a Flat Plate with Adverse Pressure Gradient," NACA TN 639, Washington, D.C.

Walsh, M.J. (1980) "Drag Characteristics of V-Groove and Transverse Curvature Riblets," in Viscous Flow Drag Reduction, ed. G.R. Hough, pp. 168-184, AIAA Prog. in Astro. & Aero. 72, New York.

Walsh, M.J. (1982) "Turbulent Boundary Layer Drag Reduction Using Riblets," AIAA Paper No. 82-0169.

Walsh, M.J. (1983) "Riblets as a Viscous Drag Reduction Technique," AIAA J. 21, pp. 485-486.

Walsh, M.J., and Lindemann, A.M. (1984) "Optimization and Application of Riblets for Turbulent Drag Reduction," AIAA Paper No. 84-0347.

Walsh, M.J., and Weinstein, M. (1978) "Drag and Heat Transfer on surfaces with Small Longitudinal Fins," AIAA Paper No. 78-1161.

Whitcomb, R.T. (1956) "A Study of the Zero-Lift Drag-Rise Characteristics of Wing-Body Combinations Near the Speed of Sound," NACA Report No. 1273.

Wilkinson, S.P., Anders, J.B., Lazos, B.S., and Bushnell, D.M. (1988) "Turbulent Drag Reduction Research at NASA Langley: Progress and Plans," Int. J. Heat and Fluid Flow 9, pp. 266-277.

Wooldridge, C.E., and Muzzy, R.J. (1966) "Boundary-Layer Turbulence Measurements with Mass Addition and Combustion," AIAA J. 4, pp. 2009-2016.

Wuest, W. (1961) "Survey of Calculation Methods of Laminar Boundary Layers With Suction in Incompressible Flow," in Boundary Layer and Flow Control, Vol. 2, ed. G.V. Lachmann, pp. 771-800, Pergamon Press, New York.

Yajnik, K.S., and Acharya, M. (1978) "Non-Equilibrium Effects in a Turbulent Boundary Layer Due to the Destruction of Large Eddies," in Structure and Mechanisms of Turbulence, Vol. 1, ed. H. Fiedler, pp. 249-260, Springer-Verlag, New York.

THE LOW FREQUENCY OSCILLATION IN THE FLOW OVER A NACA0012 AIRFOIL WITH AN "ICED" LEADING EDGE

K.B.M.Q. Zaman and M.G. Potapczuk
National Aeronautics and Space Administration
Lewis Research Center
Cleveland, Ohio 44135

Abstract

The unusually low frequency oscillation in the wake of an airfoil, studied in [1], is explored experimentally as well as computationally for a NACA0012 airfoil with a "glaze ice accretion" at the leading edge. Experimentally, flow oscillations are observed at low frequencies that correspond to a Strouhal number of about 0.02. This occurs in the angle of attack range of 8° to 9°, near the onset of static stall for this airfoil. With a Navier-Stokes computation, "limit-cycle" oscillations in the flow and in the aerodynamic forces are also observed at low Strouhal numbers. However, the occurrence of the oscillation is found to depend on the turbulence model in use as well as the Reynolds number.

Nomenclature

c	airfoil chord
C_l	lift coefficient
C_d	drag coefficient
C_m	moment coefficient about 0.5c
M	Mach number
f_s	shedding or oscillation frequency
R_c	chord Reynolds number
St_s	Strouhal number, $f_s c \sin\alpha / U_\infty$
$u'(f)$	one-dimensional, longitudinal velocity spectrum
U_∞	free-stream mean velocity
α	angle of attack

1. Introduction

An unusually low frequency oscillation in the flow over a LRN(1)-1007 airfoil was studied experimentally as well as computationally in [1], over the R_c range of 15×10^3 to 300×10^3. The phenomenon involved a naturally occurring periodic switching between stall and unstall around the onset of the static stall condition. The frequency was considered low as the corresponding Strouhal number, based on the cross-stream length scale and the free stream velocity, was only about 0.02, an order of magnitude lower than that experienced in the commonly observed bluff-body shedding (see also [2]). Even though the mechanism of frequency selection has remained unresolved, the measurements established that the phenomenon was indeed aerodynamic in origin. Any connection to a standing acoustic wave, structural resonance or unsteadiness associated with the tunnel was ruled out. The flow oscillations imparted very large unsteady forces to the airfoil. Thus, investigations leading to a clearer understanding of the phenomenon are well in order from the point of view of applications involving unsteady aerodynamic loads, e.g., in stall flutter of blades and wings.

In [1], unsteady oscillations were also observed computationally that had striking similarities with the experimental results. In an independent study [3], symptoms of the phenomenon were also encountered in a Navier-Stokes computation for the flow over a NACA0012 airfoil with "glaze ice accretion." A glaze ice accretion develops during flight conditions just below the freezing temperature. These shapes are typified by large "horns" (see the simulated shape in Figure 1(b)). This led to a wind tunnel experiment with an airfoil model having the same cross-sectional shape as used in [3]. Lift characteristics showed the onset of static stall around $\alpha \approx 7°$ with this airfoil. Slightly above this α, wake oscillations were observed resulting in an unambiguous peak in the velocity spectra, although not as sharp as in the "LRN" airfoil case, at a Strouhal number of about 0.02!

The objective of the present paper is to summarize these results. Salient features of the phenomenon observed with the "LRN" airfoil are reviewed first. The experimental results obtained with the "iced" airfoil are then presented, followed by a discussion of limited computational results. As it will be observed, even though the computation seems to capture the essence of the phenomenon, the results vary widely depending on the turbulence model in use and the Reynolds number.

2. Experimental Procedure

The experiments were carried out in a low speed wind tunnel which has been described in detail elsewhere [1,4]. Figure 1(a) is a photograph of the test section as set up during a somewhat similar experiment [4]. The test section is 76 cm (wide) by 51 cm (high). The maximum speed is about 16 ms^{-1}, however, some of the data were obtained using a smaller drive motor yielding a maximum speed of about 10 ms^{-1}. The free-stream turbulence intensity is less than 0.1 percent, but could be raised to about 0.4 percent by inserting a 30-mesh screen about 21 cm upstream

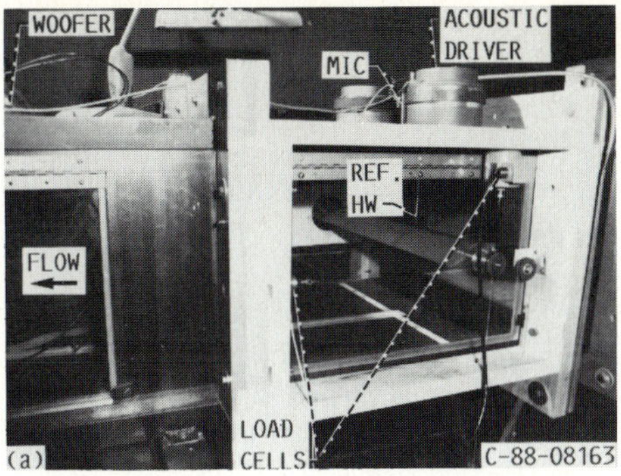

FIGURE 1a. - WIND TUNNEL TEST SECTION.

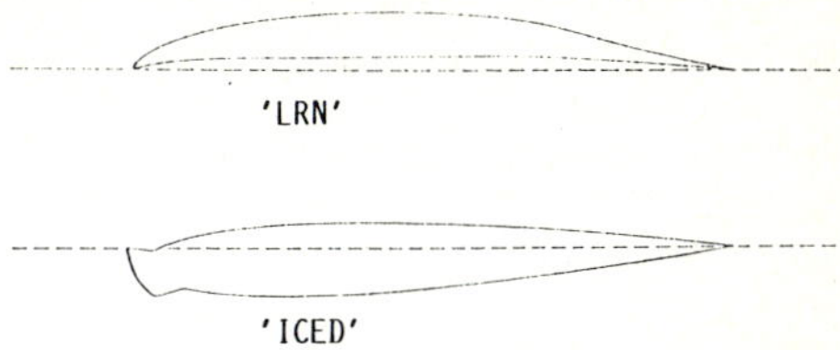

FIGURE 1b. - AIRFOIL CROSS-SECTIONS. UPPER: LRN(1)-1007, LOWER: NACA0012 WITH 'GLAZE ICE ACCRETION'.

of the airfoil support. Two-dimensional airfoil models were used in the experiment. Data from the LRN(1)-1007 and a NACA0012 airfoil with simulated "glaze ice shape" at the leading edge are presented. Both are of nominal chord of 12.7 cm and of aspect ratio of about 6. These two airfoils, whose cross sections are shown in Figure 1(b), are referred to simply as the "LRN" and the "iced" airfoils, respectively. The photograph in Figure 1(a) shows the "LRN" airfoil mounted in the tunnel.

Standard hot-wire anemometry was used for velocity measurements. A Nicolet 660B analyzer was used for spectrum analysis. There was provision for acoustic excitation of the flow, and an automated angular positioning device was used during the measurement of the lift and the drag [4]. The airfoils were supported at both ends at mid-chord. The coordinate origin is at the airfoil mid-chord and the axis of the test section. x, y and z denote streamwise, vertical and spanwise coordinates, respectively. The angle of incidence α is measured with respect to the (dashed) chord lines as shown in Figure 1(b).

3. Results and Discussion

3.1 Summary of results of [1]: Key results from [1], for the "LRN" airfoil, are shown in Figures 2 to 5. The wake velocity spectra for different α are shown in Figure 2. At large α spectral peaks are observed at relatively higher frequencies which correspond to $St_s \approx 0.2$. This is due to the familiar bluff-body shedding involving the asymmetric Kárman vortex street. Around the onset of static stall, however, at about $\alpha = 15°$ in this case, the low frequency oscillation is observed, the corresponding St_s being an order of magnitude lower. It is this low St_s oscillation -- its origin, significance as well as its contrast to the bluff-body shedding -- that has been addressed in [1].

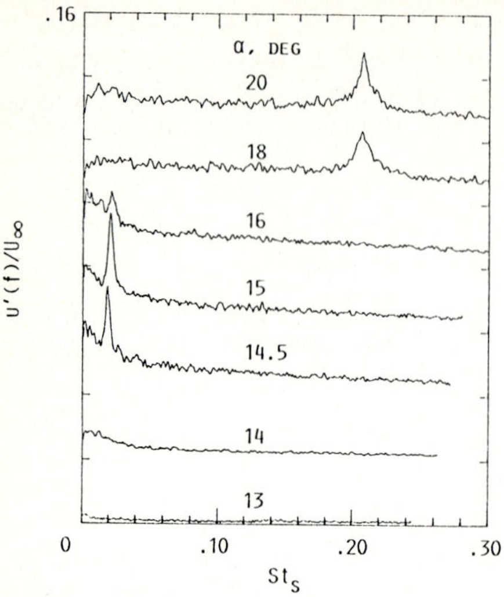

FIGURE 2. - WAKE VELOCITY SPECTRA FOR 'LRN' AIRFOIL. $R_c = 10^5$, $u'_\infty/U_\infty =$ 0.4 PERCENT; MEASUREMENT AT $x/c = 1.5$, $y/c = 0.15$, $z = 0$. SPECTRA TRACES ARE STAGGERED SUCCESSIVELY BY ONE ORDINATE DIVISION.

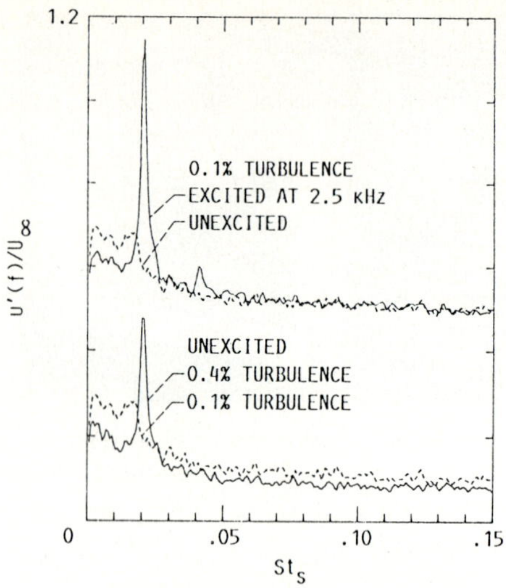

FIGURE 3. - WAKE VELOCITY SPECTRA FOR 'LRN' AIRFOIL AS IN FIGURE 2, SHOWING ENHANCEMENT OF $St_s \approx 0.02$ OSCILLATION. $R_c = 10^5$, $\alpha = 15^0$. UPPER PAIR OF TRACES SHIFTED BY TWO ORDINATE DIVISIONS.

The occurrence of the low St_s oscillation was found to be rather illusive and sensitive to ambient conditions. This is delineated by the data in Figures 3 and 4. The lower pair of spectra in Figure 3 shows that with the normal tunnel operation yielding a turbulence intensity of about 0.1 percent, the periodic oscillation does not take place; however, it does when the free-stream turbulence is raised with the help of the screen inserted upstream of the test section. Note that the oscillation was observed previously in a Langley tunnel in which the free-stream turbulence was also higher -- about 0.25 percent [2]. The upper pair of spectra in Figure 3 demonstrate that the low frequency oscillation can also be precipitated if acoustic excitation in a certain high frequency range is applied.

On the other hand, when the oscillation takes place naturally, e.g., in the flow with the higher free-stream turbulence, acoustic excitation in a relatively higher frequency range is found to eliminate it. An example is shown by the pair of spectra in Figure 4. The ranges of excitation frequency augmenting or suppressing the $St_s \approx 0.02$ oscillation have been documented for various conditions in [1]. The higher frequency excitation producing the suppression effect shown in Figure 4 appears to involve "acoustic tripping" of the separating boundary layer. It is conjectured that the boundary layer around the region of separation is turned fully turbulent by the acoustic excitation -- in which case the low frequency oscillation does not take place. Thus, on the one hand the phenomenon does not seem to take place in clean flows yielding laminar boundary layers, while on the other, it also disappears when the separating boundary layer is fully turbulent. Therefore, a certain transitional state near the separation region seems to be a requisite for the low frequency oscillation to take place [1].

Also included in [1] are the computational results for the "LRN" airfoil. C.L. Rumsey had previously observed a similar low frequency oscillation computationally for a NACA0012 airfoil at $\alpha = 18°$ [5]. The computations were then carried out for the Reynolds number and airfoil shape as in the experiment. The results were found to capture the essential features of the experimental observation. The computed flow field agreed in overall features with corresponding measurements. The computed force fluctuations also agreed generally with the experimental results, as shown in Figure 5. Despite the difference in the Strouhal number, the amplitudes and the details of the fluctuations are found to have close similarities. Aspects of Rumsey's computation will be discussed further in the following.

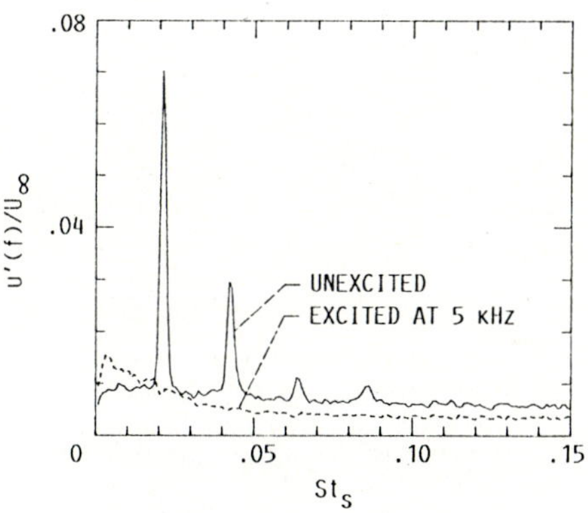

FIGURE 4. - WAKE VELOCITY SPECTRA FOR 'LRN' AIRFOIL AS IN FIGURE 2, SHOWING SUPPRESSION OF $St_s \approx 0.02$ OSCILLATION. $R_c = 0.75 \times 10^5$, $\alpha = 15^0$, $u'_\infty/U_\infty = 0.4$ PERCENT.

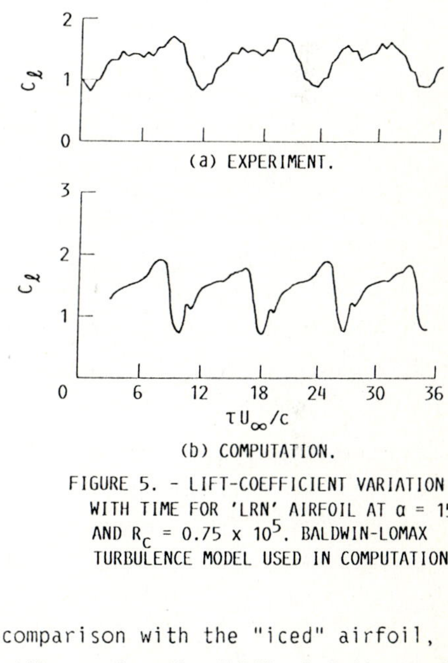

FIGURE 5. - LIFT-COEFFICIENT VARIATION WITH TIME FOR 'LRN' AIRFOIL AT $\alpha = 15^0$ AND $R_c = 0.75 \times 10^5$. BALDWIN-LOMAX TURBULENCE MODEL USED IN COMPUTATION.

3.2 Present experimental results: For comparison with the "iced" airfoil, the lift, drag and moment coefficient variation with α for the "LRN" airfoil are documented in Figure 6. Note that the low frequency oscillation takes place around 15° (within a narrow range of α) where the forces and the moment go through a large change due to stall.

Corresponding aerodynamic data for the "iced" airfoil are shown in Figure 7. In comparison to the "LRN" airfoil, the stall occurred at a much lower angle of attack of about 7°. No hysteresis was observed around the stall angle. It is slightly above this α where this airfoil exhibited a similar low frequency oscillation.

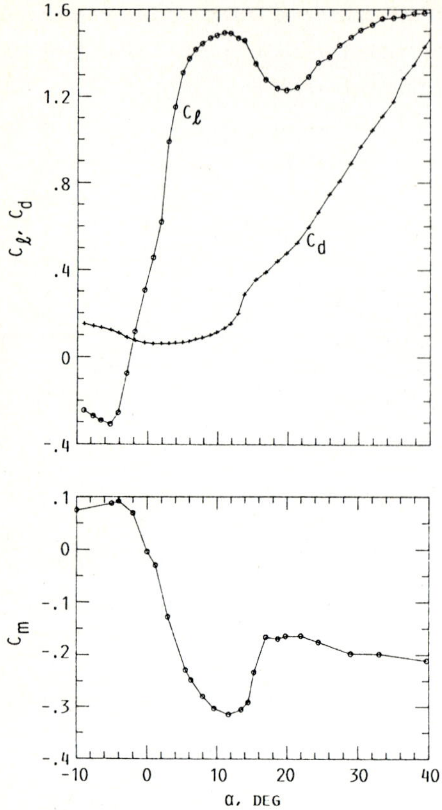

FIGURE 6. - C_ℓ, C_d AND C_m VERSUS α FOR THE 'LRN' AIRFOIL. R_c = 0.75 x 10^5 AND u'_∞/U_∞ = 0.4 PERCENT.

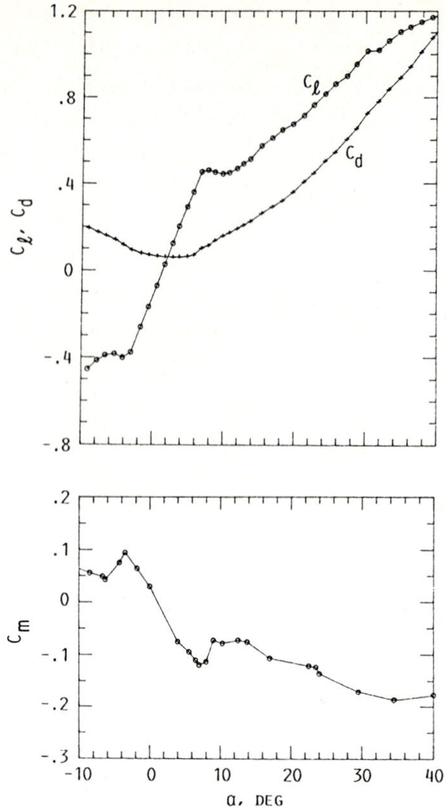

FIGURE 7. - C_ℓ, C_d AND C_m VERSUS α FOR THE 'ICED' AIRFOIL. R_c = 0.75 x 10^5, u'_∞/U_∞ = 0.4 PERCENT.

Wake velocity spectra similar to those in Figure 2 are presented in figure 8 for the "iced" airfoil. The oscillation is observed to occur around the α-range of 8° to 9°. Note that in this case the spectral peak is not as sharp as in the "LRN" airfoil case, but the tendency towards an oscillation around $St_s = 0.02$ is unambiguous. The oscillation at three different Reynolds numbers at $\alpha = 8.5°$ is documented in Figure 9. $R_c = 1.25 \times 10^5$ corresponds to the highest speed obtainable in the tunnel.

The effect of acoustic excitation on the low frequency oscillation over the "iced" airfoil was briefly studied in the present experiment. An attempt was made to augment the oscillation as in the "LRN" airfoil case (Figure 3). However, the spectral peak could not be made much sharper under the acoustic excitation. Increasing the free-stream turbulence with the screen also did not make a significant difference. However, the suppression of the low frequency oscillation, as shown in Figure 4 for the "LRN" airfoil, could be achieved. An example is shown in Figure 10.

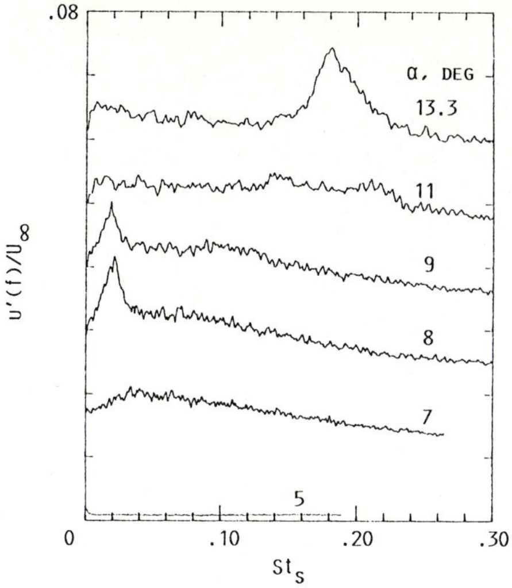

FIGURE 8.- WAKE VELOCITY SPECTRA AS IN FIGURE 2 FOR THE 'ICED' AIRFOIL; $R_C = 10^5$. TRACES ARE STAGGERED BY ONE DIVISION.

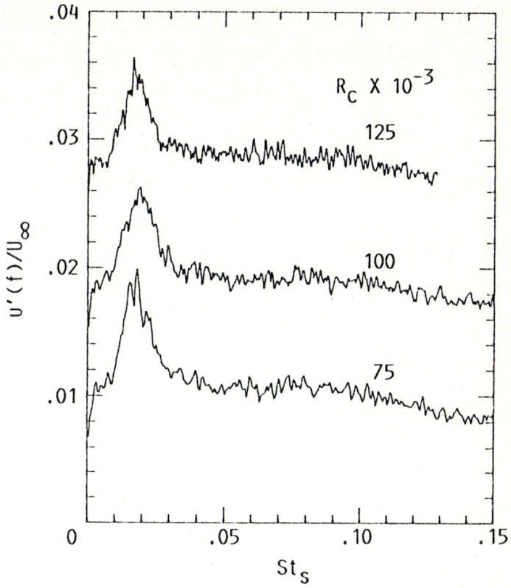

FIGURE 9. - WAKE VELOCITY SPECTRA AS IN FIGURE 2 FOR THE 'ICED' AIRFOIL AT $\alpha = 8.5^0$ FOR INDICATED R_C. TRACES ARE STAGGERED BY ONE MAJOR DIVISION.

The data in Figures 7 to 10 lead us to believe that the oscillation taking place around $\alpha \approx 8.5°$ with the "iced" airfoil is morphologically the same as that studied in [1]. The differences observed with respect to the effect of acoustic excitation are not surprising and reinforces the notion that the phenomenon is very sensitive to ambient perturbations affecting the state of the separating boundary layer.

It was observed in [1] that the phenomenon seems to occur with airfoils exhibiting "trailing edge stall" or "leading edge stall accompanied by a separation bubble", but not with ones exhibiting the "abrupt leading edge stall". The computations and experiments suggest that the "iced" airfoil belong to the second of the aforementioned stall categories (leading edge stall accompanied by a separation bubble), thus, corroborating the above observation. Note that the "LRN" airfoil is a borderline case between the first and the second categories of stall. An Wortman airfoil, on the other hand, is clearly characterized by the third category of stall, and thus, does not exhibit the low frequency oscillation [1].

3.3 Similar observations by others: It appears that the low frequency oscillation phenomenon has gone practically unrecognized over the several decades of airfoil research. To our knowledge, there are only a few references in the literature regarding similar observations. B.M. Jones (1934; see [1]) and Farren (1935; [6]) apparently encountered the phenomenon. Figure 11(a) reproduces the essential feature of Farren's data in this connection. Farren was actually developing a fast response balance and experimenting with oscillating airfoils. In his report, also included was the normal force variation with time for a R.A.F. 28 airfoil held fixed at $\alpha = 14°$. This airfoil was previously observed by B.M. Jones to undergo violent fluctuations near stalling conditions. The average period of oscillation (Figure 11(a)) was observed by Farren to correspond to 13 chords of travel. This converts to $St_s = 0.019$.

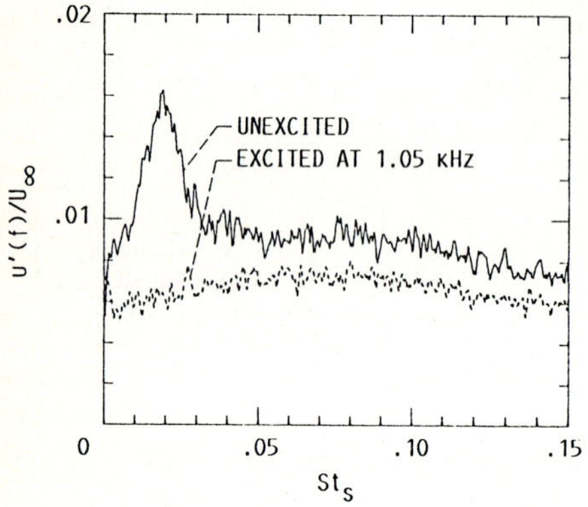

FIGURE 10.- WAKE VELOCITY SPECTRA AS IN FIGURE 4 FOR THE 'ICED' AIRFOIL, SHOWING SUPPRESSION OF $St_s \approx 0.02$ OSCILLATION. $\alpha = 8.5^0$, $R_c = 10^5$.

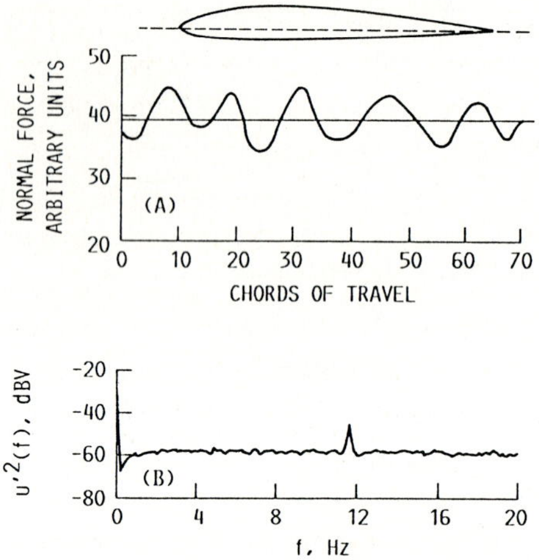

FIGURE 11. - (A) NORMAL FORCE VARIATION FOR R.A.F. 28 AIRFOIL AT $\alpha = 14^0$ AND $R_c = 10^5$ FROM REF. [6]. (B) VELOCITY SPECTRA MEASURED SHORTLY DOWNSTREAM OF THE 'HORN' OF AN 'ICED' AIRFOIL AT $\alpha \approx 6^0$ AND $R_c = 1.5 \times 10^6$ FROM REF. [7].

At NASA Lewis there is a research program addressing the problem of airfoil leading edge ice accretion and its remedies. The program includes computation and experiments on the aerodynamic characteristics of such airfoil sections both in-house and under grants. Bragg and Khodadoust [7], under a grant, obtained wind tunnel data on the aerodynamic characteristics of a NACA0012 airfoil with simulated ice shapes. Recently, for a "glaze ice" shape, (same as in the present study), they also observed a low frequency flow oscillation. Figure 22 of their paper is reproduced as Figure 11(b). The data represent u'-spectrum measured near the upper edge of the shear layer, shortly downstream of the "ice horn", for $\alpha \approx 6°$. (Note that "$\alpha = 4°$" quoted in their paper is measured with respect to the chord of the basic NACA0012 airfoil. α, measured with respect to the chord as shown in Figure 1(b), is about 2° larger). There is a clear peak in the spectrum, and this corresponds to $St_s \approx 0.016$. The authors did not find such spectral peak at other locations over the airfoil and, therefore, commented that it should represent a local shedding from the ice horn and not a global bubble oscillation. However, their observations were preliminary, and it would appear that the same phenomenon, as addressed here, has been encountered by them. Their data, however, represent a significantly higher R_c of 1.5×10^6, indicating that the phenomenon may not be characteristic of low Reynolds number airfoils only. Other relevant observations in the literature, including the characteristic frequencies in dynamic stall and stall flutter, have been discussed in [1].

3.4 Computational result: A two-dimensional Navier-Stokes code coupled with a C-mesh grid generation code was used for the computations, the details of which can be found in [3]. A "modified mixing length" (MML) turbulence model was used in the computation. An unsteady flow with "periodic vortex shedding" was noted around $\alpha = 8°$. The lift coefficient variation with time, computed for the same flow conditions of [3], is shown in Figure 12(a). Similarities can be observed with the C_l variation for the "LRN" airfoil (Figure 5). However, when the Baldwin-Lomax turbulence model was used in the present computation the oscillations damped out in the resultant flow field. This is shown in Figure 12(b).

Here, let us note that in the computations by Rumsey that resulted in low frequency oscillation for the "LRN" airfoil (Figure 5), the Baldwin-Lomax turbulence model was used. Rumsey [5] also observed similar low frequency oscillations for a NACA0012 airfoil at $\alpha = 18°$ and $R_c = 10^6$.

The algorithms used in [3] and [5] are basically similar. The "MML" is an algebraic turbulence model which attempts to avoid some of the difficulties encountered with Baldwin-Lomax turbulence model when applied to separated flows. Details of the model will be discussed in the second author's dissertation, which is forthcoming. The "turbulent viscosity" (μ_t) obtained from the two turbulence models are quite different. Typically, the "MML" model obtains μ_t/μ_l values on the order of 10^3 while in a comparable flow the Baldwin-Lomax model obtains values on the order of 10^4, where μ_l is the laminar viscosity. It is probable that the "transitional state" of the separating boundary layer, conjectured in [1] to be necessary to produce the low frequency oscillation, may be characterized by certain values of the ratio μ_t/μ_l. However, a systematic trend is far from clear at this point and will require further investigation.

The computations were also carried out for the "iced" airfoil, at $\alpha = 9°$, $R_c = 10^5$ and $M = 0.10$, to approximately match the present experimental conditions. The results have added further to the confusion. The "MML" turbulence model, which yielded the oscillation at the higher R_c, resulted in damped oscillations. The corresponding C_l variation is shown in Figure 13(b). On the other hand, when the flow was assumed laminar for this case, the computation yielded a low frequency oscillation! This is shown in Figure 13(a). The St_s values corresponding to the oscillations in Figures 12(a) and 13(a) turn out to be about 0.011 and 0.008, respectively. Let us note, for comparison, that Rumsey's computation for both airfoils ("LRN" and NACA0012) yielded an $St_s \approx 0.03$. The experimental and the computational results are summarized in Tables 1 and 2, respectively.

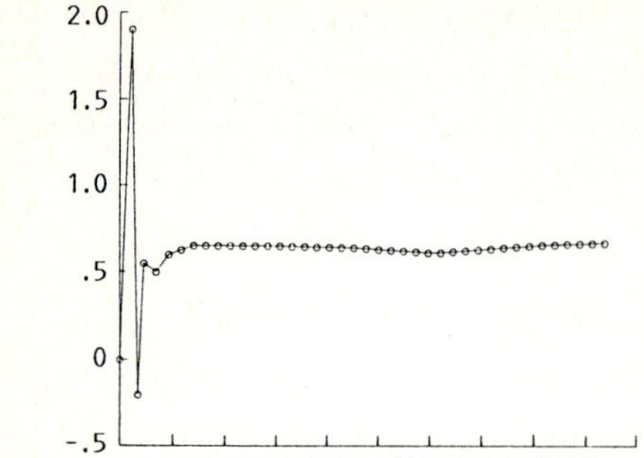

(b) BALDWIN-LOMAX TURBULENCE MODEL.

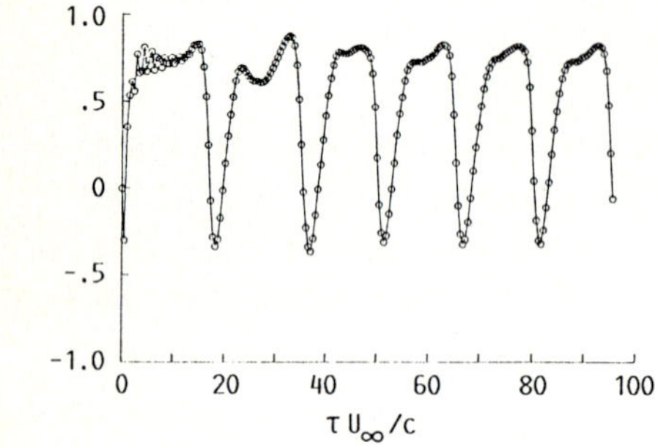

(a) "MML" TURBULENCE MODEL.

FIGURE 12. - COMPUTED LIFT-COEFFICIENT VARIATIONS WITH TIME FOR THE 'ICED' AIRFOIL AT A HIGHER R_c. $R_c = 1.4 \times 10^6$, M = 0.12, $\alpha = 10^0$.

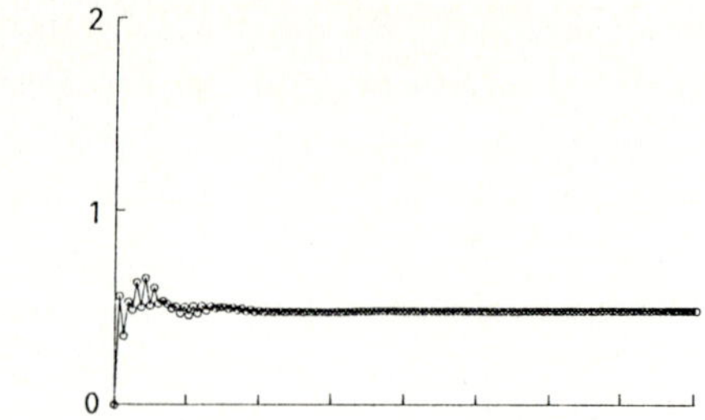

(b) "MML" TURBULENCE MODEL.

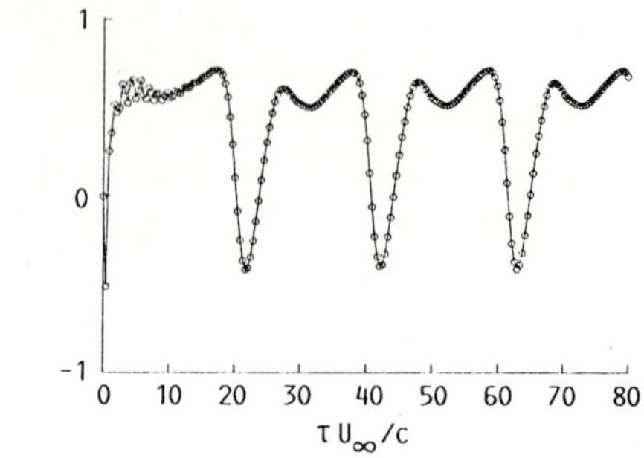

(a) LAMINAR FLOW.

FIGURE 13. - COMPUTED LIFT-COEFFICIENT VARIATIONS WITH TIME FOR THE 'ICED' AIRFOIL AT THE R_c COVERED IN THE EXPERIMENT. $R_c = 10^5$, M = 0.1, $\alpha = 9^0$.

TABLE 1. - EXPERIMENTAL RESULTS ON THE LOW FREQUENCY OSCILLATION

Airfoil	α	$R_c x 10^{-3}$	c, cm	St_s	Reference
R.A.F.28	14	100	15.2	0.019	[6]
"LRN"	15	40 to 140	10.2	0.02	[2]
"LRN"	17	15 to 25	7.3	0.025	[1]
"LRN"	15	50 to 150	12.7	0.02	[1]
"LRN"	16	100 to 300	25.4	0.033†	[1]
NACA0012 (w/trip)	14	70	10.2	0.022	[1]
Wortman	15	50 to 150	12.7	Damped	[1]
"Iced"	6	1500	55	0.016	[7]
"Iced"	8.5	75 to 125	12.7	0.02	Present

†Higher mean velocity due to large blockage may have resulted in higher St_s.

TABLE 2. - COMPUTATIONAL RESULTS

Airfoil	α	Turbulence model	$R_c x10^{-3}$	M	St_s	Reference
NACA0012	18	Baldwin-Lomax	1000	0.3	0.03	[5]
NACA0012	18	Laminar	1000	0.3	0.16	[5]
"LRN"	15	Baldwin-Lomax	75	0.3	0.03	[1]
Wortmann	15	Baldwin-Lomax	75	0.3	Damped	[1]
"Iced"	10	Baldwin-Lomax	1400	0.12	Damped	Present
"Iced"	10	MML	1400	0.12	0.011	Present and [3]
"Iced"	9	MML	100	0.10	Damped	Present
"Iced"	9	Laminar	100	0.10	0.008	Present

4. Concluding Remarks

The experimental evidence gathered in this paper suggest that the low frequency oscillation may indeed be a frequent occurrence with various airfoils. Its occurrence with "glaze ice accretion" is certainly a possibility. Thus, the leading edge ice accretion may not only be detrimental for aerodynamic performance but should also be of concern in terms of large unsteady loads associated with the phenomenon. The phenomenon, however, occurs in a narrow range of angle of attack around the onset of static stall. The acoustic excitation effect results also suggest a possible remedy, which needs to be explored further. The accumulated experimental evidence points towards a Strouhal number of about 0.02 associated with the phenomenon, although there are scatter in the data. With respect to the computations, questions remain in the application of turbulence models to separated flows. Nevertheless, it is felt that the essence of the phenomenon can be captured computationally with certain combinations of turbulence model, Reynolds number and airfoil shape. Obviously, a detailed computational study is in order not only to explore the validity of the algorithms but also to shed light into the still unresolved mechanism of this curious phenomenon.

References:

[1] "A natural low frequency oscillation of the flow over an airfoil near stalling conditions", K.B.M.Q. Zaman, D.J. McKinzie and C.L. Rumsey, J. Fluid Mech., 202, pp. 403-442, 1989.

[2] "Effect of acoustic excitation on the flow over a low-Re airfoil", K.B.M.Q. Zaman, A. Bar-Sever and S. M. Mangalam, J. Fluid Mech., 182, pp. 127-148, 1987.

[3] "Numerical analysis of a NACA0012 airfoil with leading edge ice accretions", M.G. Potapczuk, AIAA Paper No. 87-0101, 1987.

[4] "Control of "Laminar Separation" over Airfoils by Acoustic Excitation", K.B.M.Q. Zaman and D.J. McKinzie, AIAA Paper No. 89-0565, 1989.

[5] "A computational analysis of flow separation over five different airfoil geometries at high angles-of-attack", C.L. Rumsey, AIAA Paper No. 87-0188, 1987.

[6] "The reaction on a wing whose angle of incidence is changing rapidly. Wind tunnel experiment with a short period recording balance", W.S. Farren, Report & Memorandum 1648, Aeronautics Laboratory, Cambridge, 1935.

[7] "Experimental measurements in a large separation bubble due to a simulated glaze ice accretion", M.B. Bragg and A. Khodadoust, AIAA Paper No. 88-0116, 1988.

DETACHMENT OF TURBULENT BOUNDARY LAYERS WITH VARYING FREE-STREAM TURBULENCE AND LOWER REYNOLDS NUMBERS

J. Leith Potter, R. Joel Barnett, Costas E. Koukousakis, and Carl E. Fisher

Mechanical Engineering, Vanderbilt University, Nashville, TN 37235

SUMMARY

Experiments were conducted to determine if free-stream turbulence scale affects detachment of turbulent boundary layers. In consideration of possible interrelation between scale and intensity of turbulence, the latter characteristic also was varied and its role was evaluated. Flow over a 2-dimensional airfoil in a subsonic wind tunnel was studied with the aid of hot-wire anemometry, liquid-film visualization, a Preston tube, and static pressure measurements. Profiles of velocity, relative turbulence intensity, and integral scale in the boundary layer were measured. Detachment boundary was determined for various angles of attack and free-stream turbulence.

The free-stream turbulence intensity and scale were found to spread into the entire turbulent boundary layer, but the effect decreased as the airfoil surface was approached. When the changes in stream turbulence were such that the boundary layer velocity profiles were only slightly changed, detachment location was not significantly affected by the variations of intensity and scale. Pressure distribution and laminar or turbulent state remained the key factors in determining detachment location.

New data on the detachment of turbulent boundary layers were obtained. The range of flow conditions made it possible to evaluate the best-known rapid methods for predicting turbulent detachment at lower Reynolds numbers than previously reported. This revealed that approximations and empirical constants based on data for high Reynolds numbers and strongly adverse pressure gradients should not be uncritically accepted for predicting detachment of flows at low Reynolds numbers.

INTRODUCTION

This investigation was undertaken to learn if the scale of turbulence in the free stream has any influence upon the location at which turbulent boundary layer detachment occurs in an adverse pressure gradient. Both free-stream turbulence intensity and scale recently have been reported to influence local skin friction coefficient (refs. 1-7). Thus, in view of the linkage of that factor with separation location provided by Townsend's analysis (ref. 8), it is implied that both free-stream turbulence intensity and scale may affect separation. If so, there would be concern about the varying turbulence scales in wind tunnels, airbreathing propulsion units, and numerous other internal flows. In addition, the possibility of optimizing flow-manipulating devices such as vortex generators would seem feasible. Flow separation and related aerodynamic phenomena are of great importance in many fields of fluids engineering.

In consideration of the expected need to investigate turbulence intensities that would be rather high in the context of wind tunnel flows, this investigation has been focused largely on turbulent boundary layers. Although laminar detachment can occur at low Reynolds numbers, in circumstances of practical interest it will frequently be followed by transition of the unstable separated shear layer and subsequent reattachment as a turbulent boundary layer. The latter may separate still further downstream if it is exposed to a sufficiently strong positive pressure gradient. Thus, it is possible to have both laminar and turbulent separations within rather short streamwise distances.

This investigation was confined to subsonic, essentially two-dimensional flows. The approach was, first, to determine if free-stream intensity and scale variations affect detachment and then to define the parameters that controlled the separation process.

TECHNICAL APPROACH

Experimental Configuration

A subsonic wind tunnel with its associated pressure transducer and hot-wire anemometry systems and a 2-dimensional airfoil comprised the principal experimental tools used in this investigation. A complete description of the experimental apparatus and techniques is given in ref. 9.

Figure 1 illustrates the airfoil and spoiler used in this research. The upper surface has static pressure orifices located at the nominal stagnation point and at 1.27-cm intervals downstream.

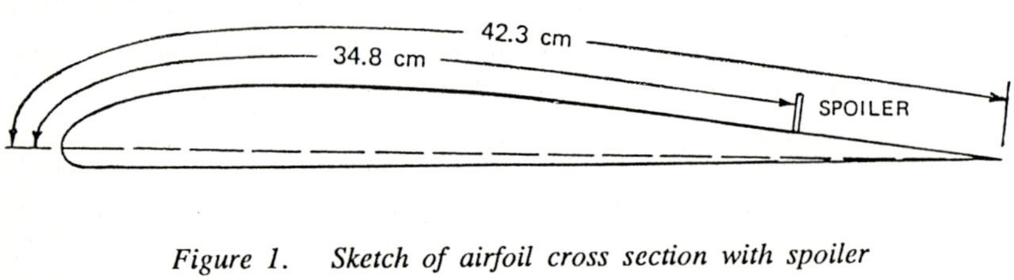

Figure 1. Sketch of airfoil cross section with spoiler

Angles of attack were varied. The spoiler was present in all cases discussed in this paper. It produced a uniform, essentially two-dimensional separation at relatively low airfoil angles of attack.

Free-stream turbulence characteristics were determined at a point in the empty tunnel where the leading edge of the airfoil was later installed. Measurements were made in the boundary layer on the airfoil at selected stations immediately upstream of the separation region. Care was taken to assure fully-developed turbulent boundary layers.

Following Simpson (ref. 10) and others, separation is here taken to mean the entire process of breakaway of a boundary layer from its bounding surface. It is essential to note that separation, like boundary layer transition, occupies a spatial zone and detachment may occur at different stations within the separation zone at any instant in time. This can have important implication in connection with the interpretation of patterns in liquid surface films or boundary layer profiles measured with probes that reflect a timewise integration or averaging of the phenomenon.

Definitions of Turbulence Characteristics

Relative longitudinal turbulence intensity is defined as

$$T' = (\overline{u^2})^{0.5}/U$$

where u = fluctuating streamwise velocity component

U = mean streamwise velocity

References 11-12 , for example, contain detailed discussions of integral scales, so only the particular approach followed in the present work is described. If the fluctuating velocity component u is measured at times t and (t + r), a correlation coefficient may be defined as

$$R_u = \overline{u(t) \cdot u(t+r)}/[\overline{u^2}(t) \cdot \overline{u^2}(t+r)]^{0.5} \quad (1)$$

A quantity defined as

$$t = \int_0^\infty R_u \, dt \quad (2)$$

is multiplied by the mean velocity, U, to give a length I_x which represents the typical streamwise dimension of the energy- containing eddies in the flow. This is the longitudinal integral scale used hereafter. Consideration of other coordinate directions and velocity components leads to the other integral scales. In practice the integration in Eq. (2) is terminated where R_u first becomes zero. A catalog of free-stream parameters, T'_∞ and $I_{x\infty}$, was compiled for five turbulence-producing grids located at various distances upstream of the test section.

The previous investigations of the combined effect of both intensity and scale on local skin friction or Stanton number (cf. refs. 1-7) led to several empirically determined correlation parameters. These parameters collapsed skin friction or Stanton number data to single curves with moderate scatter, within the scope or range of conditions covered in the referenced studies. They all suggest that the effects of intensity and scale are opposite, i.e., skin friction increases with intensity but decreases as scale increases. In regard to surface shear, both influences are weak at levels of turbulence normally encountered in aeronautics.

RESULTS

Detachment Locations

Figure 2 is an example of the results obtained by the liquid-film technique for locating detachment. All records displayed the general features outlined here:

(a) A region approximately 1 cm in streamwise extent where the film was cleaned from the airfoil surface. In view of the film being pushed downstream to (b), the mean surface shear apparently was in the downstream direction and sufficient to overcome the adverse pressure gradient.

(b) A highly repeatable, narrow, 3-mm-wide, spanwise liquid band apparently marking the mean detachment location. Varying liquid film thicknesses, viscosities and drying times did not alter results. Band (b) may be regarded as the station where averaged positive or downstream surface shear is balanced by negative or upstream shear and the adverse pressure gradient. It has been taken as the detachment location. The dividing streamline, on average, originates here.

(c) An area several cm in streamwise extent where relatively low and random surface shear is evident. When the paint in this region was not overflowed by paint from other regions, it remained smooth and uniform while it dried.

(d) High repeatable, roughly 1-cm region at the foot of the spoiler where the surface is cleansed of paint by energetic vortical flow. The paint is swept upstream to the rear boundary of region (c).

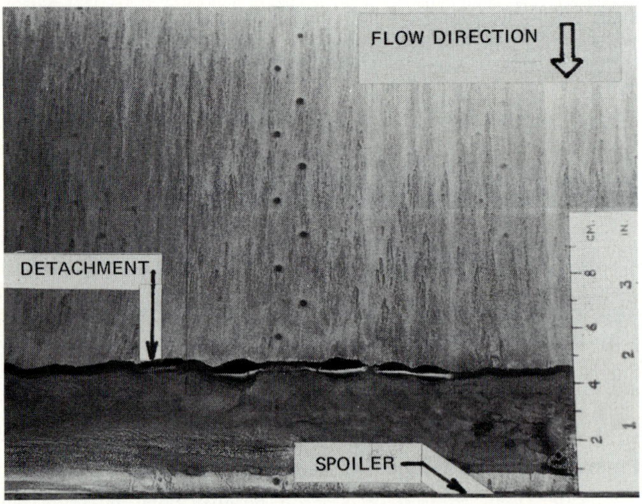

Figure 2. Liquid-film data on the separation zone

Interaction of Free-Stream and Boundary Layer Turbulence

Profiles of U/U_e, T', and I_x were measured at the stations x = 22.1 cm and x = 29.7 cm for different levels of turbulence in the free stream (T'_∞). The former station is well upstream of the separation zone, while the latter station is immediately upstream of the average detachment. The subscript, e, designates boundary layer edge.

It was noted that change of T'_∞ from 0.16% to 3.16% had only a small effect on thickness and shape of the boundary layer at either chordwise station. Figures 3-5 display the profiles of

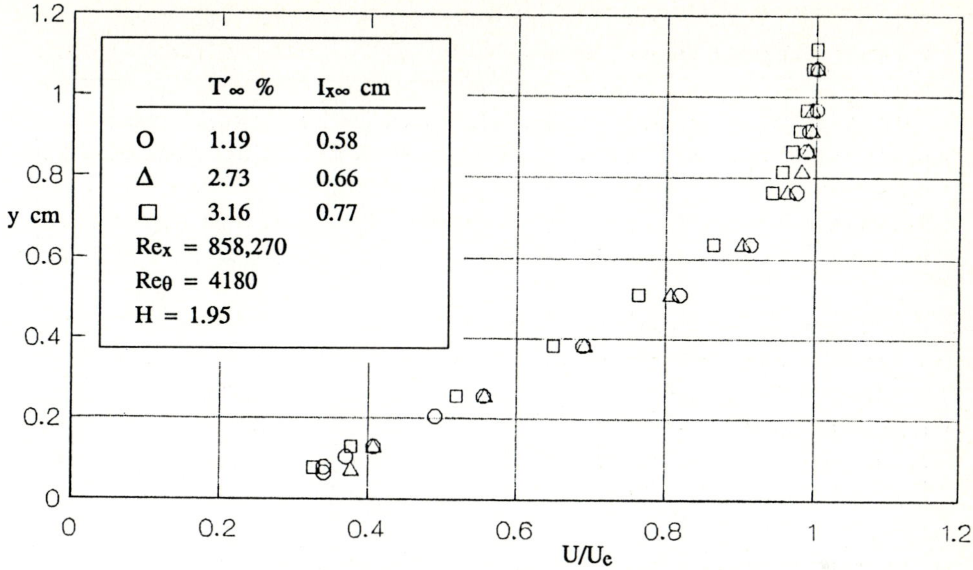

Figure 3. Boundary layer velocity profiles for varying free-stream turbulence: x = 29.7 cm

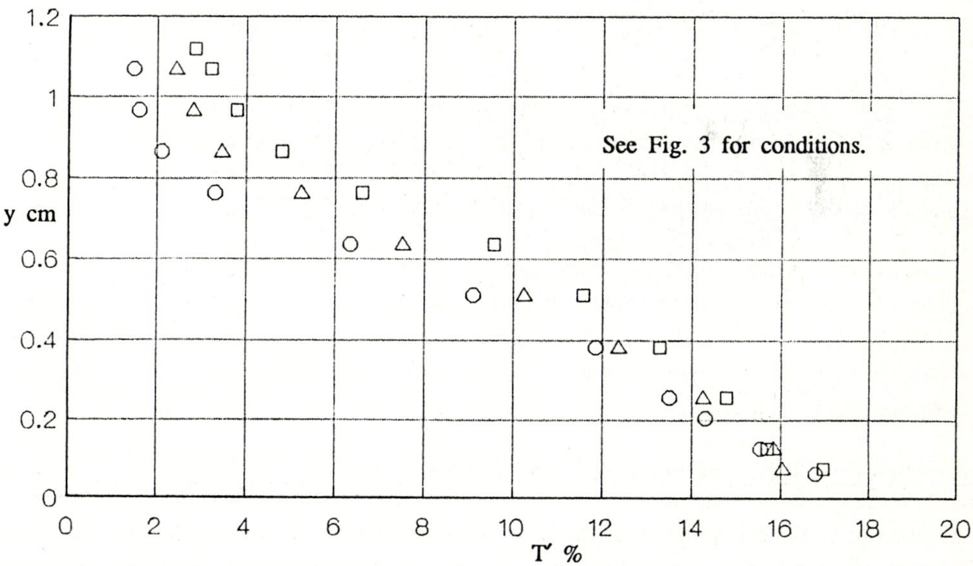

Figure 4. Boundary layer turbulence intensity profiles for varying free-stream turbulence: x = 29.7 cm

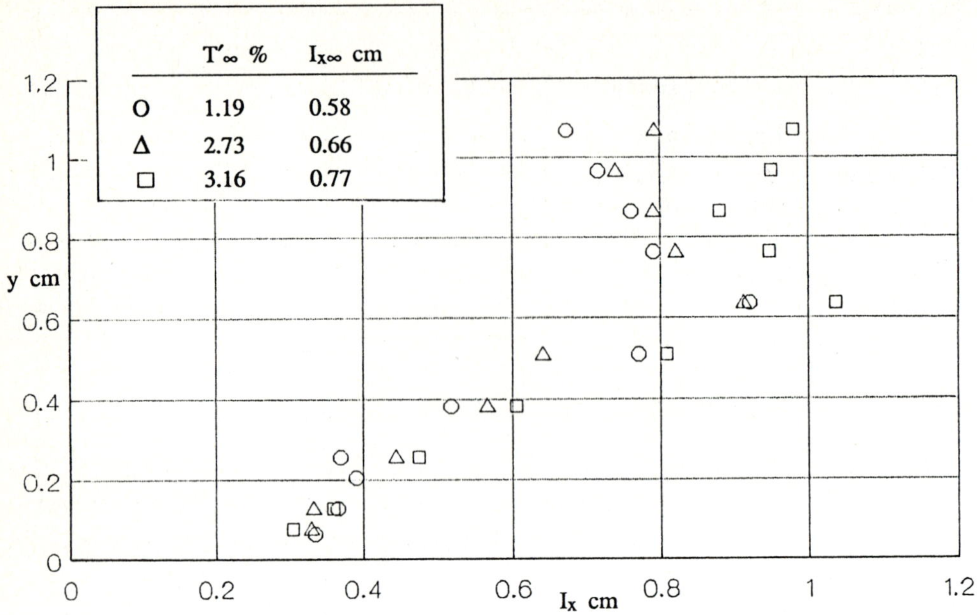

Figure 5. Boundary layer integral scale profiles for varying free-stream turbulence: x = 29.7 cm

U/U_e, T′, and I_x found at x = 29.7 cm. At this station (approximately at detachment), the situation is different in some respects from the x = 22.1 cm case. Boundary layer thickness increased markedly in comparison to the data for x = 22.1 cm, and the velocity profile displays considerable retardation caused by the adverse pressure gradient.

The response of the boundary layer profiles to free-stream turbulence intensity and scale shown in Figs. 4 and 5 is similar to that found for those two parameters at x = 22.1 cm. These figures show that the feed-in of free-stream turbulence characteristics weakens as the surface is approached and has reduced influence near the solid boundary where very high turbulence arises even when free-stream intensity is low.

A significant point well illustrated in Fig. 4 is that T′ near the solid surface in the separation zone is much greater than the free-stream turbulence intensity. That strongly implies that only very high levels of free-stream turbulence may have any possibility of affecting separation. Of course, very great turbulence will also affect boundary layer growth and thereby affect the pressure distribution. The latter factor is dominant in regard to detachment and may easily overshadow any direct effect of changes in turbulence intensity within the boundary layer.

Figure 5 shows that integral sales within the boundary layer had a characteristic profile that contained scales lower than free-stream near the wall and greater than free-stream in the outer part of the boundary layer. There was a consistent increase of I_x as $I_{x\infty}$ increased, but the effect was small deep within the boundary layer, i.e., as y approached zero.

When examining the profile of I_x; some consideration should be given to the nature of the autocorrelation procedure and the degree of uncertainty. A repeatability of approximately ± 10% is indicated by $I_{x\infty}$ measured in the free stream on different days under the less favorable signal - to - noise conditions corresponding to low free- stream turbulence. However, conditions near the outer edge of turbulent boundary layers are such as to impose greater scatter of I_x values. It should also be recalled that I_x will decrease as the local mean velocity decreases because of the direct relation between I_x and U assumed in determining I_x.

Detachment and the Influence of Free-Stream Turbulence

Tables 1 and 2 and Figs. 6 and 7 present detachment station, x_d, at corresponding T'_∞ and $I_{x\infty}$, as determined in this investigation of subsonic, turbulent boundary layers in typical adverse pressure gradients on a two-dimensional airfoil. Figures 6 and 7 correspond to angle of attack, α = 5 deg and 50.5 m/s. Typical surface static pressure distributions across the separation zone are shown in these figures. The "detachment" label in these figures is placed where the paint band (b) was located. It will be noted that the pressure coefficient did not reach a constant value at the designated detachment, but it did become constant a short distance downstream. Tables 1 and 2 present data selected so that either T'_∞ or $I_{x\infty}/\delta$ was nearly constant while the other parameter varied. This format makes it easier to see how or if each turbulence parameter independently influenced detachment point. The airfoil was at 5 deg. angle of attack and 50.5 m/s conditions when the data of Tables 1 and 2 were recorded.

$I_{x\infty}/\delta_{.955}$	T'_∞%	x_d cm
0.67	0.46	29.7 ± 0.25
0.67	1.19	29.7 ± 0.25
0.67	2.73	29.7 ± 0.25
0.56	0.31	29.7 ± 0.25
0.56	0.38	29.7 ± 0.25
0.56	2.85	29.7 ± 0.25

Table 1. Effect of free-stream turbulence intensity upon detachment location with constant turbulence integral scale

In all figures, the pressure coefficient is the conventional one, i.e.,

$$C_p = (p - p_\infty)/q_\infty$$

$T'_\infty\%$	$I_{x\infty}/\delta_{.955}$	x_d cm
0.30	1.64	29.7 ± 0.25
0.30	1.32	29.7 ± 0.25
0.30	1.23	29.7 ± 0.25
0.30	0.63	29.7 ± 0.25

Table 2. Effect of free-stream integral scale upon detachment location with constant turbulence intensity

where p = local static pressure and p_∞, q_∞ = free-stream static and dynamic pressure, respectively.

Some of the methods for predicting detachment location use the pressure coefficient, $(p - p_o)/q_o$, where p_o, q_o are the local static and dynamic pressures, respectively, at the station of minimum pressure, i.e., at the upstream beginning of the rising pressures leading to detachment. Hereafter, the subscript o will denote conditions at x_o, the point of minimum pressure on the airfoil surface, and the modified pressure coefficient based on p_o and q_o will be designated K_p.

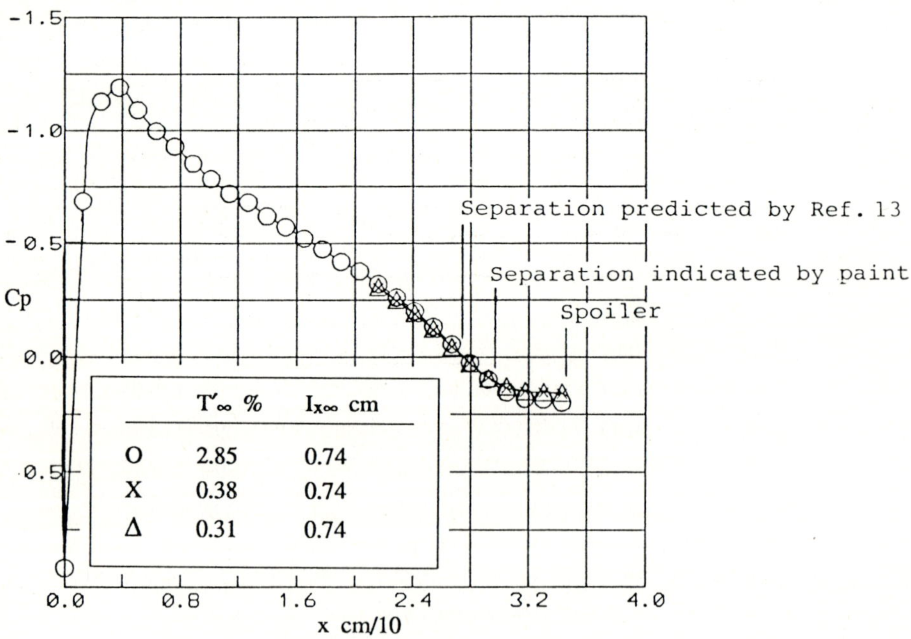

Figure 6. Pressure coefficient and liquid-film results for varying turbulence intensity

The range of T'_∞ adequately represents wind tunnel flows and the turbulence of some other types of internal flows. As planned, the free-stream integral scales are approximately equal to the boundary layer thicknesses upstream of the separation zone. Several previous investigators have concluded that this should lead to the maximum influence on surface shear stress. The boundary layer thickness grew rather rapidly between the mid-chord and detachment stations. Thus, $I_{x\infty}$ generally fell between the values of $\delta_{.995}$ at x = 22.1 cm and x = 29.7 cm in these experiments. The ratio $I_{x\infty}/\delta_d$ given in the tables is based upon $\delta_{.995}$ at x = 29.7 cm, which is approximately the detachment station in all cases. If $\delta_{.995}$ at x = 22.1 cm were used, all ratios would be approximately doubled because the boundary layer thickness grew rapidly toward detachment. The subscript, .995, designates a value of y where U = .995 U_e.

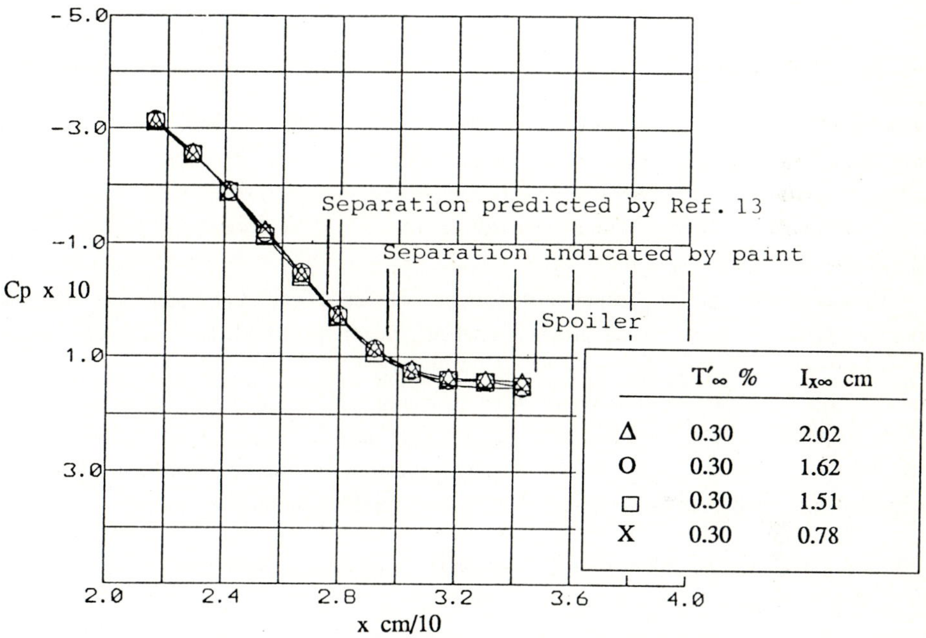

Figure 7. Pressure coefficient and liquid-film results for varying integral scale

The evidence in the tables and figures seems decisive; in the range of T'_∞ and $I_{x\infty}$ investigated, neither parameter had a significant effect on x_d. All measurements were repeatable to a very satisfactory degree, and no anomalies were observed.

Comparisons with Detachment Prediction Methods

A second objective in this work was the comparison of the detachment results with predictions by Townsend's and Stratford's methods, refs. 8 and 13, respectively. Inasmuch as Reynolds numbers were relatively low in this investigation, there was an opportunity to learn how the better-known predictive methods perform under that condition. These results are presented in Table 3.

					x_d cm			
Run	$\alpha°$	U_∞ m/s	T'_∞%	$l_{x\infty}$ cm	Exp.*	Strat.1	Strat.2	Townsend
396B	1	50.5	0.16	1.35	29.6	28.3	29.9	30.8
404A	3	50.5	0.16	1.35	29.7	27.1	27.7	28.7
404B	9	50.5	0.16	1.35	27.9	27.9	27.9	-
374A	1	50.5	1.19	0.58	29.7	27.4	30.1	30.8
374B	1	64.2	1.17	0.64	30.2	27.7	29.9	-
363	1	49.8	2.73	0.66	29.7	28.5	28.8	29.3
403	1	51.1	3.16	0.77	30.5	28.4	30.0	30.7

*Uncertainty ± 0.25 cm

Table 3. Comparison of experimental and predicted detachment locations

The experimental airfoil pressure distributions were used when calculating x_d. This point is important because if experimental pressure gradients begin to "relax" or "anticipate" detachment, that will influence the predicted value of x_d. However, Figs. 6 and 7 show that experimental x_d almost always fell upstream of any significant relaxation of the pressures. Thus, the comparisons seem to be generally valid. In cases where actual x_d is unknown, care must be exercised. Usually, a best estimate or calculation of the non-separating pressure distribution must be obtained, and then the predictive method may be based on that "uncontaminated" condition.

It is implied by the gradual relaxation of pressure gradient at experimental x_d that the leading edge of the detachment region is characterized by a very shallow concave curvature of the displacement boundary layer. This is also indicated by results obtained with various probes and flow visualization techniques used by the present investigators, cf. Fig. 3.

In Table 3, the designations Strat. 1 and Strat. 2 refer, respectively, to the version of Stratford's method originally presented in ref. 13 and the modified method proposed by Cebeci, et al. in ref. 14. The modified version incorporates a small change in the critical value of Stratford's separation criterion. In both cases, the initial length of laminar flow and favorable pressure gradient that existed on the forward portion of the airfoil in these experiments is taken into account. When the modification recommended by Cebeci, et al. is made, Table 3 shows excellent agreement between Strat. 2 and experiment.

Inasmuch as the surface shear stress coefficient, C_{fo}, appears directly in Townsend's method for predicting x_d (ref. 8), a brief examination of the effect of increased C_{fo} caused by elevated free-stream turbulence was carried out. The experimental C_{fo} values obtained in the present study by use of a Preston tube fall very close to the Simonich and Bradshaw curve given in ref. 3. It was found that, at the levels of turbulence dealt with in this case, the influence of slightly elevated C_{fo} upon x_d is negligible.

The calculations by Townsend's method generally agree closely with the results obtained by Stratford's method, as demonstrated in Table 3. However, it was observed that the Townsend method tends to predict x_d slightly higher than the experimental value when Reynolds numbers are lower, or C_{fo} is higher. Two key relations in that method are expressible as follows:

$$(dK_p/dx)_d/(dK_p/dx)_i = R = 0.24$$

and

$$(x_o - x')(dK_p/dx)_o >> \sqrt{C_{fo}}$$

where x' is the effective origin of a fully turbulent boundary layer and $K_p = (p - p_o)/q_o$.

It was discovered that to force agreement of Townsend's predictions and experimental data, it was necessary to make R a function of C_{fo}. Furthermore, the values that R must assume become higher than 1.0 when $C_{fo} > 0.0047$. This corresponds roughly to $Re_o < 3 \times 10^5$ for low-speed turbulent boundary layers. Considering that $R > 1$ is unrealistic, it seems clear that not only the above relation involving R but also the second relation involving pressure gradient and C_{fo} become questionable at lower Reynolds numbers. This constitutes a warning that empirical coefficients selected on the basis of data for high Reynolds numbers and strongly adverse pressure gradients should not be uncritically accepted when low Reynolds numbers and moderate pressure gradients are involved.

ACKNOWLEDGMENT

This work was largely supported by NASA Research Grant NAG-1-483, W. D. Harvey, Technical Officer.

REFERENCES

1. Hancock, P. E.: Effect of Free-Stream Turbulence on Turbulent Boundary Layers. Ph.D. Thesis, Imperial College, University of London, 1978.

2. Meier, H. U.: The Response of Turbulent Boundary Layers to Small Turbulence Levels in the External Free Stream. ICAS Paper No. 76- 05, The Tenth Congress of the International Council of the Aeronautical Sciences, Oct. 1976.

3. Simonich, J. C. and Bradshaw, P.: Effect of Free-Stream Turbulence on Heat Transfer Through a Turbulent Boundary Layer. *ASME Journal of Heat Transfer*, vol. 110, Nov. 1978, pp. 671-677.

4. Meier, H. U. and Kreplin, H.-P.: Influence of Freestream Turbulence on Boundary-Layer Development. *AIAA Journal*, vol . 18, Jan. 1980, pp. 11-15.

5. Hancock, P. E. and Bradshaw, P.: The Effect of Free-Stream Turbulence on Turbulent Boundary Layers. *ASME Journal of Fluids Engineering*, vol. 105, Sept. 1983, pp. 284-289.

6. Blair, M. F.: Influence of Free-Stream Turbulence on Turbulent Boundary Layer Heat Transfer and Mean Profile Development, Part 1 - Experimental Data and Part 2 - Analysis of Results. *ASME Journal of Heat Transfer,* vol. 105, Feb. 1983, pp. 33-47.

7. Castro, T. P.: Effects of Free-Stream Turbulence on Low Reynolds Number Boundary Layers. *ASME Journal of Fluids Engineering*, vol. 106, Sept. 1984, pp. 298-306.

8. Townsend, A. A.: The Behaviour of a Turbulent Boundary Layer Near Separation. *Journal of Fluid Mechanics*, vol. 12, April 1962, pp. 536-554.

9. Potter, J. L., Barnett, R. J., Fisher, C. E., and Koukousakis, C. E.: The Influence of Free-Stream Turbulence on Separation of Turbulent Boundary Layers in Incompressible Two-Dimensional Flow. NASA CR- 180638, 22 Dec. 1986.

10. Simpson, R. L.: A Review of Some Phenomena in Turbulent Flow Separation. *ASME Journal of Fluids Engineering*, vol. 103, Dec. 1981, pp. 520-533.

11. Townsend, A. A.: *Structure of Turbulent Shear Flow.* University Press, Cambridge, 1956.

12. Hinze, J. O.: *Turbulence.* McGraw-Hill, New York, 1959.

13. Stratford, B. A.: The Prediction of Separation of the Turbulent Boundary Layer. *Journal of Fluid Mechanics,* vol. 5, 1959, pp. 1-16.

14. Cebeci, T.; Mosinskis, G. J.; and Smith, A. M. O.: Calculation of Separation Points in Incompressible Turbulent Flows. *AIAA Journal of Aircraft,* vol. 9, Sept. 1972, pp. 618-624.

Wind-Tunnel Investigations of Wings with Serrated Sharp Trailing Edges

P.M.H.W. Vijgen
High Technology Corporation
28 Research Drive
Hampton, VA 23666

C. P. van Dam
University of California
Davis, CA 95616

B. J. Holmes and F. G. Howard
NASA Langley Research Center
Hampton, VA 23665

Abstract

Exploratory wind-tunnel force measurements are presented for two wing geometries with small-scale planar and nonplanar serrated trailing-edge devices (chord-Reynolds numbers ranged from $1.0 - 3.7 \times 10^6$). The planar serrated trailing-edge extensions reduced the drag at conditions when trailing-edge separation occurred at low angles of attack. The introduction of serrations reduced or eliminated the drag penalty due to the small (1-2 percent of the chord length) nonplanar trailing-edge flaps, while maintaining the effects of increase in camber. The presence of streamwise vortices immediately downstream of the serrated trailing edges is believed to have favorably affected the boundary-layer flow approaching the trailing edge and the near-wake development, resulting in reduced pressure (form) drag.

Introduction

The detailed shape of the trailing-edge of airfoils significantly affects the aerodynamic characteristics of finite- and infinite-span wings. According to inviscid theory, the camber shape near the trailing edge and the trailing-edge shape itself determine the total amount of lift generated (Kutta condition). In a viscous flow, the shape of the trailing edge affects the magnitude of pressure (form) drag, in addition to affecting the confluence of the viscous flow layers over the upper and lower surface into the trailing wake. Depending on the Reynolds number, the trailing-edge shape can affect the oscillatory character of the trailing wake and the feedback coupling with the boundary-layer flow approaching the trailing edge. Also, depending on the Mach number and the Reynolds number, the airfoil trailing-edge shape can affect the location and strength of the terminating shock waves, and, hence, the wave drag incurred by the lifting surface.

The trailing edges of aircraft wings and empennages of most aircraft are straight when viewed in the spanwise direction. By contrast, the trailing edges of the fins and wings of many aquatic animals and birds show a (small-scale) three-dimensional serrated geometry (Refs. 1 and 2). One may observe such serrated shapes along the trailing edges of the fins of small fish as well as large whales (i.e., over a large range of Reynolds numbers based on the average fin chord). Many birds, such as the owl, show a distinctly serrated trailing-edge wing shape (Ref. 2). The precise function of such spanwise-periodic trailing-edge shapes is not well understood. For the owl, where a silent approach to its prey is imperative, the presence of serrated wing edges might affect the character of vortex shedding into the wake and alter or attenuate the acoustic signature of the bird. Leading-edge serrations, resembling leading-edge barb feathers on the owl's wing, have been shown to affect both the radiated sound as well as the high-angle-of-attack separated-flow and stalling behavior of wings (Refs. 3-5). Spanwise periodic vortex generators at the trailing edge have been used to control or inhibit the unsteady Von Kármán vortex shedding and the generation of sound behind bluff bodies (Ref. 6).

To reduce the base pressure drag of wings with thick blunt trailing-edge shapes, two-dimensional and three-dimensional splitter plates of various geometries have been shown effective to increase the base pressure (Refs. 7-11). In particular, segmenting the trailing edge into a spanwise discontinuous geometry prevents primary vortex shedding from the blunt edge. Gai and Sharma (Ref. 10) first investigated the application of triangular sawtooth-shaped serrations to generate streamwise vortices in the wake behind a blunt trailing edge.

Recently, the effect of axial (streamwise) vorticity in plane free-shear layers has been extensively studied by Lasheras and Meiburg et al. (Refs. 12-14), who investigated the three-dimensional development and interaction of Von Kármán and streamwise vortices in free-shear flow behind a sharp-edged splitter plate. Introduction of vertical or horizontal spanwise-periodic perturbations in the free-shear flow generated by a sinusoidally corrugated (rippled) splitter plate or a splitter plate with a spanwise sinusoidally indented flat trailing edge, resulted in the generation of axial counterrotating vortex filaments which interacted with the spanwise primary vortices (Ref. 12). In Lasheras' experiments, the spanwise wavelength of the serrated (indented) trailing edge varied between 1.5 - 8.0δ (δ is the total initial thickness of the wake), while the depth of the serration varied between 0.15 - 0.7δ (Ref. 13).

Werle et al. studied sinusoidally corrugated trailing-edge shapes of splitter plates in laminar or turbulent free-shear flows to generate spanwise periodic axial vortex arrays in the flow (Ref. 15). Compared to Lasheras et al., the lobe height of their convoluted trailing edge is much larger than the wake thickness. Application of such a large-scale corrugated trailing edge to a thick airfoil shape (thickness-to-chord ratio, t/c = 0.21, chord Reynolds number, R_c = 110,000) resulted in a significant increase in maximum lift, as well as a decrease in drag at high lift, due to the delay in trailing-edge separation of the turbulent boundary layer (Ref. 16). The corrugated trailing-edge shape apparently provided three-dimensional relief of the streamwise adverse pressure gradient that would normally have caused the turbulent boundary layer to separate prematurely (Ref. 16). Finally, unpublished drag measurements by Howard (NASA Langley, 1985) suggest a reduction in drag due to the addition of planar serrated trailing edges to a wing at chord-Reynolds numbers up to 250,000. From these experiments it appears that streamwise vortices immediately behind the trailing edge may favorably influence the shear-layer development both upstream and downstream of the trailing edge.

Small two-dimensional trailing-edge flaps mounted at the airfoil trailing edge perpendicular to the airfoil pressure side, have been reported to produce favorable changes in the lift-curve slope, the maximum attainable lift, and the nose-down pitching moment (Refs. 17 and 18). The observed changes are caused by the increase in camber (Ref. 19) due to the small so-called "Gurney" flaps (Ref. 17). In addition to a significant increase in lift, Liebeck reported a reduction in drag with a Gurney flap with depth-to-chord ratio d/c = 0.0125 attached to a thick symmetric airfoil (Ref. 17). The reduction in drag is attributed to a reduction in wake-momentum deficit in comparison to the conventional trailing-edge shape, caused by the formation of smaller separation bubbles at the trailing edge of the airfoil with the Gurney flap (Ref. 17). Such flow structure has recently been confirmed in flow-visualization studies by Newhart and Pendergraft at low Reynolds numbers (Ref. 20). Roesch and Verillet observed increases in lift but also a small increase in drag for a Gurney flap with d/c = 0.0125 while an increase in the depth of the nonplanar trailing-edge flap to d/c = 0.05 resulted in a large drag penalty at small lift coefficients (Ref. 18).

The studies conducted thus far have not addressed the effect of planar, serrated, trailing-edge shapes on the aerodynamic characteristics of aircraft wings with sharp trailing edges at higher chord-Reynolds numbers. This paper presents wind-tunnel measurements of wings equipped with planar serrated trailing-edge geometries. In addition, results are given of tests of wings equipped with two-dimensional (solid) Gurney flaps, comparable to the devices reported in Refs. 17 and 18, and sawtooth-shaped serrated Gurney flaps. The serrated Gurney flap is a novel effort (Ref. 21) to combine the favorable effect on lift of the solid Gurney flap with the effects of a three-dimensional vertically-indented trailing edge on the shear-layer development.

Experiments

Exploratory wind-tunnel experiments were conducted in the 14-By 22-Ft Subsonic Tunnel and the 30-By 60-Ft Wind Tunnel at the NASA Langley Research Center to measure the effects of planar and nonplanar trailing-edge serrations on the aerodynamic characteristics of high-aspect-ratio wings. This section briefly presents the test conditions and the geometries of the wing planforms and the small trailing-edge additions investigated in the experiments.

Figure 1 shows the geometry of the full-span high-aspect-ratio wing-body model used in the 14-By 22-Ft wind-tunnel experiment, and figure 2 provides a photograph of this model (Ref. 22). The baseline untwisted wing had an aspect ratio of 12.56 and a reference wing area of 13.06 ft^2. The section shape of the wing in the freestream direction was the NASA NLF(1)-0414F airfoil (Ref. 23). The airfoil trailing-edge had a finite thickness of 0.05 in. and had sharp corners. Boundary-layer transition in the present tests was fixed at 5 percent of the chord on the upper and lower surface along the complete wing span. Based on the reference chord $\bar{c}$ (mean aerodynamic chord) of 1.10 ft for the baseline

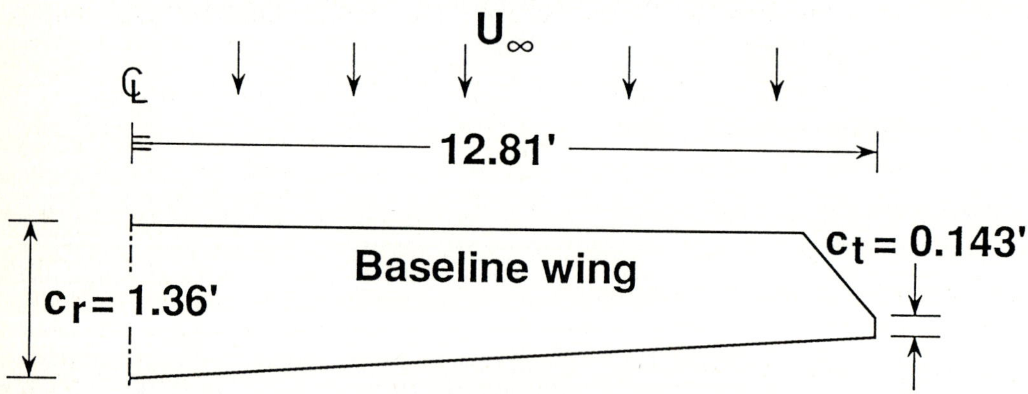

Fig. 1. Planform geometry of the baseline wing geometry tested in the NASA-Langley 14-By 22-Ft Subsonic Tunnel.

Fig. 2. Photo of the full-span baseline model installed in the NASA-Langley 14-By 22-Ft Subsonic Tunnel.

wing, R_c was 1.1×10^6. Location of transition at 5-percent chord at this Reynolds number resulted in flow separation (beyond nondimensional chord location $x/c = 0.8$) at angles of attack greater than about 1 deg for the baseline wing. The implications of this turbulent separated flow region are discussed later in this paper. The test procedures and accuracies are similar to those stated in Ref. 22.

Both planar and nonplanar trailing-edge extensions were studied using the wing model depicted in Fig. 1. Three types of planar trailing-edge shapes were attached to the lower surface of the baseline wing outboard of the nondimensional spanwise location $\eta = y/(b/2) = \pm\, 0.22$ (Fig. 3a). (Note that the centerbody extended to $\eta = \pm\, 0.09$. The 0.01 in. thick extensions were fixed to the lower surface of the wing using double-sided adhesive tape. The solid planar strip extended 0.5 in. behind the trailing edge of the airfoil, resulting in an increase in projected wing area of 0.42 ft^2 (3.2 percent); the scallop- and triangular-shaped serrated trailing-edge strips were attached in such a fashion that the projected area of the extension did not change (Fig. 3a). Defining the depth d of the serrations as the normal distance between the

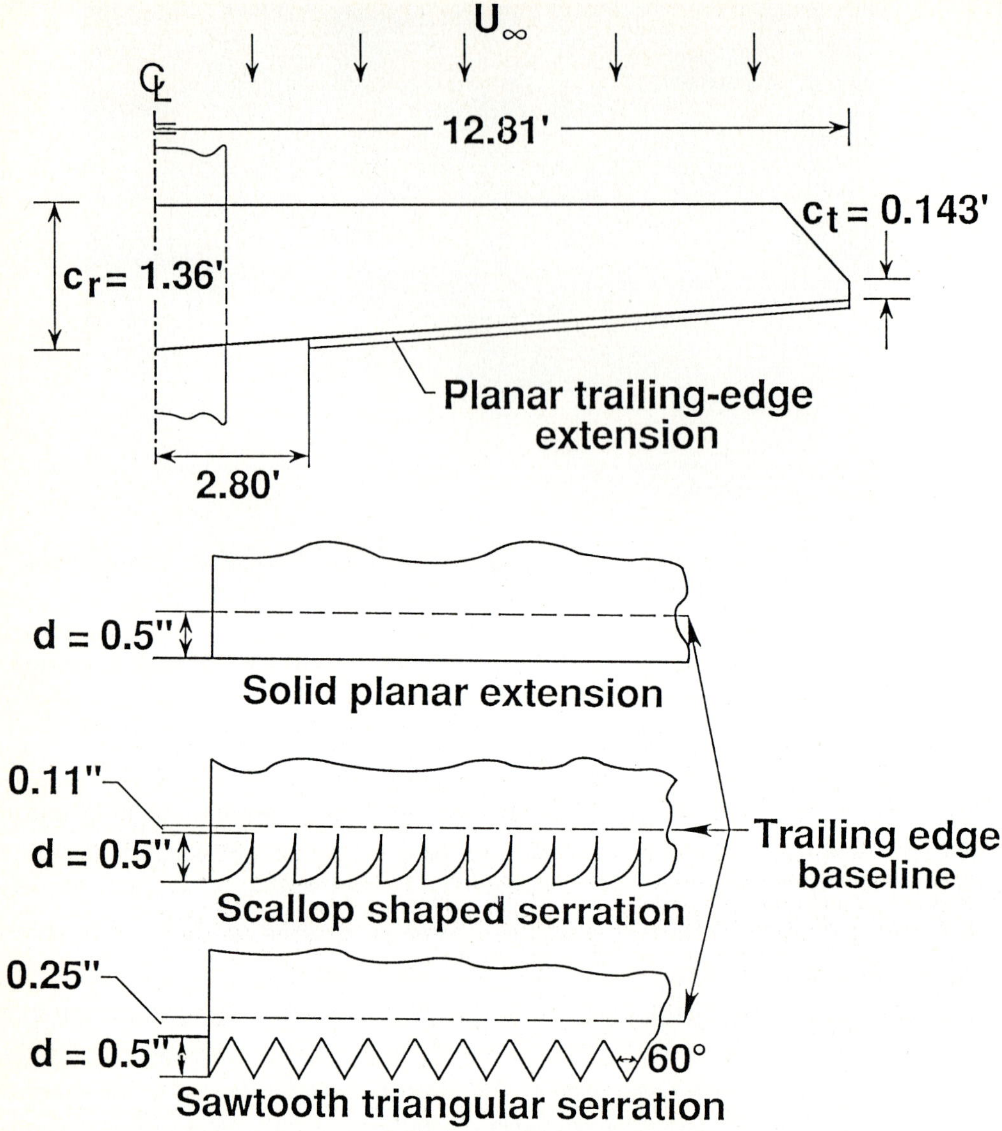

a. Planar solid and serrated configurations.

Fig. 3. Geometry and installation of the trailing-edge devices on the full-span wing (Fig. 1).

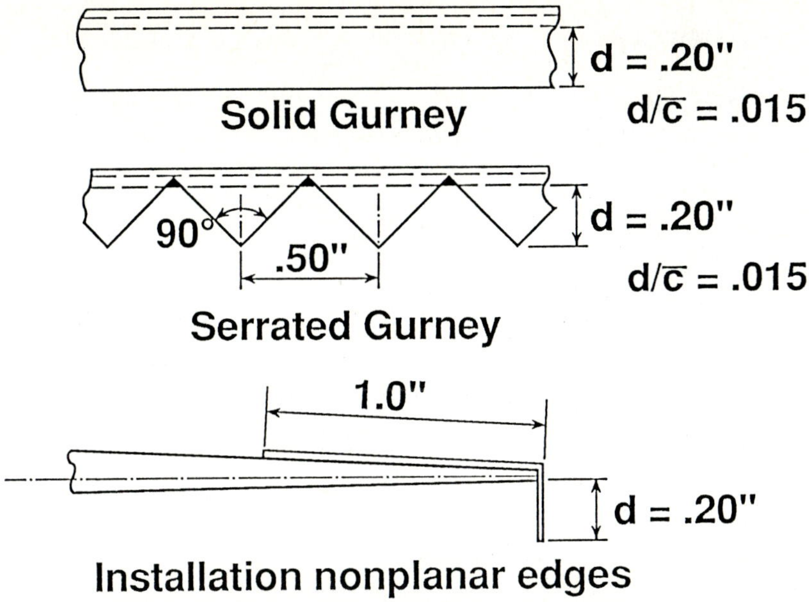

b. Nonplanar solid and serrated configurations.

Fig. 3. Concluded.

Fig. 4. Photo of the semi-span model tested in the NASA-Langley 30-By 60-Ft Wind Tunnel. (Wing trailing edge is equipped with planar serrations).

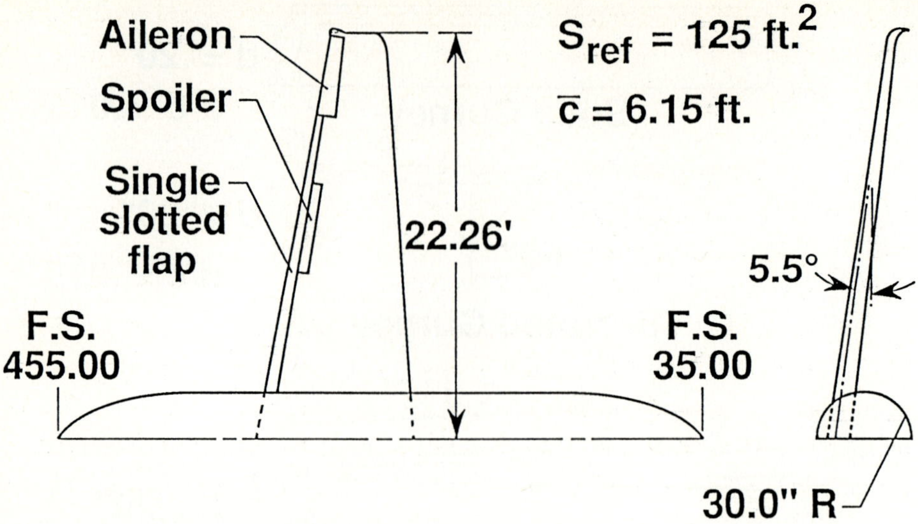

a. Baseline wing-body model.

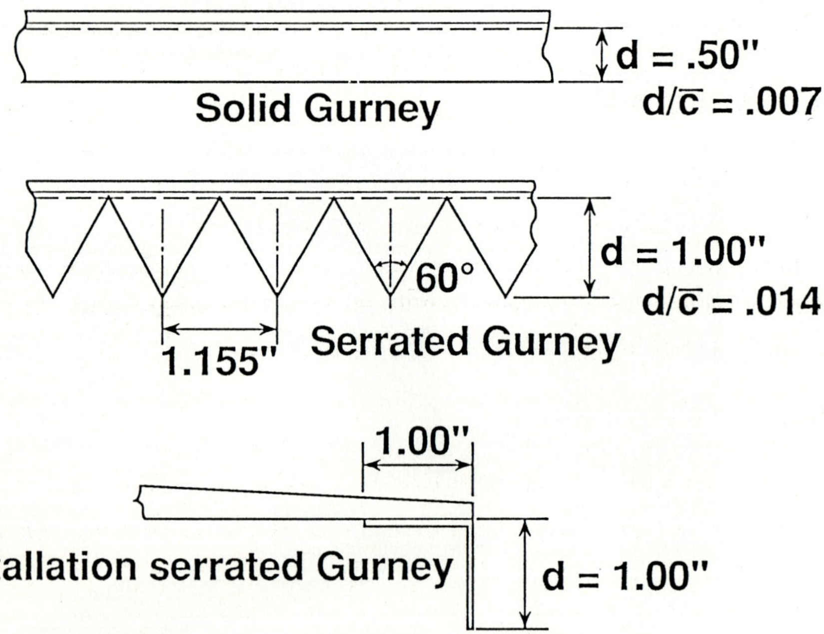

b. Nonplanar solid and serrated trailing-edge devices.

Fig. 5. Model geometry and trailing-edge devices tested in the 30-By 60-Ft Tunnel.

vertex and the base of each indentation (Fig. 3a), the depth-to-chord ratio was $d/\bar{c} = 0.038$. Figure 3b shows the geometry of the solid and serrated Gurney flaps investigated in the 14-By 22-Ft tunnel experiment. Both full-span (outboard of $\eta = \pm 0.09$) nonplanar flaps had a depth of 0.20 in. ($d/\bar{c} = 0.015$), while the acute angle of the serrations was 90 deg. The frontal area of the solid Gurney flap was slightly more than twice the area of the serrated flap. The Gurney flaps of thickness 0.025 in. were attached to the upper surface of the wing using double-sided adhesive tape (Fig. 3b).

Figure 4 shows the full-scale, semi-span, wing-body model installed in the 30-By 60-Ft Tunnel. The model geometry, depicted in Fig. 5a, was previously used to investigate the characteristics of a business-jet type wing (Ref. 24). The baseline wing (Fig. 5a) had a reference semi-span area of 125 ft^2, a semi-span aspect ratio of 3.96 while $\bar{c}$ = 6.15 ft. The section shape of the wing was the NASA HSNLF(1)-0213 airfoil (Ref. 25). The rectangular airfoil trailing-edge shape had sharp corners and a thickness of about 0.125 in. at $\bar{c}$. Data are presented here with transition strips installed at the 5-percent upper-surface and the 10-percent lower-surface chord location. Aerodynamic data were obtained for a Reynolds number of 3.67×10^6 (based on $\bar{c}$). Reference 25 presents details of the test procedures in the tunnel. Figure 5b shows the geometry of the solid and the serrated nonplanar Gurney flaps attached to the pressure side of the wing along the full semi-span. Both Gurney flaps (thickness 0.032 in.) had equal frontal area (0.84 ft^2), while the serrated Gurney flap had a depth-to-chord ratio $d/\bar{c}$ = 0.014. The acute angle of the triangular serration was 60 deg.

Experimental Results

The aerodynamic coefficients for the wing-body geometries tested were reduced using the indicated reference (baseline) wing areas and $\bar{c}$. In the 14-By 22-Ft Tunnel experiment, the moments of the wing-body configuration are presented with respect to a fixed point on the model centerline, located longitudinally at the 0.25 $\bar{c}$ location of the wing (Ref. 22).

Figure 6 shows the lift, drag and pitching-moment characteristics of the configurations with the planar serrations as compared to the baseline wing and the wing with the solid extension for the 14-By 22-Ft Tunnel experiments. Data for the planar extensions were obtained up to an angle of attack of 14 deg, where the maximum lift coefficient $C_{L_{max}} \simeq 1.25$ was reached for the baseline geometry (Fig. 6a). The configurations with the trailing-edge extensions

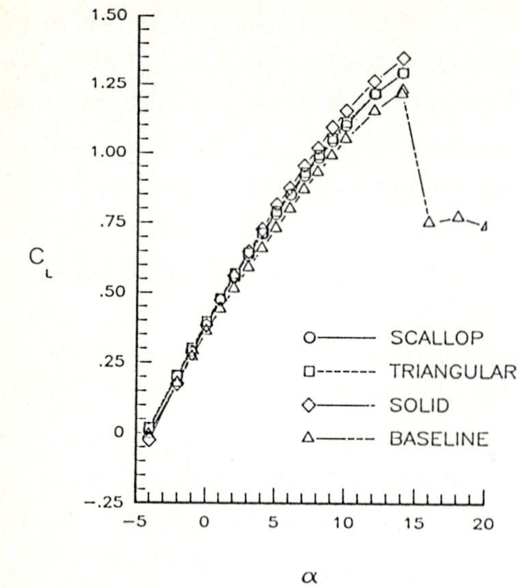

a. Lift curves.

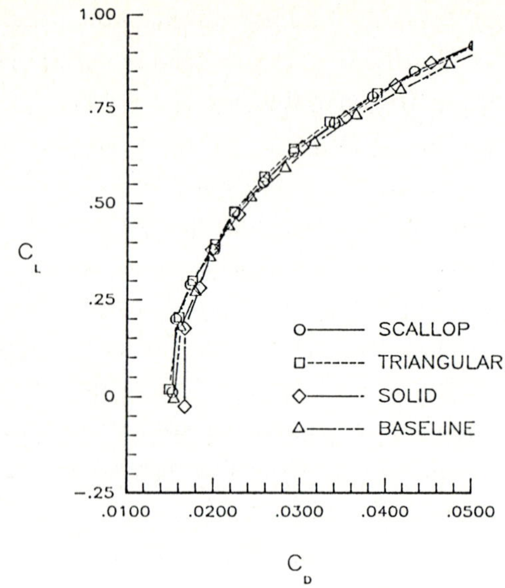

b. Drag polars.

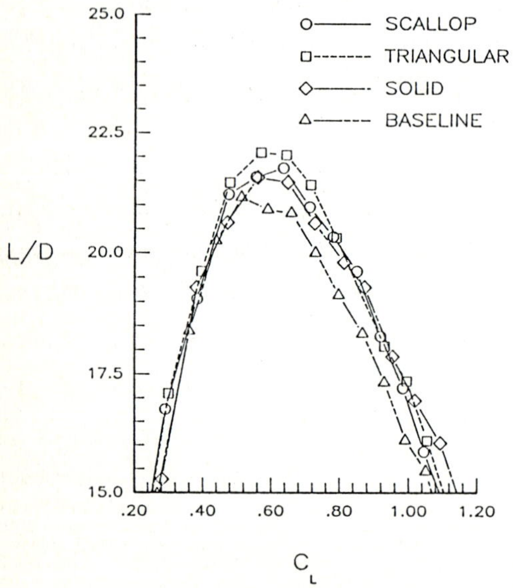

c. Lift-to-drag ratios.

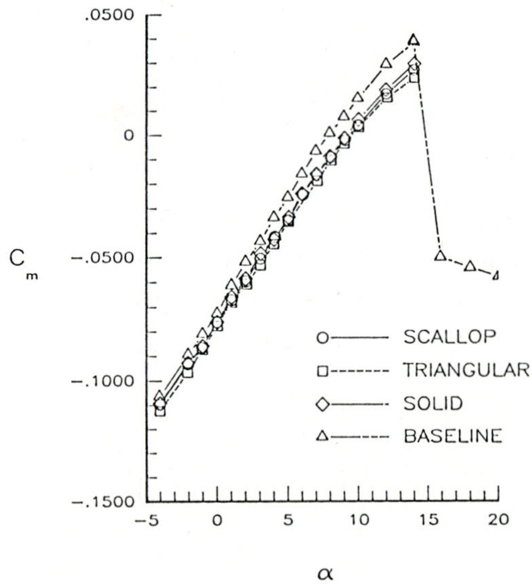

d. Pitching-moment curves.

Fig. 6. Experimental results of the full-span wing with the planar trailing-edge devices, $R_c = 1.1 \times 10^6$, $(x/c)_{trip} = .05$.

showed an increase in lift-curve slope C_{L_α} which is caused by the increase in projected area and a change in effective camber shape due to changes in the boundary-layer development. At $\alpha = 14$ deg, C_L increased by 4- and 8-percent for the configurations with the serrated and the solid extensions, respectively. The magnitude of the measured reduction in C_{L_α} for the serrated geometries, compared to the solid extension, is larger than can be attributed to the "filling" in of the serrations by the viscous fluid and the associated small reduction in effective aspect ratio. C_{L_α} for the baseline wing flattened at $\alpha = 1$ deg, while C_{L_α} for the extended geometries became nonlinear for $\alpha \geq 4$ deg (Fig. 6a). Limited flow-visualization studies using tufts for the baseline geometry indicated the presence of separated flow near the trailing edge at $\alpha \geq 1$ deg.

The scalloped and triangular-serrated trailing-edge configurations had less drag than the baseline wing for $C_L < 0.3$ and $C_L > 0.5$ (Fig. 6b), in particular, the serrations showed a decrease in drag of 5-10 counts for $C_L < 0.3$ (Fig. 6b). Compared to the configuration with the solid addition, both serrated configurations showed a large decrease in drag (10-20 counts) at small lift coefficients (Fig. 6b), whereas the maximum lift-to-drag ratio $(L/D)_{max}$ increased from about 21.5 to 22.1 for the triangular serration (Fig. 6c). $(L/D)_{max}$ occurred near $C_L = 0.65$ for all configurations with the trailing-edge extensions. The $(L/D)_{max}$ for the baseline wing was 21.0 (Fig. 6c), thus the maximum lift-to-drag ratio increased by 5 percent as a result of the triangular serrations. Finally, the serrated geometries exhibited nearly identical pitching-moment behavior as the solid extension (Fig. 6d).

Figure 7 provides the measured characteristics of the baseline wing with and without the nonplanar solid and serrated Gurney flaps (Fig. 3b). The addition of the small Gurney flaps increased the camber of the airfoil, and resulted in an increase in lift ΔC_L at a given angle of attack. At $\alpha = 0$ deg, ΔC_L is nearly 0.15 and 0.30 due to the addition of the serrated and solid Gurney flaps, respectively (Fig. 7a). Note that the frontal area of the serrated Gurney flap is about one half of the frontal area of the solid flap for the full-span model. At the lowest angles of attack (Fig. 7a), an increase in C_{L_α} was observed caused by the changes in the flow field at the trailing edge of the wings with the flaps. Near $\alpha = 1$ and 2 deg, respectively, the lift curve of the baseline wing and the wing with the two-dimensional flap became nonlinear; however, the wing with the serrations displayed this behavior at a slightly higher angle of attack (α = 3-4 deg). Whereas the serrated Gurney flap showed the same lift-curve gradient as the baseline wing at higher angles of attack ($\alpha > 5$ deg), the

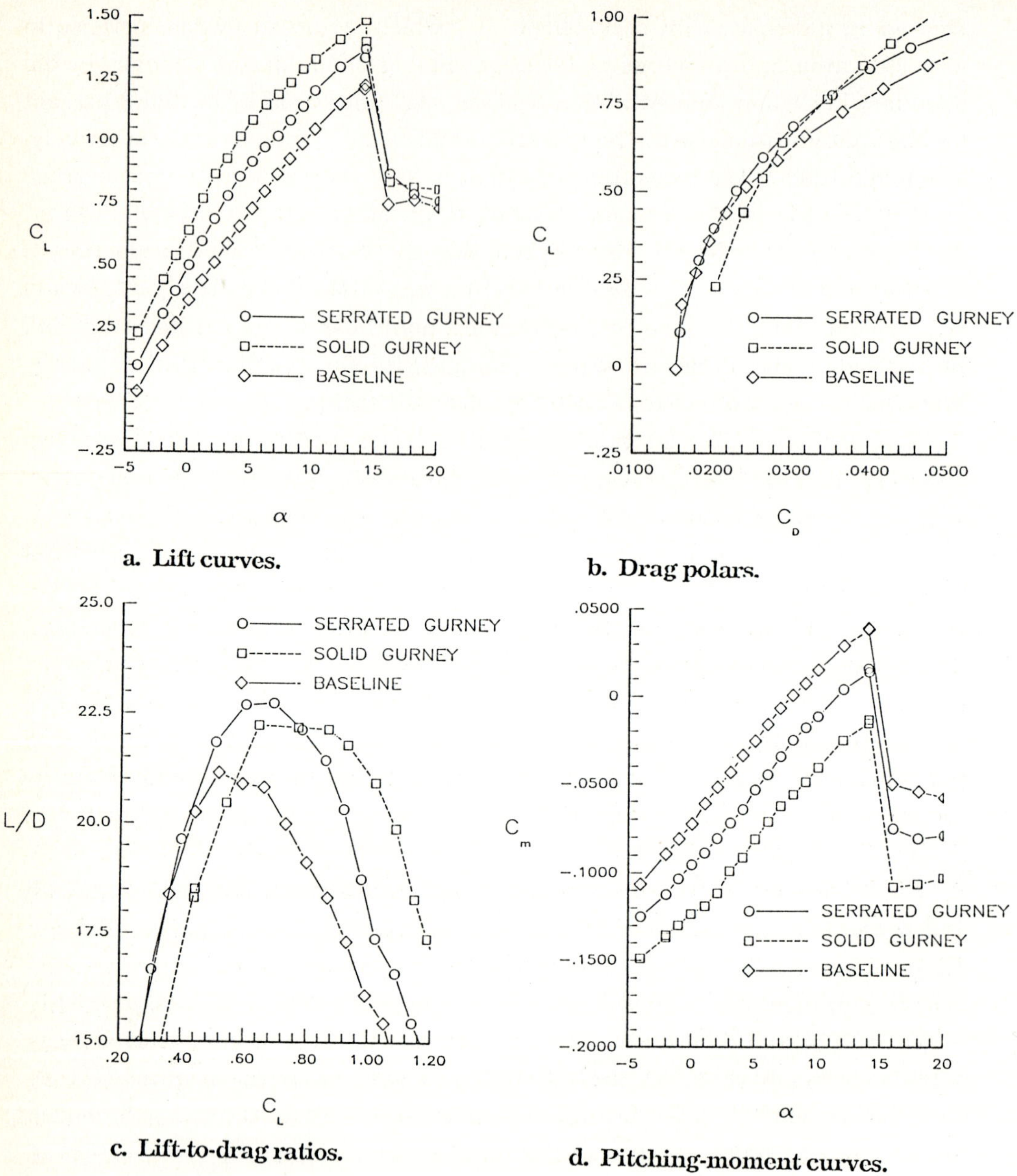

a. Lift curves.

b. Drag polars.

c. Lift-to-drag ratios.

d. Pitching-moment curves.

Fig. 7. Experimental results of the full-span wing with the nonplanar trailing-edge devices, $R_c = 1.1 \times 10^6$, $(x/c)_{trip} = .05$.

solid flap showed a smaller lift-curve slope (Fig. 7a). At $\alpha = 14$ deg, $C_{L_{max}} \simeq$ 1.45 and 1.35 for the solid and serrated Gurney flaps, respectively, as compared to about 1.20 for the baseline configuration.

Figures 7b and 7c show the effects of the Gurney flaps on drag. The model-sting mounting system precluded measurements at lower lift coefficients than indicated in Figure 7b. The solid Gurney flap had a large increase in drag for $C_L < 0.5$; for example, at $C_L = 0.25$ an increase in drag coefficient $\Delta C_D \simeq$ 0.0030 (nearly 20 percent) was measured. Due to the camber effect, the solid Gurney flaps had less drag than the baseline wing for $C_L > 0.6$; the serrated flap had less drag for $C_L > 0.4$. However, unlike the solid Gurney flap, the serrated configuration had nearly identical drag as the baseline wing for $C_L < 0.4$ (Fig. 7b). From Fig. 7c, $(L/D)_{max}$ for the solid and the serrated Gurney flaps increased from 21.0 for the baseline wing to about 22.25 and 22.75, respectively (an improvement of 6-8 percent). The lift coefficient at which the maximum lift-to-drag ratio occurred increased from 0.55 to 0.65 (Fig. 7c). The effect of the camber change due to the Gurney flaps on the pitching-moment curves is given in Fig. 7d. At $\alpha \simeq 2$ deg ($C_L \simeq .85$), the pitching-moment-curve slope C_{m_α} changed abruptly for the solid Gurney flap (Fig. 7d). The change in C_{m_α} can be explained by the sudden occurrence of flow separation near the trailing edge of the airfoil, causing a forward shift of the aerodynamic center as the angle of attack is increased. The model with the serrated flap did not show this behavior.

The effects of the solid and serrated Gurney flaps were also tested in the 30-By 60-Ft Tunnel experiment. The lift and drag characteristics of the semi-span wing with the nonplanar flaps (Fig. 5b) are summarized in Fig. 8. Both flap geometries increased both the lift-curve slope and the maximum lift coefficient in comparison to the baseline wing (Fig. 8a). C_{L_α} increased by nearly 10 percent, while $C_{L_{max}}$ reached about 1.55. The model with the serrated Gurney flap incurred less drag than the model with the solid Gurney at low lift conditions (Fig. 8b). However, the serrated configuration showed an increase in minimum drag coefficient $C_{D_{min}}$ of approximately 7.5 percent in comparison to the baseline wing. As C_L increased above 0.50-0.60, the serrated and the solid Gurney flap had lower drag than the baseline wing, as might be expected from the increase in camber.

Finally, results obtained for planar 60-deg sawtooth wing extensions in the 30-By 60-Ft tunnel experiment indicated that at cruising conditions the planar serrated trailing edges had marginally lower or equal drag and showed a

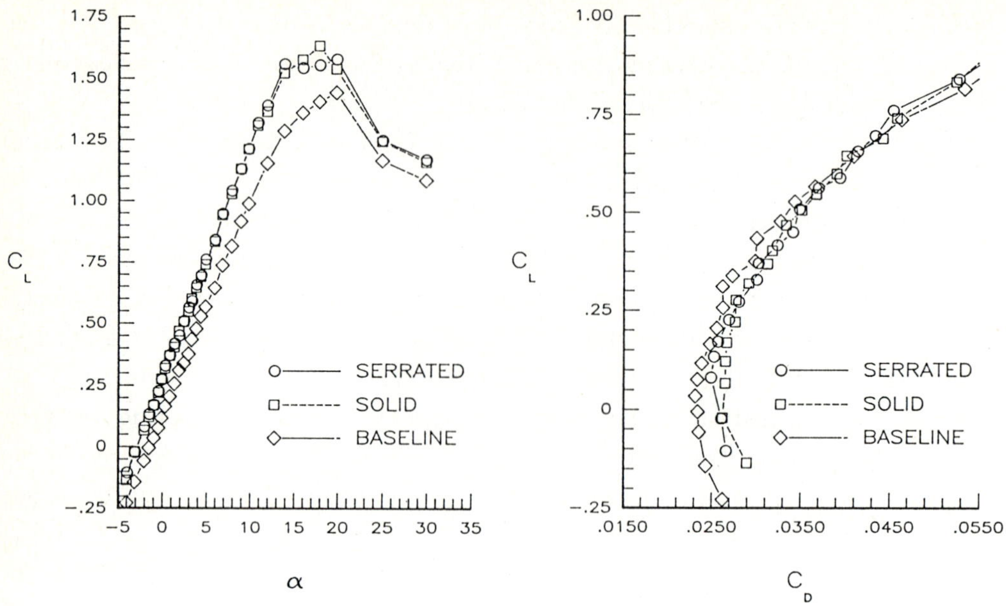

a. Lift curves. **b. Drag polars.**

Fig. 8. Experimental results of the semi-span wing with the nonplanar trailing-edge devices, $R_c = 3.67 \times 10^6$, $(x/c)_{trip} = .05/.10$ (upper/lower surface).

small increase in $(L/D)_{max}$ compared to the baseline and the solid-extension configuration. Thus, the addition of the planar serrations did not result in an increase in drag.

Discussion

The application of planar triangular serrations for the full-span model with transition fixed at 0.05*c* indicated a noticeable reduction in drag at lower angles of attack, as well as a small increase in $(L/D)_{max}$ compared to the baseline wing with the solid planar extension. The reduction in drag due to the planar serrations for the full-span model were obtained for flow conditions where premature turbulent separation occurred along the upper surface of the airfoil. The streamwise vortices generated by the planar serrated trailing edges in the near wake favorably affected the development and separation location of the turbulent boundary layer at the Reynolds number of the test. Entrainment of higher-momentum boundary-layer flow towards the surface

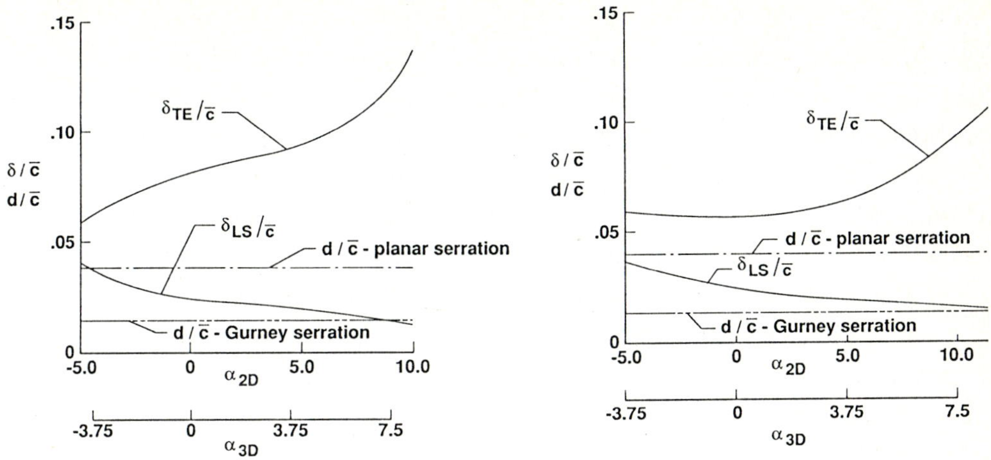

a. Full-span model; $R_c = 1.1 \times 10^6$, $(x/c)_{trip} = .05$.

b. Semi-span model; $R_c = 3.67 \times 10^6$, $(x/c)_{trip} = .05/.10$ (upper/lower surface).

Fig. 9. Estimated total-wake thickness (δ_{TE}) and lower-surface boundary-layer thickness (δ_{LS}).

near the trailing edge, as well as mixing of higher-momentum flow into the wake region when boundary-layer separation occurred may account for the reduction in pressure drag. A reduction in pressure drag due to the effects of the serrations on trailing-edge separation seems to explain the observed decrease in drag at lower angles of attack for the full-span model and the increase in $(L/D)_{max}$. Results for the semi-span model at higher chord-Reynolds numbers in the 30-By 60-Ft Tunnel did not indicate a definite reduction in drag for the serrated planar trailing edges. The absence of a noticeable effect of planar serrations in these tests without premature trailing-edge flow separation, supports the hypothesis that the serrations favorably affected boundary-layer separation and, hence, reduced form drag in the 14-By 22-Ft Tunnel experiment.

Calculations were performed using a two-dimensional integral boundary-layer method (Ref. 26) to estimate the boundary-layer thickness near the trailing-edge for comparison with the geometric sizes of the serrated trailing edges. Figure 9 presents the predictions by the boundary-layer method as a function of the airfoil angle of attack α_{2D}. (For comparison, approximate angles of attack α_{3D} are provided for the wings with finite span, assuming no wing twist). The attached-flow boundary-layer calculations were continued up to the trailing edge, even when the onset of turbulent separation was predicted on the upper surface, as was the case for the full-span wing when α_{2D} exceeded 0 deg. From figure 9a it can be determined that the depth-to-total-boundary-layer thickness d/δ_{TE} of the planar serrations in the 14-By 22-Ft experiment varied from $d/\delta_{TE} \simeq 0.35 - 0.65$ for the angle-of-attack range of interest (δ_{TE} is the predicted sum of the boundary-layer thicknesses of the upper and the lower surface near the trailing edge). Figure 9b indicates that in the 30-By 60-Ft experiment, $d/\delta_{TE} \simeq 0.40 - 0.70$. Thus, the nondimensional depths of the planar serrations studied here are comparable and fall within the range studied by Lasheras (Ref. 13).

The application of solid two-dimensional Gurney flaps in both experiments demonstrated the large effect of camber changes on lift due to these small-scale trailing-edge devices, in agreement with the measurements reported in Refs. 17 and 18. The camber effect due to the solid Gurney flap resulted in less drag at higher C_L's and a higher $(L/D)_{max}$. However, unlike the results reported by Liebeck (Ref. 17), the solid Gurney flaps produced more drag at low angles of attack for the wing geometries tested here. Serration of the Gurney flap resulted in a decrease in drag compared to the solid flap. For the full-span model (Fig. 2), the addition of the serrated Gurney flap did not increase $C_{D_{min}}$ of the baseline. From Fig. 9, it can be concluded that the depth of the serrated Gurney flaps was such that $d/\delta_{LS} < 1$, where δ_{LS} is the boundary-layer thickness near the trailing edge of the pressure side, at low angles of attack in both experiments. That is, the thickness of the turbulent boundary-layer near the lower surface at the trailing edge was larger than the depth of the serrations in both cases. The absence or reduction of pressure-drag with the serrated Gurney flaps suggests that the serrations had a favorable effect on the wake development immediately downstream and, particularly in the test in the 14-By 22-Ft Tunnel, on the turbulent boundary-layer development upstream of the trailing edge.

After the wind-tunnel experiments, a water-tunnel study was conducted to visualize the flow field near solid and serrated Gurney flaps (Ref. 20). A triangularly serrated Gurney flap (74-deg acute angle) was modelled after the geometries described in this paper. The water-tunnel results, obtained at Reynolds number of the order of 10,000, provide evidence that the serrated Gurney flap generates near-streamwise vortices which are stretched and entrained by the spanwise Von Kármán vortex system shed from the trailing-edge geometry (Ref. 20), in a fashion that resembles the observations by Lasheras (Ref. 13). Most importantly, the water-tunnel study showed that the separation point of the boundary layer along the upper surface moves downstream as a result of the serrated Gurney flap (Figs. 10 and 15 in Ref. 20). The effect of the serrated Gurney flap on boundary-layer separation supports the hypothesis that the introduction of streamwise vortices behind the serrated trailing edges affected the development of the (separated) boundary layer along the upper surface and, hence, allowed reduction of pressure (form) drag encountered by the wing in the 14-By 22-Ft tunnel experiment.

Concluding Remarks

Exploratory wind-tunnel force measurements have been conducted to study the effects of small-scale serrated planar and nonplanar trailing-edge devices on the aerodynamic characteristics of medium-to high-aspect-ratio unswept wings with sharp trailing edges. The results showed that a planar serrated sawtooth-shaped trailing-edge reduced the drag at low angles of attack and increased $(L/D)_{max}$ for a wing with trailing-edge boundary-layer separation. The addition of a two-dimensional (solid), small, nonplanar, Gurney flap to the trailing edge of the wings resulted in the expected increase in lift and nose-down pitching moment, but also showed a pressure-drag penalty. The novel introduction of serrations to the Gurney flap reduced or eliminated this drag penalty.

The favorable drag effect of the planar and the nonplanar serrated devices are thought to be the result of reduced pressure (form) drag. The presence of nearly streamwise (axial) vortices immediately downstream of the serrated edges is believed to favorably affect the boundary-layer flow approaching the trailing edge by inducing higher-momentum flow into the flow near the surface and in the wake immediately behind the wing. Recent water-flow studies demonstrate the large effect of streamwise vortices on the development of the near wake and the boundary-layer separation point upstream of the trailing edge.

Additional wind-tunnel experiments are needed to measure the effect of planar and nonplanar serrated trailing edges on the boundary-layer and wake flow of wings at moderate to high Reynolds numbers (e.g., $R_c = 1 - 10 \times 10^6$). Such investigations could contribute to the understanding of the effects of streamwise vortices on the viscous flow field near the serrated trailing-edge region of lifting surfaces. Layouts of wing trailing edges that combine planar and nonplanar serrated-edge devices to control the near-wake vortex structure throughout the angle-of-attack range are presented in Ref. 21.

Acknowledgements

The authors would like to thank Mr. David Hahne and Mr. Tom Emerson for their assistance during the experiments in the 30-By 60-Ft Wind Tunnel at NASA Langley. The effort of the first author was supported under NASA Grant NAG1-345 and Contract NAS1-18240.

References

1. Norman, J. R.; and Fraser, F. C.: Giant Fishes, Whales and Dolphins. Putnam, London, 1937.
2. Lighthill, M. J.: Aerodynamic Aspects of Animal Flight. In: Swimming and Flying in Nature. Vol. 2 (eds, Wu, T. Y., Brokaw, C. J., and Brennen, C.), pp. 423-491, Plenum, New York, 1975.
3. Hersh, A. S.; and Hayden, R. E.: Aerodynamic Sound Radiation from Lifting Surfaces with and without Leading-Edge Serrations. NASA CR-114370, 1971.
4. Soderman, P. T.: Aerodynamic Effects of Leading-Edge Serrations on a Two-Dimensional Airfoil. NASA TM-X-2643, Sept. 1972.
5. Blick, E. F.; Watson, D.; Belie, G.; and Chu, H.: Bird Aerodynamic Experiments. In: Swimming and Flying in Nature, Vol. 2 (eds, Wu, T. Y.; Brokaw, C. J.; and Brennen, C.), pp. 939-952, Plenum, New York, 1975.
6. Kuethe, A. M.: Effect of Streamwise Vortices on Wake Properties Associated with Sound Generation. J. Aircraft, Vol. 9, No. 10, Oct. 1972, pp. 715-719.
7. Nash, J. F.; Quincey, V. G.; and Callinan, J.: Experiments on Two-Dimensional Base Flow at Subsonic and Transonic Speeds. ARC R&M No. 3427, Jan. 1963.
8. Bearman, P. W.: Investigations of the Flow Behind a Two-Dimensional Model with a Blunt Trailing Edge and Fitted with Splitter Blades. J. Fluid Mech. (1965), Vol. 21, Part 2, pp. 241-255.
9. Tanner, M.: A Method for Reducing the Base Drag of Wings with Blunt Trailing Edge. The Aeronautical Quarterly, Vol. 23, Feb. 1972, pp. 15-23.
10. Gai, S. L.; and Sharma, S. D.: Experiments on the Reduction of Base Drag of a Blunt Trailing-Edge Aerofoil in Subsonic Flow. Aeronautical J., May 1981, pp. 206-210.
11. Sharma, S. D.: Influence of Yaw and Incidence on Base Drag of Rectangular Wings. Z. Flugwiss. Weltraumforsch. 11, 1987, pp. 19-22.
12. Meiburg, E.; and Lasheras, J. C.: Experimental and Numerical Analysis of the Formation of Streamwise Vortices in the Plane Wake behind a Flat Plate. 6th Symposium on Turbulent Shear Flows, Toulouse, France, Sept. 7-9, 1987. Proceedings: University Park, PA, Pennsylvania State University, 1987, pp. 16-1-1 to 16-1-6.

13. Lasheras, J. C.; and Choi, H.: Three-Dimensional Instability of a Plane Free Shear Layer: An Experimental Study of the Formation and Evolution of Streamwise Vortices. J. Fluid Mech., Vol. 189, 1988, pp. 53-86.
14. Meiburg, E.; and Lasheras, J. C.: Experimental and Numerical Investigation of the Three-Dimensional Transition in Plane Wakes. J. Fluid Mech., Vol. 190, 1988, pp. 1-37.
15. Werle, M. J.; and Paterson, R. W.: Flow Structure in a Periodic Axial Vortex Array. AIAA Paper 87-0610, Jan. 1987.
16. Werle, M. J.; Paterson, R. W.; and Presz, W. M., Jr.: Trailing-Edge Separation/Stall Alleviation. AIAA J., Vol. 25, No. 4, 1987, pp. 624-626.
17. Liebeck, R. H.: Design of Subsonic Airfoils for High Lift. J. of Aircraft, Vol. 15, No. 9, Sept. 1978, pp. 547-561.
18. Roesch, Ph.; and Verillet, A.: New Designs for Improved Aerodynamic Stability on Recent Aerospatiale Helicopters. Vertica, Vol. 6, 1982, pp. 145-164.
19. Moriarty, J. A.; and Tuck, E. O.: Thin Aerofoils with High-Incidence Flaps or Blunt Trailing Edges. Aeronautical J., March 1989, pp. 93-99.
20. Neuhart, D. H.; and Pendergraft, O. C., Jr.: A Water Tunnel Study of Gurney Flaps. NASA TM 4071, Nov. 1988.
21. Vijgen, P.M.H.W.; Howard, F. G.; Bushnell, D. M.; and Holmes, B. J.: Serrated Trailing Edges for Improved Lift and Drag of Lifting Surfaces. NASA-Langley Pending Disclosure of Invention, LAR 13870-1-CU, Aug. 31, 1987.
22. Vijgen, P.M.H.W.; van Dam, C. P.; and Holmes, B. J.: Sheared-Tip Aerodynamics: Wind-Tunnel and Computational Investigation. J. Aircraft, Vol. 26, No. 3, March 1989, pp. 207-213.
23. McGhee, R. J.; Viken, J. K.; Pfenninger, W.; Beasley, W. D.; and Harvey, W. D.: Experimental Results for a Flapped Natural-Laminar-Flow Airfoil with High Lift-Drag Ratio. NASA TM 85788, May 1984.
24. Hahne, D. E.; Jordan, F. L.; Davis, P. J.; and Muchmore, C. B.: Full-Scale Semi-Span Tests of an Advanced NLF Business-Jet Wing. SAE Paper 871860, Oct. 1987.
25. Viken, J. K.; Viken, S. A.; Pfenninger, W.; Morgan, H. L.; and Campbell, R. L.: Design of the Low-Speed NLF(1)-0414F and the High-Speed HSNLF(1)-0213 Airfoils with High-Lift Systems. In: Research in Natural Laminar Flow and Laminar-Flow Control, ed. Hefner, J. N. and Sabo, F. E., NASA CP-2487, 1987, Part 3, pp. 637-671.
26. Stevens, W. A.; Goradia, S. H.; and Braden, J. A.: Mathematical Model for Two-Dimensional Multi-Component Airfoils in Viscous Flow. NASA CR-1843, 1971.

LOW REYNOLDS NUMBER AIRFOIL DESIGN FOR SUBSONIC COMPRESSIBLE FLOW

by

Robert H. Liebeck

Douglas Aircraft Company

McDonnell Douglas Corporation

Long Beach, California

ABSTRACT

An airfoil design study has been conducted to examine the capability of providing high lift in the Reynolds number range of 0.5 to 5.0 million at Mach 0.4. This relatively moderate Mach number produces significant compressibility effects when design lift coefficients in excess of 1.5 are desired. Four example airfoils have been developed with various thickness ratios and degrees of aft loading. The MIT ISES airfoil analysis code has been used to theoretically predict performance including drag rise characteristics. In addition, as a calibration, the ISES code was applied to airfoils where wind tunnel results were available, and agreement with the data was very good.

INTRODUCTION

The purpose of this study has been to develop an array of low Reynolds number airfoil designs for application to the high-altitude long-endurance (HALE) class of airplanes that is currently being studied by the Air Force. These airfoil designs will provide a data base for the design and sizing of HALE vehicles.

The airfoil design specification set by the Air Force was:

$1.0 < C_{l_{des}} < 1.8$

$0.5 \times 10^6 < RN < 5.0 \times 10^6$

$0.25 < M < 0.40$

$t/c > 15\%$

$c_{m_{c/4}}$ "low" (preferably)

As a purely theoretical study, no wind tunnel verification of the predicted performance of the newly designed airfoils has been established. Nevertheless, the state-of-the-art methods that have been employed in analyzing the airfoils have shown excellent correlation with experimental results for earlier airfoils that were wind tunnel tested. A presentation and discussion of these comparisons is offered in the following section.

APPROACH AND RESULTS

AIRFOIL ANALYSIS CODE CALIBRATION

As a preliminary task, the Massachusetts Institute of Technology ISES [1] airfoil analysis computer code was acquired by Douglas, and test cases were run on airfoils where reliable wind tunnel data were available. The appeal of this recently developed (1986) code is that it can predict airfoil performance at low Reynolds numbers where laminar separation bubble formation is virtually inevitable. Typically, such bubbles cause an increase in drag and possibly a reduction in maximum lift. The ISES code not only predicts the existence of said bubbles, it also predicts the drag and maximum lift with the bubble present.

In addition to comparisons with wind tunnel data, the predictions of the ISES code were compared with those of the Douglas MADAAM airfoil analysis code [2]. The latter has been well calibrated in the low Reynolds number regime; however, it cannot predict the effects of laminar separation bubbles on airfoil performance.

The ability of the ISES code to predict drag rise due to compressibility was investigated first. Here the classic data for the NACA 0012 symmetric airfoil [3] was used for comparison with the ISES prediction. The results are shown in Figure 1 for lift coefficients of 0.0 and 0.4. ISES appears to slightly overpredict the drag level, but the code is remarkably accurate for its prediction of drag rise. There exists, however, an area of concern regarding compressibility effects at low Reynolds numbers. The data of Reference 3 were obtained at a Reynolds number of 9.0×10^6, which is hardly within the regime of interest of the present study. Unfortunately, no reliable data on drag rise at lower Reynolds numbers were available. It is possible that if the shock location was nearly coincident with the laminar bubble, an interaction could occur that the ISES code would be unable to predict. This reservation should be considered when viewing the drag rise predictions at low Reynolds numbers.

Next, a basic evaluation of the ISES code capability at low Reynolds numbers was conducted. Reference 1 gives some example analysis comparisons for Douglas airfoils LNV109A and LA203A, which show excellent agreement of both drag polars and

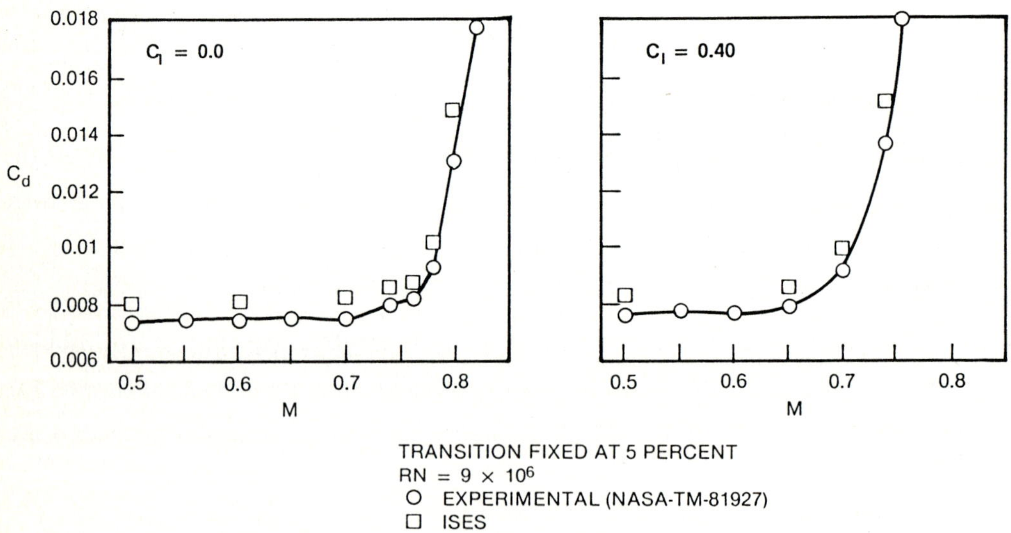

FIGURE 1. COMPARISON OF DRAG RISE PREDICTION FOR NACA 0012 AIRFOIL

chordwise pressure distributions. In the present study, the code was applied to some other existing Douglas low Reynolds number airfoils, and the results for airfoils LA2573A and LA5104E are given in Figures 2 through 5.

Airfoil LA2573A is particularly vulnerable to laminar separation bubble formation because of the relatively short rooftop of its upper surface pressure distribution. (A short rooftop results in a relatively low local Reynolds number at the rooftop peak, and laminar bubble formation is dependent on the value of this parameter.) The chordwise pressure distribution comparison shown in Figure 2 demonstrates that the ISES code accurately predicts the occurrence and extent of the bubble. Drag polar comparison (Figure 3) indicates that the ISES code accurately predicts the total drag penalty due to the laminar bubble at the mid and lower C_l range. (Comparison with the MADAAM prediction, which does not consider the laminar bubble, indicates the level of the drag penalty of the bubble.) This example represents a relatively severe test where a short rooftop is coupled with a very low Reynolds number, and the drag prediction accuracy is very good. Maximum lift is slightly underestimated.

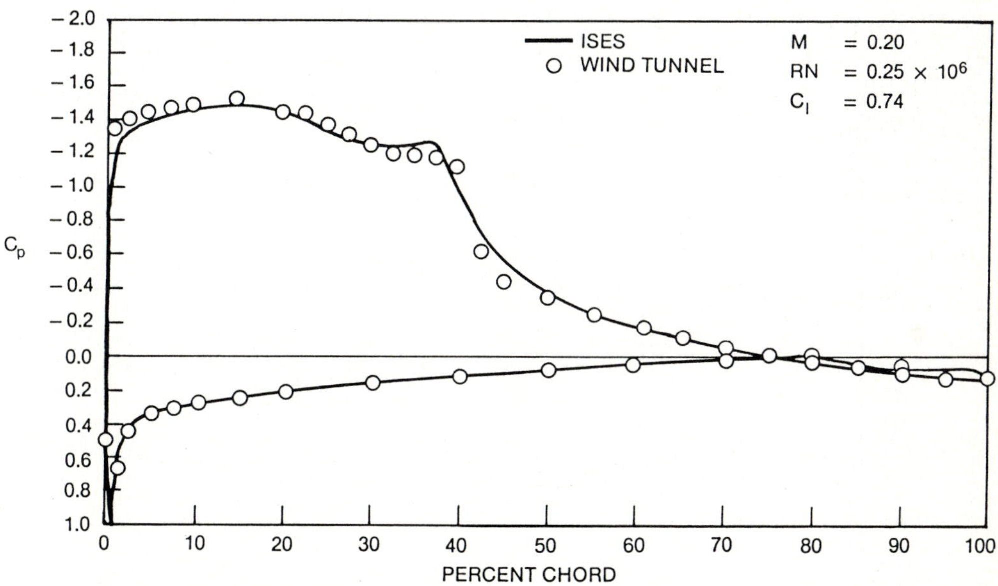

FIGURE 2. COMPARISON OF PRESSURE DISTRIBUTION PREDICTION FOR LA2573A AIRFOIL

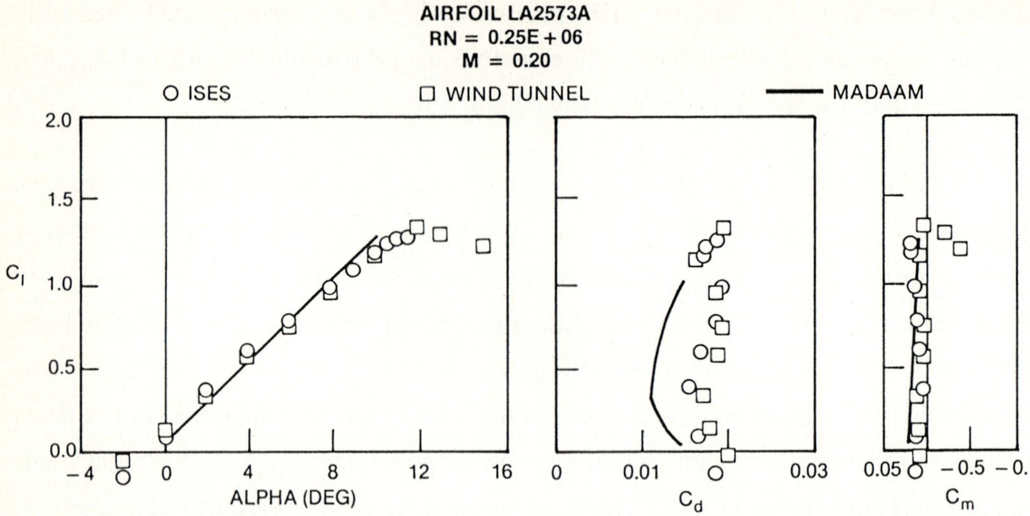

FIGURE 3. COMPARISON OF PERFORMANCE PREDICTION FOR LA2573A AIRFOIL

Airfoil LA5104E is a representative high-lift design for a Reynolds number of 0.50×10^6 with an upper surface rooftop of moderate length. Figure 4 shows very good agreement of the chordwise pressure distribution predicted by ISES, and again excellent prediction of the laminar bubble itself. The drag polar comparison of Figure 5 also shows excellent agreement with the experimental data, and again there is an underprediction of $C_{l_{max}}$.

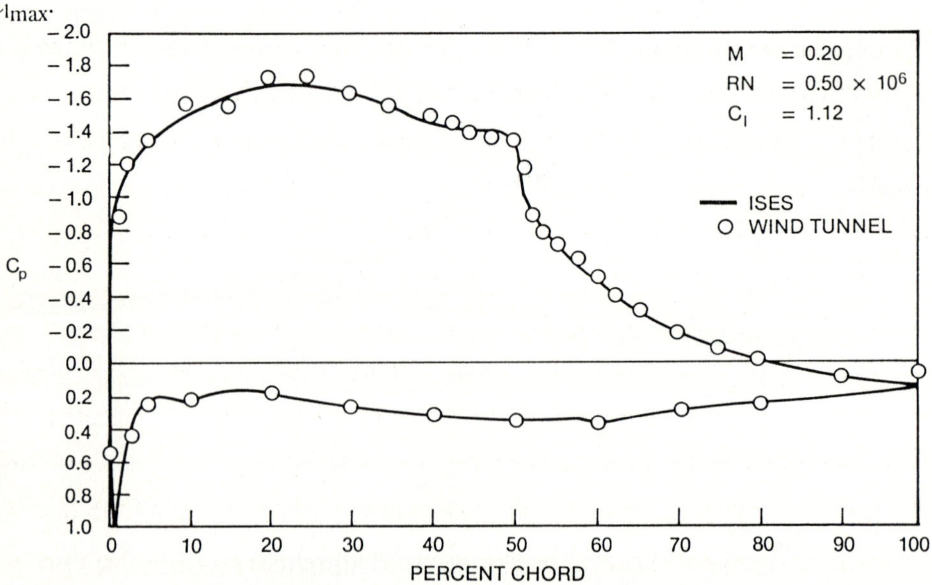

FIGURE 4. COMPARISON OF PRESSURE DISTRIBUTION PREDICTION FOR LA5104E AIRFOIL

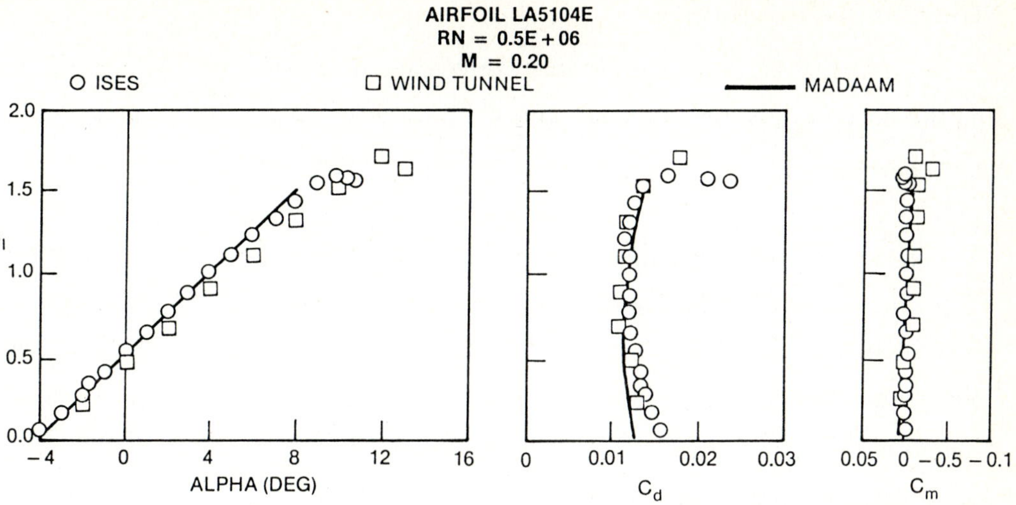

FIGURE 5. COMPARISON OF PERFORMANCE PREDICTION FOR LA5104E AIRFOIL

In the course of the present study, there were some cases where the ISES code refused to start and/or successfully converge. This does not mean that the airfoil being analyzed will not perform properly at the particular condition where the code is intractable. Rather, it is suspected that this simply represents an idiosyncrasy of a young computer code, which will ultimately become robust as it is refined with experience and time.

Before closing the subject of ISES code calibration, some discussion of one of the input parameters is in order. Prediction of transition in the free shear layer of the laminar separation bubble depends on the assumed value of the exponent n in the e^n stability criterion. (This relates to the amplification of unstable disturbances which ultimately produce transition.) Experience has shown that a value of n on the order of 9 is proper for flows with very low free-stream turbulence, and this is the recommended value for use in the ISES code. However, if the flow calculation is being compared with experimental data where the free-stream turbulence is not necessarily low (say turbulence greater than 0.1 percent), the value $n = 9$ may not be appropriate. This problem is currently being studied by first generating wind tunnel data for varying levels of turbulence and comparing the predictions of the ISES code for varying values of n. While beyond the scope of the present study, preliminary results indicate a correlation between n and free-stream turbulence level with the "best" value of n decreasing with increasing turbulence. In the present study, $n = 9$ was used for all of the calculations.

NEW AIRFOIL DESIGNS

Based on the Air Force specification, three candidate airfoils were designed. The first two have low pitching moments, and they are distinguished by their design Reynolds numbers: L101B, $RN_{des} = 1.0 \times 10^6$ and L505B, $RN_{des} = 5.0 \times 10^6$.

For the specified Reynolds number range, airfoil L101B is the most conservative design, and airfoil L505B is the most optimistic. The appeal of L505B was that it was thicker than L101B. It was left to the theoretical analysis to establish whether or not L505B would perform properly at the lower portion of the Reynolds number range. In addition to the two low pitching moment airfoils, a design with a small amount of aft loading was prepared: LC111A, $RN_{des} = 1.0 \times 10^6$.

This airfoil has a significant negative pitching moment (–0.17) compared with the moments of the first two airfoil designs (–0.05). However, airfoil LC111A is a "conservative" airfoil designed at a Reynolds number of 1.0×10^6, and is slightly thicker than airfoil L505B, designed at 5.0×10^6. The geometry and design pressure distributions of the three new airfoils are given in Figures 6 through 8.These three airfoils provide a representative data base in response to the Air Force specification.

An airfoil designed at lower Reynolds number will typically improve in performance at higher Reynolds numbers, but the converse is not necessarily true. For example, it is possible that an airfoil designed at 5.0×10^6 will suffer a substantial increase in drag due to a large laminar separation bubble (10- to 15-percent chord) when it is operated at 1.0×10^6. Experience has shown that the theoretical method used to design the airfoils is somewhat conservative itself. Consequently, it was decided to attempt successful operation of 5.0×10^6 designs at 1.0×10^6.

Pitching moment does not affect an airfoil's ability to perform in terms of lift and drag, but it can have a profound effect on the lift and drag of the complete airplane. Airfoil LC111A was developed by relaxing the pitching moment requirement to demonstrate the level of performance improvement in terms of airfoil thickness and Reynolds number "conservatism" that might be achieved. Configuration-dependent phenomena

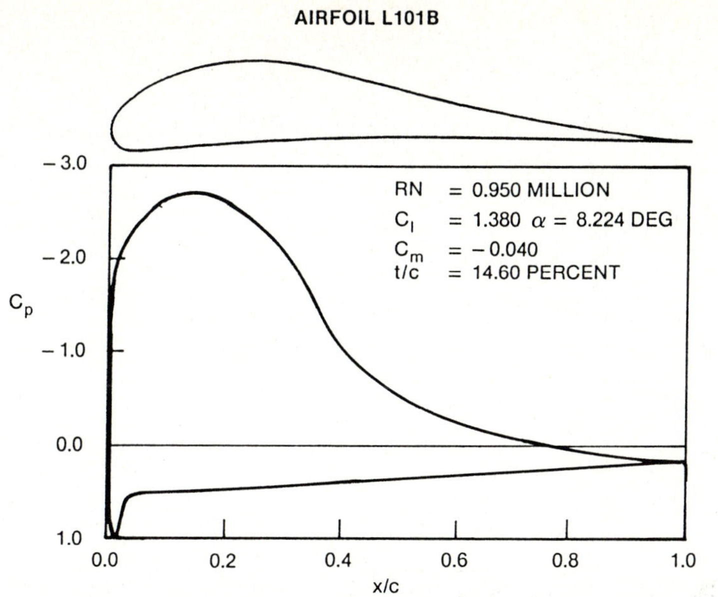

FIGURE 6. AIRFOIL L101B GEOMETRY AND DESIGN PRESSURE DISTRIBUTION

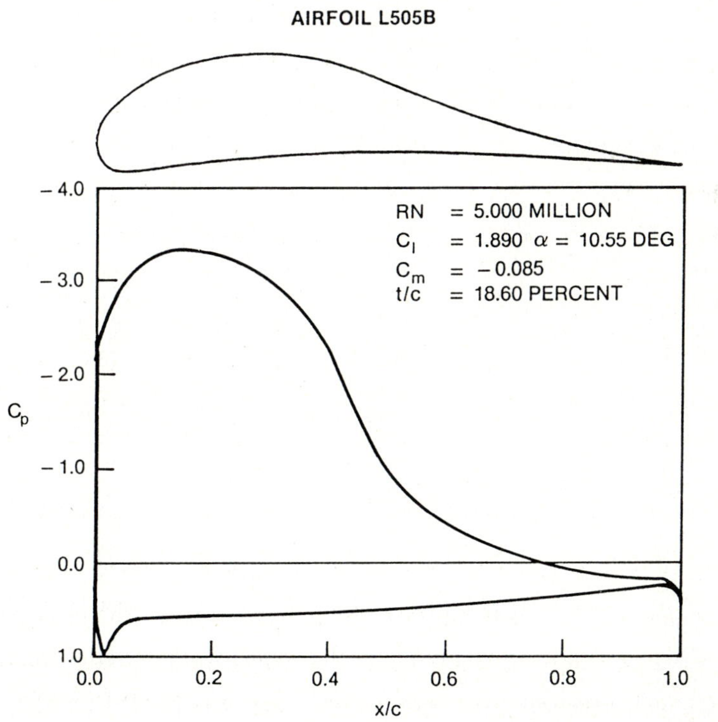

FIGURE 7. AIRFOIL L505B GEOMETRY AND DESIGN PRESSURE DISTRIBUTION

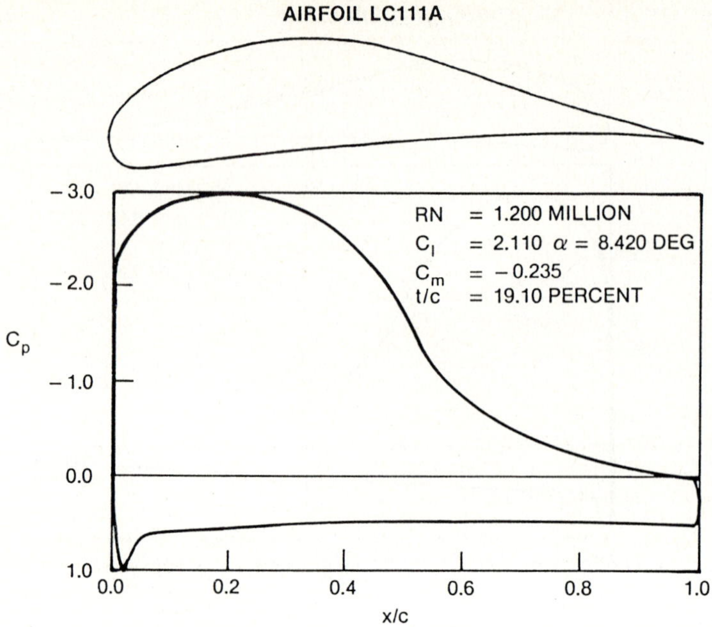

FIGURE 8. AIRFOIL LC111A GEOMETRY AND DESIGN PRESSURE DISTRIBUTION

such as trim drag must ultimately be considered when applying any of these airfoils to an actual airplane design.

The theoretically predicted performance of the new airfoils L101B, L505B, and LC111A is given in Figures 9 through 11, respectively. Results from both the Douglas MADAAM and MIT ISES codes are shown for Reynolds numbers of 0.5, 1.0, 2.0, and 5.0×10^6. MADAAM predictions are shown as solid lines, and ISES predictions are shown as discrete symbols. There were some cases where the ISES code failed to provide analysis solutions (e.g., airfoil LC111A at RN $= 5.0 \times 10^6$) as discussed earlier, but these should not be taken as indicative of the airfoil's failure to perform. The data at RN $= 0.5 \times 10^6$ show considerable disagreement between the MADAAM and ISES predictions for drag. This is a consequence of the significant laminar separation bubble effect not being accounted for by the MADAAM code, and the ISES prediction should be regarded as correct.

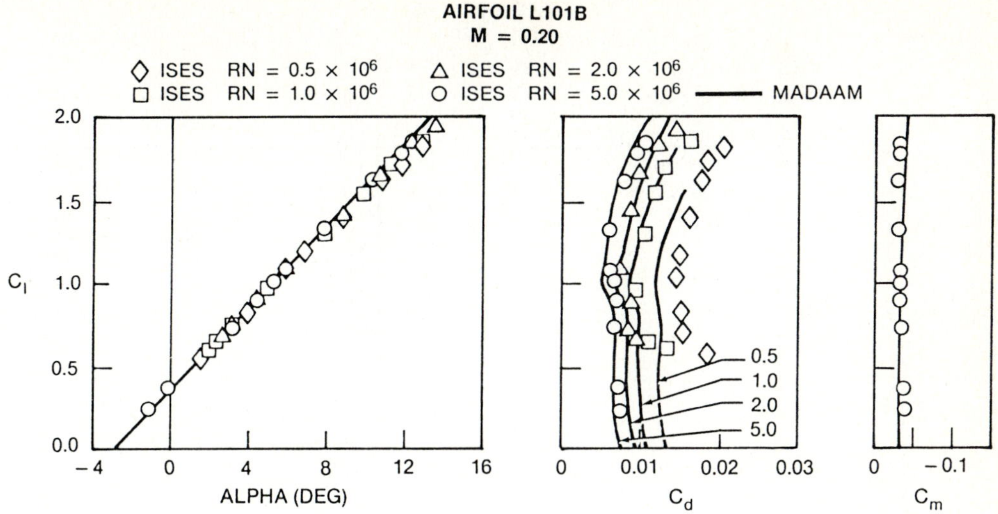

FIGURE 9. AIRFOIL L101B PERFORMANCE

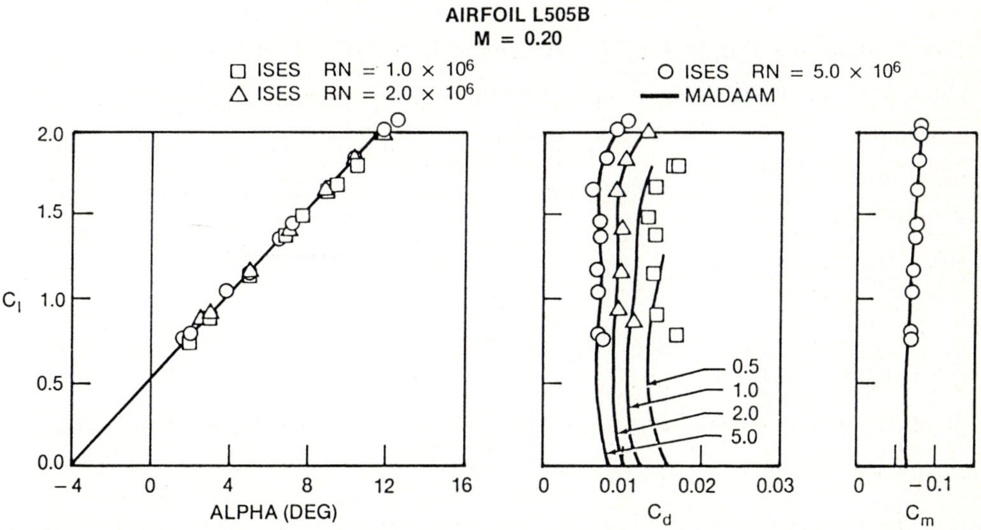

FIGURE 10. AIRFOIL L505B PERFORMANCE

The MADAAM and ISES predictions shown in Figure 11 disagree on angle of zero lift and pitching moment. Aft-cambered airfoils typically experience a significantly modified Kutta condition because of the rapid thickening of the boundary layer on the upper surface near the trailing edge. This is unaccounted for in the MADAAM program, which calculates both lift and pitching moment inviscidly. A similar result appears in the analysis of airfoil LC120A, shown in Figure 18.

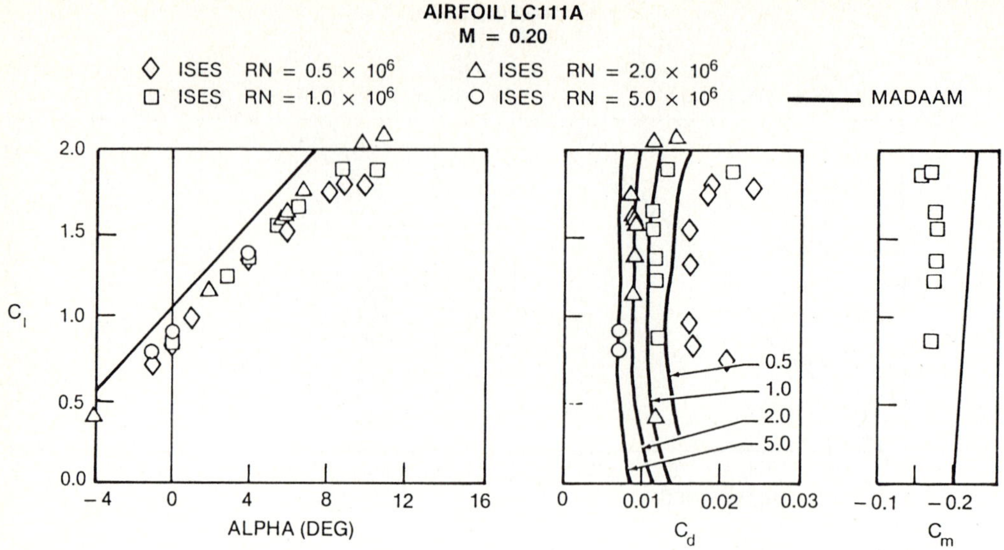

FIGURE 11. AIRFOIL LC111A PERFORMANCE

Compressibility drag rise results predicted by the ISES code for airfoils L101B, L505B, and LC111A are given in Figures 12 through 14, respectively. Based on the comparison with the test data for the NACA 0012 airfoil discussed earlier, the results for the new airfoils should be reliable and representative of how the airfoils will in fact perform — barring any unusual effects of shock/laminar bubble interaction. However, the

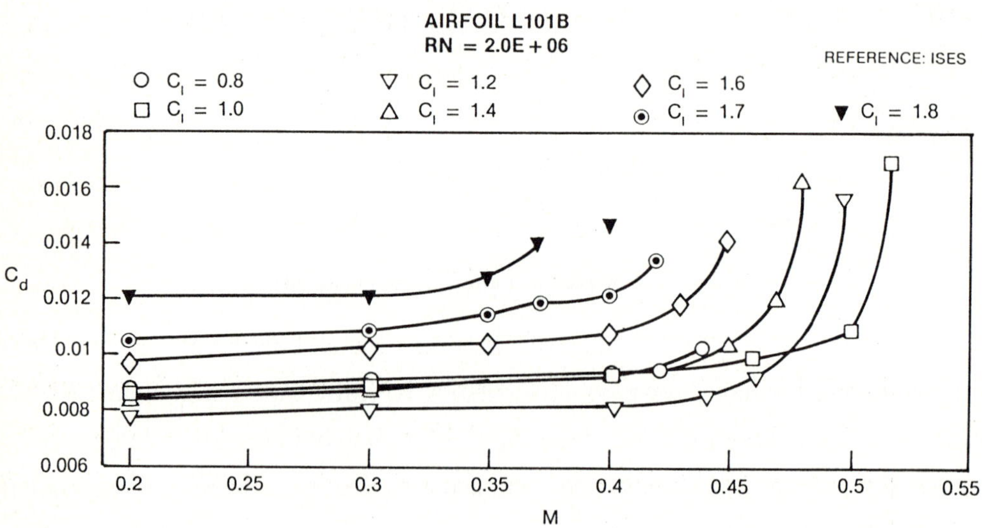

FIGURE 12. AIRFOIL L101B DRAG RISE

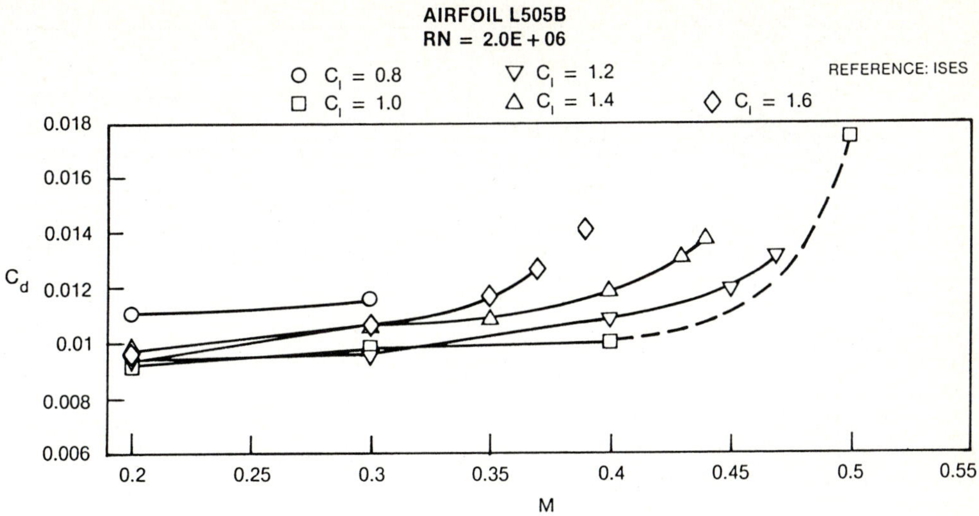

FIGURE 13. AIRFOIL L505B DRAG RISE

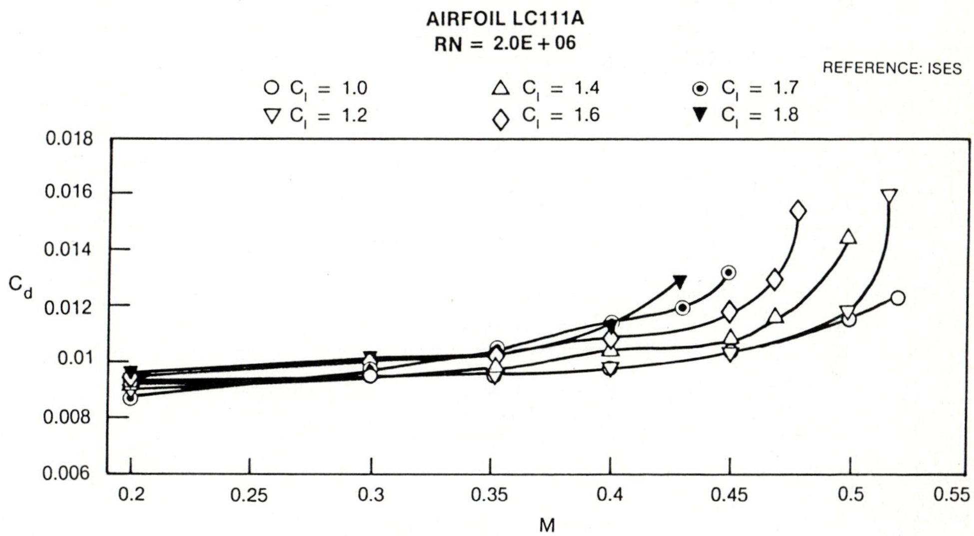

FIGURE 14. AIRFOIL LC111A DRAG RISE

calculated chordwise pressure distributions shown in Figures 15 and 16 for airfoils L101B and LC111A, respectively, appear to indicate that the shock is in fact coincident with the laminar bubble. This was true for all of the airfoils when their rooftop regions became supercritical. Nevertheless, the ISES code did not predict separation when this phenomenon occurred, as indicated by the successful pressure recovery shown in the examples of Figures 15 and 16.

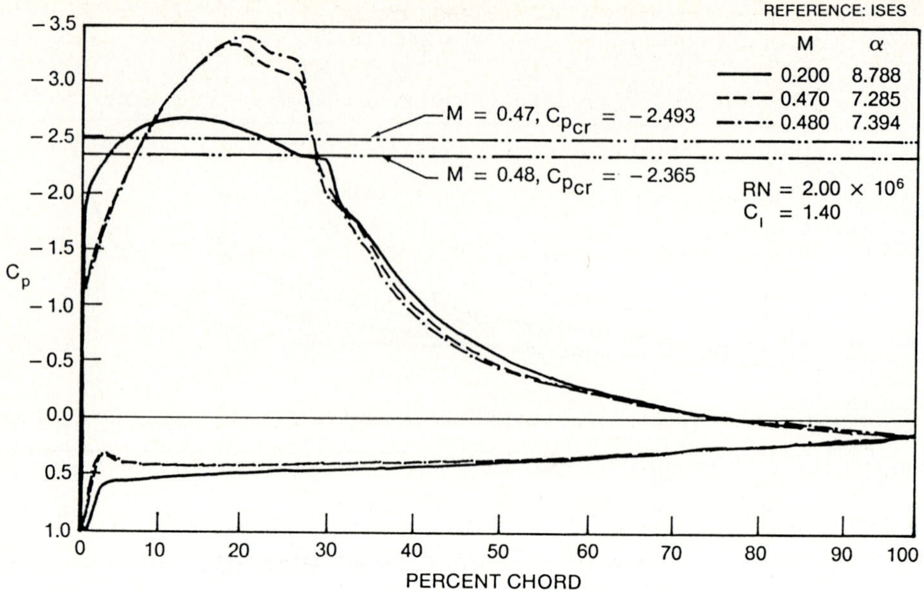

FIGURE 15. AIRFOIL L101B PRESSURE DISTRIBUTIONS AT CONSTANT C_l = 1.40

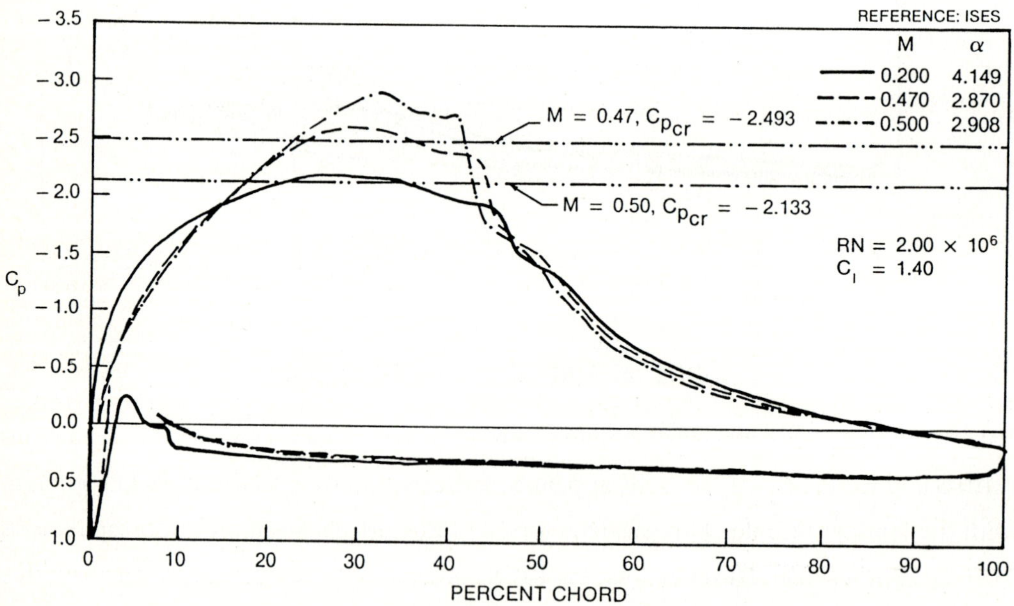

FIGURE 16. AIRFOIL LC111A PRESSURE DISTRIBUTIONS AT CONSTANT C_l = 1.40

The portion of the specification that called for proper airfoil performance up to Mach numbers as high as 0.40 was in effect met by the three new airfoils just described. Figures 12 through 14 indicate that all three airfoils resist drag rise below M = 0.4 at a C_l of 1.4, and airfoil LC111A meets this requirement at C_l = 1.8. "Retreating" to aft loading is almost irresistible to the aerodynamicist when faced with compressibility requirements. Consequently, an additional airfoil was designed with extended upper surface rooftop pressure distribution and moderate aft camber: LC120A, $RN_{des} = 1.0 \times 10^6$.

The geometry and design pressure distribution are given in Figure 17, and performance and drag rise are given in Figures 18 and 19. Airfoil LC120A lacks the $C_{l_{max}}$ capability of airfoil LC111A; however, it demonstrates improved compressibility characteristics for application at cruise Mach numbers greater than 0.4. Also, the extended rooftop of this airfoil results in a higher local Reynolds number at transition, which in turn accounts for its very good performance at RN = 0.5×10^6.

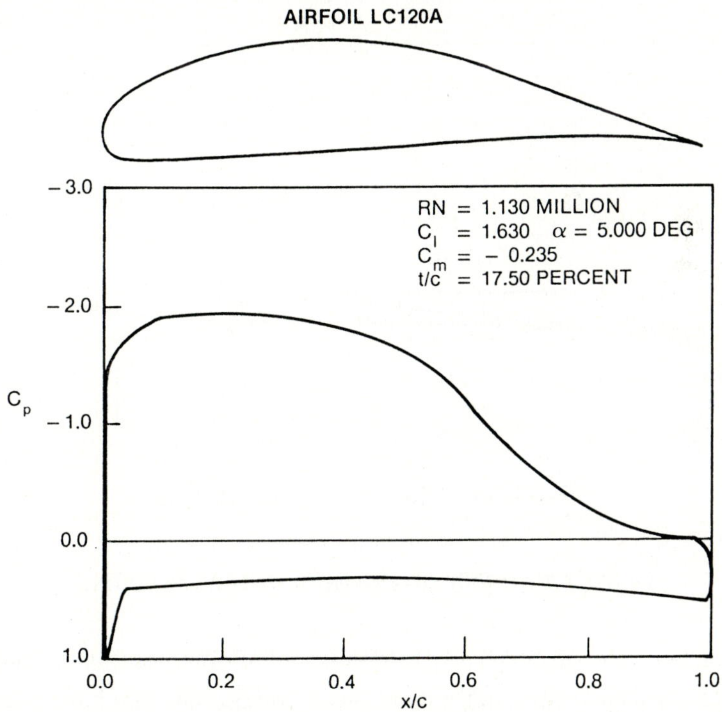

FIGURE 17. AIRFOIL LC120A GEOMETRY AND DESIGN PRESSURE DISTRIBUTION

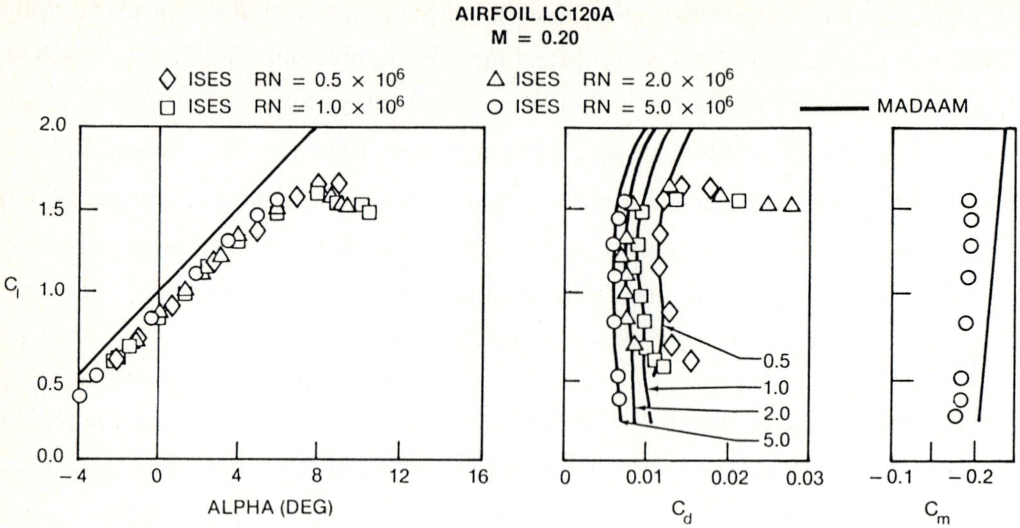

FIGURE 18. AIRFOIL LC120A PERFORMANCE

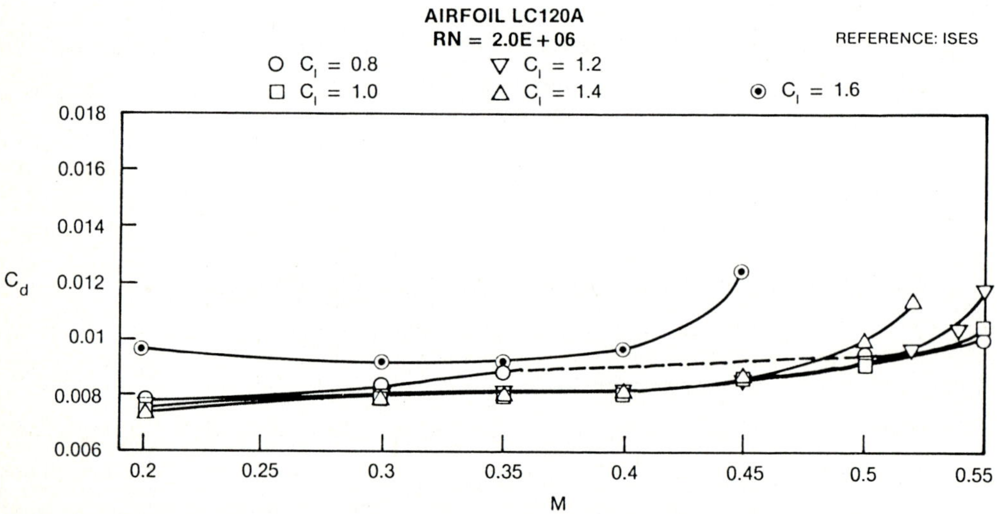

FIGURE 19. AIRFOIL LC120A DRAG RISE

CONCLUSIONS AND RECOMMENDATIONS

This study, while based purely on theoretical analyses and predictions, should provide a realistic appraisal of the capability of the airfoil designs that have been developed herein. Calibration of the ISES code indicates that its predictions are reliable and accurate. Again, those cases where ISES solutions failed to start and/or converge should not be regarded as indicative of the airfoil's failure to perform, but rather as an idiosyncrasy of the code itself. Conversely, successfully converged solutions are regarded as valid.

The confidence expressed above is tempered by one area of uncertainty: the interaction of the laminar separation bubble with a local shock wave as compressibility effects begin to become significant. It is possible that the ISES code prediction of no penalty for the shock being coincident with the laminar bubble is valid — at least for the class of airfoils considered here. The Stratford pressure recovery theoretically begins with an infinite pressure gradient, and the condition presented by the shock simply emulates this, assuming a "healthy" turbulent boundary layer at the onset of the pressure rise. If somehow the interaction between the would-be laminar bubble and the shock is such that a proper turbulent boundary layer is created, all of this might work.

The study of transonic flow at low Reynolds numbers is relatively embryonic, and experimental data are limited. Consequently, it is recommended that the next step should be some basic wind tunnel testing in this flow regime. Emphasis should be on identifying fundamental flow characteristics first, followed by the evaluation of various airfoils.

Some comments on the philosophy of airfoil design for the flow regime considered in this study are in order. Reynolds number was the dominant similarity parameter and Mach number was secondary. Designing to meet the requirement of high lift at RN = 0.5×10^6 and avoiding the penalties of laminar separation bubbles became the primary constraint. Once this was achieved, the resulting airfoils typically showed good performance at Mach 0.4. However, if the Mach number requirement was increased (say to 0.55), it is unlikely that this situation would apply. Here, the standard compressibility trades between C_l and thickness would begin to control the design with all of the low Reynolds number constraints still present. It is interesting to note that both low Reynolds number performance and compressibility-induced drag rise are enhanced by extending the length of the upper surface rooftop pressure distribution — for mutually exclusive fluid mechanical arguments (higher local Reynolds number at transition and weaker shock due to lower rooftop level, respectively). All of this suggests that pursuing the development of low Reynolds number airfoils for operation at higher subsonic Mach numbers would offer an interesting challenge.

ACKNOWLEDGEMENT

This work was sponsored by the United States Air Force under the direction of Charles F. Patterson, Wright Aeronautical Laboratories, AFWAL/FIAD, Wright-Patterson AFB, Ohio. The author also acknowledges the assistance of Peter P. Camacho of Douglas Aircraft Company for his work in designing and analyzing the airfoils.

REFERENCES

1. Drela, M. A. and Giles, M. B., Viscous Inviscid Analysis of Transonic and Low Reynolds Number Airfoils, AIAA Paper No. 86-1786, June 1986.

2. Callaghan, J. G. and Beatty, T. D., A Theoretical Method for the Design and Analysis of Multi-Element Airfoils, AIAA Journal of Aircraft, Vol. 9, No. 12, December 1972.

3. Harris, C. D., Two-Dimensional Aerodynamic Characteristics of the NACA 0012 Airfoil in the Langley 8-Foot Transonic Pressure Tunnel, NASA TM81927, April 1981.

Computation of Viscous Unsteady Compressible Flow About Profiles

K. Dortmann
Aerodynamisches Institut der RWTH Aachen
GFR

1. Introduction

An essential aspect of aerodynamics is the time-dependent behaviour of flows. In this context the separation of boundary layers is of great importance. Depending on the similarity parameters the flow can become unstable, which often leads to selfinduced periodic flows.

The numerical simulation of these problems requires the solution of the Navier-Stokes equations. In addition grid spacing and time steps have to be chosen sufficiently small, so that vortex shedding for example at trailing edges of profiles is not disturbed or even completely suppressed [1]. The consequence is a considerable computational effort. With the developement of a more powerful computer generation these questions could be taken up.

The objective of this investigation is the numerical simulation of unsteady flows about airfoil sections. Since the flow considered is separated from the profiles, even at zero angle of attack, it can be expected that the order of accuracy of the discretization is crucial. To study the numerical influence on the solution two difference approximations are compared and their truncation errors are discussed. The use of central differences in space requires artificial damping, of which the influence on the solution of a model equation and on the solution of a complex separated flow is examined. Finally numerical and experimental data will be compared.

2. Governing Equations

Compressible viscous flows are described by the components of the velocity vector u and v, the thermodynamic quantities density ρ, pressure p and temperature T, the material quantities dynamic viscosity μ, bulk viscosity $\hat{\mu}$, thermal conductivity coefficient λ and the specific heats c_p and c_v. The relation between the first five quantities is given by the conservation equations for mass, momentum and energy formulated in curvilinear time-dependent coordinates $\tau(t)$, $\xi(x,y,t)$, $\eta(x,y,t)$, where external forces and heat sources are neglected:

$$Sr_0 \frac{\partial}{\partial \tau}(J\vec{q}) + \frac{\partial}{\partial \xi}\vec{\bar{E}} + \frac{\partial}{\partial \eta}\vec{\bar{F}} = 0 \qquad (2.1)$$

with

$$\vec{q} = (\rho, \rho u, \rho v, E)^T$$

$$\vec{\bar{E}} = \frac{\partial y}{\partial \eta}\left(\vec{E}_K - \frac{1}{Re_0}\vec{E}_D - \vec{q}\frac{\partial x}{\partial \tau}\right) - \frac{\partial x}{\partial \eta}\left(\vec{F}_K - \frac{1}{Re_0}\vec{F}_D - \vec{q}\frac{\partial y}{\partial \tau}\right)$$

$$\vec{\tilde{F}} = -\frac{\partial y}{\partial \xi}(\vec{E}_K - \frac{1}{Re_0}\vec{E}_D - \vec{q}\frac{\partial x}{\partial \tau}) + \frac{\partial x}{\partial \xi}(\vec{F}_K - \frac{1}{Re_0}\vec{F}_D - \vec{q}\frac{\partial y}{\partial \tau}) \quad .$$

Therein $J^{-1}$is the Jacobian of the transformation. The terms $\vec{q}\frac{\partial x}{\partial \tau}$ and $\vec{q}\frac{\partial y}{\partial \tau}$ take into account the movement of the curvilinear system relativ to the Cartesian. $\vec{E}_K$ and $\vec{F}_K$ are the vectors of the mass, momentum and energy fluxes. $\vec{E}_D$ and $\vec{F}_D$ contain the viscous forces, dissipation and thermal conductivity. In the present study laminar flow and perfect gas is assumed. The variables have been scaled with the stagnation quantities $\overline{\rho}_0$, $\overline{a}_0$, $\overline{T}_0$ likewise $\overline{\mu}_0$ and $\overline{\lambda}_0$. For the nondimensional time t a reference time $\overline{t}_{ref}$ was chosen and for the independent variables x and y a characteristic length $\overline{l}$, such as the chord lenght of an airfoil. With these parameters Strouhal, Reynolds and Prandtl number are defined:

$$Sr_0 = \frac{\overline{a}_0}{\overline{t}_{ref}\overline{l}} \quad , \qquad Re_0 = \frac{\overline{\rho}_0\overline{a}_0\overline{l}}{\overline{\mu}_0} \qquad \text{and} \qquad Pr_0 = \frac{\overline{\mu}_0\overline{c}_p}{\overline{\lambda}_0} \quad . \tag{2.2}$$

3. Boundary Conditions

On the surface of the profile boundary conditions are determined by the no slip condition and by the the assumption of an adiabatic wall. Necessary pressure information on the airfoil is deduced from the normal momentum equation.

On the farfield boundaries, which are located far away from the profil (15 l), viscous forces and heat conduction can be neglected. Then the Navier-Stokes equations change from a mixed parabolic-hyperbolic system into the Euler equations, a pure hyperbolic system. Assuming further a locally onedimensional flow an approximation for the boundary values can be found by means of the eigenvalues λ_k^i and the characteristic variables c_k^i with $k = \xi$ and $k = \eta$, respectively [2]:

$$c_{kR}^i = \begin{cases} c_{k\infty}^i & \text{for } \lambda_k^i < 0 \\ c_{kI}^i & \text{for } \lambda_k^i > 0 \end{cases} \qquad i = 1(1)4 \quad . \tag{3.1}$$

This approximation correspond to an extrapolation of the characteristic variables from outside the integrational domain, subscript "∞", or from the interior, subscript "I", onto the boundary, subscript "R". System (3.1) specifies the primitive variables ρ_R, u_R, v_R and p_R, by which the conservative variables can be assembled.

4. Method of Solution

The steady-state operator of equation (2.1) was discretized with central difference approximations. Two different schemes have been considered: a node-centred scheme, abbreviated with "NC", and a cell-vertex-centred scheme [3] ("CVC$_H$"). With the latter scheme only the steady-state operator of the Euler equation was approximated:

$$\frac{\partial}{\partial t}\vec{q}_{Konv} = -\frac{1}{J\,Sr_0}(\frac{\partial}{\partial \xi}\vec{\tilde{E}}_K + \frac{\partial}{\partial \eta}\vec{\tilde{F}}_K) \quad . \tag{4.1}$$

The diffusive terms were discretized according to scheme "NC":

$$\frac{\partial}{\partial t}\vec{q}_{Diff} = \frac{1}{J\,Sr_0}\left(\frac{\partial}{\partial \xi}\vec{\vec{E}}_D + \frac{\partial}{\partial \eta}\vec{\vec{F}}_D\right) \tag{4.2}$$

The superposition of (4.1) and (4.2) results in the complete approximation of the steady-state operator of the Navier-Stokes equations. Both schemes are second order accurate in Cartesian grids with constant spacing. However in curvilinear coordinates the accuracy is reduced to first order due to stretching and skewing of the grid. Table 4.1 shows the truncation errors of a simple differential operator in stretched grids with $\Delta x_+ = x_{i+1} - x_i$ and $\Delta x_- = x_i - x_{i-1}$.

scheme	truncation error of $f_x + v_0 f_y$
"NC"	$\frac{1}{2}(\Delta x_+ - \Delta x_-)\,f_{xx} + \frac{1}{2}(\Delta y_+ - \Delta y_-)\,v_0 f_{yy} +$ $+ o(\Delta x^2, \Delta y^2)$
"CVC$_H$"	$\frac{1}{2}(\Delta x_+ - \Delta x_-)(f_x + v_0 f_y)_x +$ $+ \frac{1}{2}(\Delta y_+ - \Delta y_-)(f_x + v_0 f_y)_y + o(\Delta x^2, \Delta y^2)$

Table 4.1

Substituting the derivatives in space of equation (2.1) by their difference formulation L_Δ yields the semi-discrete approximation [4]:

$$\frac{\partial}{\partial \tau}\vec{q} = L_\Delta(\vec{q}) \quad . \tag{4.3}$$

This set of ordinary differential equations is solved by an integration formula of the Runge-Kutta type:

$$\begin{aligned} \vec{q}^{(n+1)} &= \vec{q}^{(n)} + \sum_{r=0}^{m-1} \Theta_r \vec{k}_r^{(n)} \\ \vec{k}_r^{(n)} &= \Delta\tau L_\Delta\Big(\vec{q}^{(n)} + \sum_{s=0}^{r-1} \lambda_{r,s}\, \vec{k}_s^{(n)}\Big) \quad . \end{aligned} \tag{4.4}$$

The original goal of these methods was to increase the accuracy of the integration by repeated (m-times) evaluations of the right hand side of (4.3). Looking at the truncation error in space, which is at best of second order, offers a different strategy. One part of the parameters Θ_r and $\lambda_{r,s}$ is chosen such, that second order accuracy in time is guaranteed. The other free parameters are used to maximize the region of stability [5]. The following five point integration formula [6], which was used for this investigation, is an example of such an optimized procedure:

$$M = \begin{pmatrix} \lambda_{0,0} & \lambda_{0,1} & \cdots & \lambda_{0,m-1} \\ \lambda_{1,0} & & & \\ \vdots & & & \vdots \\ \lambda_{m-1,0} & & & \lambda_{m-1,m-1} \\ \Theta_0 & \Theta_1 & \cdots & \Theta_{m-1} \end{pmatrix} = \begin{pmatrix} 0 & 0 & 0 & 0 & 0 \\ \frac{1}{4} & 0 & 0 & 0 & 0 \\ 0 & \frac{1}{6} & 0 & 0 & 0 \\ 0 & 0 & \frac{3}{8} & 0 & 0 \\ 0 & 0 & 0 & \frac{1}{2} & 0 \\ 0 & 0 & 0 & 0 & 1 \end{pmatrix} \quad . \tag{4.5}$$

The accuracy of $o(\Delta\tau^2)$ is controlled by the parameters Θ_4 and $\lambda_{4,3}$. Neglecting the truncation error in space the truncation error in time reads:

$$\varepsilon_\tau = \tfrac{1}{48}\Delta\tau^2 L_\Delta\left[\left(\tfrac{\partial}{\partial \vec{q}}L_\Delta\right)^2 - 2L_\Delta\tfrac{\partial^3}{\partial \vec{q}^3}L_\Delta\right](\vec{q}) + o(\Delta\tau^3) \quad . \tag{4.6}$$

5. Analysis of the Method of Solution

According to Lax's equivalence theorem the solution of a finite-difference approximation of a properly posed initial-value problem converges to the solution of the differential equation, if the conditions of consistency and stability are satisfied.

The consistency condition is fulfilled if the difference approximations are at least of first order accurate. Both space and time discretization satisfy this condition. The superposition of these truncation errors does not produce terms of order zero, what is shown by the truncation error of the complete difference formulation:

$$\begin{aligned}\varepsilon_g = {} & \Delta h\, g(\vec{q}(\tau)) + \tfrac{1}{2}\Delta\tau\,\Delta h\left[f\tfrac{\partial}{\partial \vec{q}}g + g\tfrac{\partial}{\partial \vec{q}}(f + \Delta h\, g)\right](\vec{q}(\tau)) + \\ & + \tfrac{1}{48}\Delta\tau^2 f\left[\left(\tfrac{\partial}{\partial \vec{q}}f\right)^2 - 2f\tfrac{\partial^2}{\partial \vec{q}^2}f\right](\vec{q}(\tau)) + o(\Delta\tau^2\Delta h) + o(\Delta\tau^3) \quad .\end{aligned} \tag{5.1}$$

Herein $f(\vec{q}(\tau))$ represents the steady-state operator of the Navier-Stokes equations and $\Delta h g(\vec{q}(\tau))$ its truncation error.

In order to check the stability of the scheme an analysis according to von Neumann is carried out. Although the theory is restricted to systems of linear equations with constant coefficients and to pure initial-value problems, it is often applied to nonlinear initial-boundary-value problems by analysing model equations. In this case

$$\tfrac{\partial u}{\partial t} + \tfrac{\partial u}{\partial x} + a\tfrac{\partial u}{\partial y} = \tfrac{1}{Re}\left(\tfrac{\partial^2 u}{\partial x^2} + \tfrac{\partial^2 u}{\partial y^2}\right) \qquad \text{with } a = \tfrac{v_0}{u_0} \text{ and } Re = \tfrac{\rho_0 u_0 l}{\mu_0} \quad . \tag{5.2}$$

The resulting stability bounds for parameters like the time step are in general a good estimation for the nonlinear problem. By mapping the solution $u^{(n)}$ into the Fourier space the influence of the scheme on different frequencies can be studied. The amplification factor

$$G(\Delta t, k, l) = \tfrac{V(k,l)^{(n+1)}}{V(k,l)^{(n)}} \tag{5.3}$$

shows whether the amplitude of a Fourier mode is increasing or decreasing. The variables k and l are wave numbers. Their maximum is given by the smallest wave length $\lambda_x = 2\Delta x$ and $\lambda_y = 2\Delta y$ which can be resolved. Stability requires, that $G(\Delta t, k, l)^n$ is uniformly bounded. From this condition the stability criterion due to von Neumann can be derived: $R(\Delta t, k, l) \leq 1 + o(\Delta t)$ with $R(\Delta t, k, l)$ the spectral radius of $G(\Delta t, k, l)$. The critical parameters are the onedimensional Courant and mesh Reynolds numbers: $Cou_x = \frac{\Delta t}{\Delta x}$, $Cou_y = a\frac{\Delta t}{\Delta y}$, $Re_{\Delta x} = \frac{u_0 \Delta x}{\nu}$ and $Re_{\Delta y} = \frac{u_0 \Delta y}{\nu}$. Figure 5.1 shows the stability limits for $Re_\Delta = Re_{\Delta x} = Re_{\Delta y}$, $Cou = Cou_x + Cou_y$ and $a = 1$. The scheme is unstable for $Cou > 4$.

6. Numerical Damping

The numerical integration of conservation laws with central differencing can cause spurious oscillations especially in the neighbourhood of boundaries or shocks, due to the limited ability of numerical grids to resolve small wave lengths. The nonlinear interaction of Fourier modes in the convection terms can form a wave with higher or lower frequency. Generally the lower frequency does not cause any problems, but the continual cascading into higher frequencies does. The exact solution of (5.2)

$$u = V(t)e^{I(kx+ly)} \quad \text{with} \quad V(t) = V_0 V(t)_{Re} V(t)_{Ph} \tag{6.1}$$

and

$$V(t)_{Re} = e^{-\left(\frac{k^2}{Re}+\frac{l^2}{Re}\right)(t-t_0)} \quad , \qquad V(t)_{Ph} = e^{-I(k+al)(t-t_0)} \quad ,$$

shows, that on the one hand the amplitudes of the waves are damped by viscous forces, $V(t)_{Re}$, on the other hand they are carried away by phase movement $V(t)_{Ph}$. The solution of the difference approximation deviates, depending on Courant and mesh Reynolds number, for different wave angles $\Theta = k\Delta x$ and $\Phi = l\Delta y$ more or less from the exact behaviour. In figure 6.1 the amplification factor of the numerical procedure $\|G(\Theta,\Phi)\|$ and of the differential equation

$$\|G_A(\Theta,\Phi)\| = \frac{V(t+\Delta t)}{V(t)} \tag{6.2}$$

are compared for $Cou = 1,2$ and 3 (a), b) and c)) and a typical mesh Reynolds number $Re_\Delta = 10$. The dissipative behaviour of the exact and the numerical solution is in good agreement only for the low frequency modes. For increasing wave angles $\|G(\Theta,\Phi)\|$ is more and more influenced by the truncation error. Increasing the Courant number impairs also the numerical solution. Compared to the exact solution the highest frequencies are scarcely damped.

Figure 6.2 shows the ratio of phase angles of the difference approximation φ and the exact solution φ_A as a function of the wave angles Θ and Φ with

$$\varphi = \arctan\left[\frac{\Im(G(\Theta,\Phi))}{\Re(G(\Theta,\Phi))}\right] \quad \text{and} \quad \varphi_A = -(\Theta Cou_x + \Phi Cou_y) \quad . \tag{6.3}$$

The dispersive error behaves in a similar way as the dissipative error. Furthermore it can be seen, that the phase velocity vanishes for the highest frequencies. Hence pertubations are neither properly damped, nor move they away from their origin. In order to achieve smooth solutions artificial dissipative terms are added to the right hand side of equation (5.2):

$$D(u) = -\varepsilon_x \frac{\partial^4 u}{\partial x^4} - \varepsilon_y \frac{\partial^4 u}{\partial y^4} \quad \text{with} \quad \varepsilon_x = \varepsilon \frac{\Delta x^4}{\Delta t} \quad , \qquad \varepsilon_y = \varepsilon \frac{\Delta y^4}{\Delta t} \quad . \tag{6.5}$$

With this the exact solution (6.1) gets the additional factor

$$V_D(\varepsilon,\Theta,\Phi) = e^{-\varepsilon(\Theta^4+\Phi^4)} \quad . \tag{6.6}$$

For small wave angles the deviation of V_D from one is not worth mentioning and the influence on the solution can be neglected. Figure 6.3 shows the same effect of the artificial damping on the difference equations. The parameters are the same as in figure 6.2 c). ε was chosen in such a way, that $\|G(\Theta = \pi, \Phi = \pi\|$ approaches to zero. This value of ε serves as an estimation for the nonlinear problem.

In curvilinear coordinates the damping terms are used in the nonlinear form [7]:

$$D(\vec{q}) = D_\xi(\vec{q}) + D_\eta(\vec{q}) \tag{6.7}$$

$$D_k = -\Delta k \delta_k \left[\tfrac{J}{\Delta t} \varepsilon \Delta k^3 \delta_k^3 \vec{q}\right] \qquad \text{with } k = \begin{cases} \xi \\ \eta \end{cases}$$

In order to keep the computational effort per time step low $D(\vec{q})$ was evaluated only in the first of the five Runge-Kutta steps and then kept constant [8]:

$$\vec{q}^{(n+1)} = \vec{q}^{(n)} + \sum_{r=0}^{4} \Theta_r \vec{k}_r^{(n)} \tag{6.9}$$

$$\vec{k}_r^{(n)} = \Delta t \left[L_\Delta\left(\vec{q}^{(n)} + \sum_{s=0}^{r-1} \lambda_{r,s} \vec{k}_s^{(n)}\right) + D(\vec{q}^{(n)})\right] \quad .$$

Adding fourth order derivatives to the differential equation requires additional boundary conditions. In this study $\frac{\partial^2}{\partial k^2}\vec{q} = 0$ was applied to all boundaries.

7. Results

The numerical method described above shows some free parameters, for which the influence on the solution has been studied. In particular the effect of the damping coefficient ε, equation (6.7), in cennection with cell-vertex-centred scheme "CVC$_H$" on an unsteady solution will be discussed for the flow around a NACA 0012 profile at an angle of attack of $\alpha = 20°$. The similarity parameters are $Ma_\infty = 0.3$ and $Re_\infty = 2 \cdot 10^4$.

Figure 7.1 shows details of the mesh used, which has 171 grid points in ξ-direction, 81 of them on the profile, and 51 grid points in η-direction. The farfield boundary is approximately 15 chords away from the profile and the minimum grid spacing is $\Delta h = 10^{-3}$. In figure 7.2 streaklines have been plotted, which were computed with the parameters $Cou = 3$ and $\varepsilon = 0.012$. On the upper side of the profile a complex separation region can be seen, which consists of a leading edge vortex, a smaller vortex nearly in the middle of the profile, a big shedded vortex from the leading edge and a trailing edge vortex. One period later in the wake the trailing edge vortex is distorted. This effect can be explained by the different stretching of the grid in ξ- and η-direction. Across the vortex in ξ-direction there are only 3 or 4 grid points , whereas in η-direction the full resolution of the boundary layer is still available. Hence it follows, that for a given wave length the wave angle in ξ-direction is by far larger than that for the η-direction. The result is a stronger damping in ξ-direction due to the artificial damping and to the dissipative terms of the truncation error, which cannot be neglected for high frequency modes. The time-dependent behaviour of the lift is mainly determined by the vortex shedding at the leading and the trailing edge of the profile. Figure 7.3

shows the lift coefficient as a function of time in dependence of the damping coefficient ε. The curve with the weakest damping $\varepsilon = 0.012$ varies between $c_{A_{min}} = 0.54$ and $c_{A_{max}} = 1.13$. Beside of the absolute extremum, relative extrem values exist, what indicates the complex interaction between leading edge and trailing edge separation. The Strouhal number for this case is $Sr = 0.088$. By increasing the damping coefficient to $\varepsilon = 0.02$ the minimum lift rises slightly upto $c_{A_{min}} = 0.59$. The maximum declines noticeable to $c_{A_{max}} = 0.88$ and can hardly be distinguished from the relative maximum. The Strouhal number decreases to $Sr = 0.079$. Finally a further increase of the damping coefficient to $\varepsilon = 0.03$ makes the difference between absolute and relative minimum vanish and leads to a period doubling with a Strouhal number of $Sr = 0.162$. The absolute extrem values do not change.

Looking at the diagrams for the amplification factor of the model equation (5.2) with the parameters of the above described flow and a typical combination of the grid steps, $\Delta x = 0.01$ and $\Delta y = 0.001$ (figure 7.4), shows, that for $\varepsilon = 0$ there is no damping in x-direction. This lack is removed with $\varepsilon = 0.01$, a value above which smooth solutions for the nonlinear problem are possible. A further increase of the damping coefficient suppresses not only the spurious oscillations but also the high frequency modes of the solution, which have a significant influence on selfinduced unsteady flows. Hence a minimization of ε is inevitable.

Beside of flow instabilities, which can lead to complex periodic solutions, the time-dependent behaviour of a flow is often enforced by unsteady boundary conditions. Both effects may come together, if the pressure gradient of a decelerated flow has a destabilizing influence. In this study such a flow was calculated for a NACA 4412 airfoil. Starting from steady inflow conditions, $Re_\infty = 10^4$ and $Ma_\infty = 0.2$, the free stream velocity u_∞ was sinusoidally accelerated to its double value within the nondimensional time interval $0 \leq \tau \leq 0.5$ and $Sr_0 = 0.05$. Even for steady inflow conditions, at an angle of attack of $\alpha = 0°$ and with Reynolds numbers between $5 \cdot 10^3$ and $2 \cdot 10^4$ the flow separates at 60% of the chord length and a von Kármán vortex street arises. The onset of vortex shedding depends on the kind and magnitude of pertubations and cannot be predicted [9].

Figure 7.5 shows the coarsest and the densest mesh used. Both grids have in common the smallest grid step $\Delta h = 10^{-3}$ and the distance of the farfield boundary, which is approximately 15 chords. The number of grid points in ξ-direction are 141 and 512 respectively, from which 31 (199) are located in the wake. In η-direction there are 41 (64) grid points.

With the coarse grid the influence of the space discretization, using scheme "NC" or "CVC_H", on the solution was investigated. Figure 7.6 a) shows the lift coefficient as a function of time computed with scheme "NC". The lift is scaled with $\rho_\infty(\tau \leq 0)$ and $u_\infty(\tau \leq 0)$. During the acceleration the separation point moves downstream, which results in a rapid increase of the lift to a maximum of $c_A = 0.68$. The reason for the subsequent decrease can be found in the reestablishment of the trailing edge separation. For $\tau > 1.2$ the solution begins to oscillate and generates a von Kármán vortex street with $Sr = 0.209$. The lift fluctuates between $c_{A_{max}} = 0.04$ and $c_{A_{min}} = -0.002$. The parameters of the computation were $Cou = 3$ and $\varepsilon = 0.03$. Although this damping coefficient is significantly larger than the value recommended in literature $\varepsilon = 1/128$ [6], it is still within the range for steady-state computations. Figure 7.6 b) shows

$c_A(t)$ computed with scheme "CVC_H" with the same parameters. The movement of the separation point is reproduced in a similar way. The maximum value of the lift coefficient $c_A = 0.75$ is higher than that of the previous computation. But this time no periodic solution could be achieved. For $\tau \to \infty$ is $c_A = 0.03$, which is not a mean value of the solution of the previous computation. However, reducing the damping coefficient to $\varepsilon = 0.0075$ results in a periodic solution too (figure 7.7). Not only before but also after the acceleration a periodic solution exists with the Strouhal numbers $Sr = 0.177$ for $\tau < 0$ and $Sr = 0.224$ for $\tau \to \infty$. During the acceleration the lift coefficient is increasing up to $c_A = 0.64$ and for $\tau \to \infty$ it fluctuates between $c_{A_{min}} = -0.005$ and $c_{A_{max}} = 0.074$. These results show the strong influence of artificial damping and space disretization on the solution.

A comparison between numerically and experimentally determined streaklines is given in figure 7.8. The main purpose of the experimental investigation of this problem has been the visualization of the flow [10]. Good agreement was achieved with respect to the movement of the separation point. However the computed shape of the wake in fig 7.8 a) shows only the principle elements of a vortex street and the Strouhal number $Sr = 0.177$ is too low compared to the experimental one $Sr_E = 0.28$. The reason for this big difference is the coarse mesh resolution. In order to improve the simulation before and after the acceleration a mesh refinement is necessary. Figure 7.9 compares the shapes of the vortex streets computed with the coarsest and the finest mesh (figure 7.5) with $Re = 10^4$. A significantly better agreement with the experiment could be achieved. The Strouhal number of the fine mesh solution is $Sr = 0.262$.

8. Conclusions

The Navier-Stokes equations were solved for the numerical simulation of unsteady viscous flows about airfoil sections. Because of the strong influence of the numerical approach on the time-dependent solution, the truncation errors of two difference approximations for stretched Cartesian grids were discussed. In connection with overlapping C-type grids a formulation of Hall has been advantageous. As in other central difference schemes too artificial fourth order damping terms have to be added to suppress spurious oscillations. Since in unsteady flow calculations the time step is limited for reasons of accuracy, an explicit Runge-Kutta type integration formula was used.

For the flow about a NACA 0012 at an angle of attack of $\alpha = 20°$ with $Re_\infty = 2 \cdot 10^4$ and $Ma_\infty = 0,3$ it could be shown, that artificial dissipation can impair an unsteady solution by suppressing the high frequency modes. These modes play an important role in selfinduced unsteady flows. By increasing the damping coefficient the period time of the vortex shedding is prolonged and the amplitude of the lift fluctuations decreases. To get accurate solutions within the range of the numerical approximation the damping coefficient has to be minimized.

At zero angle of attack and with Reynolds numbers between $5 \cdot 10^3$ and $2 \cdot 10^4$, the flow about the NACA 4412 profile is characterized by a laminar separation. In the wake a Kármán vortex street is generated. During the accelerration of the oncoming flow the solution is stabilized, a phenomenon which could already be observed in experiment. The movement of the separation point downstream during acceleration and the reestablishment of the separation when the final velocity is reached is in good agreement with the experiment even if a rather coarse grid (141×41 points) is used. However at this resolution the Strouhal number of the wake is too low. After a mesh refinement (512×64 points) the relative error of this quantity was approximately 6%.

The numerical integration of the Navier-Stokes equations was carried out on a CYBER 205 (2 pipes). Using the parallel architecture of this computer the effort per time step and grid point of $20 - 25\,\mu$sec could be kept at a low level. Nevertheless computing time has the magnitude of CPU-hours. Hence the main target of further research will be the improvement of the numerical algorithm.

9. Literature

[1] E. Krause: "*Review of Some Vortex Relations* ", Technical Note, Computers and Fluids, Vol. 13, pp. 513-515, 1985

[2] D. L. Whitfield: "*Three-Dimensional Unsteady Euler Equation Solutions Using Flux Vector Splitting* ", Notes for Computational Fluid Dynamics User's Workshop, Tullahoma Tennessee, 1984

[3] M. G. Hall: "*Cell-Vertex Multigrid Schemes for Solution of the Euler Equations* ", Proceedings of Conference on Numerical Methods for Fluid Dynamics , Clarendon Press, Oxford, 1986

[4] R. F. Warming, R. M. Beam: "*Stability of Semi-Discrete Approximations for Hyperbolic Initial-Boundary-Value Problems II: Asymtotic Estimates* ", Computational Fluid Dynamics, G. de Vahl Davis and C. Fletcher (Editors), Elsevier Science Publishers B. V. (North-Holland), 1988

[5] P. J. van der Houwen, B. P. Sommeijer: "*Internal Stability of Explicit, m-Stage Runge-Kutta Methods* ", ZAMM Z. angew. Math. Mech., 60, pp. 479-485, 1980

[6] R. Radespiel, N. Kroll: "*Progress in the Development of an Efficient Finite Volume Code for the Three-Dimensional Euler Equations* ", DFVLR-FB-85-31, 1985

[7] R. W. MacCormack, B. S. Baldwin: "*Numerical Method for Solving the Navier-Stokes Equations with Application to Shock-Boundary Layer Interactions* ", AIAA paper 75-1, 1971

[8] A. Jameson, W. Schmidt, E. Turkel: "*Numerical Solution of the Euler Equations by Finite Volume Methods Using Runge-Kutta Time Stepping Schemes* ", AIAA paper 81-1259, 1981

[9] D. J. Tritton: "*Physical Fluid Dynamics* ", van Nostrand Reinhold Company, New York, Cincinnati, Toronto, London, Melbourne, 1977

[10] W.-B. Schweitzer: "*Experiments on Unsteady Flows About Wing Sections* ", Proceedings of the conference on low Reynolds number airfoil aerodynamics, edited by T. J. Mueller, Notre Dame, Indiana, 1985

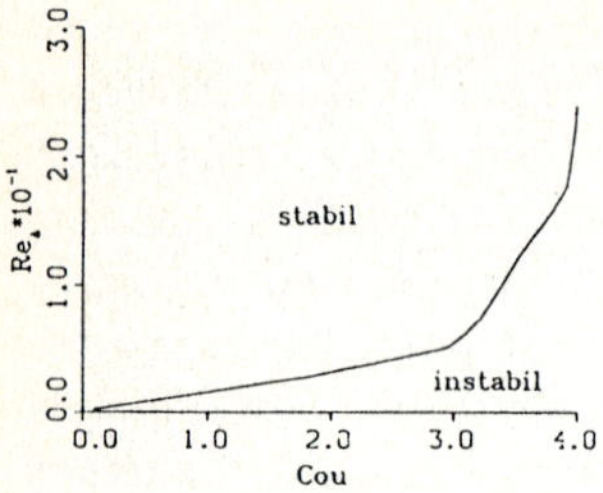

Figure 5.1: Domain of stability for central differences and integration formula (4.5).

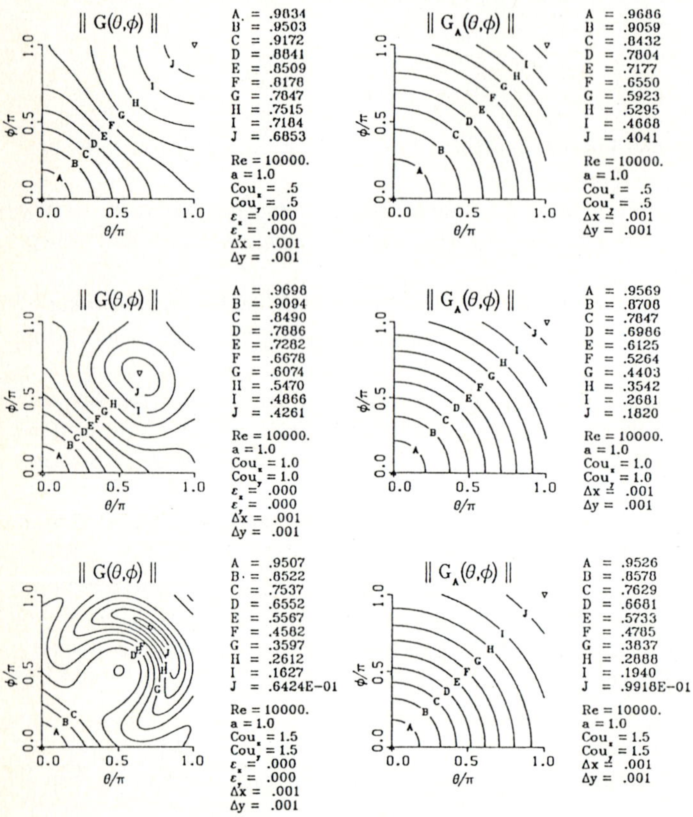

Figure 6.1: Comparison of the amplification factor of the numerical $\|G\|$ and the analytical solution $\|G_A\|$ for $Cou = 1$, 2 and 3 and $Re_\Delta = 10$.

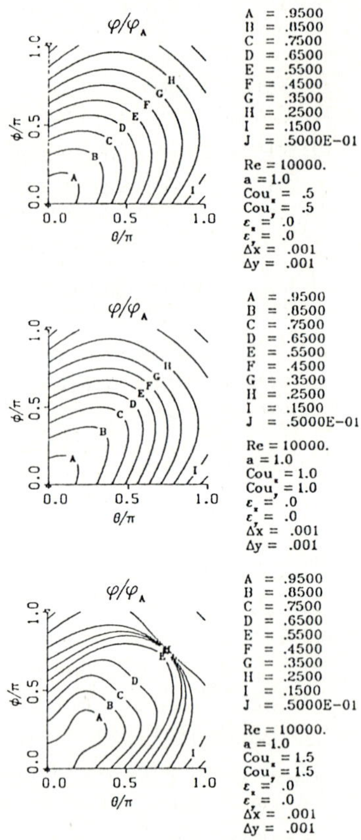

Figure 6.2: Comparison of the phase angles of the numerical φ and the analytical solution φ_A for $Cou = 1$, 2 and 3 and $Re_\Delta = 10$.

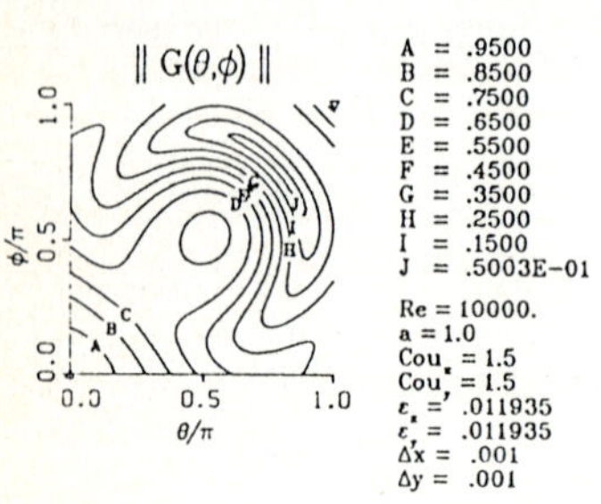

Figure 6.3: Amplification factor $\|G\|$ with maximum damping for the highest frequencies ($Re_\Delta = 10$, $Cou = 3$).

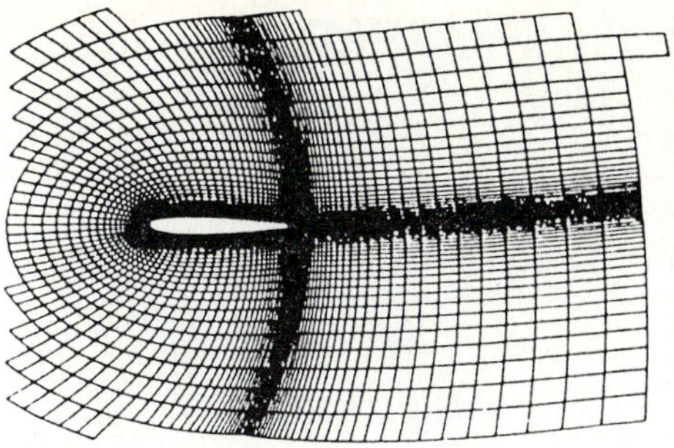

Figure 7.1: Detail of a C-type mesh for NACA 0012 with 171 grid points in ξ-direction, 81 of them on the profile, and 51 in η-direction.

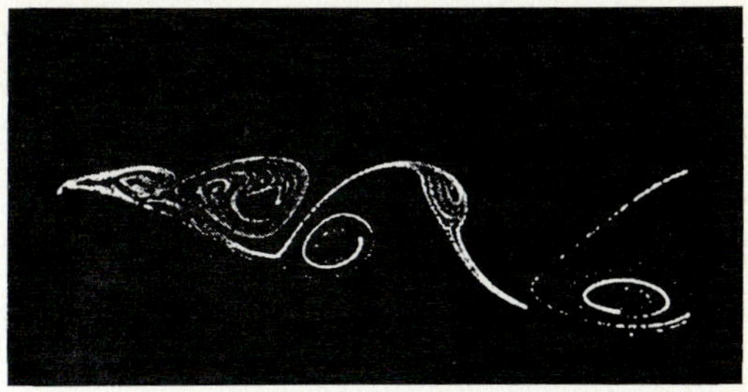

Figure 7.2: Streaklines for $Re = 2 \cdot 10^4$, $\alpha = 20°$ and $Ma = 0.3$.

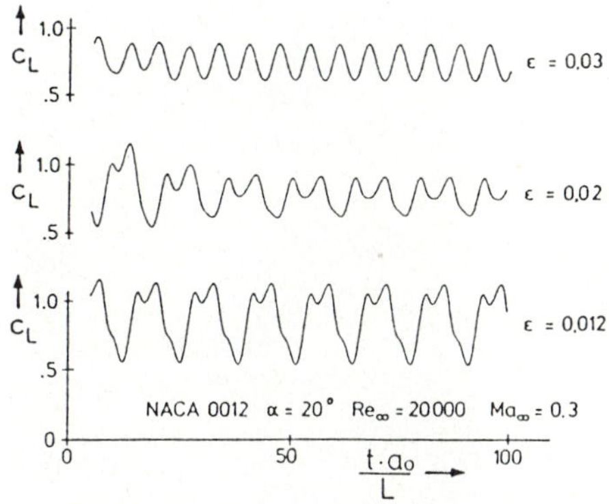

Figure 7.3: Lift coefficient as a function of time and numerical damping.

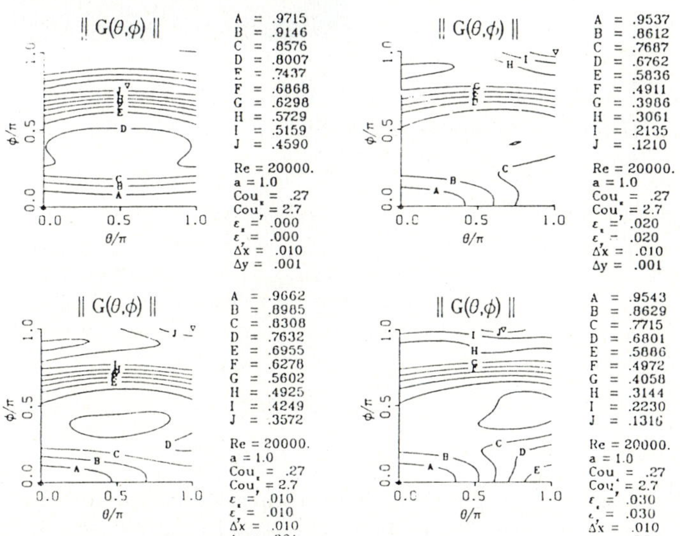

Figure 7.4: Amplification factor $\|G\|$ with $Re_{\Delta x} = 200$, $Re_{\Delta y} = 20$, $Cou = 3$ and $0 \leq \varepsilon \leq 0.03$.

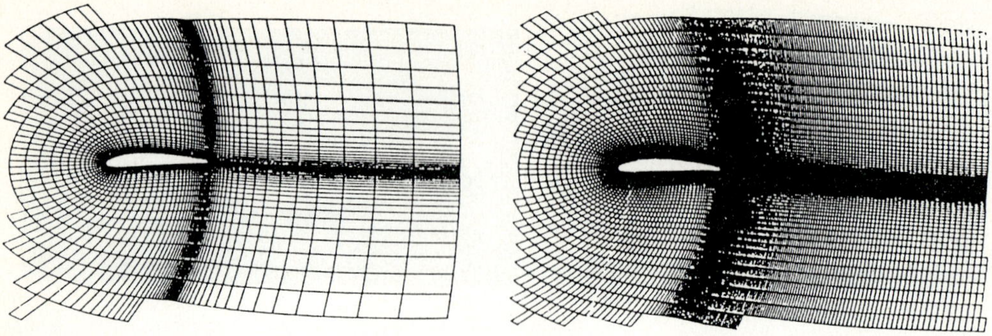

Figure 7.5: Coarsest and densest mesh for NACA 4412, with 141 (512) grid points in ξ-direction, 81 (115) of them on the profile, and 41 (64) in η-direction.

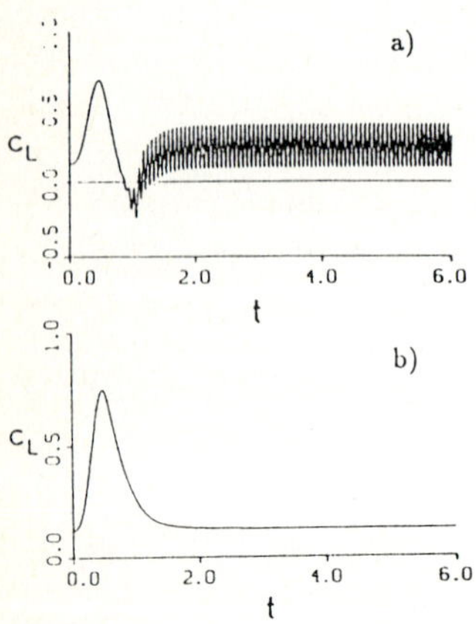

Figure 7.6: Lift coefficient as a function of time for the accelerated flow computed with scheme "NC" (a) and "CVC_H" (b). ε is in both cases 0.03.

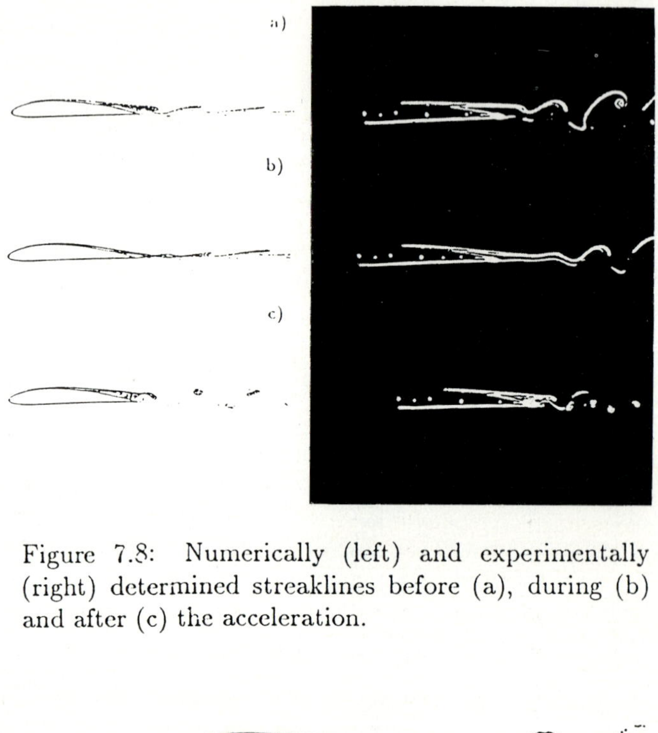

Figure 7.8: Numerically (left) and experimentally (right) determined streaklines before (a), during (b) and after (c) the acceleration.

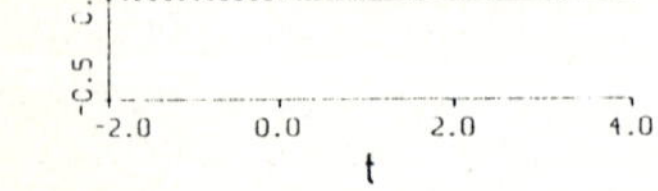

Figure 7.7: Lift coefficient as a function of time with scheme "CVC_H" and ε = 0.0075.

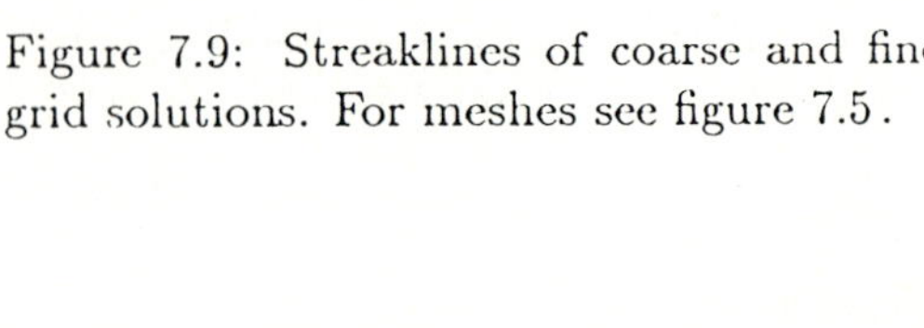

Figure 7.9: Streaklines of coarse and fine grid solutions. For meshes see figure 7.5.

COMPRESSIBLE NAVIER-STOKES SOLUTIONS OVER LOW REYNOLDS NUMBER AIRFOILS

Zuheyr Alsalihi
von Karman Institute for Fluid Dynamics
Chaussée de Waterloo, 72
1640 Rhode-Saint-Genèse,Belgium

Abstract.

Unsteady, laminar and turbulent, compressible and incompressible flow at low Reynolds numbers over a Wortmann FX63-137 airfoil is numerically modeled by solving the time-averaged two-dimensional compressible Navier-Stokes equations by means of the implicit central difference scheme of Beam-Warming. Recent improvements in efficiency,accuracy and convergence for the approximate factorization scheme are employed, including nonlinear artificial dissipation and refined body and farfield boundary conditions. C-type grids are generated by the hyperbolic grid generation technique. Turbulence is modeled by the modified algebraic turbulence model of Baldwin-Lomax and various transition criteria are tested. Results cover a range of test cases with Re_∞ =60,000 to 700,000 and M_∞ =0.01 to 0.5 for which experimental and computational results exist.

1.Introduction.

The objective of this study is to take a further step in this respect and to adequately model the requisite physics of the low Reynolds number phenomena, while still being practical for engineering applications, which is the ultimate goal.

The problems addressed in this study are of mainly two types: modeling the relevant physics and minimizing or isolating the numerical problems usually inherent in the solution procedures of the Navier-Stokes equations. The Navier-Stokes solver used in the study is already tested for various flow configurations and conditions including steady, turbulent transonic flows over airfoils with shock waves which are Stanford Conference test cases, time accurate Couette solutions and compression corners at high supersonic and hypersonic regimes. The validation of the laminar model for this class of problems is followed by the tests to improve the modeling of the special physics, such as transition. Turbulence is modeled employing a turbulence closure together with the time-averaged Navier- Stokes equations. Transition, on the other hand, can not be modeled with these equations. In case of laminar separation bubbles, the flow usually reattaches in transitional or turbulent state. Consequently the flow inside the bubble is partly laminar and partly transitional even if it reattaches fully turbulent. Rigorous modeling of turbulence for unsteady transitional flows with eddy viscosity concept can not be achieved even with multi-equation models. On the other hand, Reynolds stress models or higher order closures are not preferred on the grounds of complexity and excessive computing times. Therefore, algebraic turbulence model of Baldwin-Lomax is used with modifications for this first part of the work which is described in this paper. Transition modeling is attempted using physical arguments and reasonable accuracy is obtained within the limits of the data available for comparison.

2. Mathematical formulation.

The nondimensionalized Navier-Stokes equations written in the strong conservation form are

$$\frac{\partial \vec{Q}}{\partial t} + \frac{\partial \vec{E}(\vec{Q})}{\partial x} + \frac{\partial \vec{F}(\vec{Q})}{\partial y} = Re^{-1}\left\{\frac{\partial \vec{E}_v(\vec{Q},\vec{Q}_x,\vec{Q}_y)}{\partial x} + \frac{\partial \vec{F}_v(\vec{Q},\vec{Q}_x,\vec{Q}_y)}{\partial y}\right\} \tag{1}$$

where $\vec{Q}$ is the unknown vector of the conserved variables density, x and y momentums, and the total energy. The speed of sound is denoted by a. Inviscid fluxes in x and y directions are denoted by $\vec{E}$ and $\vec{F}$, their viscous counterparts are denoted by subscript v. Subscripts x and y denote partial differentiation.

$$\vec{Q} = \begin{pmatrix} \rho \\ \rho u \\ \rho v \\ E_t \end{pmatrix}, \vec{E} = \begin{pmatrix} \rho u \\ \rho u^2 + p \\ \rho uv \\ u(E_t + p) \end{pmatrix}, \vec{F} = \begin{pmatrix} \rho v \\ \rho uv \\ \rho v^2 + p \\ v(E_t + p) \end{pmatrix}, \vec{E}_v = \begin{pmatrix} 0 \\ \tau_{xx} \\ \tau_{xy} \\ f_4 \end{pmatrix}, \vec{F}_v = \begin{pmatrix} 0 \\ \tau_{xy} \\ \tau_{yy} \\ g_4 \end{pmatrix}$$

with

$$\tau_{xx} = \mu(4u_x - 2v_y)/3 \qquad f_4 = u\tau_{xx} + v\tau_{xy} + \mu Pr^{-1}(\gamma-1)^{-1}(a^2)_x$$
$$\tau_{xy} = \mu(u_y + v_x) \qquad g_4 = u\tau_{xy} + v\tau_{yy} + \mu Pr^{-1}(\gamma-1)^{-1}(a^2)_y$$
$$\tau_{yy} = \mu(4v_y - 2u_x)/3$$

Pressure is related to the conservative flow variables Q by the equation of state

$$p = (\gamma - 1)\left[E_t - \frac{1}{2}\rho(u^2 + v^2)\right] \tag{2}$$

The equations are mapped onto a generalized body-fitted coordinate system with uniform grid spacing so as to facilitate the difference representations and to handle the boundaries without interpolations. The so-called C-type of grid in the physical plane, which wraps around the airfoil, is created using the *hyperbolic grid generation* technique (see Alsalihi [1]).

Under the transformation, $(x,y,t) \rightleftharpoons (\xi,\eta,\tau)$, the Eq. (1) can still be written in the strong conservation form maintains the strong conservation form (see Peyret and

Viviand [2]). The unknown vector $\hat{\vec{Q}}$ is expressed as $\hat{\vec{Q}} = \vec{Q}/J$, where J is the Jacobian of the transformation. The x and y derivatives are transformed to the ξ and η derivatives by the chain rule expansion (see Pulliam [3], Müller [4]).

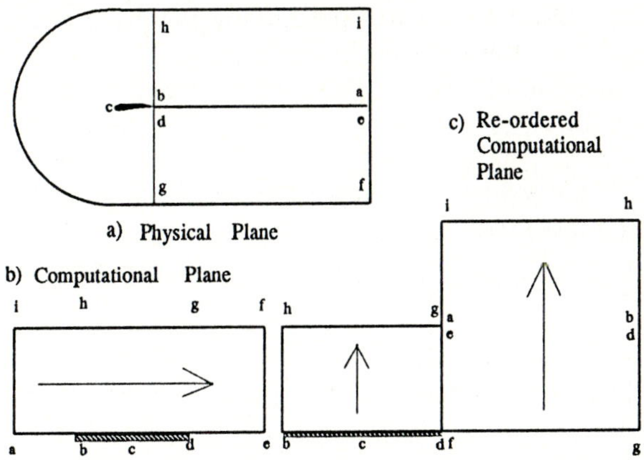

Figure 1. Physical and computational planes

Physical boundary conditions are applied on the airfoil surface, along the far field boundary, which behaves either as the inflow or outflow depending on the angle of attack, and along the outflow boundary (see Fig. (1)). The outflow boundary located downstream of the airfoil is treated differently from the far field boundary because of the presence of viscous effects due to the wake. At the far field boundary viscous effects are neglected. As it is seen in Fig. (1), in addition to these *physical* boundary conditions there are boundaries in the *computational* plane which are due to the wake cut and is purely *mathematical.* Classical treatment has been to average the extrapolated values along the wake cut from upper and lower points. A different way of integration, at the expense of a more complicated solver, is followed in the present work. The computational $\{\xi, \eta\}$ plane is re-transformed into the $\{\xi', \eta'\}$ plane as shown in Fig. (1c). Therefore, in this new computational plane the wake region is one block in the η' direction. As will be clarified in section 3, the operations in the η direction are transformed to the η' direction and the boundaries encountered in this direction are physical far field boundaries. Hence, the wake points are treated exactly in the same manner as other interior points. In this new configuration the ξ coordinate remains unchanged.

No-slip boundary conditions, i.e., $u = v = 0$, are applied on the solid walls. Wall temperature is either specified or evaluated by the specified wall heat flux. Adiabatic wall conditions are specified in this work. In implicit boundary condition application, the equations are simplified with the physical boundary conditions and are solved directly to give the conserved variables on the wall (Pulliam [3], Müller [4], Beam and Warming [5], Thomas [6], Steger [7]).

Stretched grids are usually used to place far field boundaries far away from body surfaces and therefore viscous effects are assumed to be negligible. Therefore, inviscid characteristic like boundary conditions, locally one dimensional Riemann invariants, are used at the outer far field boundary (see Pulliam [3]). For lifting airfoils in subsonic free stream, circulation at the far field boundary is accounted for to first order (Salas et al. [8], Pulliam [3]) by imposing a compressible potential vortex solution which is added as a perturbation to the free stream quantities. The vortex correction logic can be modified to produce boundary conditions which allow one to compute the angle of attack for a given lift [3].

The above outlined procedure can not be applied to the outflow region downstream of the airfoil where the wake viscous effect can not be neglected. Nevertheless, the signal propagation in the inviscid part of the outflow boundary is approximately modeled by the following, [4]

$$\frac{\partial^2 \rho}{\partial \xi^2} = 0, \quad \frac{\partial^2 (\rho u)}{\partial \xi^2} = 0, \quad \frac{\partial^2 (\rho v)}{\partial \xi^2} = 0, \quad p = p_\infty \tag{3}$$

3. Description of the numerical algorithm.

A two-parameter family of explicit and implicit schemes written in a finite volume formulation is given as

$$\Delta \hat{Q}^n_{j,k} = \frac{\lambda^\xi \theta \Delta t}{(1+\omega)} \frac{\partial \Delta \hat{Q}^n_{j,k}}{\partial t} + \frac{\Delta t}{(1+\omega)} \frac{\partial \hat{Q}^n_{j,k}}{\partial t} + \frac{\omega}{1+\omega} \Delta \hat{Q}^{n-1}_{j,k} \tag{4}$$

Here $\Delta \hat{Q}^n_{j,k} = \hat{Q}^{n+1}_{j,k} - \hat{Q}^n_{j,k}$ and Δt denotes the time step [9]. This two-parameter family of algorithms contains first and second order implicit as well as explicit schemes. The scheme is temporally second order if $\theta = \omega + \frac{1}{2}$ and first order otherwise. For time accurate calculations temporally second order three-point backward differentiation ($\theta = 1, \ \omega = 1/2$) is used [9]. The spatial accuracy of the scheme depends on the form of the numerical flux functions. An equation for $\hat{Q}^{n+1}$ is obtained by substituting the the transformed system (1) into the Eq.(4). However, to solve for $\hat{Q}^{n+1}$ one normally needs to solve a set of nonlinear algebraic equations iteratively. One way of avoiding this is to linearize the implicit operator and solve the linearized form by other means. Following the same procedure as in references [10] and [11], a conservative linearized alternating direction implicit (ADI) form of Eq.(4) is obtained. In the linearization of the viscous fluxes the procedure of Steger [7] is followed. Let the grid spacing be denoted by $\Delta\xi$ and $\Delta\eta$ such that $\xi = j\Delta\xi$ and $\eta = k\Delta\eta$. Also let $\lambda^\xi = \frac{\Delta t}{\Delta \xi}$ and $\lambda^\eta = \frac{\Delta t}{\Delta \eta}$. The final two-step algorithm using an intermediate unknown $E*$ is as follows

$$\left[I + \frac{\lambda^\eta \theta}{1+\omega} \left(\tilde{H}^\eta_{j,k+\frac{1}{2}} - \tilde{H}^\eta_{j,k-\frac{1}{2}} \right) + \frac{\lambda^\eta \theta}{(1+\omega) Re_\infty} \frac{\partial M}{\partial \eta} \right]^n E^* = -\frac{\lambda^\eta}{1+\omega} \left[\tilde{F}^n_{j+\frac{1}{2},k} - \tilde{F}^n_{j-\frac{1}{2},k} \right]$$

$$-\frac{\lambda^{\eta}}{1+\omega}\left[\tilde{G}^{n}_{j,k+\frac{1}{2}}-\tilde{G}^{n}_{j,k-\frac{1}{2}}\right]+\frac{\Delta t}{(1+\omega)Re_{\infty}}\left[\frac{\partial\tilde{E}_v}{\partial\xi}+\frac{\partial\tilde{F}_v}{\partial\eta}\right]+\frac{\omega}{1+\omega}\Delta\hat{Q}^{n-1}_{j,k}, \quad (5)$$

$$\left[I+\frac{\lambda^{\xi}\theta}{1+\omega}\left(\tilde{H}^{\xi}_{j+\frac{1}{2},k}-\tilde{H}^{\xi}_{j-\frac{1}{2},k}\right)+\frac{\Delta t\theta}{(1+\omega)Re_{\infty}}\frac{\partial N}{\partial\xi}\right]^{n}E^{n}=E^{*}, \quad (6)$$

$$E^{n}=\Delta\hat{Q}^{n};\quad \hat{Q}^{n+1}=\hat{Q}^{n}+E^{n} \quad (7)$$

where for the finite volume formulation

$$\tilde{H}^{\xi}_{j+\frac{1}{2},k}E^{*}=\frac{1}{2}\left[\hat{\hat{A}}_{j+1,k}E^{*}{}_{j+1,k}+\hat{\hat{A}}_{j,k}E^{*}{}_{j,k}\right]E^{*} \quad (8)$$

Where A and B are the inviscid flux Jacobians, M and N are the viscous flux Jacobians. Note that the Jacobians are the derivatives of the *Cartesian fluxes* with respect to $\vec{Q}$. The expressions for A, B, M and N matrices, as well as for the viscous terms of the right hand side of Eq.(5) can be found in [3], [4], [5]. The inviscid numerical flux vector in ξ direction is

$$\bar{F}_{j+\frac{1}{2},k}=\frac{1}{2}\left[\left(\frac{\xi_x}{J}\right)_{j+\frac{1}{2}}(F_{j+1,k}+F_{j,k})+\left(\frac{\xi_y}{J}\right)_{j+\frac{1}{2}}(G_{j+1,k}+G_{j,k})+\Phi_{j+\frac{1}{2}}\right] \quad (9)$$

Where vectors F and G are the Cartesian inviscid fluxes. Similar expression is obtained in η direction replacing ξ by η. Dissipation vector in Eq.(9) is given as

$$\Phi_{j+\frac{1}{2}}=-\bar{\epsilon}^{(4)}\left(Q_{j+2}+Q_{j+1}+Q_{j}+Q_{j-1}\right)$$

$$\bar{\epsilon}^{(4)}=\kappa\Delta t\sigma_{j+\frac{1}{2}},\qquad \sigma_j=\left(|U|+|V|+c\left(\sqrt{({\xi_x}^2+{\xi_y}^2)}+\sqrt{({\eta_x}^2+{\eta_y}^2)}\right)\right)J_{j,k}$$

The spectral radii of the inviscid Jacobian matrices A and B is denoted by σ and κ is a constant (see section 5). Similar expressions are obtained for the η direction. Note that no dissipation is included in the implicit matrix inversions (left hand sides of Eqs.(5) and (6)).

One begins the iteration by first constituting the ξ derivatives in the right hand side of Eq.(5) in the ξ configuration of the computational plane (Fig.1b). Secondly, *over the airfoil region* the η derivatives of the right hand side of the Eq.(5) and the implicit portion (the left hand side of Eq.(5)) are formed. The so-called η-sweep is performed in this computational plane. Then the data is ordered to the $\eta\prime$ configuration Fig. (1c). The η derivatives of the right hand side and the block matrices of the left hand side of the Eq.(5) are formed *in the wake region* for the second η sweep. The resulting $E*$ is re-ordered to the ξ configuration Fig. (1b), and the left hand side block matrices of the Eq.(6) are formed and solved in the *inner field, i.e., excluding the wake cut line.* Finally, another ξ sweep is performed in the direction of the flow *along the upper wake line.* The iteration is completed by updating the solution (Eq. (7)) and applying the *explicit* boundary conditions, i.e., the far field boundary condition.

4. Turbulence model.

The expressions for the total viscosity and thermal energy conductivity which replace the μ and k are given as $\mu_e = \mu + \mu_T$ and $k_e = k + k_T$. Where μ_T and k_T are the turbulent viscosity and thermal conductivity.

The equilibrium model used is the highly modified Baldwin-Lomax model [12] patterned after the Cebeci-Smith model. This model is a two-layer model and removes the necessity for finding the edge of the boundary layer of its parent model. The *standard model* is expressed in terms of an inner and outer eddy viscosity function as

$$\mu_T = \begin{cases} \mu_{inner}; & \text{if } y \leq y_c \\ \mu_{outer}; & \text{otherwise.} \end{cases}$$

where y_c is the first point at which μ_{inner} exceeds μ_{outer}. (For the details of the model see Baldwin and Lomax [12]). The modeling constants have been determined by requiring agreement with the Cebeci [13] for *constant pressure boundary layers at transonic speeds.* In the present study this standard model is modified by taking into account the strong pressure gradients and the low Reynolds number region following the transition to turbulent flow. Granville [14], expressed the constants C_{cp} and C_{kleb} as functions of the modified Clauser pressure-gradient parameter by examining the experimental velocity similarity laws. The Clauser constant K is expressed as a function of the displacement-thickness Reynolds number to take into account the low Reynolds number effects by applying the correction given by Cebeci [15].

Another problem area that needs modification is inherently related to the airfoil geometry: the model *does* take into account the *wall bounded* and *wake* flows. However, *near* the trailing edge the flow is strongly affected by the wall (or rather the trailing edge). Therefore, near the trailing edge the flow is modeled as if the trailing edge is a perturbance. This fact is first noticed by Bradshaw [16], who noted that for y distances approximately of the order of 0.2δ , the flow can be described as "self-similar " for the perturbations of velocity and shear stress. This idea is incorporated into the model by replacing the Prandtl mixing length by a characteristic length $y_{c_{te}}$ in an eddy viscosity expression similar to Cebeci formulation as follows

$$\mu_{T_{te}} = \rho \kappa U_{\tau}{}^{te} y_{c_{te}}, \quad with \quad y_{c_{te}} = 0.03(x - x_{te})$$

The quantity $U_{\tau}{}^{te}$ is the shear velocity at the trailing edge. This eddy viscosity is used downstream the trailing edge for a distance of 7 trailing edge boundary layer thicknesses as the inner eddy viscosity. Furthermore, to provide a smooth transition from inner to outer eddy viscosity in a given profile the the following combined eddy viscosity is used everywhere [17]

$$\mu_T = \mu_{outer} tanh(\mu_{inner}/\mu_{outer})$$

The model also provides a mechanism which attempts to simulate the transition by setting μ_T equal to 0 everywhere in a profile for which the maximum tentatively computed value of μ_T is less than a threshold value μ_{mutm}=14.

In order to simulate the effect of the upstream history, particularly after the separation, the following relaxation model is used [18].

$$\mu_T = \mu_{T-eq} - (\mu_{T_{eq}} - \mu_{T_i})exp[-(x - x_i)/\lambda_r], \quad where \quad \lambda_r = C_r\delta_0$$

is a relaxation length parameter as a function of the distance from a selected point (denoted by subscript i). Typically, it is selected slightly upstream of separation, δ_0 denotes the boundary layer thickness at that location, $\mu_{T_{eq}}$ is the local equilibrium value and C_r is a constant taken as C_r=50. The results reported here use the above formulation, although many other variations have been tested.

5.Results and discussion.

In order to diagnose and solve or at least to be aware of the numerical effects, main parameters and problem areas of the numerical method are tested. These results will be discussed mainly for *unsteady* flows as they are more sensitive to these numerical effects than the steady flows.

The present algorithm is highly vectorized and parallelized on an ALLIANT FX/8 mini supercomputer with 3 parallel computing elements and 12 MB core memory. Computational speed averages approximately 320μs/grid point/iteration, and the main memory required in words is about 280 x the meshsize. Reduction in computational cost is possible for steady computations by updating the viscosity, effective turbulence viscosity, convergence check and the viscous flux Jacobians every set number of iterations (about 10 or 20).

In order to be able to analyze the effect of the grids used, an extensive grid sensitivity study is performed. In all the cases a Wortmann FX63-137 airfoil is computed at typical 10 degree angle of attack at Reynolds number 700,000 and Mach number 0.01. Turbulence model used is the above described Baldwin-Lomax model with the model's original transition criterion.

A 185x65 grid was developed with 120 points on the airfoil, a maximum grid extend of L_{max}=25c, and a minimum spacing normal to the wall of $\Delta n = 1.0x10^{-6}$c. With this minimum spacing at least three points were found to be in the viscous sub-layer for the given conditions.

In order to analyze the effect of the location of the far field, various outer boundary locations were tested with the identical interior grid point distribution (number of points and spacing near the body). In these calculations far field circulation correction is used.

Since this correction is correct only for steady calculations, a smaller time step must be used. The time operator in the equations is split and each *physical time step* is treated as the steady state. Therefore, a time relaxation for that state is done by using a *numerical time step*. This also reduces the computation time. With the use of this point vortex correction, 25 chord-length distance is found to give acceptable results for the extreme low Mach number and high angle of attack . The drag coefficient is less sensitive to the variation in this distance. As for the normal grid resolution, at least three to five points are required in the viscous sublayer. Adequacy of the grid density is checked by inspecting the velocity profiles and the flow resolution especially near the leading edge. A grid, generated with the *hyperbolic grid generation* technique and used as the standard grid (for most of the results reported here) is shown in Fig. (2).

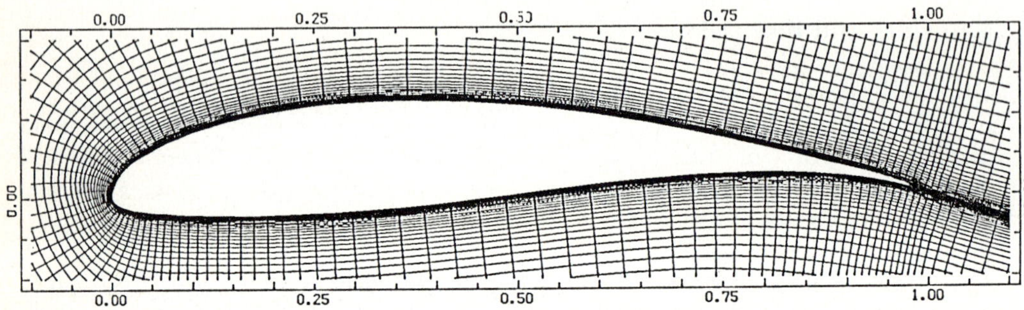

Figure 2. The near-body detail of the 185x65 grid

Numerical dissipation is particularly necessary in separated flow regions and is a potential problem especially in case of separation. The size of separation bubbles can be altered or separation may be completely prevented by the numerical dissipation. Many possible choices available in the literature,([3], [4], [5], [6], [19], [20]), are tested. The best results are obtained by combining the one given by Beam and Warming, [5], with the nonlinear dissipation of Jameson [20]. The only empirical entry κ is optimized by trial and error. However, once fixed, its value is kept constant for the class of problems presented in this report. *No second order explicit or implicit dissipation is used.* It is also found that grid refinement is another (and more acceptable) alternative to *increasing* the dissipation in case of unresolved physical flow regions due to insufficient grid resolution.

The method is tested for laminar flows in order to avoid any modeling problem of phenomena like transition and turbulence. In order to ensure laminar flow, a Wortmann FX63-137 airfoil is computed at Re=60,000 and Mach number 0.01. The leading edge details of the flow at 0 and 14 degrees angle of attack are shown in Fig. (3).

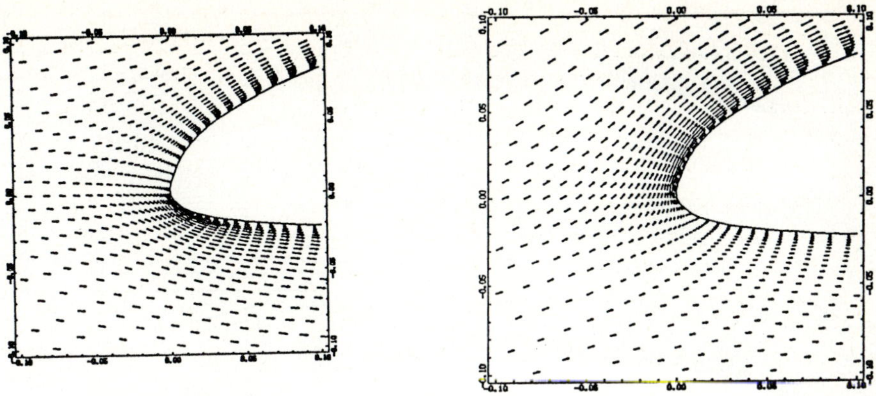

Figure 3. The leading edge details at zero and 14 degrees of angle of attack, laminar flow.

The peculiar feature of the Wortmann airfoil, the stagnation point *on the upper surface* as noted in ref. [21], is also evident in the present computation. The movement of the stagnation point towards the lower surface can be seen in the same figure.

Long separation bubbles encountered especially at lower Reynolds numbers are shown in Fig (4). for 0 and 5 degrees of angle of attack.

At 0 degree angle of attack, the suction-side laminar boundary layer separates near the mid-chord station. Unstable flow behavior in the separated region then appears downstream of the trailing edge. Because of the special geometry of the airfoil, the separated flow does not attach, whereas on the pressure-side it does. At 5 degrees angle of attack, the physical picture does not change significantly, but the pressure-side bubble moves toward the trailing edge. The upper-surface separation point moves forward. Also noticeable is the fact that the vortices created are swept over the airfoil and shed there of. This point is more clearly observed in Fig. (5), where the evolution of the flow at the same conditions and at 0 degrees of angle of attack is shown. For a NACA 23009 airfoil, basically the same features are observed in the smoke photographs of Fisher et. al. [22].

In the low Reynolds number regime considered in this study, flow is usually dominated by either separation bubbles or, depending on the angle of attack and the Reynolds number, massive separation. Since massive separation is also caused by the bursting of a separation bubble, separation bubble is known to play the key role [23], [24], [25], [26]. Therefore, attention is focused on the mechanics of the bubble and its effects on the flow structure, including transition. Assuming the simpler, and in this class of flows more probable case of laminar separation, the transition is modeled by using the results of the classical inviscid stability theory [16]. Despite the limiting assumptions, the results of the theory have been confirmed by many experiments (see, for example Wortmann [27]).

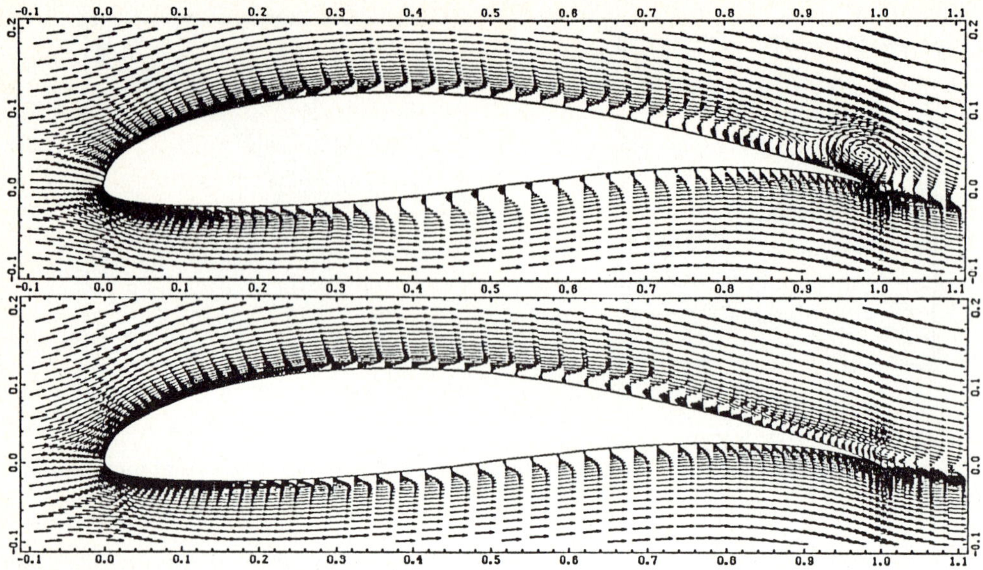

Figure 4. Laminar flow detail near the body.

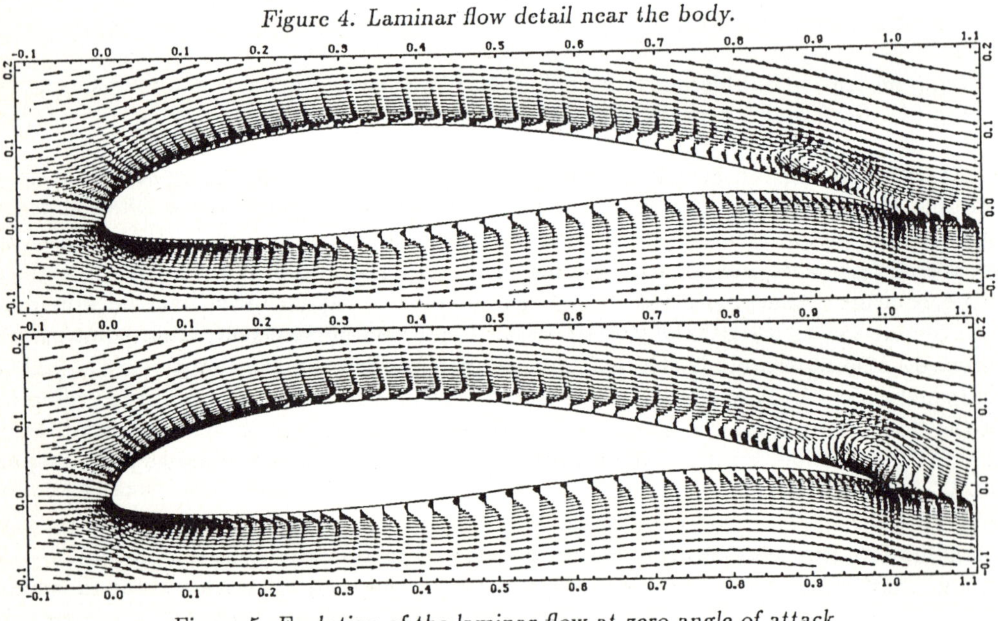

Figure 5. Evolution of the laminar flow at zero angle of attack.

Therefore, in the calculations it is assumed that the transition takes place at the point where the mean velocity profile has an inflection point. Shown in Fig. (6) are the computations over a Wortmann FX63-137 airfoil at low Mach number (0.01) and at Reynolds number 700,000. This is a case for which there is experimental data as well as viscous-inviscid interaction calculations of Williams [28]. In Fig. (6a), the angle of attack is 0 degree and the experimental transition location on the upper surface for this case is given as about 0.6 chord. The computed location of the *first* velocity profile with inflection point is seen to coincide with this position.

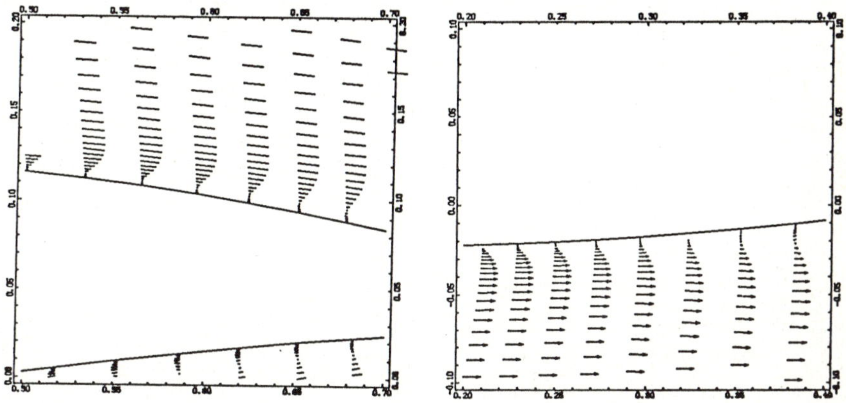

Figure 6. Location of transition point a) upper surface , b) lower surface.

The interaction theory predicts a separation bubble at this locations, whereas no bubble is encountered on the upper surface in the computations and experiment. Furthermore, the interaction theory predicts the bubble positions at about 0.5 chord position. Same agreement is obtained for the lower surface of the airfoils. Predicted bubble position on the lower surface for the cases discussed above are shown in Fig. (6b). The experimental bubble position is given as about 0.4 chord point. The Baldwin-Lomax turbulence model transition criterion predicts the transition as 0.7 and 0.5 chords for 0 degree angle of attack, The same trend is observed for the angle of attack values up to 10 degrees with the decreasing accuracy. The accuracy of the turbulent predictions with this criterion is tested for the limited data available by computing the global aerodynamic force coefficients. Lift coefficient is compared with the calculations of Kothari et Anderson [21] and the experimental data of [26] in Fig. (7). The experiment is incompressible and the computational data of [21] is obtained with Mach 0.5 and Reynolds 100,000.

The comparison reveals the fact that the quantitative accuracy of the low Mach number results are acceptable at this relatively low Reynolds number. Furthermore, the flow is believed to be transitional and therefore, the reasonable agreement is taken as encouraging for the transition criterion described above. At 0 degree angle of attack, for which the only computational data is given, the transition predicted by the Baldwin-Lomax model is 0.4 and 0.2 chord points on the upper and lower surfaces, respectively. The same case is computed at M=0.5 and the results obtained showed the fact that at this Mach number, the effect of compressibility is not significant but not altogether negligible as the lift at 0 incidence is increased by approximately 2%. The same case is re-computed for purely laminar flow and found to be *unsteady*, whereas the turbulent computation indicates completely attached flow. The transition criterion predictions are about 0.3 on both surfaces. Also of importance is the differences between the results of two different facilities.

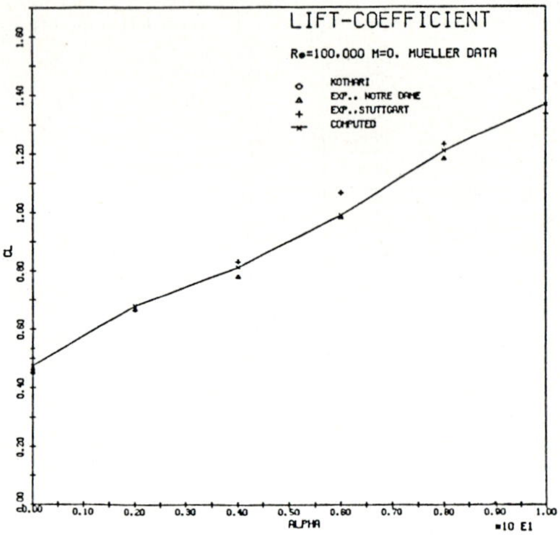

Figure 7. Lift coefficient comparison ([21], [26]).

A more definitive and severe case is the incompressible flow at Reynolds number 700,000 for which both experiment and interaction method data is available. This Reynolds number is also a more realistic value for the practical RPV (remotely piloted vehicles) applications. The lift and lift-drag comparisons are shown in Figs. (8) and (9).

The first point to be noted is the repeated discrepancy between two two experiments. The computed lift coefficients are in good agreement with the interaction theory and with the experiment. However, the lift-drag curve shows a pronounced difference in the predictions of the two theoretical methods. On the other hand, the computations are seen to be slightly but consistently overpredicting the drag at a given lift. The interaction theory also shows the same tendency. This may be attributed to the deficiency of the turbulence model employed. As the angle of attack increases, the discrepancy becomes more pronounced. These results underline the fact that in case of massive separation, the equilibrium models become increasingly inaccurate. The relaxation model described in this report proves to be insufficient in the presence of strong positive gradients.

The pressure coefficient comparison at eight degree angle of attack is of interest in two points. Primarily, the transition is reasonably well predicted when compared with the experiment; and secondly, the presence of a sharp suction peak at the leading edge observed in the computations. The computation shows separation on both sides of the airfoil and this is also observed experimentally. This pressure peak, which is closer to the theoretical data than the experimental, is believed to be of numerical origin. Grid resolution is more pronounced at high angles of attack. This is due to the fact that the separation point on the suction-side moves forward as the angle of attack increases. However, because of the particular shape of the airfoil, an accurate description of the leading edge region is limited by the practical memory considerations.

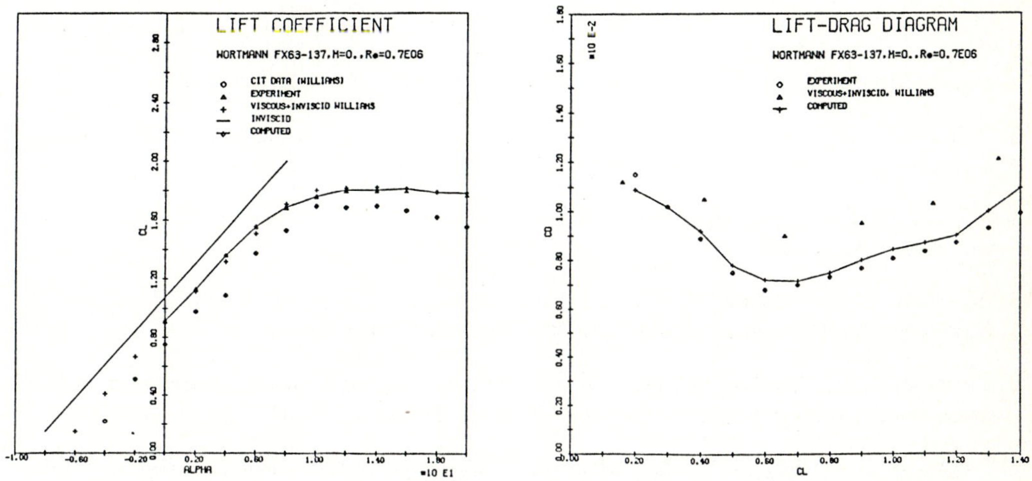

Figure 8. Comparison for lift coefficient a) and lift-drag coefficient b) [28].

6 Concluding remarks.

This paper has described the preliminary phase of an on-going research towards a better modeling of the low Reynolds number airfoil flows.

The results indicate that the numerical features can be as important as the physical modeling considerations when the accuracy of the solution is concerned, in the modeling of complex flows with the Navier-Stokes equations. In case of laminar separation bubble caused transition, the proposed simple criterion is found to work reasonably well. However, this point needs further test and also, lacks the versatility to include the environmental effects such as the free stream turbulence level. Turbulence modeling can be the solution to include such effects. It is known that in case of small separated regions equilibrium models give better results than the one or two-equation models. However, in case of large separated regions, the latter class of equations produce better results. As for the transition, only at least two-equation models can produce nonlinear instabilities which resemble the transition but without any physical meaning. Additionally, the low Reynolds number flow turbulence, in the sense of turbulent Reynolds number based on the kinetic energy, is another difficult physical problem to model because of the effect of the molecular viscosity on the turbulence. Therefore, following these arguments, in the second phase of the research modeling with a modified two-equation k-ϵ low Reynolds number turbulence is already in progress.

References

1 Alsalihi, Z., "Two-Dimensional Hyperbolic Grid Generation", VKI-TN 162, October 1987

2 Peyret, R., Viviand, H., "Computation of Viscous Compressible Flows Based on the Navier-Stokes Equations", AGARD-AG-212, 1975

3 Pulliam, T. H., "Solution Methods in Computational Fluid Dynamics", Von Karman Institute LS-1986-02, January 1986

4 Müller, B., "Navier-Stokes Solution for Hypersonic Flow over an Indented Nose-tip",AIAA Paper 85-1504, 1985

5 Beam, R. M., Warming, R. F., "An Implicit Factored Scheme for the Compressible Navier-Stokes Equations", AIAA J., Vol. 16, April 1978

6 Thomas, P. D., "Numerical Method for Predicting Flow Characteristics and Performance of Nonaxisymmetric Nozzles - Theory", NASA Cr-3147, Sept. 1979

7 Steger, J. L., "Implicit Finite-Difference Simulation of Flow About Arbitrary Two-Dimensional Geometries", AIAA J., Vol. 16, No. 7, July 1978.

8 Thomas, J., Salas, M., "Far field Boundary Conditions for Transonic Lifting Solutions to the Euler Equations", AIAA Paper 85-0029, 1925.

9 Anderson, D. A., Tannehill, J. C., Pletcher, R. H., "Computational Fluid Mechanics and Heat Transfer", Mc Graw Hill, 1984.

10 Yee, H. C., "Linearized Form of Implicit TVD Schemes for the Multidimensional Euler and Navier-Stokes equations", Computers and Mathematics with applications, Vol. 12A, No.4/5,1986.

11 Yee, H. C., "High-Resolution Shock-Capturing Schemes for Inviscid and Viscous Hypersonic Flows", NASA tm 100097, April 1988.

12 Baldwin, B. S., Lomax, H., "Thin layer Approximation and Algebraic Model for Separated Turbulent Flows", AIAA Paper No. 78-257, January 1978.

13 Cebeci, T., "Calculation of Compressible Turbulent Boundary-Layers with Heat and Mass Transfer", AIAA Paper 70-741, presented at AIAA 3rd Fluid and Plasma Dynamics Conference, Los Angeles, Calif., June 1970.

14 Granville, P. S.,"Baldwin-Lomax factors for Turbulent Boundary-Layers in Pressure Gradient, AIAA J., Vol., 25, No., 12, December 1987

15 Bradshaw, P.,"Prediction of the Turbulent Near-Wake of a Symmetrical Aerofoil", AIAA J., Vol., 8, No., 8, August 1970

16 Schlichting, H.,"Boundary-Layer Theory",Mc Graw Hill, 1968

17 Johnston, L. J., "Finite-Difference Methods for the 3D Turbulent Boundary-Layer Equations", VKI-LS-1989-04, March 1989

18 Shang, J. S., Hankey, W. L., "Numerical Solution of the Navier-Stokes Equations for Compression Ramp", AIAA Paper 75-3, January 1975.

19 Müller, B., "Calculation of Separated Laminar Supersonic Flows over Blunt Bodies of Revolution at zero Angle of Attack", ESA-TT-953, 1985.

20 Jameson, A., Schmidt, W., Turkel, E., "Numerical Solutions of the Euler Equations by Finite Volume Methods Using Runge-Kutta Time-Stepping Schemes", AIAA Paper 81-1259, presented at the AIAA 14th Fluid and Plasma Dynamics Conference, Palo Alto, 1981.

21 Kothari, A. P., Anderson, J. D. "Flows over Low Reynolds Number Airfoils", AIAA Paper No. 85-0107, presented at the AIAA 23rd Aerospace Sciences Meeting, January 1985.

22 Fisher, S. S., Abbitt, J. D., O'Leary, G. W. O., "Separation Bubble Behavior for a NACA 23009 Airfoil", Conference on Low Reynolds Number Aerodynamics Proceedings, June 1985.

23 Mueller, T. J., Batill, S. M., "Experimental Studies on a Two-Dimensional Airfoil at Low Reynolds Number", AIAA-80-1440 1980.

24 Pohlen, L. J., Mueller, T. J., "Boundary-Layer Characteristics of the Eppler 61 Airfoil at Low Reynolds Numbers", Journal of Aircraft, Vol. 21, No. 9, Sept. 1984

25 Arena, A.V., Mueller, T. J., "Laminar Separation, Transition, and Turbulent Reattachment near the Leading Edge of Airfoils",AIAA J., Vol. 18, No. 7, July 1980

26 Bastedo, W. G. Jr., Mueller, T. J., "Performance of Finite Wings at Low Reynolds Numbers", Conference on Low Reynolds Number Airfoil Aerodynamics, June 1985

27 Wortmann, F. X.,"Fifty Years of Boundary-Layer Research",Braunschweig, 1955

28 Williams, B. R., "The Calculation of Flow About Aerofoils at Low Reynolds Number With Application to Remotely Piloted Vehicles", Aerodynamics at Low Reynolds Numbers International Conference Proceedings, Vol. III, October 1986

SHOCK/TURBULENT BOUNDARY LAYER INTERACTION IN LOW REYNOLDS NUMBER SUPERCRITICAL FLOWS

G. R. Inger
Iowa State University

Abstract

This paper describes a theoretical study of the effects of Reynolds number on nonseparating transonic shock/turbulent boundary layer interaction, with emphasis on the low Reynolds number regieme $10^3 \leqslant Re_{\delta}^{*} \leqslant 10^6$. A unique non-asymptotic triple deck theory of the interaction is used which is specifically designed to treat this regieme, where classical high Reynolds number asymptotic theory is inapplicable. The results of a comprehensive parametric study are given showing how the local interactive pressure, displacement thickness and skin friction distributions each depend on Reynolds number and how and why this dependence differs markedly from the predictions of high Re asymptotics. For example, under conditions of interest to the low Reynolds number aerodynamicist, it is shown that the size of the interaction zone does not even approximately scale with the local boundary layer thickness, but instead follows its own interactive scaling law inherent to the compressible viscous flow involved.

1. INTRODUCTION

The appearance of compressibility effects in a low Reynolds Number flow field (which can occur, for example, on high lift wings and their slots and flaps, on fast helicopter blades, or within airbreathing engines on high altitude transonic/supersonic aircraft) can significantly alter the viscous flow physics and scaling laws of the resultant aerodynamics. This alteration becomes especially pronounced when zones of supersonic flow terminated by a normal shock appear, accompanied by intense local shock-boundary layer interaction.[1-2] Such interaction significantly alters not only the local transonic flow but also the flow well downstream within the boundary layer, thus affecting the global aerodynamic properties of lift, drag and pitching moment. These interaction effects become particularly influential at low Reynolds numbers; indeed, they introduce new "low Reynolds number" effects that onset at higher Re than, and may actually overshadow, the usual incompressible low Re effects associated with transition and/or separation. In such applications, it is therefore important that any aerodynamic flow field analysis include a fundamentally sound account of the interaction effects and how they scale with Reynolds number.

This paper describes a basic theory of the local shock/boundary layer interaction for the case of unseparated 2-D transonic flow with a turbulent boundary layer. Our approach is based on a triple deck model of the interaction zone (see Fig. 1) which is specifically designed to treat the lower Reynolds number regime $10^3 < Re_\delta \lesssim 10^6$ where the results of classical large Reynolds number asymptotic theory are inapplicable.

2. NONASYMPTOTIC TRIPLE-DECK THEORY

2.1) Rationale

Because it is the foundational framework of the present study, a brief outline of the triple-deck approach and the advantages of its non-asymptotic version will first be given. We consider small disturbances of an arbitrary incoming turbulent boundary layer due to its interaction with a weak external shock. At those Reynolds numbers of aerodynamic interest it is known that the disturbance field has a triple-decked structure (Fig. 1) consisting of an outer inviscid flow region above the boundary layer which contains the incident shock wave, an intermediate deck of rotational-inviscid disturbance flow occupying the outer 90% or more of the incoming boundary layer thickness, and an inner sublayer adjacent to the wall containing both turbulent and laminar shear stress disturbances, which accounts for the interactive skin friction perturbations (and hence possible incipient separation) plus most of the upstream influence of the interaction. The middle deck couples the outer interactive field to the response of the inner deck while itself modifying the disturbance field, while the slow viscous flow in the thin inner deck reacts very strongly to the pressure gradient disturbances imposed by these overlaying decks.

The present study is based on a unique and experimentally well-validated non-asymptotic version[3] of this triple deck model which is specifically designed to treat the entire practical Reynolds number regime. This is made possible by including (among other features) the lateral pressure and streamline slope variations across the middle deck, the interactive changes in the eddy viscosity model within the inner deck adjacent to the wall, and the powerful incoming boundary layer shape factor effect on the interactive disturbance field in both these decks.

2.2) Incoming Turbulent Boundary Layer Profile

The incoming undisturbed 2-D turbulent boundary layer is characterized by the classical Law of the Wall/Law of the Wake structure. It is modeled by Walz's[4] composite analytical expression for the velocity profile $U_o(y)$ combined with an adiabatic wall reference temperature method-correction for compressibility, allowing arbitrary nonequilibrium values of its shape factor H_i. Thus if we let Π be Coles' (incompressible) Wake Function, $\eta \equiv y/\delta$ and denote for convenience $R \equiv .41\, Re_{\delta^*}/[(1+\Pi)(T_W/T_e)^{1+\omega}]$ and $\lambda \equiv .41(2Te/TwCf_o)^{\frac{1}{2}}$ with $\omega = .76$ and $\gamma = 1.4$ for a perfect gas, $U_o(y)$ is described by

$$\frac{U_o}{U_e} = \lambda^{-1}\left[\left(\frac{R}{1+R}\right)\eta^2(1-\eta) - 2\pi + 2\pi\,\eta^2\cdot(3-2\eta) + \ln\left(\frac{1+R\eta}{1+R}\right) - (.215+.655R\eta)e^{-3R\eta}\right] \quad (1)$$

subject to the following condition linking Π to C_{f_o} and Re_{δ^*}:

$$\lambda = 2\pi + .215 + \ln(1+R) \quad (2)$$

In addition, the defining integral relations for $_i^*$ and $_i^*$ yield the following relationship linking the wake parameter to the resulting incompressible shape factor $H_i = (\delta_i^*/\theta_i^*)$:

$$\frac{H_i - 1}{H_i} = \frac{2}{\lambda} \cdot \left(\frac{1 + 1.59\pi + .75\,\pi^2}{1 + \pi} \right) \quad (3)$$

Equation (1) for $\eta > .10$ yields a Law of the Wake behavior with $U_o/U_e \rightarrow 1$ and $dU_o/dy \rightarrow 0$ as $\eta = 1$, while for very small η, it assumes a Law of the Wall-type behavior consisting of a logarithmic term that is exponentially damped out into a linear sublayer profile $U/U_e = R\eta$ as $\eta \rightarrow 0$. From the adiabatic temperature-velocity relationship $T_o(y) = T_w + (T_e - T_w)UY_o^2/U_e^2$ the associated Mach number profile $Mo(y) \simeq U_o/\sqrt{\gamma RT_o}$ may then be determined.

Equations (1) - (3) provide a very general and accurate profile model in terms of three important physical quantities: the shock strength (M_1), the displacement thickness Reynolds number Re_{δ^*} and the Wake function Π that reflects the prior upstream history of the incoming boundary layer. The resulting relationship of the incompressible shape factor H_i, to the Wake Function as a function of Reynolds number for a typical $M_1 = 1.20$ transonic flow is illustrated in Fig. 2. It is seen that Hi_1 approaches a limiting value of unity as $Re_\delta \rightarrow \infty$ but that this approach is very gradual, especially for wake function values larger than zero (slightly favorable and adverse pressure gradient upstream flow histories).

2.3) Outer/Middle Deck Formulation

The analysis of this region rests on the key simplifying assumption that for non-separating interactions the turbulent Reynolds shear stress changes are small enough to have only a negligible back-effect on the mean flow properties along the interaction zone; hence this stress is taken to be "frozen" along each streamline at its appropriate value in the undisturbed incoming boundary layer. This approximation is supported by both asymptotic analysis and detailed experimental studies of a non-separating shock turbulent boundary layer interaction.[5] Significant local turbulent shear stress disturbances are thus essentially confined to a thin sublayer within the Law of the Wall region (see below) where the turbulence rapidly adjusts to the local pressure gradient.

The middle deck disturbance field caused by a weak shock is thus one of small rotationally-inviscid perturbation of the incoming nonuniform turbulent boundary layer profile $M_o(y)$ as governed by the equations

$$\frac{\partial}{\partial y}\left[\frac{v'(x,y)}{U_o(y)}\right] = \frac{1-M_o^2(y)}{\gamma M_o^2(y)} \cdot \frac{\partial(p'/p_o)}{\partial x} \quad (4)$$

$$\frac{\partial u'}{\partial x} = -\frac{\partial p'/\partial x}{\rho_o(y)\,U_o(y)} - \frac{dU_o}{dy} \cdot \frac{v'}{U_o} \quad (5)$$

$$\frac{\partial^2 p'}{\partial y^2} - \frac{2}{M_o}\frac{dM_o}{dy}\frac{\partial p'}{\partial y} + \left[1-M_o^2 - \frac{2\,u'M_o^2}{U_o}\right]\frac{\partial^2 p'}{\partial x^2} = 0 \qquad (6)$$

where for a short-ranged interaction we may neglect the streamwise variation of the undisturbed turbulent boundary layer properties $U_o(y)$, $\rho_o(y)$ and $M_o(y)$. The outer deck disturbances are governed by Eqs. (4) - (6) specialized to the uniform flow case $U_o = U_{oe}$, $M_o = M_{oe}$; they are matched to the middle deck solutions by the physical conditions that both v'/U_{oe} and p' be continuous across $y = \delta_o$.

2.4) Inner Disturbance Shear Stress Deck

This very thin layer lies well within the Law of the Wall region of the incoming turbulent boundary layer profile and contains all the wall region turbulent stress perturbation effects. Further, for the weak interactions considered the sublayer disturbance flow is assumed to be a small perturbation upon the incoming boundary layer but one which still contains all the physically-important effects of streamwise pressure gradient, streamwise and vertical acceleration, and both laminar and turbulent disturbance stresses. Finally, for adiabatic flows the undisturbed and perturbation flow Mach numbers are both so small that the sublayer density perturbations may be neglected and the corresponding modest compressibility effect on the undisturbed profile treated by the reference temperature method.[6]

The disturbance field is thus governed by the equations

$$\frac{\partial u'}{\partial x} + \frac{\partial v'}{\partial y} = 0 \qquad (7)$$

$$U_o\frac{\partial u'}{\partial x} + v'\frac{dU_o}{dy} + (\rho_{w_o}^{-1})\frac{dp'_w}{dx} = \frac{\partial}{\partial y}\Big(\nu_{w_o}\frac{\partial u'}{\partial y} + \underbrace{\epsilon_{T_o}\frac{\partial u'}{\partial y} + \epsilon'_T\frac{dU_o}{dy}}_{2\,\epsilon_{T_o}\frac{\partial u'}{\partial y}}\Big) \qquad (8)$$

$$\tau_o(y) = \text{const.} = \tau_{w_o} = [\mu_{w_o} + \rho_{w_o}\epsilon_{T_0}(y)]\frac{dU_o}{dy} \qquad (9)$$

where ρ_{w_o} and μ_{w_o} are evaluated at the adiabatic wall recovery temperature and the Van Driest-Cebeci eddy viscosity model with damping constant A = 26 and $y+ = (y\sqrt{\tau_{w_o}/\rho_{w_o}})/\nu_{w_o}$ provides

$$\epsilon_T = [.41y\,(1-e^{-y+/A})]^2\frac{\partial u}{\partial y} \qquad (10)$$

It can be seen that inclusion of the eddy viscosity perturbation has exactly doubled the turbulent shear stress disturbance term.

These equations are solved subject to the wall boundary conditions $U_o(0) = u'(x,o) = v'(x,o) = 0$ plus an initial condition $u'(-\infty,y) = 0$ requiring that all interactive disturbances vanish far upstream of the shock. Furthermore, at some distance $_{SL}$ sufficiently far from the wall, we require that $\partial u'/\partial y \to 0$ with u' matching to the inviscid solution u'_{inv} from the bottom of the middle deck.

2.5) Solution Methodology

The foregoing triple deck equations can be solved by an operational method to obtain all the essential physics of the viscous interaction field for non-separating flows.[3] Following Fourier Transformation of Eq. (3) w.r.t. x, the resulting middle deck pressure is governed by an ordinary differential equation in y that can be solved numerically quite efficiently for the input turbulent boundary layer profile Mo(y). Then the Fourier inversion process via the calculus of residues can be employed to find the physical-plane pressure distribution.[3] Upstream of the shock, this yields for the wall pressure

$$p'_w \simeq \text{CONST.} \times \Delta p \; e^{x/\ell u} \tag{11}$$

where ΔP is the overall pressure jump and ℓu is the characteristic upstream influence distance determined by the appropriate residue pole in the Fourier plane. In two opposite extremes, an analytical expression can in fact be obtained for ℓu. Thus in the very high (asymptotic) Reynolds number limit where $\ell u/\delta o \ll 1$, it is found that ℓu scales on the sonic height y_s within the incoming boundary layer:

$$\ell u \simeq \frac{4}{\pi}\int_0^{y_s}\sqrt{1-M_o^2(y)}\,dy \tag{12}$$

which has a very weak explicit dependence on Reynolds number (see Fig. 4). On the other hand, at lower Reynolds numbers where $\ell u/\delta o \gtrsim 0(1)$, ℓu instead is governed by the highly Reynolds number-dependent relation

$$\ell u \simeq \frac{Me_1^2 \cdot I_o}{\sqrt{Me_1^2 - 1}} + \frac{\sqrt{Me_1^2 - 1}\cdot I_1}{Me_1^2} \tag{13a}$$

in terms of the profile-dependent integrals

$$I_o \equiv \int_{Y_{weff}}^{\delta_o} \left\{\left[1 - M_o^2(b)\right]/M_o^2(b)\right\}db \tag{13b}$$

$$I_1 \equiv \int_0^{\delta_o} M_o^2(b)db \tag{13c}$$

The parameter y_{weff} here is the effective inviscid wall shift or displacement thickness of the underlying inner deck, and is highly Re-dependent (see Fig. 4).

The corresponding Fourier transformation of the inner deck problem, followed by the introduction of new inner deck variables and a y-scaling defined by Inger[3], yields a set of ordinary differential equations in a universal form that can be solved and tabulated once and for all. Among other properties, this in turn yields the following result for the deck's displacement thickness:

$$y_{weff} = .766\left(\frac{\mu_{w_o}^2\,\ell u}{\rho_{w_o}\tau_{w_o}}\right)^{1/3} H(T) \tag{14}$$

where the eddy viscosity function H depends on the interactive turbulence parameter

$$T = (.41)^2 \frac{\rho_{o_w} \tau_{w_o}}{\mu_{ow}^2} \left(\frac{\mu_{ow}^2 \ell u}{\rho_{o_w} \tau_{w_o}} \right)^{2/3} \tag{15}$$

as plotted in Fig. 3. The simultaneous solution of Eqs. (12) and (13) for ℓ_u and y_{weff} implements the coupling of the middle and inner decks. The companion result for the upstream disturbance skin friction is found to be

$$\tau_w' \simeq -.72\, p_w'(N)\; f_p(X) \left[\frac{\rho_{w_o} \ell_u^2 \tau_{w_o}}{\mu_{wo}^2} \right]^{-1/3} S(T) \tag{16a}$$

where

$$f_p(x) \equiv \left[\frac{2}{3} \frac{\ell_u \, (p_w')^{3/2}}{\int_{-\infty}^{x} (p_w')^{3/2} dx} \right]^{2/3} \tag{16b}$$

and S(T) is another interactive-turbulence effect function, also plotted in Fig. 3. From the foregoing, it is obvious that Fig. 3 is an important result of the present general turbulent shear-disturbance inner deck treatment; it gives a unified account of the inner interactive physics over the entire Reynolds number range from quasi-laminar behavior at $T \ll 1$ (lower Reynolds numbers) to the opposite extreme of wall turbulence-dominated behavior at $T \gg 1$ pertaining to asymptotic theory at very large Reynolds numbers where the inner deck thickness and its disturbance field become vanishingly small.

A computer program has been constructed to carry out the foregoing solution method: it involves the middle-deck disturbance pressure solution coupled to the inner deck by means of the effective wall shift combined with an upstream influence solution subroutine (the corresponding local total interactive displacement thickness growth and skin friction are also obtained). This provides a very general fundamental description of the boundary layer in terms of the three arbitrary parameters M_1, Re_{δ^*} and Π. Numerous detailed comparisons[8] of the predictive results with experiment have shown that it gives an excellent account of all the important interactive features over a wide range of Mach-number, Reynolds number and shape factor conditions.

3. APPLICATION TO A STUDY OF LOW REYNOLDS NUMBER INTERACTIONS

A major advantage of the present nonasymptotic triple-deck theory is its applicability to a wide range of practical Reynolds numbers including those that are far below (by four to five order of magnitude) the values appropriate to classical asymptotic theory. We have exploited this by carrying out a detailed parametric study of the scale (Reynolds number) effect on nonseparating transonic interactions in the heretofore-unexplored lower Reynolds number regieme pertinent to the applications mentioned in the Introduction. Some of the interesting features of the results will now be discussed.

Figures 5a, b and c illustrate the predicted influence of Reynolds number on the non-dimensional distributions vs. x/δ_0 of wall pressure, displacement thickness growth and local skin friction, respectively. Whereas these properties became independent of Re_δ at very large Reynolds number such that the size of the interaction scales with δ_0 in this limit, it is seen that this is decidedly not true at "low" Reynolds numbers; rather, with decreasing $Re_{\delta *}$ below about 10^6, the interaction progressively spreads out and thickens well beyond simple proportionality according to its own inherent scaling law (see below).

Fig. 6 summarizes results for the important upstream influence property of the interaction by plotting $\ell u/\delta_0$ versus Re_δ with the wake function as a parameter. Several important conclusions can be drawn from this Figure. First, the functional dependence on Re_δ changes significantly at lower Re to one which yields a much stronger increase in $\ell u/\delta_0$ with decreasing Re_δ. When compared to the corresponding predictions based on neglecting the inner deck turbulent eddy viscosity perturbation effect that are also indicated in Fig. 6, this change in Re-dependence is seen to be due to the growing importance of this effect (not included in high Re asymptotic theory) at lower Re_δ. Second, the wake function is seen to also have an important direct effect of significantly increasing $\ell u/\delta_0$, roughly equivalent to that of decreasing Re_δ. Indeed, this wake function influence may be regarded as a "low Reynolds number effect" in view of the fact that it is totally excluded in high Reynolds number asymptotic theory (which is based on the limiting value $Hi_1 = 1.0$ pertaining to the infinite Reynolds number limit, wherein the wake component completely vanishes: see Fig. 2).

Further insight as to the role of Reynolds number is brought out in Fig. 7 where $\ell u/\delta_0$ from the full solution is again plotted vs. Re_δ but now in terms of comparisons with the predictions from either high Reynolds number theory (Eq. 12) or the "lower Reynolds number" Theory of Eq. 13, for three different values of the Π function. Clearly, the use of high Reynolds number asymptotics at moderate to low Reynolds numbers ($Re_\delta < 10^8$) is not only incorrect in its predicted Re-dependence but also quantitatively underestimates the upstream influence by hundreds of percent for all Π values; indeed, its approach toward even marginal validity with increasing Re_δ is very slow.

4. CONCLUDING REMARKS

Further work on the present subject appears warranted in several respects. The first would be a study of incorporating the present interaction model into a more global flow field prediction method for treating supercritical low Reynolds number flow fields. The second is a more detailed investigation of incipient separation at low Reynolds numbers as it depends on the local flow Mach number and wake function (or shape factor).

REFERENCES

1. Katary, M. "An Experimental Study of the Development of a Supersonic Zone Near the leading Edge of an Airfoil Oscillating in Subsonic Flows," AIAA Paper 83-2133, Aug., 1983.
2. Fung, Y. and L. W. Carr, "The Effects of Compressibility on Dynamic Stall," AIAA Paper 88-3541, 1988.
3. Inger, G. R. "Nonasymptotic Theory of Unseparated Turbulent Boundary Layer Interaction," in Numerical and Physical Aspects of Aerodynami Flows (T. Cebeci, Ed.), Springer-Verlag, N.Y. 1983.
4. Walz, A., Boundary Layers of Flow and Temperature, M.I.T. Press, 1969, pp. 113-116.
5. Rose, W. C., and Childs, M. E., "Reynolds Shear Stress Measurements in a Compressible Boundary Layer within a Shock Wave-Induced Adverse Pressure Gradient," JFM 65, 1, 1974, pp. 177-188.
6. Burggraf, O. R., "The Compressibility Transofrmation and the Turbulent Boundary Layer Equation," Journal of Aerospace Scie. 29, 1962, pp. 434-39.
7. Lighthill, M. J., "On Boundary Layers and Upstream Influence; II. Supersonic Flow Without Separataion," Proc. Royal Soc. A 216, 1953, pp. 578-587.
8. Inger, G. R., "The Modular Application of a Shock/Boundary Layer Interaction Solution to Supercritical Viscous Inviscid Flow Field Analysis," in Computational Methods in Viscous Flows, Vol. 3, Pineridge Press, U.K., 1984, pp. 475-512.

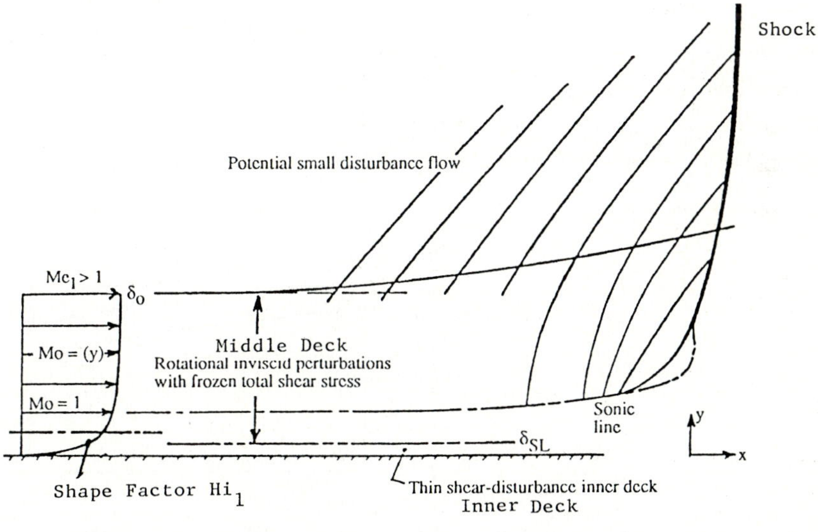

Figure 1. Triple-deck structure of interaction zone.

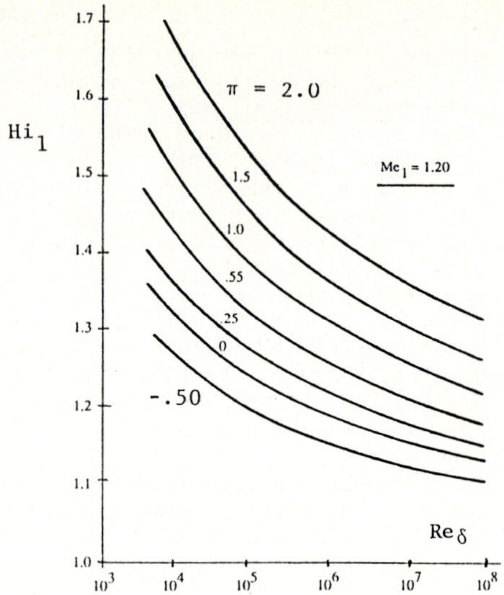

Fig. 2

Incompressible shape factor versus Reynolds number with the wake function as parameter.

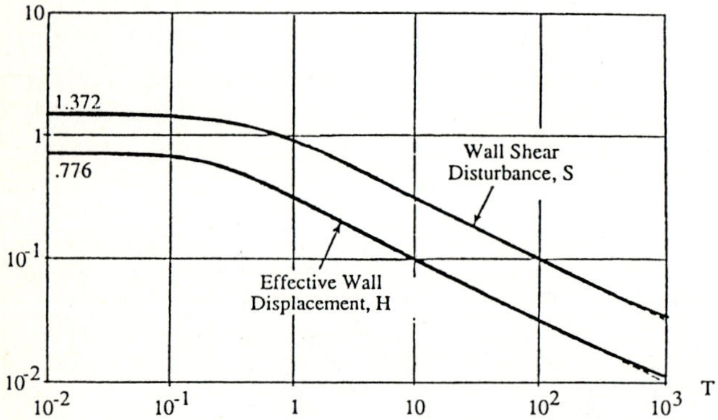

Fig. 3
Turbulent interaction parameter effect on interactive displace. thickness and skin friction

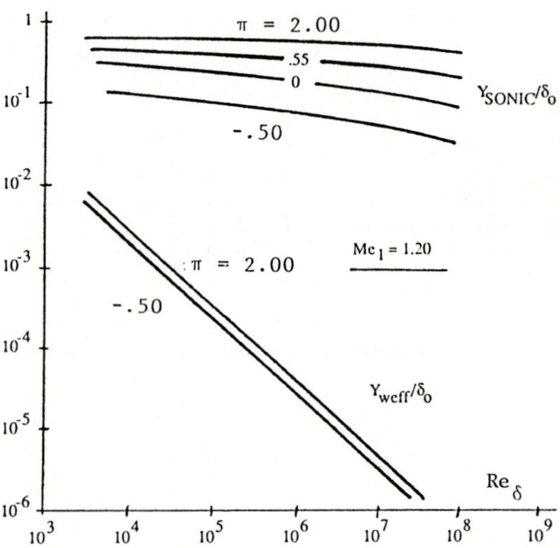

Fig. 4

Nondimensionalized inner deck thickness and sonic height variations with Reynolds number with wake function as parameter.

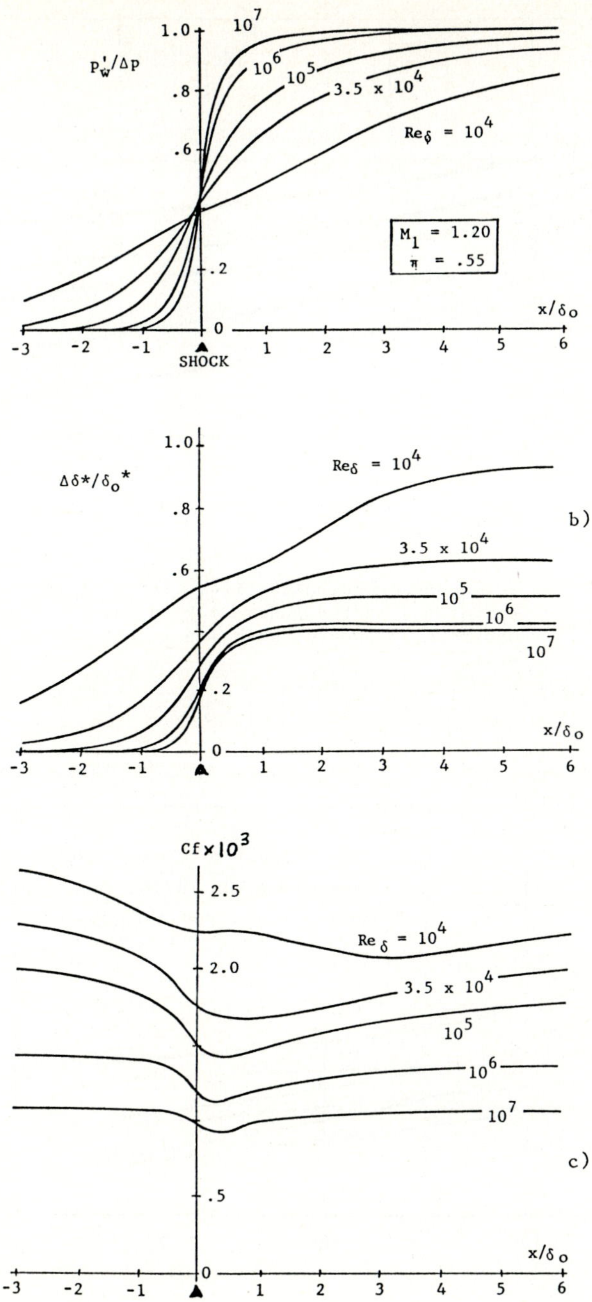

a) Wall Pressure

b) Displacement Thickness

c) Local Skin Friction Coefficient

Fig. 5 Reynolds Number effects on typical Interactive Properties ($Me_1 = 1.20$, $\pi = .55$)

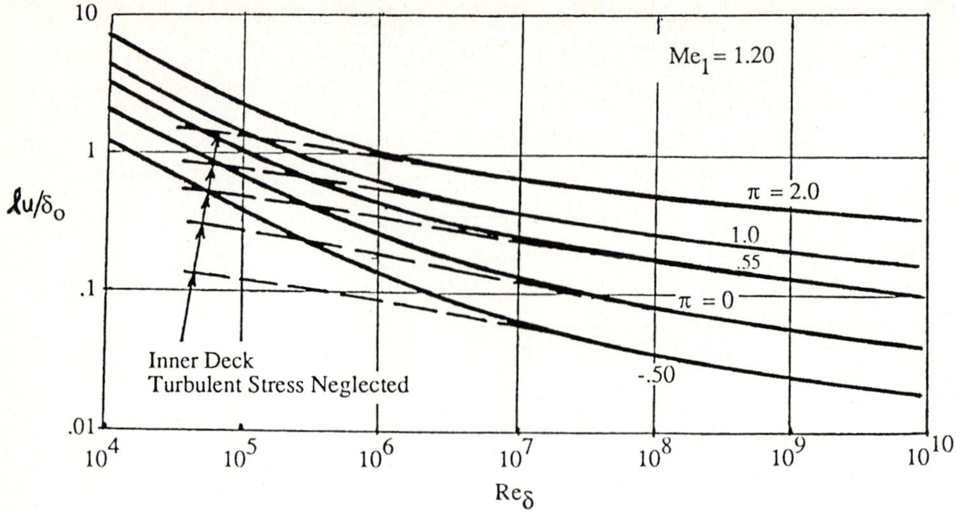

Fig. 6 Influence of Turbulent Eddy Viscosity Perturbations in the Inner Deck on the Upstream Influence Variation with Reynolds Number

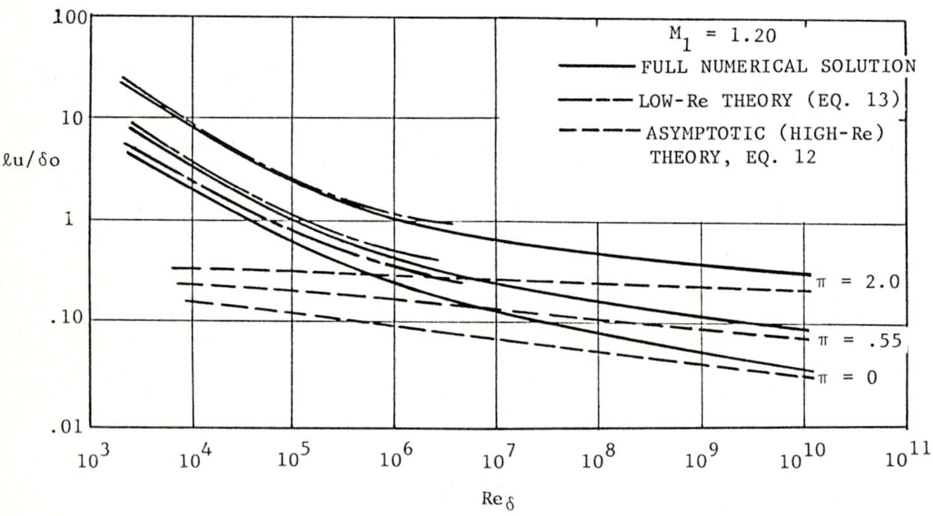

Fig. 7 Comparison of full numerical solution with results from asymptotic (high Re) and low-Re theory.

SUMMARY OF EXPERIMENTAL TESTING OF A TRANSONIC LOW REYNOLDS NUMBER AIRFOIL

Peggy L. Toot
U.S. Naval Research Laboratory
Washington, D.C. 20375

BACKGROUND

The U.S. Navy is developing small, transonic speed unmanned aircraft for electronic warfare missions. Typically, these small vehicles have wing chords of less than one inch, and with flight speeds of Mach 0.8 to 1.2, operate with chord Reynolds numbers (RN_c) less than 500,000. For such low Reynolds numbers, viscous effects might be expected to dominate the character of the boundary layer, despite the compressibility effects. The design and analysis for one such vehicle, developed by NRL, identified the need for additional LRN transonic test data, and inspired these experiments.

The purpose of the experimental testing was to establish a transonic low Reynolds number (LRN) 2-D airfoil data base for verifying computational analysis as well as providing insight into the aerodynamics of this flight regime.

The RN_c region of interest was determined to be below 1,000,000 at speeds of 0.7 to 1.2 Mach. The biconvex circular-arc airfoil was chosen since it was easy to computationally model. Also, its sharp leading edge encourages leading edge shock attachment, thus making the transition to supersonic flow clearly defined.

This research was conducted in three phases over a five year period. Phase I involved designing the test model, establishing test parameters, familiarization with the wind tunnel and taking Schlieren data. The most interesting test condition was then quantified with Laser Doppler Velocimetry

(LDV). Phase II included detailed LDV measurements over a range of conditions to provide data needed to compute the aerodynamic forces on the airfoil. Phase III involved the design of a pressure model and the subsequent measurement of pressure distributions for comparison of this measured force data with that computed during Phase II.

WIND TUNNEL FACILITY

Wind tunnel features crucial for this LRN transonic research included the need for a low turbulence intensity, ventilated walls to eliminate shock reflections, continuous operation near Mach 1, Schlieren equipment, two-component LDV equipment, and relatively inexpensive operation.

The Notre Dame Transonic Smoke Tunnel (NDTST) easily met these requirements, and had high availability, so it was selected for all Phases of testing. It is an indraft tunnel which draws ambient air from inside the laboratory and exhausts to the outside. The 4 inch by 4 inch test section has a 6% ventilation area on the upper and lower walls and solid removable side walls. It is capable of continuous operation in excess of 8 hours at up to Mach 1.3. Its 150:1 inlet contraction ratio downstream of 11 inlet screens insured excellent quality smoke streamlines which qualitatively indicated low turbulence intensities.

PHASE I RESEARCH - SCHLIEREN AND LIMITED LDV TESTS

Approach

Test conditions were selected to reproduce well known flow phenomena that exists for 2-D airfoils operating in the transonic regime including expansions and compressions, trailing edge flow separation, boundary layer/shock interaction, and unsteadiness.

Schlieren flow visualization gave high resolution results for locating chordwise locations of flow phenomena of interest. A Hi-Cam prism type camera shooting over 1000 frames per second recorded the unsteady nature of the shock-boundary layer interaction to determine the existence of periodicity. Still photographs were taken to record time-averaged conditions at each test case and to select conditions for limited Laser Doppler Velocimetry (LDV) measurements.

The Phase I test procedure was divided into three parts. First, the tunnel was calibrated, the range of operating conditions established, and transonic conditions of interest identified. Second, Schlieren photographs were taken which provided shock and boundary layer information for comparison with computational results, and to select one case for LDV measurement. Third, LDV measurement provided values for horizontal and vertical velocities near the airfoil for the preselected test case, as well as providing another tunnel calibration reference.

Phase I Wind Tunnel Model - Airfoil Selection

Several desired features drove the selection of the airfoil: a classical shape with a sharp leading edge to encourage leading edge shock attachment and thus precisely defining the upper limit of the transonic region, a mathematically simple configuration for incorporating into computational codes, an easy to fabricate 2-dimensional wind tunnel model, no surface slope discontinuities which would force boundary layer transition, and structural integrity for the expected dynamic forces. A biconvex, circular-arc airfoil with a 10% thickness to chord ratio, having sharp leading and trailing edges was selected.

The size of the model was driven by the requirement to minimize blockage in the tunnel and to achieve the preferred Mach and Reynolds number range. The method used to compute Reynolds number is detailed in NACA report 1135, Appendix B. (Ref. 1)

A 1.0 inch wing chord was determined as the maximum acceptable size based on allowable wind tunnel blockage. Consequently, the RNc values were ideal, being 415,000 for M = 0.7 and 712,000 for M = 1.2.

Model Fabrication

Four brass airfoils, readily fabricated by lathe turning, were built. During initial tests they exhibited bending deflections that clouded the Schlieren data and two were severely damaged. Four stainless steel replacement airfoils were machined using an electrical discharge (ED) technique. Although expensive, these new airfoils were structurally adequate and dimensionally precise.

Phase I Schlieren Test Results

Details of the first Schlieren tests are described in Reference 2. The highlight of these tests was the high speed motion pictures which clearly showed a periodicity to the apparent unsteadiness of the shock/boundary layer interaction. Considerable motion of the shock accompanied the unsteadiness of the separating boundary layer. The frequency of the periodicity, measured on the basis of the frame rate of the film, was estimated to be approximately 1000 Hz. This value is comparable to the 1300 - 1400 Hz values for free transition and 1100 - 1500 Hz values for fixed transition on a 14% biconvex airfoil measured at Cranfield Institute of Technology. (Ref. 3) Based on a comparison of a previous wind tunnel acoustic survey and the frequencies estimated for the periodicity, it appeared that the periodicity was not driven by the wind tunnel.

The test condition having the greatest combination of flow phenomena was observed to be the 6° angle of attack, Mach = 0.845, Re_c = 477,000 case. This case had expansion/compression waves, a strong shock, a thick separated unsteady boundary layer, and supersonic flow across 50% of the airfoil. This case was selected for the LDV survey.

Laser Doppler Velocimetry Test Results

The LDV system description and the details of the test procedure are given in Ref.2, Section 8. Since this system was newly installed at the time of these tests, the data storage and system control software was not yet fully operational. Hence, the LDV measurements were sufficiently laborious to preclude more than one test case.

First, LDV-measured velocities (and computed Mach numbers) were taken on the longitudinal centerline both with and without the airfoil in the test section. The data closely agreed with the pressure calibration measurements. Following this calibration, the selected case was tested.

The LDV data and plots of the horizontal and vertical velocity components are shown in Ref. 2. The plots clearly showed transition between subsonic and supersonic flow. A plot of the sonic lines versus chord revealed excellent correlation with the corresponding Schlieren photograph.

A major limitation of the LDV data was the inability to determine the absolute direction of the velocity components. For example, the vertical velocity components near the body had small magnitudes, but varied greatly in value with respect to time for each point. Although likely, it is unresolvable whether or not these velocities were changing directions with respect to time as would be expected for periodic unsteady flow. Another limitation with the LDV measurements was the difficulty in seeding the flow with a smoke streamline lined up precisely with the control point. Filling the entire section with smoke caused unacceptable scattering of the laser.

PHASE II - DETAILED LDV MEASUREMENTS

Test Approach

The range of test conditions for Phase II testing included Mach numbers from 0.72 to 1.35. For each Mach number four angles of attack: 0°, 2.5°, 4° and

6° were tested. The first part of Phase II consisted of using LDV for freestream Mach number calibration ahead of the airfoil. Very close agreement was achieved with previous pressure-based calibrations in the upstream subsonic region. Mach numbers measured in the subsonic flow downstream of the airfoil differed from the pressure measurements due to slightly different reference locations. A detailed explanation of the results are given in Reference 4.

LDV measurements were then taken in the flow field surrounding the airfoil for freestream Mach numbers of 0.72, 0.78, 0.84, 1.03, and 1.35. Mach number was mapped at multiple streamwise and vertical locations. In addition, unsteady regions were identified by the relative root-mean-square (RMS) velocities. Plots of these results are shown in Reference 4, also.

Phase II Results

Due to light reflections from the airfoil, LDV measurements could not be made in the region closer to the airfoil than $y/c = \pm 0.20$. In both the $M = 0.72$ and $M = 0.78$ cases, even at 6° angle of attack, the freestream Mach was too low for supersonic flow to develop anywhere in the flow field. In the $M = 0.84$ case, at 6° angle of attack, the flow decelerated near the leading edge, and then accelerated over the upper surface to $M = 1.22$, near the three-quarter chord location. At this point the Mach number decreased sharply into the subsonic region indicating the presence of a strong normal shock. A comparison of the plotted data with the Schlieren photo for the same condition is shown in Figure 1.

For the remaining cases, $M = 1.03$ and 1.35, LDV measurements were made at zero angle of attack. The flow over the airfoil for the $M = 1.03$ case was supersonic except for a small portion near the leading edge. In the case of $M=1.35$, the flow was entirely supersonic.

The primary goal of these experiments was to determine the aerodynamic forces on the airfoil using the flow field information gathered by the LDV. An attempt to calculate the profile drag was made by the wake momentum

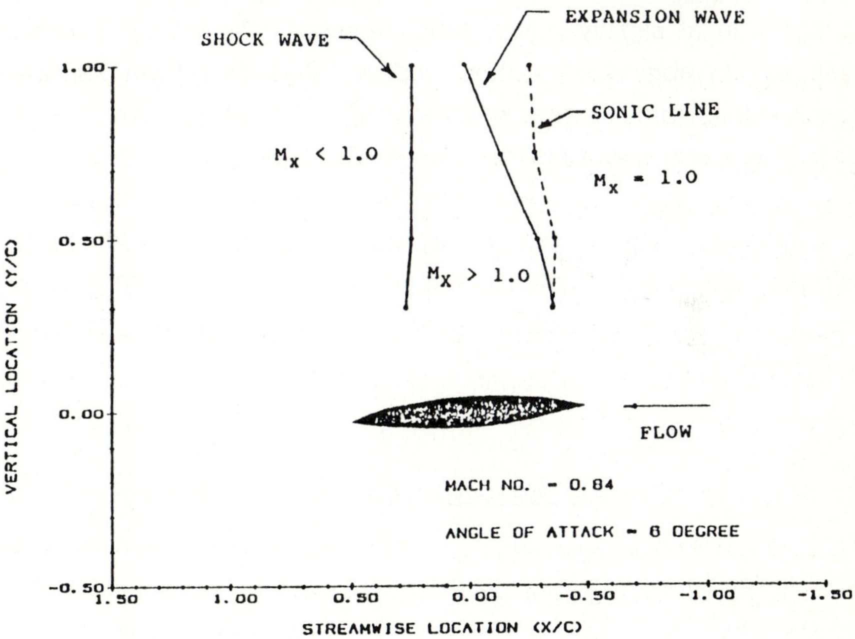

Figure 1. Comparison of Schlieren Photography to LDV Result for Mach = 0.84, Angle of Attack = 6 deg.

method for the M = 0.78 case at three angles of attack: 0°, 3° and 6°. Only the M = 0.78 condition could be tested since high ambient temperatures and humidity during these tests caused condensation build-up in the test section at higher speeds.

Details of the data reduction methods and the determination of uncertainties in Phase II are detailed in Reference 4. For the M = 0.78 case, there was no apparent shock induced boundary layer separation; therefore, the primary contributions to drag were expected to be caused by skin friction and viscous separation. Drag values calculated by the wake momentum method appeared close to expected values for skin friction drag, leading one to deduce that the LRN effects were minimal. Unfortunately, the accuracy of the measurements due to experimental uncertainty does not allow validation of this deduction and the drag values remain inconclusive. It has been shown experimentally by Shapiro (Ref. 5) that the drag forces increase by an order of magnitude at slightly higher Mach numbers (M = 0.9). These larger drag forces are, in part, due to shock induced boundary layer separation. More accurate drag force measurements might therefore have been made had higher freestream Mach numbers been attainable.

Lift coefficient data is presented in Reference 4. The variation of lift coefficient with angle of attack agreed with theory and previous experiments. The values of lift coefficient were in the expected range, considering uncertainties.

PHASE III - PRESSURE DISTRIBUTION MEASUREMENTS

Phase III Model

A 10% thick circular-arc biconvex airfoil pressure model was developed for detailed pressure surveys. Initially, the requirement for a tiny model with micro pressure taps was technically challenging and prohibitively expensive. Thanks to fabrication technology advances which occurred during the life of this research program, an affordable 1 inch chord, 4 inch span airfoil pressure model was produced in stainless steel. Ten pressure taps each were ED machined into both the upper and lower surface (Fig. 2).

Test Procedure

The twenty pressure taps were connected to a mercury manometer board via capillary tubing to a scanivalve system. Data was collected at the following freestream Mach numbers: 0.72, 0.78, 0.84, and 1.0. Five angles of attack were studied at each Mach number: -6°, -3°, 0°, 3° and 6°. Data was taken and recorded directly into a MacIntosh computer. The freestream Mach numbers, ambient pressures, and temperatures were recorded before and after each set of tests to check for variation. In addition, testing was performed during the winter months to minimize condensation.

Test Results

The data was reduced for pressure coefficient and Mach number values. The results agreed well with those obtained in both Phase I and Phase II. As before, the flow remained subsonic over the airfoil for the 0.72 and 0.78 cases at 0° angle of attack. The M=0.84 case typifies the detail obtained with these pressure measurements. For this case, the flow became supersonic at 30% chord and reached its maximum Mach number of 1.13 at about 55% chord. A shock or compression was observed near 80% chord. Further details of the results are described in Reference 6.

CONCLUSION

Despite the tedious nature of setting up and maintaining the LDV equipment, its nonintrusive nature makes it an excellent method for sensitive flowfields such as those encountered at low Reynolds numbers and those with unsteady and/or compressibility effects. The pressure test results correlated very well with those obtained in the LDV, thus validating both techniques.
The results of these experiments established a data base useful for

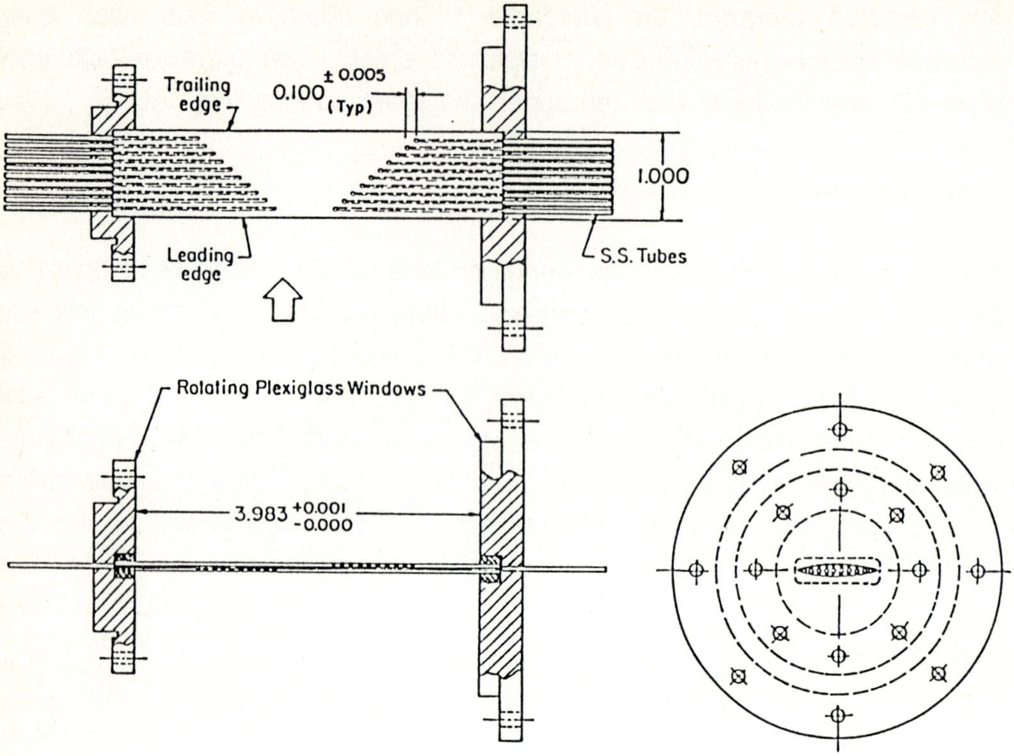

Figure 2. Three View Drawing of 10% Thick Biconvex Airfoil With 20 Pressure Ports

computational aerodynamics of 2-D low Reynolds number transonic flows. Classical transonic 2-D airfoil flow conditions were reproduced in each test phase despite the expected LRN viscous effects. Therefore, these experiments showed that compressibility still dominates the character of the flow, even at the low chord Reynolds numbers.

The shock boundary layer interaction observed in test Phase I was evaluated as a characteristic of only circular-arc airfoils having t/c between 10% and 14%. It is suspected that the shock on the circular-arc airfoil oscillates due to the smoothness of the surface. Therefore, a 10%-14% thick circular-arc type airfoil is not recommended for use in the transonic low Reynolds number region.

Thus, a flat plate airfoil with a 45° wedge leading edge was selected for a canard on NRL's transonic vehicle. The vehicle was wind tunnel tested to look for evidence of shock/boundary layer interaction induced flutter in the canard. These canard tests were deemed mandatory since the 1000-1400 Hz boundary layer shock interaction frequency exhibited by the circular arc airfoil was very near the canard structure's torsional natural frequency. Fortunately, no evidence of canard flutter was observed.

Future testing of various LRN airfoil shapes at transonic speeds would be useful as well as desirable to establish a broader data base for future advanced unmanned vehicles.

REFERENCES

1. Ames Research Staff, "Equations, Tables and Charts for Compressible Flow," NACA Report 1135, 1953

2. Foch,R.J., "Transonic Low Reynolds Number Wind Tunnel Testing of a Two-Dimensional Biconvex Circular-Arc Airfoil," Masters Thesis, University of Maryland, Dec.,1985.

3. Mohan,S.R., "Periodic Flows on Rigid Airfoils at Transonic Speeds," Cranfield Institute of Technology Synopsis, 27 Feb.,1984

4. Good, M.T., "Transonic Biconvex Airfoil Experiments Employing Laser Doppler Velocimetry and Schlieren Flow Visualization," Masters Thesis, University of Notre Dame, Feb.,1986

5. Shapiro,A.H., The Dynamics and Thermodynamics of Compressible Fluid Flow, Vol.II, New York: John Wiley & Sons, Inc., 1953

6. Ng,T.T., "Pressure Distribution Experiment on a Transonic Biconvex Airfoil," University of Notre Dame, Final Report N00174-87-K-2037, Oct.,1988

The Design of a Low Reynolds Number RPV

S. Siddiqi, R. Evangelista, T. S. Kwa †
AS&M Inc., 107 Research Drive, Hampton, Virginia 23666.
† Aviation Advanced Technology Applications, Orlando, Florida.

Nomenclature

a	lift curve slope	NASA LaRC	NASA Langley Research Center
ac	aerodynamic center	NLF	Natural Laminar Flow
AR	aspect ratio	R_e	Reynolds Number
b	wing span	R_c, R_u	chord & unit Reynolds Number
c	wing chord	S	wing area
cg	center of gravity	V	flight speed
C_L	lift coefficient	WT	Wind Tunnel
C_{Di}	induced drag coefficient	α	angle of attack
$C_{D\infty}$	drag coefficient	α_a	absolute angle of attack
e	Oswald efficiency factor	α_g	geometric angle of attack
LRN	Low Reynolds Number	Γ_o	center span circulation
LE/TE	Leading/Trailing Edge of airfoil	μ	microns, 10^{-6} *metres*
LTPT	Low Turbulence Pressure Tunnel		

I ABSTRACT

This paper covers the ongoing design process for a long endurance Remotely Piloted Vehicle. The 45 *pound* RPV must operate in the 25 – 50 *knot* speed range at low altitudes. Specially designed LRN airfoils were used for efficient aerodynamics. A wing of $AR = 23$ was selected which will give an estimated $\frac{L}{D}|_{\max}$ of 27. Wing tip feathers will be used to reduce the induced drag. A comparison between the computed aerodynamic predictions and wind tunnel results is given. The predicted handling quality and performance results that have been obtained so far are given. The structural challenges in designing a light weight structure for the wing and control surfaces are only outlined as this phase is not yet complete.

II THE LAURA PROJECT

The low altitude unmanned research aircraft (*LAURA*) project is a study sponsored by the NRL and the ONR with the participation of other government agencies and of the aeronautical industry. The details of the LAURA project are reported in [Foch 1986]. The aim of the research project is to investigate and flight test the potential of LRN aerodynamic technology. The project has a standard fuselage, engine, propeller, and landing gear to which the different wing and control surface configurations will be fitted. The design specifications for the LAURA are given in Table I which also shows an artist's impression of the RPV being designed.

TABLE I AIRCRAFT DESIGN SPECIFICATIONS

Gross Weight: 45 lb.
Wing/tail weight: 18 lb.
Ultimate load factor: +6, -4 g's
Cruise speed range: 20-50 kt.
Best rate of climb at gross weight: 400 fpm.
Stability requirements: positive static margin (stick-fixed)
Flutter free up to an airspeed of 150 fps.

Wing Configuration:
Wing section: LRN-1-1010
Aspect Ratio (AR): 23.2
Mean Aerodynamic Chord (MAC): 0.895 ft.
Taper ratio (λ): 0.5
Twist: 1/2 -3° linear (washout at tip)

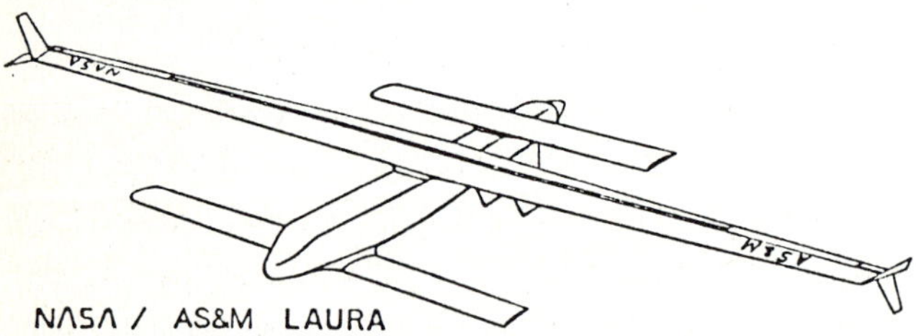

Dimensions:
Wingspan: 20 ft.
Wingchord: 1.15 ft.
Planform area(s): 17.24 sq.ft.
Fuselage length: 6.67 ft.

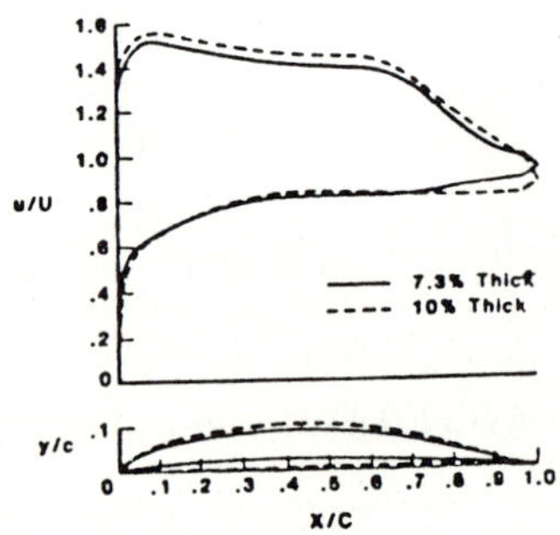

Fig. 1 LRN-1-1010,LRN-1-1007

The four aircraft concepts being evaluated are:

- JVW/ACM Industries - joint wing configuration
- Advanced Aeromechanisms Corp. - hinged wing configuration
- Locus - tandem wing configuration
- NASA/AS&M - three surface configuration

The standard initial sizing process based on the LAURA mission is as follows.

[1] The wing area is sized by the required low speed capability. Then for V=25*kts* and C_L=1.23 $S=17.24ft^2$.

[2] The engine selected gives sufficient power for the high speed capability. A power of approximately 2.3 hp is required for an $L/D = 6$ at $V = 50kts$ assuming the propeller efficiency is about 50%.

[3] The 400 fpm climb requirement is easily met with this available power. A climb rate of approximately $600fpm$ can be obtained at $30kts$ assuming an $L/D = 16$.

The overall aerodynamic design efficiency goal can be quantified as follows. The LRN optimized airfoil has a 2D $\frac{L}{D}|_{max}$ of approximately 100 for R=250,000. A wing designed with this airfoil will then have a L/D of approximately 50. Wind tunnel measurements for the fuselage gave a minimum $C_{D\infty}$=.012 based on the wing area. Assuming that the horizontal and vertical tail are totally <30% of S their C_{D_∞}=.007 assuming flow without separation at R_c=150,000 (which is possible with the use of turbulators). The overall $\frac{L}{D}|_{max}$ possible for such a configuration then is approximately 28. Trim, separation and interference drag and the loss of wing planform efficiency will reduce the overall L/D. The design challenge is then to aim to get an $\frac{L}{D}|_{max}$ as close to 28.

The design aim was to keep the wing weight below 10 lbs so leaving 8 lbs for the control surfaces. For such low wing loaded aircraft a major structural problem is to ensure stiffness in torsion to prevent skin buckling especially near the root. The majority of the wing weight is then in the skin (estimated to be 70% of the total wing weight). The structural configuration chosen for the wing was a single cell torsion box extending from the LE to $40\%c$, where the main spar was located. This solution was chosen rather than using a circular tube to serve as the torsion box and spar because the torsional stiffness is proportional to the square of the enclosed area of the torsion cell. The details of the structural design and fabrication will be reported after the fabrication is completed. Aeroelastic considerations were taken into account in the design procedure to unload the tips when they are under g loading. This also gives a dihedral effect that improves with g loading.

The aerodynamic problems associated with LRN RPV's are discussed in [Mueller 1985]. The design, wind tunnel testing and flight testing of a RPV was reported in [Stollery et al. 1988], this effort included a WT evaluation of several different airfoil sections. This RPV was designed to function in the R_e range of 300,000 – 1×10^6 and used a wing AR of about 8. Two useful papers on composite model/RPV building techniques are available. The first is by Vranas, of the NASA LaRC, this discusses building using molding techniques with vacuum bagging. The second is by Jacob, of the Aeronautical Development Establishment in India, this discusses building using molding techniques with pressure forming (this paper was published in the Proceedings of an RPV Conference sponsored by the Royal Aeronautical Society and the University of Bristol). Historical and technical information on existing RPVs can be found in sources such as [Reed 1979] while technical specification data is available in the standard reference work, Jane's All The World Aircraft. For the older RPVs such as the Mastiff, the Aquila, etc., great attention does not seem to have been paid to aerodynamic optimization.

III AERODYNAMICS

The key aerodynamic factor which can assure efficient RPVs are the newly designed laminar flow airfoils which improve the L/D for low Reynolds numbers and whose section $\frac{L}{D}|_{max}$ is approximately 100. These designs were based on LRN ideas described in [Pfenninger 1947 & 1956]. The RPV performance goals for this design were to fly efficiently in the 25–50kts speed range (R_u=250,000–500,000). The NASA LRN-1-1010 airfoil, that was designed by Evangelista specifically for the low R_e regime typical for LAURA applications (100,000<R_c<500,000), was chosen for the aircraft's main lifting surfaces. Figure 1 shows the NASA LRN-1-1010 which is a 9.8% thick airfoil and has its maximum thickness located at 36%c, its design details are reported in [Evangelista 1987]. Its design was based on the 7.3% thick NASA LRN-1-1007 airfoil, designed by Mangalam & Pfenninger to operate at high lift-to-drag ratios, especially near a C_L of about 1.0. The design details of the 1007 are reported in [Mangalam & Pfenninger 1984] and [Mangalam et al. 1986].

For the wing this airfoil choice can be evaluated with respect to the other available LRN airfoils on the basis of the figure of merit formula for wing endurance given in [Maughmer et al. 1988], which is $FOM_e=\frac{C_{L_{max}}}{C_D@C_{Lop}}$. The comparisons were made using WT data where available or with computationally predicted data.

[1] NASA LRN-1-1010 FOM_e=107 (WT data for R_c=250,000)

[2] Eppler 387 FOM_e=100 (WT data from [McGhee 1988] for R_c=300,000)

[3] Wortmann FX63-137 FOM_e=60 (WT data estimate from [Stollery 1988] for R_c=300,000)

[4] NACA 0010 FOM_e=54 (Estimate using the Drela code for R_c=250,000)

The LRN-1-1010 and the Eppler 387 are fairly close but at lower R_c the LRN-1-1010 has better performance with turbulators, whose use is discussed below.

Many older LRN airfoils have been compared in [Carmichael 1981].

Conventional airfoils exhibit rapid deterioration in their aerodynamic characteristics at low chord Reynolds numbers, especially below 500,000 because of laminar separation. The non-linear effects of laminar separation are apparent even at Reynolds numbers greater than a million as is evident from the results of [Ferris et al. 1987] for the NASA LS(1)-0013 airfoil for which careful wind tunnel measurements were made in the NASA LaRC LTPT. Ladson, also of the NASA LaRC, made measurements for the standard NACA 0012 airfoil in which the Mach Number and the R_e were varied independently, these results also showed the same characteristics. A good experimental investigation of the laminar separation bubble for airfoils is reported in [O'Mara and Mueller 1987].

When the chord Reynolds number is low, the laminar boundary-layer on the airfoil tends to be very stable and transition is not easily triggered. The inevitable adverse pressure gradient following the minimum pressure peak over an airfoil will then cause laminar separation. To avoid the airfoil drag penalties caused by laminar separation Pfenninger proposed design remedies which prevent laminar separation by forcing transition close to the onset of laminar separation. To minimize drag without penalizing lift, the pressure distribution of the airfoil was designed with the minimum pressure peak near the leading edge, this was followed by a gradual flow deceleration to about the $65\%c$ after which there was a pressure recovery to the trailing edge. The suction peak near the leading edge ensures that a reasonable C_L is generated, and the gentle adverse pressure gradient which follows destabilizes the laminar boundary-layer making it susceptible to transition. The boundary layer on the suction side of the airfoil stays laminar and attached up to the pressure recovery point at $65\%c$ so that the skin friction drag contribution is low up to this point.

The design aim for the airfoil shape were two-fold.

[1] The upper surface contour is designed for the high C_L point of the low drag regime.

[2] The lower surface contour has an undercut near the LE and the TE regions and is designed to meet the low C_L limit of the low drag region.

The pressure drag caused by the separation bubble was minimized by the use of turbulators. These devices should be placed so as to destabilize the boundary layer by promoting the growth of T-S instabilities without incurring an unacceptable large device drag penalty. Their optimum use is to cause transition just before the start of the laminar separation bubble. The skin friction drag is then minimized by ensuring the maximum possible extent of laminar flow. At the lower Reynold's numbers often several turbulators must be used. The advantage of 3D turbulators devices is that they have greater boundary layer destabilizing capability. Turbulators located on the front of the airfoil are favorable because they create instabilities but do not disrupt the laminar flow. Conversely, as they are located in a relatively high speed region of the boundary layer their device drag for a given height is greater than for devices located behind the maximum thickness point. A qualitative evaluation of the merits of different turbulators such as roughness strips, bleed holes, grooves, and three-dimensional roughness elements are discussed in [Pfenninger et al. 1988]. Operationally durable turbulators are necessary for flight operations, hence, a study was undertaken to evaluate some of the relative merits of three turbulator devices.

The experiments were run with a NASA LRN-1-1010 model in a (15" x 12" open WT operated by the Viscous Flow Branch of the NASA LaRC whose turbulence level near the test section is about 1%). Wake surveys were used to measure the airfoil drag, surveys made at different span locations gave consistent results. The available equipment and facility could not match the quality of the NASA LaRC LTPT WT in which LRN measurements have been made for the Eppler 387 and other airfoils that are reported in [McGhee et al. 1988]. This shortcoming was partially overcome by using computationally predicted data using the Drela airfoil analysis code whose results are shown in Figure 2 for the LRN-1-1010. A more detailed comparison of the Drela code computations and LRN experimental data available from the LTPT data can be found in [Evangelista et al. 1989]. For the LRN-1-1010 the Drela code predicted C_D*'s* lower than the WT results shown in Figure 3 which shows C_D *v/s* α for the NASA LRN-1-1010 for various R_e*'s*. Hence the WT results were considered to be valid and were used to compare the C_D of the *'rough'* airfoil against the baseline *smooth* case. In the tests the angle of attack α_g was measured relative to a horizontal reference line (such as the x axis shown in Figure 1). The absolute angle of attack is 6.5° more than the measured α_g. As the C_{D_∞} reduces sharply at α_g=5° natural boundary layer transition probably occurred prior to the formation of a laminar separation bubble (which was estimated to start at about 65%c for the other low C_D α values). The uncertainty of these results motivated a turbulator device study. Figure 4 a & b shows C_D v/s α_g for the smooth and rough airfoil for R_c=150,000. The effect of different turbulators for drag reduction are shown. The dashed line in Figure 4a was obtained when two 55μ thick 5*mm* wide tapes were located at 30%*c* & 45%*c*. These thin tapes do not always give repeatable results because their handling and the method by which they are installed can affect their thickness and form locally. The results of the tape located at 25%*c* are more promising, the best separation bubble control is at α_g=3°, however, its control effect is not uniform as can be seen from the α_g=5° result.

Fig. 2 LRN-1-1010, Drela Code n=9

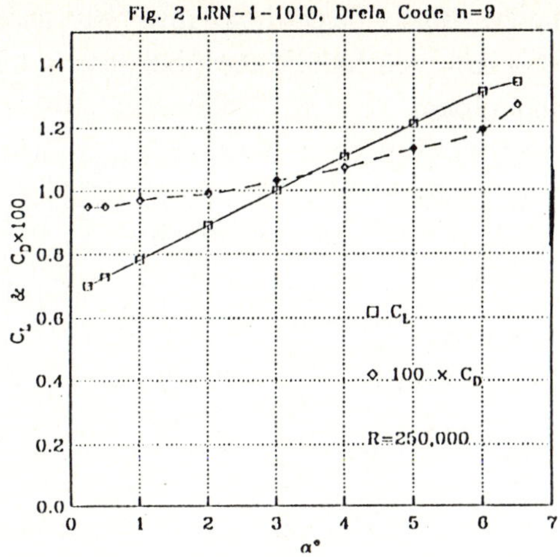

Fig. 3 WT results for the LRN-1-1010

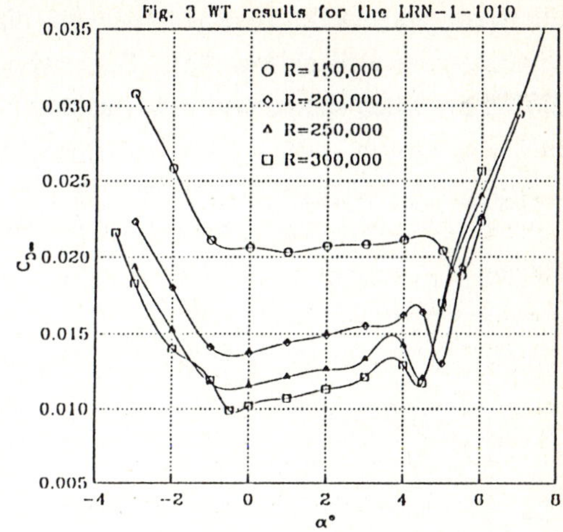

Fig 4a Turbulator Tests for LRN-1-1010

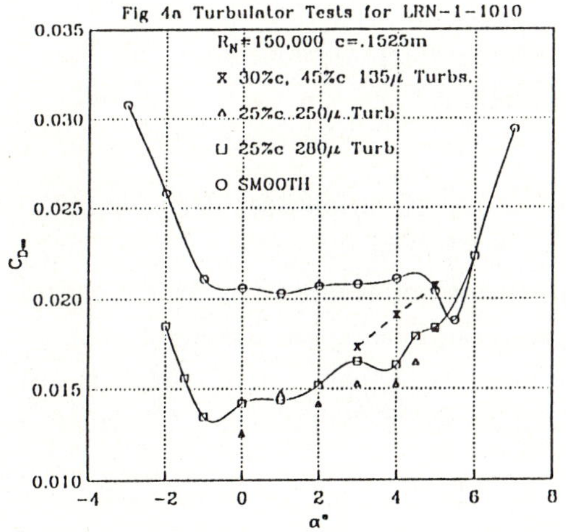

Fig 4b Turbulator Tests for LRN-1-1010

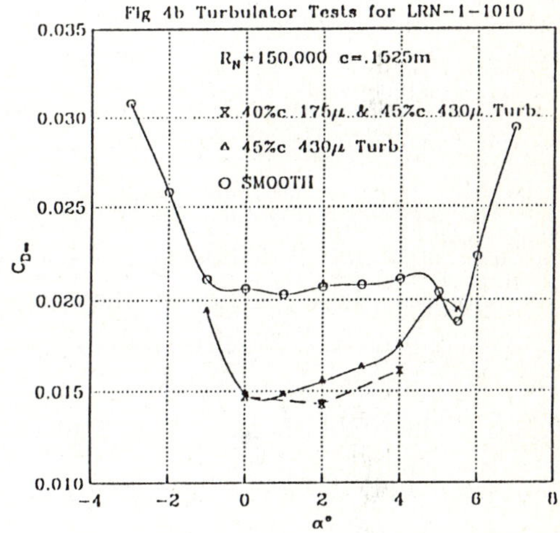

The most repeatable results were obtained with a 430μ thick *6mm* wide sandpaper strip whose LE was cut into a zig-zag shape. The results for this turbulator for an α_g=2° are shown in Figure 5. This turbulator appears to work best when located at 80%c for most other locations the smooth airfoil has a lower drag. The drag reduction probably occurs because this turbulator causes transition immediately behind it, and so it had an excessive device drag penalty when located far ahead of the separation bubble. Such thick turbulators do not work as well as the thin ones described above but are naturally less prone to installation irregularities. A comparison of the performance of two bleed hole configurations located at 50%*c* & 57%*c* for different R_c is shown in Figure 6. The bleed holes were spaced 5mm apart along the span and were *.6mm* & *.75mm* in diameter. Each bleed hole row was used separately as well as together. Tests were also run with half of the front row holes closed off when the second row was completely closed.

Fig. 5 LRN-1-1010 Turbulator Tests

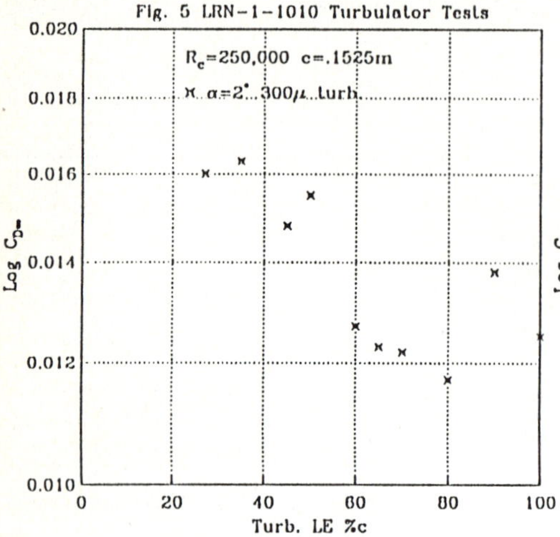

Fig 6 LRN-1-1010 Bleed Holes Tests

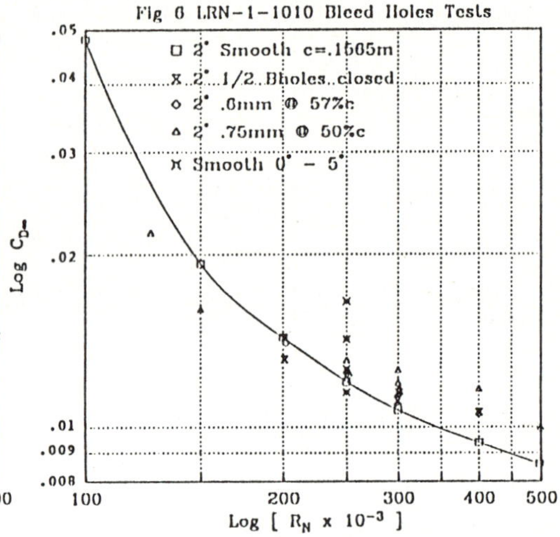

The turbulator results indicate that the LRN-1-1010 airfoil does not have severe laminar separation penalties for $200{,}000 < R_c < 400{,}000$ in the $-1° < \alpha_g < 5.5°$ range. The choice of the turbulator for the wing is perhaps not as critical as for the canard, horizontal and vertical tails which operate at about R_c=150,000 for flight at 25*kts*. They will require an operationally reliable separation bubble control which is also robust in the flight operating environment. The choice of the turbulator will be made after more WT tests are run.

IV INDUCED DRAG REDUCTION

The RPV was designed to obtain its $\frac{L}{D}\Big|_{max}$ at 25*kts*, with a C_L=1.14 for the wing and fuselage based on measured data obtained from [Mangalam et al. 1987]. The wing operates at R_c=250,000 and has an a=.092, a α_g=4.5°, its $C_{D\infty}$=.015 and its C_{Di}=.016 assuming an e=.95. It should be possible to further reduce this induced drag by improving e. Wing tip feathers were designed for this purpose by computing the lifting line vortex wake rollup with an inviscid Point Vortex Method. (Tip feathers were proposed by Pfenninger for induced drag reduction in 1943 in Zurich). For wake rollup the PVM seems to give more reliable results than a vortex lattice method, this was investigated in [Siddiqi 1987]. Figure 7 shows the rollup for the wing with two feathers, one deployed up at 50° and one downwards at –25°. The C_{Di} of the two wings was compared keeping C_L, S, b, and *AR* the same. The reduction in C_{Di} is evident from the plot of the non-dimensional downwash w on the lifting line. The inboard value w=1 means that it is equal to the value $\frac{\Gamma_o}{4s}$, the uniform downwash of an elliptically loaded lifting line. Near and on the feathers the w is much less than this value and so gives a computed C_{Di} reduction of approximately 10%.

Fig.7 $C_{Di}/\pi AR$ for tip feathers with C_L=1 AR=23.2 with wake rollup

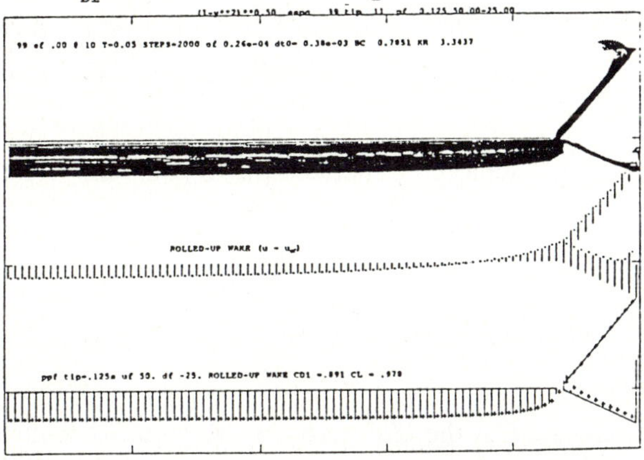

Fig.8 NACA 0008, Drela Code n=14.

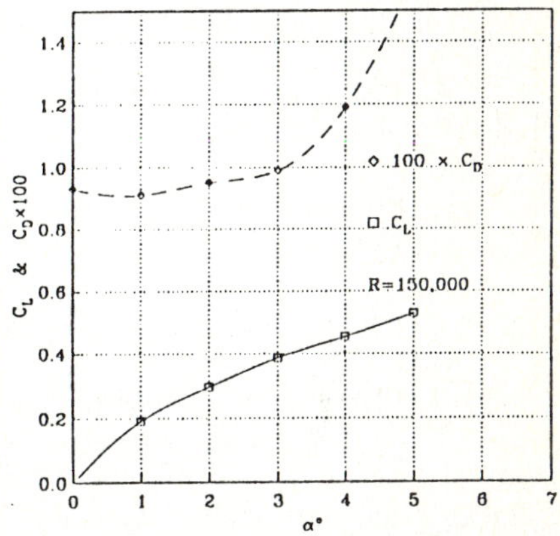

(Details of such computations will be available in a forthcoming paper). Induced drag computation studies for swept wing tips have been done by van Dam, Vijgen & Holmes, in the NASA LaRC, and in Germany, by Eppler.

The use of tip feathers or *Pfenninger Pfeathers* should give a computed C_{Di}=.0161 at C_L=1.143. Viscous effects will increase this figure but the pressure drag detriment of these effects can be minimized by well designed feather junctures. The canard chord was chosen so that at 25*kts* it has a R_c=150,000 and so a C_D=.022. Its AR=6 and so for C_L=.9 its C_{Di}=0.048 assuming an e=.9. The use of tip feathers should reduce this to about C_{Di}=.041. The overall drag penalty of the canard is less severe since S_c/S=.125 and hence, C_{Di_c}=.0051. The wake vortex rollup computations showed that a low canard position would ensure that its vortex wake passes below the wing for high α. This may contradict high R_e experience, where the interaction of the vortex wake and the wing upper surface is considered to be favorable. For the low R_e case the literature suggests that boundary layer separation may be induced on the wing upper surface by the close passage of the vortex wake. The horizontal tail was also positioned so that the wing and canard wakes passed above it for the high α cases (this was done because the tail is down loaded for this case).

V STABILITY AND CONTROL

The design specifications require a statically stable RPV which has *good* handling characteristics. A horizontal tail volume factor of .5 (where S_h/S=.15) will ensure longitudinal stability. Similarly a vertical tail volume factor of .7 (where S_v/S=.15) should give dynamically stable lateral modes. This Section concentrates on longitudinal stability, the investigation of the lateral stability characteristics is not complete and will be reported later.

Ensuring *good* handling qualities is the design challenge, it is closely tied to the RPV's dynamic stability characteristics. The three main handling features that were judged to be desirable were:

[1] Predictable and gentle stall and departure characteristics near the long endurance design speed.

[2] An acceptable flying workload for the pilot.

[3] Uncoupled longitudinal and lateral flight modes as well as uncoupled phygoid and short period modes.

The original half scale design that was tested in the University of Maryland's tunnel [Mangalam et al. 1987] was a twin boom tail configuration. While it gave adequate longitudinal stability, its parasite drag was excessive as it was oversized. The option of reducing its S_h/S=.29 was not considered because the twin booms which are anchored to the wing will have separated flow regions in their vicinity at these low R_e for high $\alpha's$, this can locally spoil the wing flow field. Hence, other longitudinal stability and control configurations were sought. The two main guidelines used in evaluating the configurations studied (aside from stability) were.

[1] Slow flight capability below the specified $25kts$.

[2] Phygoid and short period mode frequencies that are as widely separated as possible.

To ensure slow flight, the configuration should have as high a C_{Lmax} as possible. The nose down C_{Mac} is usually quite large (-.1 to -.15) for such LRN airfoils, this generally forces the use of a down loaded horizontal tail for trim which in turn compromises the wing C_{Lmax}. A partial solution to this problem is to locate the cg aft of the wing ac so that the wing C_L partially compensates for the large C_{Mac} allowing the C_{Lh} to be approximately 0. However, longitudinal stability requirements and structural torsion considerations are unfavorably affected by aft cg locations for a conventional configuration.

The following three alternate configurations were investigated and evaluated based on the two criteria stated above and on their predicted L/D values.

[1] A conventional tail aft configuration without booms.

[2] A canard configuration.

[3] A three surface configuration.

The C_D for these configurations was calculated as follows. The wing fuselage C_D was obtained from the University of Maryland's wind tunnel studies [Mangalam et al. 1987]. The wing fuselage C_{Di} is included in this figure. The 2D skin friction and pressure drag of the horizontal tail, vertical tail, and canard airfoils was estimated from computations and wind tunnel tests. The induced drag due to the canard and horizontal tail (the *trim* induced drag) was then added. The NACA 0008 was chosen for the tail airfoils and its characteristics at LRNs were computed using the Drela Code, the results are shown in Figure 8. The NASA LRN-1-1010 was also chosen for the canard airfoil and its characteristics were available from the wind tunnel tests described in Section II.

Table II gives a comparison of the stability and control characteristics of the three configurations. These evaluations were made with a modified version of the 2D linear Stability & Control code described in [Smetana 1984]. The LRN non-linear C_L, C_D, C_M characteristics were used in the analysis. The conventional tail configuration was quickly eliminated because S_h/S requirements were excessive and compromised L/D. The canard configuration was eliminated because it compromises the usable high C_L of the wing. This is because good handling qualities require the canard to stall before the wing. The $C_{L_{\max}}$ of the canard is also limited by the fact that its aspect ratio is limited by its chord which was selected to ensure a LRN limit of 150,000 for the R_c at 25*kts* (c_c = .6 ft. b_c = 3.6 ft. AR_c = 6).

The three surface configuration offers interesting handling features, the trim and control capabilities can be well separated. This configuration meets the stability and control requirements while retaining a favorable L/D. The three surface configuration should provide low speed capability by allowing C_L aircraft to be > C_L for the wing alone. For high R_e the C_{Lmax} requirement can be met with flaps but for LRNs the ΔC_{Lf} is severely limited by the laminar separation phenomenon. Hence, in effect here the ΔC_{Lf} is provided by the canard. The C_{Lh} trim penalty is minimized because there are now multiple trim solutions. A wing C_L was selected for a flight condition to ensure a high L/D. The trim C_L's for the canard & tail were then obtained by simultaneously solving for the required $C_L = \frac{W}{qS}$ and for $C_{Mcg} = 0$. An

TABLE II TYPICAL LONGITUDINAL STABILITY RESULTS

Configuration	$V_{(kts)}$	L/D	SM	$C_{L.TOT}$	$C_{L.W}$	$C_{L.C}$	$C_{L.H}$	$W_{n.sp}$	$W_{n.ph}$
Twin Boom	26	24.5	0.23	1.14	1.14	--	0.0	3.75	.73
3 surface	25	26	0.10	1.24	1.15	1.15	-.52	2.62	.68
3 surface	35	15.2	0.11	0.63	0.60	0.59	-.40	3.11	.62
3 surface	50	6.6	0.13	0.31	0.40	-.10	-.71	4.28	.48

CONFIGURATION COMPARISON

Configuration	Wing	Tail	S_h/S	S_c/S
Twin Boom	LRN-1-1010	NACA0012	0.29	--
Three Surface	LRN-1-1010	NACA0008	0.125	0.125

unusual way to gain high speed capability when the wing C_L is less than the low C_L limit of the low drag region is to down load the canard and the tail thus forcing the wing C_L back into the low drag region. This may require a symmetric airfoil for the canard as the LRN-1-1010 does not function efficiently for negative C_L. Another solution is to use cruise flaps to shift the low C_D region of the wing to lower C_L's, computed graphs for this may be found in [Pfenninger et al. 1988].

Configurations which had a *cg* aft of the wing *ac* gave a better L/D, because they require a less down load on the horizontal tail. However, based on experience a SM>12% was considered to be necessary for good handling characteristics at the low speeds. This forces the choice of *cg's* that are on or ahead of the wing *ac*. The selection of S_c/S and S_h/S requires a careful tradeoff between the friction drag and the induced drag. The selection of the chord is hence influenced by this compromise. A rather low canard & tail AR=6 had to be selected in spite of the above tradeoff, because this choice was also affected by the weight considerations. In order to minimize canard/tail weight for the given area their span had to be limited to limit torsion loads.

ACKNOWLEDGEMENT

The authors would like to thank their NASA LaRC sponsor Mr. W. D. Harvey, head of the Fluid Dynamics Branch, for his project management support. We are grateful to the Computational Fluid Dynamics Lab. of NASA LaRC for the use of its computing facilities. We appreciate being able to use the WT in the Viscous Flow Branch, NASA LaRC. We would also like to express our gratitude to the NASA LaRC machine shop personnel for their quick fabrication turnaround which helped to meet a tight WT schedule. We are also very grateful to Mr. R. Foch and Ms. P. Toot, of the NRL, for their support, technical advice and comments during the course of this design effort.

This work was partially supported under NASA Contract NAS1-18599.

REFERENCES

1. B. H. Carmichael, Low Reynolds Number Airfoil Survey, NASA CR 165803, v 1 Nov 1981.
2. R. Evangelista, C. S. Vemuru, Evaluation of an Analysis Method for Low-Speed Airfoils by Comparison with Wind Tunnel Results, AIAA 89-0266, 27th Aero Sc Meet, Reno, Nv, Jan 1989.
3. R. Evangelista, et al., Design and Wind Tunnel Test of High Performance Low Reynolds Number Airfoil, AIAA-87-2349-CP, AIAA 5th Appl Aero Conference, Monterrey, Calif., Sept. 1987.
4. R. Evangelista, Design and Evaluation of a Low Reynolds Number Airfoil, M.S. Thesis, George Washington University 1987.
5. J. C. Ferris, R. J. McGhee, R. W. Barnwell, Low-Speed Wind-Tunnel Results for Symmetrical NASA LS(1)-0013 Airfoil, NASA TM 4003, August 1987.
6. R. J. Foch, R. E. Wyatt, Low Altitude/Airspeed Unmanned Research Aircraft (LAURA) Preliminary Development, Int Conf on Aerodynamics at Low Reynolds Numbers, United Kingdom, 1986.
7. R. J. Foch, Detailed Technical Specifications for the Low Altitude/ Airspeed Unmanned Research Aircraft, U.S. Naval Research Laboratory, Washington, D.C. 1986.
8. S. M. Mangalam, W. D. Harvey, S. Siddiqi, Development of Wing and Tail Configurations for Low Altitude Unmanned Research Aircraft, SAE 871882, Aerospace Tech. Conf. and Expo., Long Beach California, Oct 1987.
9. S. M. Mangalam, et al. Transition and Separation Control on a Low Reynolds Number Airfoil Int. Conf. on Aerodynamics at Low Reynolds Numbers, London, United Kingdom 1986.
10. S. M. Mangalam, W. Pfenninger, Wind-Tunnel Tests on a High Performance Low-Reynolds Number Airfoil, AIAA-84-0628, AIAA 13th Aerodynamic Testing Conf., March 5-7, 1984.
11. M. D. Maughmer, D. M. Somers, Figures of Merit for Airfoil/Aircraft Design Integration, AIAA-88-4416, AIAA/AHS/ASEE, A/C Design Sys & Ops Meet, Atlanta, Georgia Sept 7-9, 1988.
12. R. J. McGhee, G. S. Jones, R. Jouty, Performance Characteristics From Wind Tunnel Tests of a Low Reynolds Number Airfoil, AIAA-88-0607, 26th Aero Sc Meeting, Reno, Nevada, Jan 1988.
13. R. J. McGhee, B. S. Walker, B. F. Millard, Experimental Results for the Eppler 387 Airfoil at Low Reynold Numbers in the Langley Low-Turbulence Pressure Tunnel, NASA Technical Memorandum 4062, Oct 1988.
14. M. M. O'Meara, T. J. Mueller, Laminar Separation Bubble Characteristics on an Airfoil at Low Reynolds Numbers, AIAA J., Vol. 25, Nol. 8, pp. 1033 Aug 1987.
15. T. J. Mueller, Low Reynolds Number Vehicles, AGARDograph No. 288 1985.
16. W. Pfenninger, C. S. Vemuru, Design of Low Reynolds Number Airfoils-I, AIAA-88-2572, AIAA 6th Applied Aerodynamic Conference, Williamsburg, Virginia June 1988.
17. W. Pfenninger, C. S. Vemuru, S. Mangalam, R. Evangelista, Design of Low Reynolds Number Airfoils-II, AIAA-88-3764-CP, AIAA/ASME/SIAM/APS 1st National Fluid Dynamics Congress, Cincinnati, Ohio, July 1988.
18. W. Pfenninger, Investigation on Reduction of Friction on Wings in Particular by Mean of Boundary Layer Suction, NASA TM-1181 1947.
19. W. Pfenninger, Experimental Investigation of an Airfoil with High Lift- to-Drag Ratios at Low Reynolds Numbers, Northrop BLC Report 84 (NAI 560188) 1956.
20. A. Reed, Brassey's Unmanned Aircraft, Brassey's Publishers Ltd., London, Great Britain 1979.
21 S. Siddiqi, Trailing Vortex Rollup Computations Using The Point Vortex Method, AIAA-87-2479-CP, Proc 5th AIAA Appl Aero Meet, Monterey, Ca., Aug 1987.
22. F. O. Smetana, Computer Assisted Analysis of Aircraft Performance stability & Control, McGraw-Hill Book Co., New York 1984.
23. J. L. Stollery, D. J. Dyer, The Flight Performance of an RPV Compared with Wind Tunnel and Theoretical (CFD) Results, ICAS-88-4.9.2, p1392, Proc. 16th Cong. of the Int. Council of Aero. Sc., Jerusalem, Israel, 1988.

CAPTIVE CARRY TESTING OF REMOTELY PILOTED VEHICLES

Alvin Cross
U.S. Naval Research Laboratory
Washington D.C. 20375-5000

Background

Captive carry testing is a valuable supplement to wind tunnel testing, allowing important dynamic properties to be validated while providing significant information pertaining to the flight worthiness of the vehicle. As in the early days of flight, some form of ground testing has preceded significant technology advances while minimizing risk and failure probability. The researcher is afforded the opportunity to evaluate integrated subsystems operation and handling qualities in a realistically simulated flight environment. Therefore, the U.S. Naval Research Laboratory (NRL) considers a pre-flight testing program an essential step in the development of advanced unmanned aircraft.

NRL has reached the flight test phase of the Low Altitude/Airspeed Unmanned Research Aircraft (LAURA) Program. Since LAURA involves the integration of advanced technology for aerodynamics, structures design, and flight control techniques into four distinct configurations, it was essential that these vehicles be pre-flight tested in a controlled environment. The design and fabrication of a reliable ground based test apparatus was seen as a logical step in the preparation of vehicles for flight testing.

Captive Carry Structure and Key Design Features

Figure 1 illustrates the Joined Wing LAURA configuration and this aircraft installed on the captive carry unit designed and built by NRL. This apparatus consists of a lightweight/high strength structure capable of testing vehicles

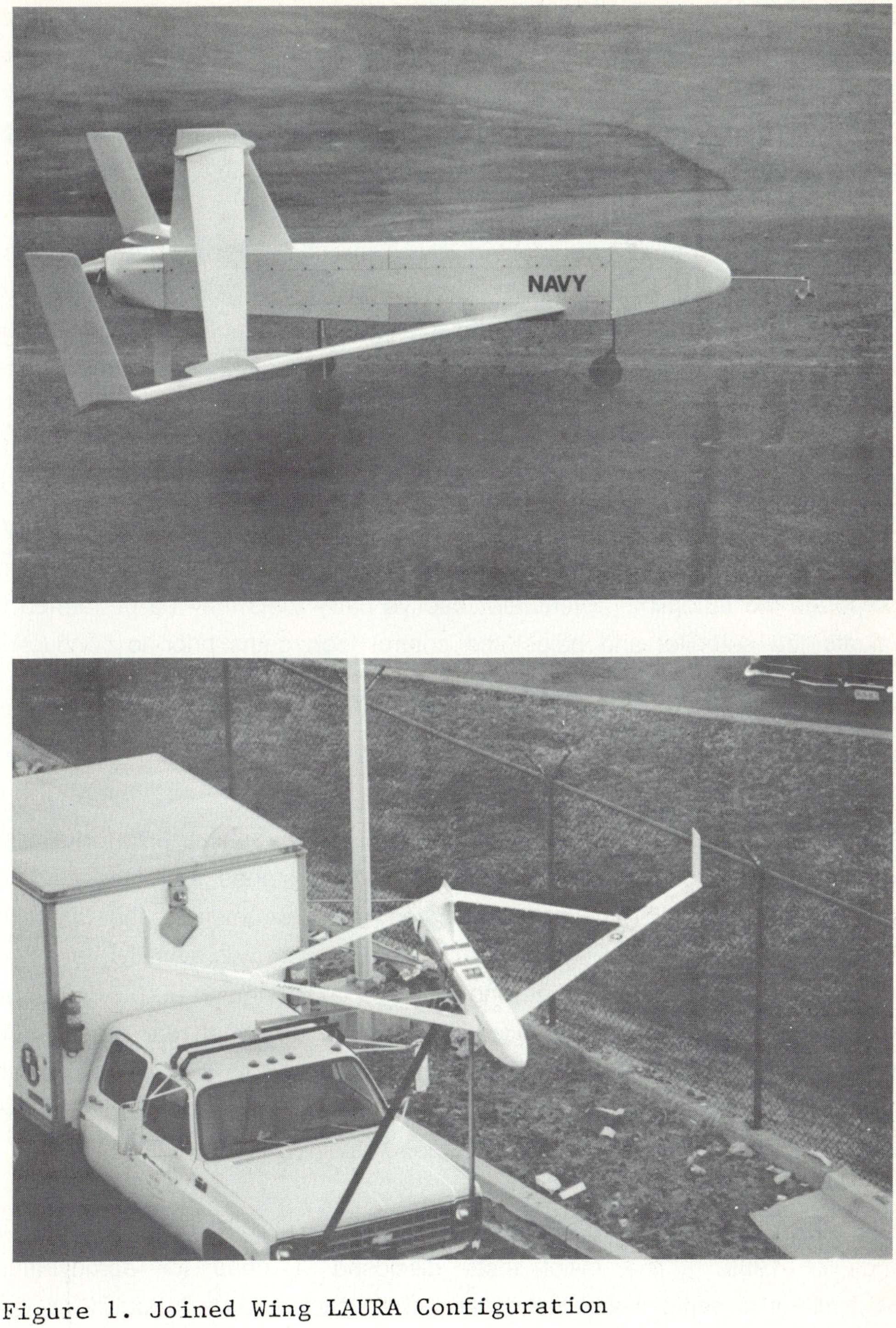

Figure 1. Joined Wing LAURA Configuration

having gross weights of 40 to 60 pounds. The current system allows +/- 20 degrees freedom of movement of the vehicle in both the pitch and yaw axis. Roll axis movement was constrained to minimize system complexity and risks to both the aircraft and the truck. This test configuration emulates the model position allowed by a classical wind tunnel balance table.

Test vehicles are fully instrumented via a ground based data acquisition system (DAS). The DAS is capable of receiving 18 channels of data at a rate of 60 bps. Sensor information includes; airspeed, magnetic heading, engine temperature and RPM, control surface positions, 3-component accelerations, angle of attack, altitude, and angular rates. With this information, a pre-selected vehicle flight path and profile can be maintained via feedback to the onboard autopilot. At this stage of the captive carry tests the autopilot is not engaged since stick-fixed directional and longitudinal stability is of primary interest for flight qualification. Also, initial LAURA flight tests will not incorporate the autopilot. Eventually, captive carry tests may be performed to evaluate the autopilot and adjust the control loop gains prior to conducting autonomous flight.

Testing and Results

The initial test aircraft was the Joined Wing LAURA configuration. Its wing set, including elevon controls and servos, was designed and fabricated by ACA Industries. NRL designed, fabricated and instrumented the fuselage. All vehicle structures and systems are those which will actually be flown. The Joined Wing LAURA main wing features a Wortmann FX63-137 airfoil with a leading edge sweep of 30 degrees and an unswept rear wing having an Eppler 201 airfoil. The wings are joined inboard of the elevons. This vehicle was tested in flight ready status with all onboard sensors, power supplies (NiCad battery packs), and fuel load (32 oz.).

Initial captive-carry tests were completed on the Joined Wing configuration in March of 1989. The initial tests disclosed a need for aerodynamic modifications to achieve satisfactory stick-fixed pitch and yaw stability as well as sufficient control power for safe flight. In addition, serious difficulties were encountered with the DAS which require correction prior to the first free-flight.

Details of these first tests follow: The test location was a 5000 ft. taxiway at the Martin State Airport, Baltimore, MD. The empty (dry) weight of the aircraft less main landing gear (removed to avoid mechanical interference with the test apparatus) was 46.84 pounds with a longitudinal c.g. location giving a static margin of 13.2% of the mean aerodynamic chord (MAC). Following assembly and mounting of the vehicle on the apparatus, experimental runs were conducted. They were of limited duration due to the early drop-out and eventual loss of data acquisition signal at approximately 400 feet of the available 5000 foot test range. The initial runs were made at speeds of 35 mph, the calculated minimum sink speed for the estimated gross weight. With all electrical and electronic systems on and engine off, the pilot (positioned in the transport truck) was unable to fly the vehicle off the rear pitch stop, which corresponds to 20° angle of attack. With full nose down elevon command the vehicle remained on the rear stop. A table of representative runs is shown in figure 2.

This run was repeated several times without any noticeable response to the pilot's inputs. The test was halted to allow a verification of c.g. position and proper balance of the boom cradle with respect to the aircraft. The runs were then immediately repeated. The aircraft continued to show no response to the elevon control inputs. Meanwhile, the DAS continued to experience reception problems and on screen data seemed to be presenting inaccurate values (of primary concern were alpha, beta, and velocity information). Figure 3 is an illustrative example of the data received. It is evident that the airspeed indicated is well below that recorded in the run table. During a particular test run, the aircraft's nose was placed fully down, -20° angle of attack, for the duration of the run, while recorded values oscillate between +9° and -1°. Additionally, it can be noted that acceptable data was recorded for a time interval of only 15 seconds (only during the period when the transport truck was in close proximity to the DAS).

RUN #	RUN PARAMETER	FLIGHT SPEED (FPS)	OBSERVATIONS
2	PITCH STABILITY	51.33	VEHICLE PITCHES UP DURING ACCELERATION. (TO REAR STOP & REMAINS IN POSITION.
3	PITCH STABILITY	51.33	GROUND CHECK OF WING BALANCE. BALLAST ADDED TO NOSE. RE-POSITION PIVOT TO NEW C.G. PITCH UP REMAINS.
5	PITCH STABILITY	51.33	NOSE PUSHES DOWN WITH FULL DOWN ELEVON, SLIGHT YAW OSCILLATIONS NOTED. STALL OF REAR WING. PITCH UP OCCURS.
7	TRIM ALPHA	51.33	TRANSMITTER @1/2 AFT STICK, VEHICLE HOLDS 0°ALPHA
9	TRIM ALPHA	51.33	OSCILLATION IN PITCH DURING RUN. DAS RECORDS 2°-5° ALPHA, PILOT IS HOLDING @-20°.
12	CNB EVALUATION	51.33	NO DAS INFORMATION. FULL LEFT & FULL RIGHT RUDDER INPUTS (+/-10°) PITCH DOWN NOTED ON COMMAND. VERY SOFT DIRECTIONAL CONTROL.
13	CNB EVALUATION	58.60	IDENTICAL RESULTS AS IN RUN #12
18	ADVERSE YAW	58.60	LEFT AILERON FULL DOWN. DAS RECORDS +6°TO -6° SWING IN BETA. AVG+5

Figure 2. Representative Captive-Carry Results for the Joined Wing LAURA (first test)

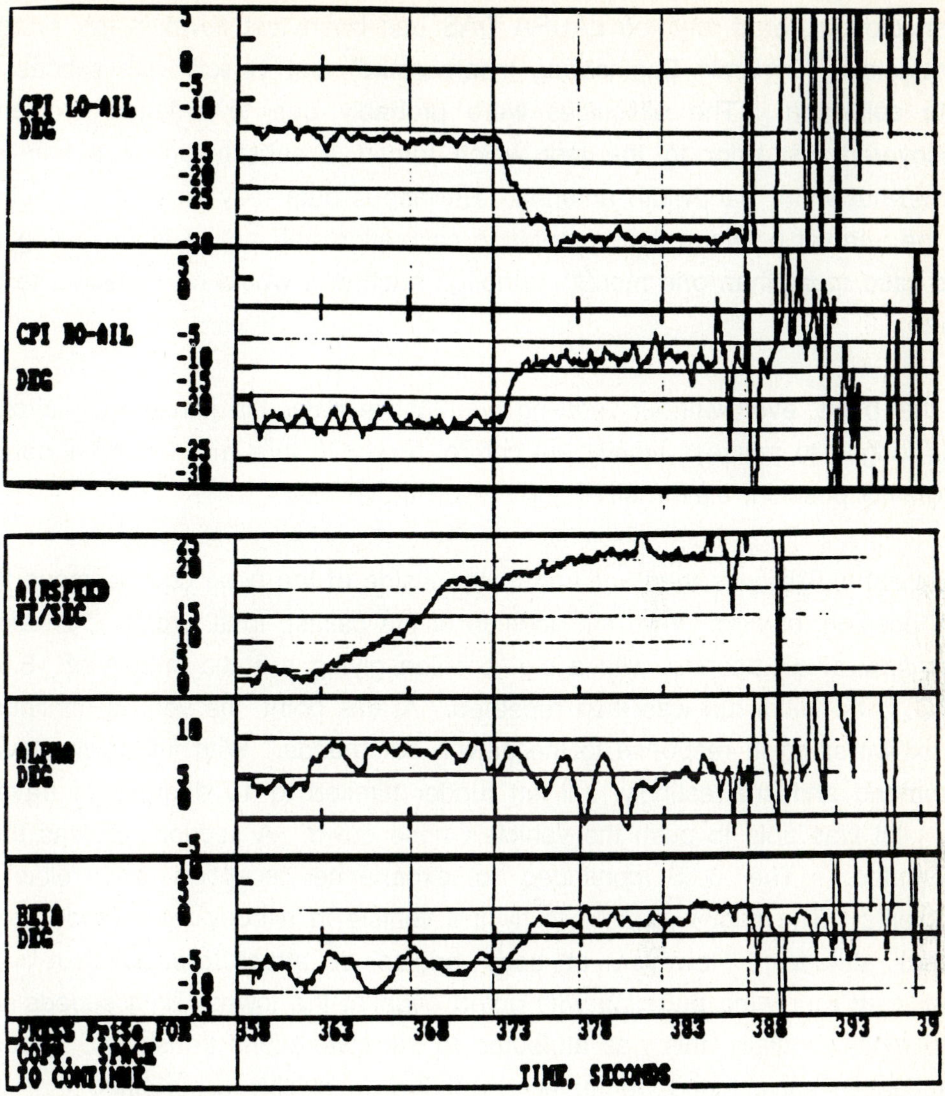

Figure 3. Sample of DAS Output

It should be noted that the LAURA DAS has been test flown in the system contractor's (Dynamic Engineering Incorporated) test vehicle with successful data collection. The difficulties were probably due to antenna problems discovered just prior to the test which forced a substitution of a smaller, somewhat wider bandwidth antenna. Erroneous data may also be a function of the interval between instrumentation calibration and actual flight testing (in this case more than one month), although such drift would be expected to be small.

At this point, even without verifying quantitative data, the joined wing aircraft was judged to possess insufficient control power to trim with the initial center of gravity position, elevon size, and elevon movement.

A ballast weight was then secured to the inside of the nose cowl to move the c.g. position forward. With the addition of the ballast, total aircraft dry weight was now 47.67 pounds, with a c.g. position giving a static margin of 18.0% MAC. The initial run was then repeated. At this point, the vehicle continued to exhibit sluggish response to the elevon commands. With full down elevon command and, interestingly, full left rudder (limited to 10 degrees of travel), the pilot was able to push the vehicle's nose down. A second run was then attempted. The DAS continued to experience reception and reliability problems. The pilot was able to fly the vehicle in a body level (body at 0° alpha) attitude. However, at body angles of attack greater than $+2^\circ$, significant rear wing buffeting was noted even at the lowest flight speeds (25 - 35 mph). Initially, this was attributed to truck vibrations transmitted through the captive carry tripod frame; however, it was quickly ascertained that the rear wing buffeting was a result of massive flow separation caused by either the wake of the forward wing or stalling of the rear wing itself. Whenever the body angle of attack exceeded approximately $+2^\circ$, the rear wing was severely vibrating and the run consequent-ly aborted. The run was repeated several times with very similar results. A series of runs was then undertaken to examine the Joined Wing LAURA configuration at higher speeds. The transport truck was accelerated to 40 and 48 mph on successive runs. With some difficulty, the pilot was able to hold the vehicle in a trimmed condition, and fly into the rear wing buffeting condition. The Joined Wing was not able to recover from this stalled condition due to its lack of pitch control power. The powerful yaw response to aileron control input during the high speed runs appeared to indicate marginal yaw stability. Yaw response to rudder inputs was also judged to be minimal.

Overall, the configuration appeared to be neutrally stable longitudinally (except at low angles of attack) and presented insufficient directional control and stability. NRL decided that relatively simple fixes could resolve these problems and allow flight testing to begin. An examination of the wind tunnel data CM versus alpha curve (fig. 4) shows the marked tendency of the vehicle to pitch-up at an alpha of +4°. This implies that the rear wing (Eppler 201) is seeing a 2-D alpha of about +10-12°, and has stalled. At the same time, it can be noted that the main wing (Wortmann FX63-137) continues to generate lift at an alpha of +4 degrees. In fact, the front wing stall break is not apparent until an alpha of +8° is attained (fig. 5). This data correlates well with available FX63-137 data (Ref. 1). The ability of the main wing to continue generating lift to an alpha of +14° (further enhanced by its sweep) exacerbates the pronounced longitudinal stability problem. Premature stall of the rear wing was felt to be a function of the rear wing incidence. A rear wing incidence of +8 degrees at the root was measured indicating a 2° error in incidence from the wind tunnel model's +6°. At this point, it was decided that lowering the rear wing incidence from +8° to +5° would sufficiently delay the stall of the rear wing. Additionally the aircraft could be trimmed at the original 13.2% MAC static margin. Based on analysis of the wind tunnel data a 13.2% static margin corresponds to trim at CLmax with the elevons at the highest (trailing edge up) angle tested in the wind tunnel.

In addressing the marginal directional stability and control (low CNB and large calculated/measured negative CLB), an increase in the effective side area was incorporated. A ventral fin with moveable rudder, and larger winglets having 2.16 times the original area as well as a lower effective center of pressure were installed. Additionally, elevon extensions were added to the configuration in order to increase the control power. The extensions effectively doubled the area of the existing elevons by doubling their chord.

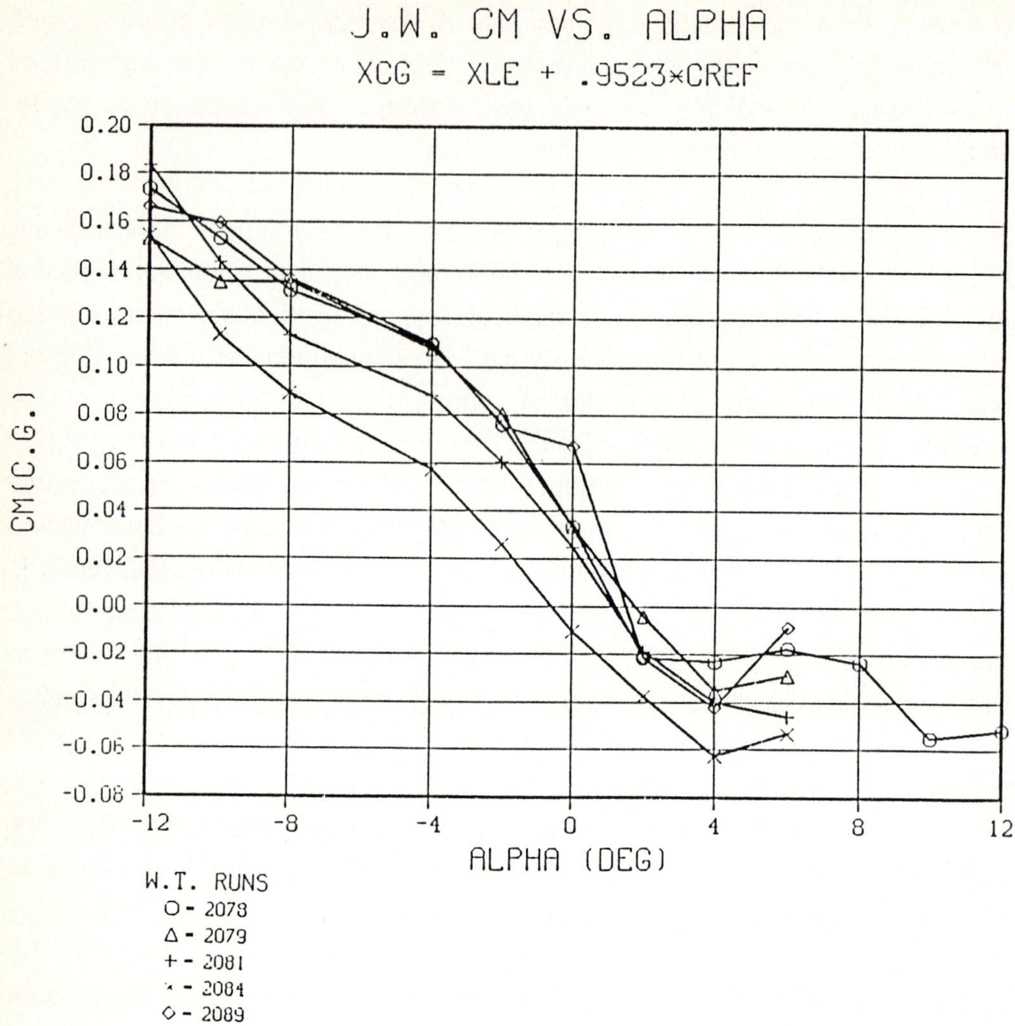

Figure 4. Joined Wing LAURA Wind Tunnel Pitching Moment Results

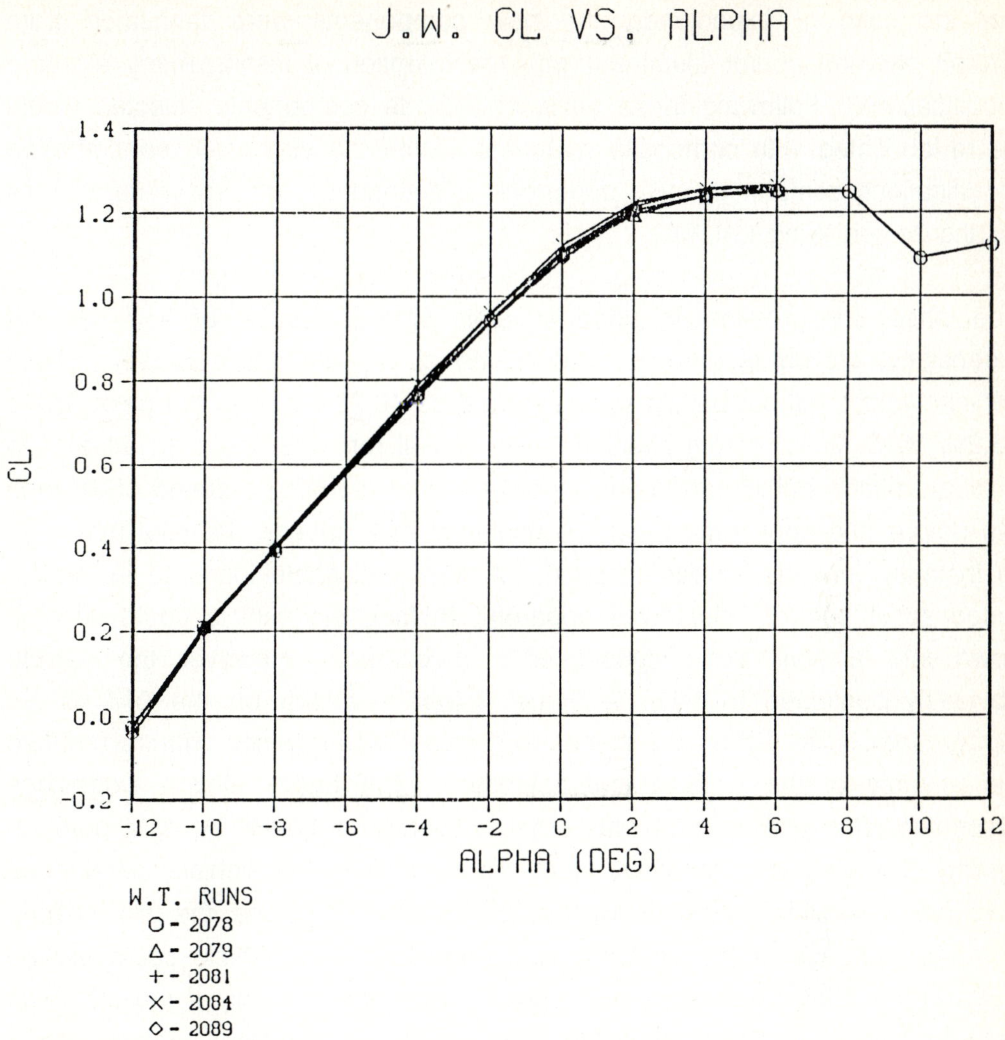

Figure 5. Joined Wing LAURA Wind Tunnel Lift Slope Results

For the sake of expediency, the new components were fabricated from aircraft plywood and/or aluminum with the intention of testing many airframe modifications. Following these tests, only those components selected would be re-fabricated with composite materials. With the suggested configuration modifications completed, NRL undertook a second set of captive-carry tests for the Joined Wing LAURA.

The base configuration included a rear wing incidence of +5°, winglet extensions, elevon extensions, and a ventral fin with active rudder. Total vehicle weight without ballast was recorded as 49.33 lbs. with the pivot at the 13.2% MAC static margin position. The initial run was again an attempt to evaluate pitch stability and control power. After reaching a speed of 36 mph (as before, the approximate vehicle minimum sink speed), the pilot was able to instantly trim the aircraft in pitch. A very noticeable increase in vehicle response to control inputs was apparent. In fact, the vehicle could now be flown into a stall and subsequently recovered. However the vehicle obviously continued to enter a "knee" (slope reversal) on the CM versus alpha curve. The elevon extensions and rear wing incidence change reduced the severity of the pitch-up characteristic. The heavy elevon extensions presented one serious problem, outer elevon/wing flutter. At this point in testing, the elevon extensions were removed and the vehicle rebalanced accordingly. Following the removal of the elevon extensions, the pilot noticed a marked change in the vehicle's stall behavior. The vehicle could still be flown into a stalled condition, but recovery was possible only if the stall was entered gently. Attempts at abrupt stall resulted in a violent pitch-up into a deep stall condition.

By utilizing a revamped antenna for the DAS, the effective telemetry range was increased slightly. However, a power failure (portable field generator) and lack of a back up power supply caused a system failure during a test run. Hard data from all previous runs to this point was lost. System modifications to address this failure mode are underway and will be properly evaluated during a follow up captive carry test prior to the initial LAURA flight test.

Following the initial alpha runs, over sized vortilons were added to the main wing (12 inches inboard of the rear-main wing joint) in an effort to energize the flow over the rear wing. This was attempted as a measure to reduce onset of separation on the rear wing, thereby allowing the vehicle to see a more gradual stall. No discernable difference in stall behavior could be noted when the alpha runs were repeated. The main wing vortilons were removed.

Next, lightweight foam elevon extensions, the same size as those previously tested, were installed. The vehicle was now found to be readily recoverable from stall when trimmed for the cruise flight condition. With the pitch control trim at the furthest aft position, the aircraft was still manageable in the stalled condition; but was significantly slower in its ability to recover since the slight additional elevon travel allowed the stall to be entered more deeply.

Having found an apparently flight worthy configuration with respect to pitch, the directional control and yaw stability concerns could now be addressed. With dual rudder (ventral rudder electronically linked to main rudder), the vehicle could now be successfully yawed in the proper direction (right rudder - negative beta). However, with the added ventral fin/rudder combination and winglet extensions (CNB due to winglet increase (predicted) from 0.034 to 0.083) the vehicle continued to exhibit a "soft" response directionally indicative of insufficient rudder control. Of significant note, the vehicle no longer exhibited a striking tendency to pitch-up with a rudder deflection, no doubt due to the greatly enhanced longitudinal stability (CMDR problem eliminated). Since the directional stability was noticeably improved, it was now felt that elevon roll inputs would no longer induce the large proverse and adverse yaw angles seen earlier. As anticipated, the vehicle performed quite well directionally with an aileron input. Little adverse yaw was observed with elevon roll deflection and at low angles of attack elevon-induced yaw was small and proverse.

Even though the aircraft now appeared to be longitudinally and directionally stable, an extra margin of safety with respect to its pitch behavior was sought to insure a successful initial test flight of the Joined Wing LAURA. Rear fuselage strakes were added to provide an extra nonlinear lifting force on the aft body, which tends to push the nose down and enhances stall recovery. With the strakes added (aircraft rebalanced), the aircraft never entered the stall condition when neutrally trimmed in pitch. With the control stick in the full aft position and full nose up pitch trim applied, the aircraft entered a mild stall but was fully recoverable even at below-stall airspeeds.

The final test of this second captive carry series of the Joined Wing LAURA involved the removal of the wing fences at the rear-main wing juncture. The aircraft's pitch stability was slightly impaired with the fence removal.

Conclusion

Having completed the second series of captive-carry testing on the Joined Wing LAURA, a flight worthy configuration has been achieved. That configuration has +8° incidence on the main wing, +5° incidence on the rear wing (with original vortilons), winglet extensions, elevon extensions (both inboard and outboard), aft fuselage strakes, and wing-juncture fences.

As evidenced by the aircraft's initial post-stall pitch instability, inability to trim in pitch, marginal directional stability, and poor directional control, the Joined Wing LAURA vehicle was not felt to be capable of sustained flight prior to captive-carry testing. The testing did indeed afford NRL the opportunity to evaluate the aircraft in a controlled, minimal risk environment. Without having to repair damaged systems and speculate as to the cause of the aircraft's failure (which the first test indicated as quite probable), flight testing efforts will be able to proceed with minimal delay. The proven value of this type testing has led NRL to incorporate the captive-carry testing into its program prior to flight testing all LAURA vehicles. Additionally, captive carry testing will be specified as a mandatory requirement for all future high value/high risk unmanned aircraft developed by NRL.

Reference

1. Aulthaus,D; Profilpolaren Fur Den Modellflug; Neckar-Verlag; Klosterring 1, 7730 Villingen-Schwenninger, 1980.

Flight Testing Navy Low Reynolds Number (LRN) Unmanned Aircraft

Richard J. Foch, Peggy L. Toot
U.S. Naval Research Laboratory
Washington, D.C. 20375-5000

Small unmanned aircraft utilizing state-of-the-art advances in low Reynolds number aerodynamics are being developed by the Naval Research Laboratory for use in Navy electronic warfare missions. Two missions in particular require vehicles exhibiting high aerodynamic performance in conjunction with stringent operational requirements. The first mission requires a long endurance, low speed, low altitude vehicle, while the second mission requires a vehicle designed for short endurance, low speed, and low altitude.

The long endurance vehicle must carry a 25 lb. payload for 24 hours endurance at 100 ft MSL altitude without exceeding a flight speed of 25 knots (42 fps). The vehicle size is limited by the requirement to fit in a 9 ft by 2 ft by 2 ft container and must be ready to launch within 20 minutes of notice. The second mission requires a vehicle which can carry a 10 - 25 lb payload for 5 minutes at 40 knots (67.5 fps) at altitudes below 100 ft MSL. The packaging constraint for this mission is much more severe. The vehicle is stored in a folded condition in a 130 mm (5 1/8 inch) by 1.7 m (66 7/8 inch) tube and is rocket launched. The system must be ready for launch within 7 seconds of launch command and deploy its tail and wing surfaces in flight. Vehicles for both these missions fly autonomously with no requirement for vehicle recovery.

Technically, the performance requirements challenge the latest low Reynolds number (LRN) technology. Meeting the long endurance mission parameters requires a vehicle with high $C_L^{3/2}/C_D$ values at high C_L for a chord Reynolds number of 250,000. The boundary layer must be insensitive to the wind gusts and water droplets, characteristic of the operating environment. Additionally, clever configurations are necessary to maximize wing area, within the packaging constraints, to optimize for the low speed flight.

Although Mission 2 is not demanding in range or endurance, wing area is restricted by available packaging volume. High lift coefficients are necessary to achieve the desired flight speeds at Reynolds numbers of approximately 350,000. Innovative configurations are being developed to maximize wing area and minimize adverse transient conditions during wing and tail deployment. Again, the boundary layer must be insensitive to wind gusts and water droplets.

Most of the vehicles developed by the Vehicle Research Section at NRL utilize LRN airfoils. The Wortmann FX63-137 has been used on several vehicles including the Long Duration Expendable Decoy (LODED). This airfoil was originally selected for its high lift performance which was documented with wind tunnel data. At the time the LODED was designed, no other high lift LRN airfoils had been wind tunnel tested, thus, newly developed LRN computational design and analysis codes were not verified by experiment. The LODED was a Mission 1 type vehicle that operated in the 210,000 to 250,000 Reynolds number range. The vehicle had a tandem wing configuration in which the front wing pivoted for pitch control and the split rear wing pivoted for roll control and pitch trim. The initial LODED incorporated upper winglets as shown in Figure 1. Roll-yaw instability was observed during initial flight testing and led the designers to modify the vehicle for larger, lower winglets to increase the directional stability while reducing the dihedral effect. The second version of the LODED, shown in Figure 2, completed several successful flight tests. LODED performance data indicated a lift to drag ratio of approximately 14:1 using lifting surfaces with aspect ratios of 10 to 11. Although the successful LODED flights proved the vehicle concept is sound for a high performance LRN vehicle, a lower cruise speed was tactically desirable without a significant increase in packaging volume. Improvements in aerodynamic performance, structures, and overall vehicle design were needed. In response, the Office of Naval Research (ONR) sponsored basic research in the LRN area to support aerodynamic advancements.

The Low Altitude/Airspeed Unmanned Research Aircraft (LAURA), currently under development by the Navy, incorporates the results of five years of LRN basic research. The LAURA consists of four different wing planforms which integrate to a common fuselage. The four configurations are the tandem wing, joined wing, hinged wing and 3-surface (Figure 3). Each planform features it own LRN airfoils to include the Eklund modified FX63-137, Calderon modified Eppler 193, Eppler 205, Eppler 201, and LRN 1010. Based on wind tunnel test results, exemplary improvements in performance are exhibited in lift to drag ratio on the order of 20:1 to 28:1, as compared to previous Mission 1 type aircraft in which L/D was on the order of 14:1. These vehicles use main lifting surfaces with aspect ratios ranging from approximately 15 to 27.

The Flying Wing Research Testbed (FLYRT), has a modified Liebeck LA2573A for its wing shape. The FLYRT was designed and fabricated as a flying testbed for the LA2573A, which features a slightly positive pitching moment about the 1/4 chord, making it attractive for flying wing configurations. Initially the FLYRT had constant chord wings with 3° dihedral. The airfoil shape was modified slightly to allow an adequate static margin for longitudinal stability and for the installation of control surfaces. Flight tests indicated moderately high values of maximum lift coefficient (C_{Lmax} = 1.1 approx.), excellent L/D, a large speed range and satisfactory handling. However, some Dutch-roll was noted. Following initial flight testing, downward canted swept wing tips [1] were fabricated and installed on the FLYRT. Trimmed flight was achieved with less "up" elevon, thus returning the airfoil to its designed shape and improving performance while reducing the Dutch rolling motion.

The Liebeck LA2573A was then used on the Pendulus vehicle (Figure 4). The Pendulus was a concept demonstrator of a Mission 2 type. The unique vehicle configuration is controlled by moving the wing with respect to the body for pitch and using the rudders for roll and yaw control. The wing folds down about the center to lie along the length of the body for packaging into the launcher tube. The Pendulus concept was successfully demonstrated from a bungee catapult launch through a controlled flight under electric propulsion.

Several observations can be made in regards to vehicle development. With recent technological advances in composite structures and LRN aerodynamics, significant performance improvements are achievable and minimum performance requirements are met with minor difficulty. The primary problem in vehicle development is with stability and control. Achieving satisfactory roll-yaw coupling in conjunction with adequate lateral and directional control proves to be the most challenging. Considerations must be made in the design of control surfaces and stabilization surfaces.

Flight testing of fully instrumented flight vehicles provides quantitative data useful for evaluating vehicle performance and validating technological advances in LRN aerodynamics. In addition, valuable data can be achieved on uninstrumented vehicles by use of video recordings, still photography and a detailed flight log.

Computational aerodynamics, real time flight simulation, and captive carry testing have been used to significantly reduce the risks of initial flight testing. This has resulted in the ability to provide cost effective demonstrations of new technologies and verify theory.

With regards to the current state-of-the-art in LRN aerodynamics, avionics and structures, the development of a second generation LAURA would be quite feasible. This vehicle would be optimized for a CL/CD greater than 30:1 at a Reynolds number of 250,000 with an approximate gross weight of 65 lbs., a 30 ft wing span and a cruise speed of 25 knots. The vehicle concept, shown in Figure 5, partly achieves its outstanding aerodynamic performance by use of a strut-braced high aspect ratio wing. Not only do the struts allow a significant wing weight reduction, but also make changing the wing incidence for roll control practical since minimal bending loads are transferred into the wing root/fuselage joint. The "salmon nose" shaped fuselage is aligned with the upwash in the flow to the wings to minimize flow separation over the body. The vertical tail is large to provide sufficient lateral-directional stability at the low design speeds. An electric motor located near the cg, (to minimize moments of inertia) provides propulsion by turning a LRN propeller via a composite shaft. For research purposes the vehicle would be equipped with sensors and real time telemetry. The resulting aircraft would not only be extremely useful for detailed LRN research, but could be utilized for a wide variety of military and civilian missions thanks to its superior flight performance for its class.

Reference

1. van Dam, C.P., "Induced-Drag Characteristics of Crescent-Moon-Shaped Wings", Journal of Aircraft, Volume 24, Number 2, February 1987, pg 115.

Fig. 1 Early LODED Configuration

LONG DURATION EXPENDABLE DECOY (LODED)

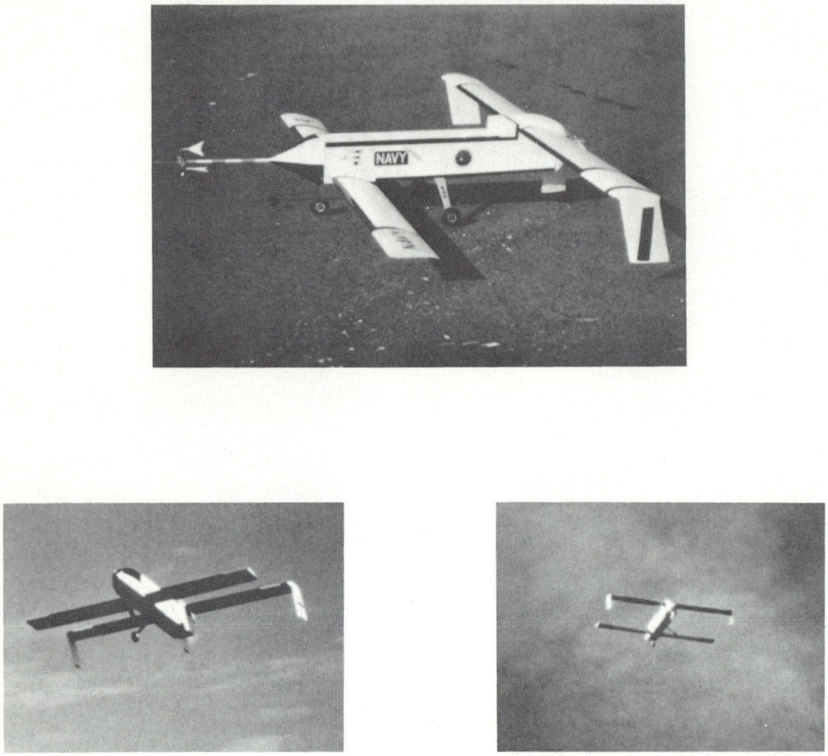

Fig. 2 LODED

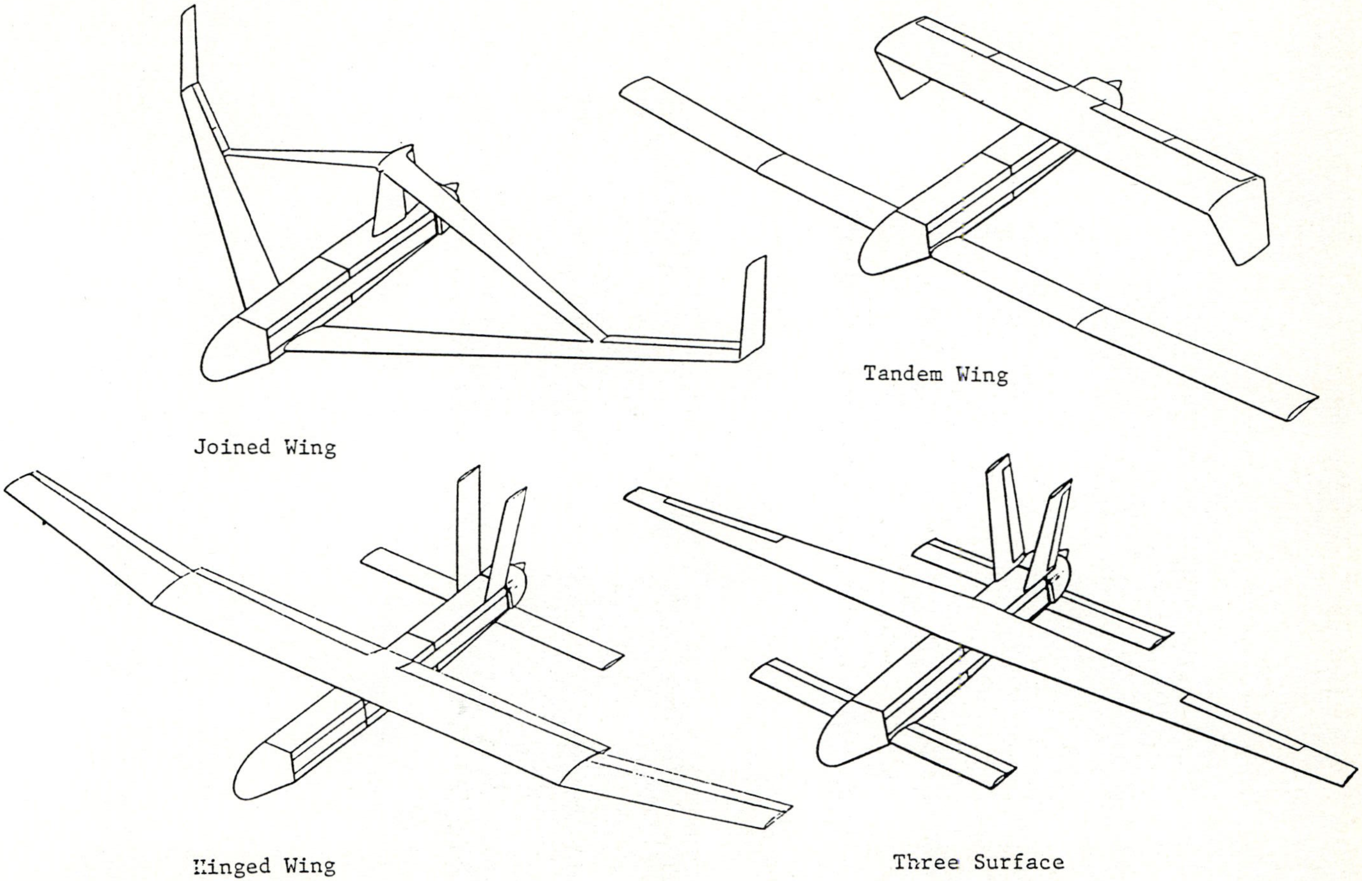

Fig. 3 LAURA Configurations

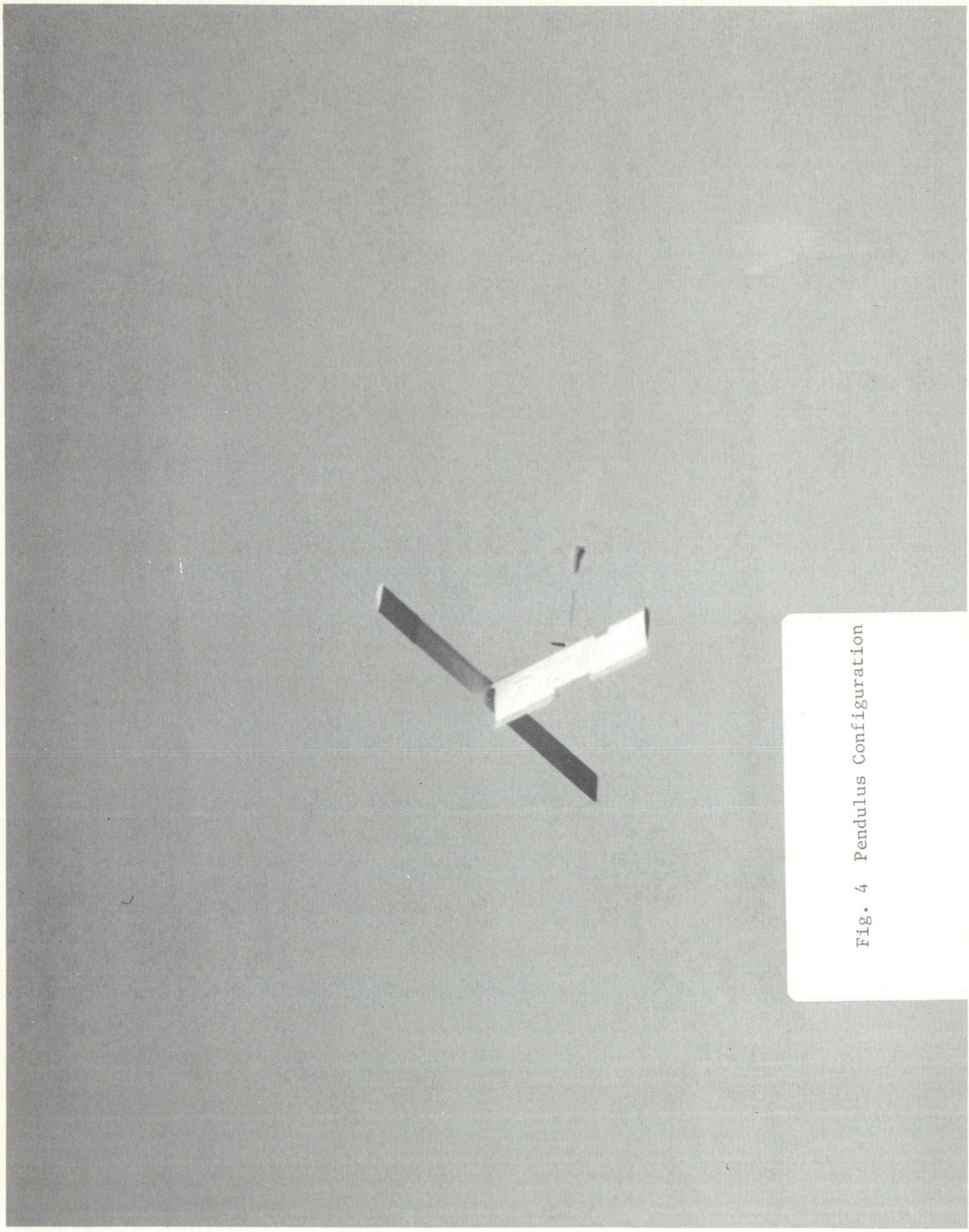

Fig. 4 Pendulus Configuration

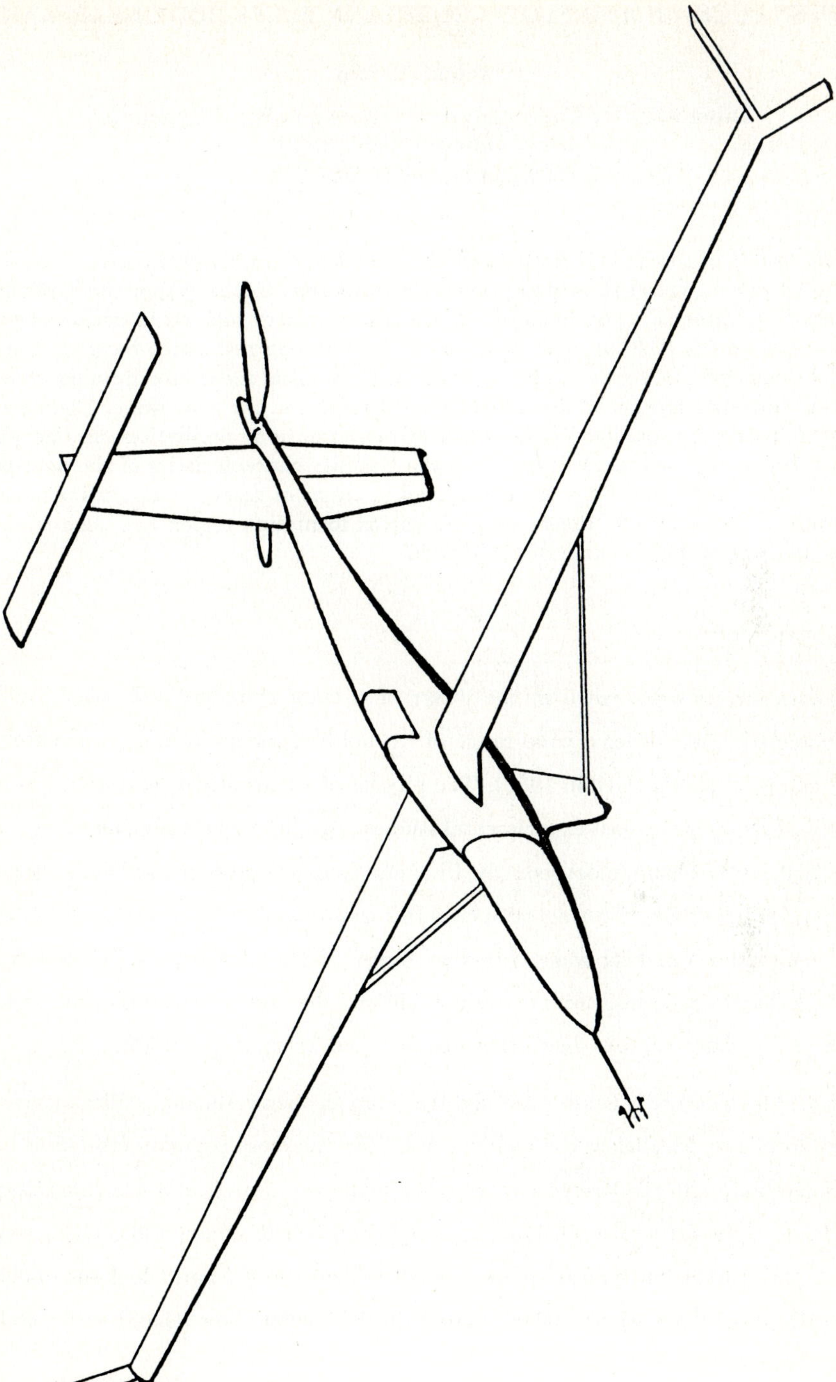

VORTEX LOCK-ON AND FLOW CONTROL IN BLUFF BODY NEAR-WAKES

Owen M. Griffin

Laboratory for Computational Physics and Fluid Dynamics
Naval Research Laboratory
Washington, DC 20375-5000

The results of recent experiments show the concept of universal vortex wake similitude to be more general than was previously supposed. In this paper these similarity arguments are directed as an example to the case of vortex lock-on in oscillatory flow. This form of vortex lock-on exhibits a particularly strong resonance between the flow perturbations and the vortices, and provides one potential means for modification and control of the basic formation and stability mechanisms in the near-wake. The essential features of active and passive wake control by means of body oscillations and geometrical alterations are discussed briefly. Then the stability characteristics of the near-wake, i.e. absolute versus convective, are considered in terms of recent computations and experiments. As an example, the flow in the vortex formation region has been shown to be absolutely unstable to small disturbances.

INTRODUCTION

Vortex streets are formed in the wakes of circular cylinders and other bluff, or unstreamlined, bodies over a wide range of Reynolds numbers from approximately 50 to 10^6 and even higher (Griffin 1982). The physics of vortex street formation has been the focal point for many past experimental studies, e.g. Gerrard (1966), Bearman (1965, 1967), Griffin (1981) and, most recently, Unal and Rockwell (1988), and Williamson and Roshko (1988), because of the importance of the near-wake flow to the eventual evolution of the overall near- and far-wake vortex patterns. Modern high-speed computers and direct numerical simulation techniques have allowed the vortex formation process to be studied numerically in great detail (Karniadakis and Triantafyllou 1989).

If a bluff structure is flexible and lightly damped, then resonant oscillations can be excited normal or parallel to the incident flow. For the more common cross flow oscillations the body and the wake have the same frequency near one of the characteristic frequencies of the structure (Griffin and Ramberg 1974; Bearman 1984). This coincidence of the vortex and vibration frequencies is commonly termed lock-on. Lock-on also occurs when the body oscillates in-line with the incident flow (Griffin and Ramberg 1976).

The phenomenon of lock-on occurs also when the incident mean flow has a sufficiently large periodic component superimposed upon it (Barbi et al. 1986; Armstrong et al. 1986, 1987). In this case the cylinder remains stationary, but the vortex lock-on modifies the character of the near-wake flow. There is a complete equivalence between this case and in-line oscillations of the cylinder (Griffin and Ramberg 1976) when the acoustic wavelength is long compared to the cylinder's diameter. The introduction of an appropriate sound field also can cause lock-on to occur (Blevins 1985). All of these external disturbances are potential means for active control of the bluff body near-wake flow (Rockwell 1987). Passive control of the shedding process can be accomplished, for example, by geometric alterations such as a wake splitter plate (Bearman 1965; Roshko 1954).

This paper further investigates the question of flow in the near-wakes of bluff bodies. Vortex shedding in oscillatory flow is employed here as a case study. This is an important case which has not been considered previously.

THE UNIVERSAL WAKE STROUHAL NUMBER

Roshko (1954, 1955) and Bearman (1967) originally showed that a characteristic non-dimensional group of parameters for scaling of the wakes of bluff bodies could be derived by applying relatively simple physical arguments. The most recent formulation (Griffin 1978, 1981) is a universal wake Strouhal number St^* for vortex shedding based upon measured parameters of the bluff body near-wake.

If one considers two shear layers a distance d' apart with the velocity just outside the layers equal to U_b, the mean velocity at separation, then a wake Strouhal number can be defined as

$$St^* = \frac{f_{so}d'}{U_b} = St\left(\frac{U}{U_b}\right)\left(\frac{d'}{d}\right). \tag{1}$$

The characteristic frequency f_{so} associated with the configuration is assumed to be proportional to the ratio U_b/d'. Here the usual Strouhal number of the cylinder is

$$St = \frac{f_{so}d}{U}, \tag{2}$$

where d is the cylinder diameter and U is the incident flow speed. When Bernoulli's equation is applied to the flow just outside the boundary layer at separation, the base pressure coefficient is

$$C_{pb} = \frac{2\,(p_b - p_\infty)}{\rho U^2} = 1 - \left(\frac{U_b}{U}\right)^2. \tag{3}$$

If the base pressure parameter $K = U_b/U$ is introduced, then

$$K^2 = 1 - C_{pb} \tag{4}$$

and

$$St^* = \frac{St}{K}\left(\frac{d'}{d}\right). \tag{4a}$$

A wake Reynolds number Re^* is defined in a corresponding way as

$$Re^* = \frac{U_b d'}{\nu} = ReK\left(\frac{d'}{d}\right), \tag{5}$$

where $Re = Ud/\nu$ is the usual free-stream Reynolds number.

Bearman (1967), in one early formulation of a universal Strouhal number, showed that St^* was dependent only upon the pressure drag coefficient C_D, the base pressure coefficient C_{pb} $(or K)$, and the Strouhal number St. The correlation between drag, base pressure and shedding frequency can be represented by the linear relation

$$St\ C_D = a + bK, a = -0.48 \pm 0.03, b = +0.50 \pm 0.02, \tag{6}$$

which was found to be optimal by Griffin (1981). Further discussion of this approach is given in Griffin (1989).

VORTEX LOCK-ON IN OSCILLATORY FLOW: A CASE STUDY

The recent experiments of Armstrong et al. (1986, 1987) and Barbi et al. (1986) were conducted to study the problem of vortex lock-on for a cylinder in a flow consisting of a steady flow with a periodic component superimposed upon it. The earlier experiments of Hatfield and Morkovin (1973) had attempted to study the same problem, but were inconclusive because the flow perturbation amplitude and frequency were too low to cause lock-on. The results obtained by Barbi et al. and Armstrong et al. show some remarkable similarities with the earlier experiments of Griffin and Ramberg (1976), which were conducted to examine vortex lock-on for a cylinder oscillating in-line with an incident uniform flow. These results are compared here on a common basis.

The vortex lock-on regime measured by Barbi et al. is compared with those of Griffin and Ramberg in Fig. 1. The vertical axis represents two different measures of the perturbation amplitude. For the experiments of Griffin and Ramberg the peak-to-peak amplitude of cylinder displacement is given by 2a/d. And for the experiments of Barbi et al. the normalized "peak-to-peak" incident velocity perturbation is given by $2\Delta U/\omega d$. The horizontal axis is the ratio of the vibration frequency f and the Strouhal

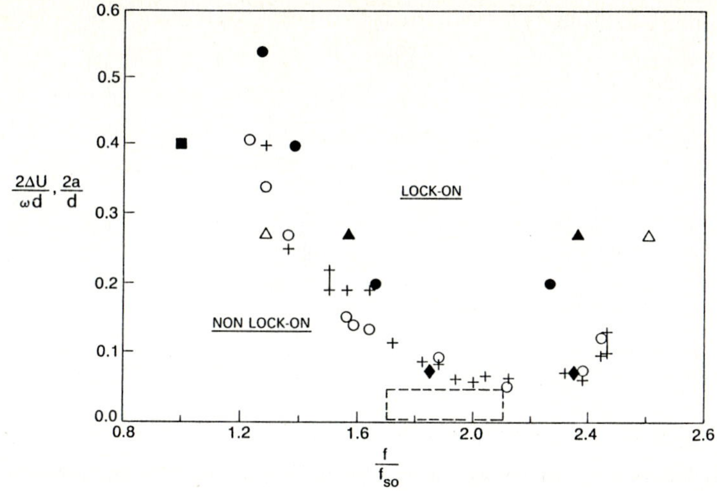

Figure 1 Limits of the lock-on regime as a function of amplitude and frequency for in-line oscillations and flow perturbations. In-line oscillations: +, ○, $Re = 190$, Griffin and Ramberg (1976); ▲, $Re = 80$ △, $Re = 4000$, Tanida et al. (1973); ●, $Re = 100$, Tatsuno (1972). Flow perturbations: ◆, $Re = 3000$; ■, $Re = 40000$, Barbi et al. (1986). For region enclosed by the dashed lines, see Fig. 4.

frequency f_{so} of a stationary cylinder. The two types of the external disturbance are identical in this case. Also shown are the cylinder vibration results of Tanida, Okajima and Watanabe (1973) and of Tatsuno (1972), reproduced from the paper by Griffin and Ramberg. The dashed lines enclose the results of Armstrong et al. (1986, 1987) which are shown on an expanded scale in Fig. 2. Vortex lock-on and crossflow oscillations usually occur at the shedding frequency f_{so} over a cycle for pairs of vortices. For in-line oscillations and flow perturbations the lock-on occurs near twice the Strouhal frequency, $f = 2f_{so}$, since the fluctuating drag force is in the flow direction and an individual vortex is shed over each cycle.

There is generally good agreement between the bounds of the lock-on regime for the two different types of external disturbance or flow control, though there is some scatter at the highest amplitudes. This is most likely due to Reynolds number effects, as noted by Barbi et al. The latter experiments were conducted at Re between 3,000 and 40,000, whereas the results of Tanida et al., Tatsuno, and Griffin and Ramberg were conducted at Re between 80 and 4,000. The overall differences are minimal.

The results obtained by Armstrong et al. are plotted in Fig. 2, where the vertical and horizontal axes have been scaled in the same way as in the previous figure. The original results of Armstrong et al. had been plotted in terms of the rms velocity u' and

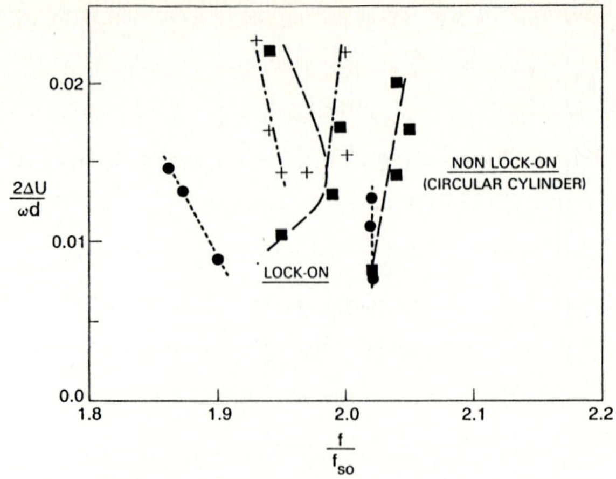

Figure 2 Limits of the lock-on regime as a function of amplitude and frequency for flow perturbations; data from Armstrong et al. (1986). ●, circular cylinder, ■, flat plate, +, D-section cylinder; $Re = 15000 - 35000$.

the reduced velocity $U/f_{so}d$. Three body shapes were investigated–a circular cylinder, a D-section cylinder, and a vertical flat plate. The regimes of lock-on versus non lock-on noted in Fig. 2 pertain only to the circular cylinder. It is clear that the circular cylinder, with free separation points, has a lock-on range of about twice the breadth of the two bodies with fixed separation points. This basic difference in the lock-on behavior for the two types of bluff bodies previously was discussed by Bearman (1984) for the case of body oscillations only. This again points to the basic similitude of the vortex resonance over still a wider variety of external disturbances. Further discussion and additional results are given by Griffin (1989).

BLUFF BODY NEAR-WAKE GEOMETRY

The base pressure coefficient C_{pb} is influenced by the flow perturbations in much the same manner as in the case of cylinder vibrations. For the stationary cylinder the base pressure coefficient is near $C_{pb} = -1.44$; this value, though somewhat low for a circular cylinder, is in reasonable agreement with the results of West and Apelt (1982) for a comparable wind tunnel blockage ratio of 9 percent. When the flow perturbation was largest, the base pressure was decreased to $C_{pb} = -1.85$ at the point of maximum resonance, a reduced velocity of $U/f_{so}d = 2.5$ (half the Strouhal value). The measured vortex formation region length l_f was reduced by this level of perturbation to 0.9d from 1.2d, the value measured for the unperturbed flow (Armstrong et al., 1987).

Lesser decreases in C_{pb} were measured for smaller levels of the flow perturbation, but there was an overall dependence upon reduced velocity $U/f_{so}d$. The mean drag coefficient C_D increased from 1.28 to 1.52 for the perturbed flow as compared to the unperturbed flow. The base pressures of the flat plate and D-section bodies also were decreased by the introduction of the incident flow perturbation. But the decrease was only half of that measured for the circular cylinder at the same perturbation amplitude, which further shows the effect of free versus fixed separation points on the vortex resonance. These experiments were conducted at Reynolds numbers between 15,000 and 35,000, and the base pressure coefficients of all three stationary bodies in the unperturbed flow were effectively constant over this range.

The wake width d' at the end of the vortex formation region of a bluff body depends directly on the base pressure coefficient or the velocity ratio K (Griffin, 1981). For a cylinder vibrating normal to the incident flow, the variation of both the measured base pressure and wake width with the frequency ratio f/f_{so} over the lock-on regime shows the same resonant behavior. Both $-C_{pb}$ and d' increase to a maximum value and then gradually decrease as the upper limit of the lock-on range of frequencies is reached. Only the base pressure variation was measured by Armstong et al., but the wake width can be estimated as a function of $U/f_{so}d$ using the wake similitude relationships summarized earlier in the paper and measured values of St and C_{pb} (or K). For the range of Reynolds numbers corresponding to the experiments of Armstrong et al., the wake Strouhal number St^* is essentially constant at a value of 0.16, as shown in Fig. 5 of Griffin (1981), so that

$$d'/d = (St^*/St)K$$

from Eq. (4a). The wake widths for the circular cylinder, obtained using $St^* = 0.16$ and measured values of St and K, are plotted as a function of $U/f_{so}d$ in Fig. 3. The predicted dependence of d' on the reduced velocity in Fig. 3 is essentially the same as the measured wake width dependence for the vibrating cylinder given in Fig. 3 of Griffin (1981).

The wake widths for the three cylinders employed by Armstrong et al. and the circular cylinder employed by Barbi et al. are plotted in Fig. 4. The other results for a host of cylinder shapes and vibration conditions have been reproduced from Griffin (1981). There is generally good agreement between the new perturbed and steady flow results and the earlier data, with two exceptions being the circular cylinder results of Armstong et al. which are displaced to the right of the overall trend of the results. This

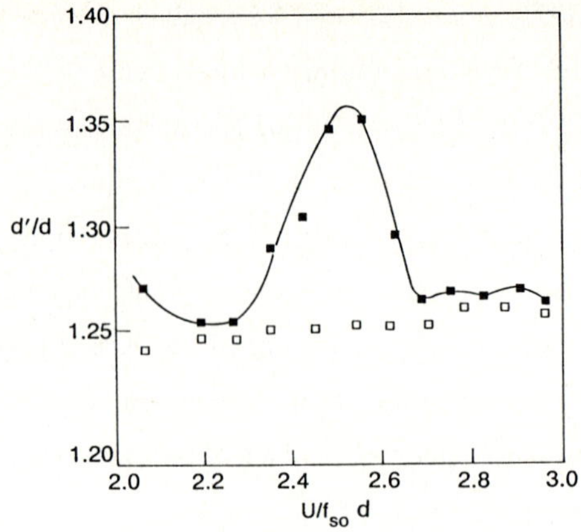

Figure 3 Wake width d'/d, predicted by equation (4a), as a function of the reduced velocity $U/f_{so}d$; from Armstrong et al. (1987). □, steady mean flow; ■, perturbed flow, $2\Delta U/\omega d = 0.014$.

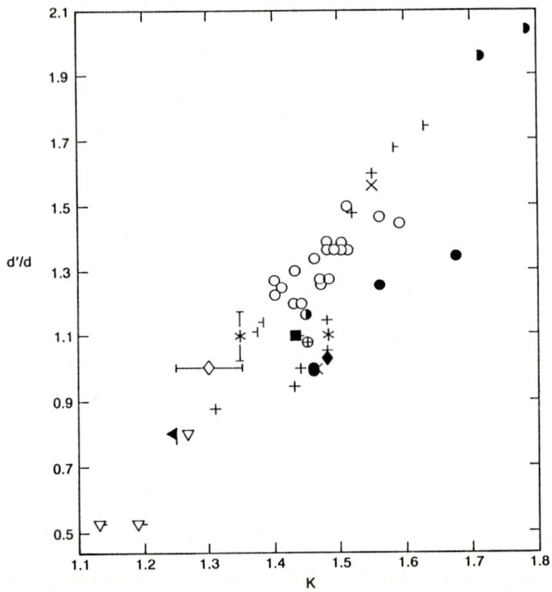

Figure 4 Wake width d'/d as a function of the base pressure parameter or velocity ratio K. Legend as in Griffin (1981), except for ◐, circular cylinder, Barbi et al. (1986); ●, circular cylinder, ⊢, flat plate, ◗, D-section cylinder, Armstrong et al. (1986).

departure from the overall trend is due to the much reduced base pressure measured on the circular cylinder in those experiments, as previously mentioned. In the experiments of Armstong et al. the circular cylinder and flat plate base pressures were virtually the same under otherwise unvarying conditions of blockage, incident flow, axial uniformity, etc. One might expect the base pressure coefficient for a circular cylinder at the Reynolds numbers studied to be closer to the values of $C_{pb} = -1$ to -1.1 measured by Barbi et al. and others at the same Reynolds numbers.

The downstream vortex spacing or wavelength is a useful diagnostic for the state of the spatial structure and development of the flow, which cannot be ascertained from dynamic system temporal characteristics such as phase plane analysis. Measurements of the spacing for a variety of in-line and cross flow oscillations, and also for stationary cylinders were reported by Griffin and Ramberg (1976). These can be compared to the direct numerical simulations of Karniadakis and Triantafyllou (1989), wherein the vortex spacing was employed to assess the spatial state of the flow as compared to phase plane diagrams of the streamwise and cross stream components of the wake velocity to assess the temporal state.

Two examples from Karniadakis and Triantafyllou are shown in Fig. 5. The upper instantaneous streamline pattern corresponds to the unforced wake at $Re = 100$ while the lower pattern corresponds to a forced wake (cross flow oscillations) at $a/d = 0.10$ and $f/f_{so} = 0.75$. For the unforced wake $\lambda = 5d$ and for the forced wake $\lambda = 7d$, an increase of forty percent. Comparable measurements were made at $Re = 190$ by Griffin and Ramberg (1976). The results are summarized in Table 1. The cylinder oscillations were in-line with the flow at near twice the Strouhal frequency (as in Fig. 1) and for the cases shown a single vortex was shed during each oscillation cycle. Thus the basic forced wake pattern shared many of the basic features of the wake with cross flow oscillations as shown by the flow visualization photograph in Fig. 6. The measured changes in the forced wake vortex spacing correspond directly with those from the direct numerical simulation; for $f < 2f_{so}$ the wavelength is increased while for $f > 2f_{so}$ the wavelength is decreased as shown in Table 1. Extrapolating the measured results in the table to the case shown in Fig. 5 using a least-squares straight line given by Griffin and Ramberg (1976), the vortex spacing is $\lambda = 6.2d$. This compares reasonably well with the computed results of $\lambda = 7d$, though it is somewhat less. The measured vortex spacing for the stationary cylinder at $Re = 190$ ($\lambda = 4.9d$) is virtually identical to the computed value at $Re = 100(\lambda = 5d)$. The vortex spacings computed by Karniadakis and Triantafyllou are compared with measured spacings in the range of Reynolds numbers from 100 to 1000 in Fig. 7. The computations fit nicely with the overall trend of the measured data, which show only a slight dependence on Reynolds number.

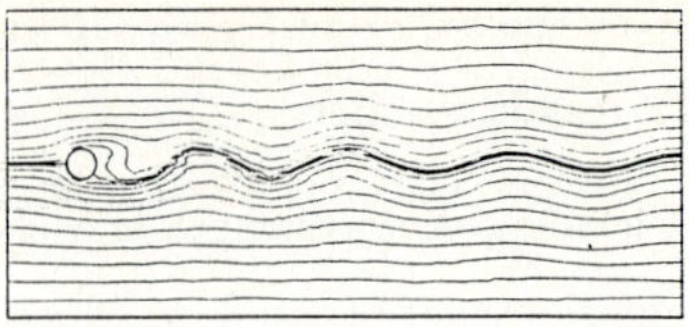

(a)

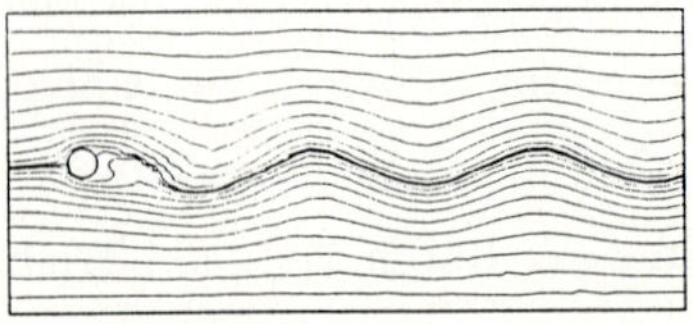

(b)

Figure 5 (a) Instantaneous streamlines at $Re = 100$ for natural shedding. The wavelength of the vortex street is approximately five diameters. (b) Instantaneous streamlines at $Re = 100$ and excitation frequency, $f_e = 0.75 f_s$ (a lock-in state). The wavelength of the vortex street is approximately seven diameters; from Karniadakis and Triantafyllou (1989).

Table 1

Longitudinal vortex spacing or wavelength in the near-wake of a circular cylinder vibrating in-line with an incident uniform flow; adapted from Griffin and Ramberg (1976).

Vibration frequency, $f(Hz)$	Frequency ratio, f/f_{so}	Vortex spacing, λ/d	Relative change, $\Delta\lambda/\lambda$	Vortex convection speed, $\frac{1}{2}f\lambda/U$
		Reynolds number = 190		
69.2	1.88	5.2	+0.07	0.94
73.6	2.00	4.9	0	0.93
75.6	2.06	4.7	-0.05	0.91
78.9	2.14	4.7	-0.04	0.96
80.4	2.18	4.4	-0.09	0.92
				Average = 0.93

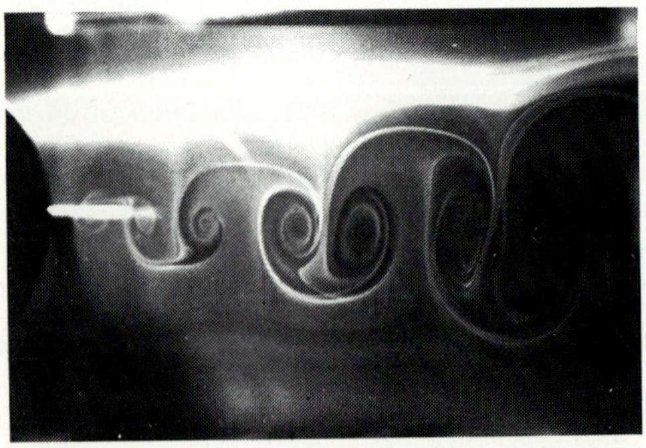

Figure 6 Flow visualization of the near-wake of a circular cylinder oscillating in line with an incident flow at $Re = 190$, $f/f_{so} = 1.76$. The wavelength of the vortex street is approximately 5.5 diameters.

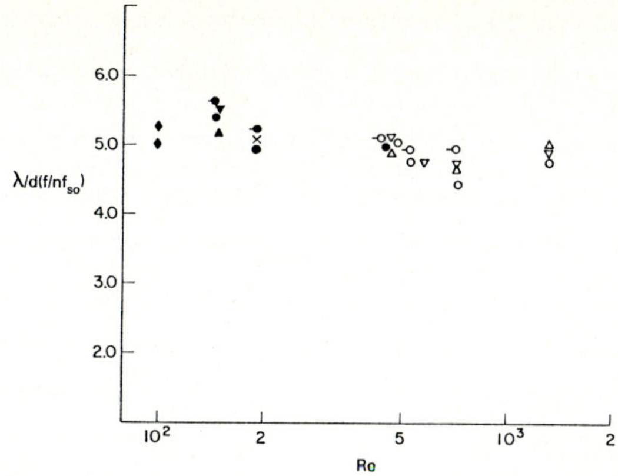

Figure 7 Longitudinal vortex spacing $\lambda/d(f/nf_{so})$ as a function of Reynolds number Re. All of the measurements were made in the wakes of vibrating cylinders and cables. ○, $f/nf_{so} = 1$; △, $f/nf_{so} = 0.9$; ×, $f/nf_{so} = 1.07$; ▽, $f/nf_{so} = 1.1$. ♦, $f/nf_{so} = 0.75, 1.00$, computations (Karnidakis and Triantafyllou 1989). Open symbols refer to vibrating cables, closed symbols to stationary and vibrating cylinders. Data points at $Re = 190$ correspond to in-line vibration ($n = 2$); all other data correspond to cross-flow vibration ($n = 1$); from Griffin and Ramberg (1976).

Several measurements of the vortex phase or convection speed, i.e. the speed at which the vortex cores move downstream, are given in Table 1. Though there is some scatter, the data generally are grouped around the average value of $U_\phi = 0.93U$. This corroborates the assertion of Karniadakis and Triantafyllou that both forced and non-forced periodic vortex wakes have the same basic non-dispersive character.

DISCUSSION OF THE RESULTS

The similarity relations for bluff body near-wakes which have been established over the years have been extended here and in Griffin (1989) to yet another case of both fundamental and practical importance - - a bluff body in a perturbed flow consisting of a mean flow with an oscillatory component superimposed on it. And this newest case has been shown to be fundamentally identical, under appropriate conditions, to the lock-on or vortex resonance of a cylinder oscillating in line with an incident uniform flow. The vortex lock-on in oscillatory flow exhibits a particularly strong form of resonance, with a relative perturbation amplitude of $2\Delta U/\omega D = 0.014$ producing a reduction in base pressure from C_{pb} = -1.44 to -1.85, or 22 percent, for a circular cylinder (Armstrong, et al. 1986, 1987). For a circular cylinder oscillating in cross flow, a peak-to-peak vibration amplitude of $2a/d = 0.20$ to 0.30 is required to provide a comparable reduction in C_{pb} (Stansby, 1976). Conditionally-averaged wake velocity measurements by Armstong et al.

(1987) have indicated that the strength of the vortices was increased by 29 percent and the spacing was decreased by 25 percent for the perturbation levels of their experiments.

These are remarkable modifications of the near-wake flow for such a relatively small perturbation amplitude. Thus, seemingly small perturbations of the basic near-wake flow can produce large changes in vortex strength, base pressure, and drag on a bluff circular cylinder or other cross-section. Modification and control of the basic formation or instability mechanisms of the wake can provide a means for making substantial changes in the near-wake vortex pattern, and possibly even the middle- and far-wake patterns as well (Cimbala et al. 1988).

Karniadakis and Triantafyllou (1989) have characterized the state of the forced and unforced vortex wakes by means of a state diagram as sketched in Fig. 8. At a small, but finite, amplitude the transitions corresponding to the upper and lower limits of the lock-on are given by two bounding frequencies; within these limits only periodic lock-on states exist. Two quasi-periodic regions develop above and below the lock-on regime, together with chaotic states in narrow regions immediately adjacent to the lock-on boundaries. In the terminology of chaos and nonlinear dynamics, these quasi-periodic states are called "resonant horns".

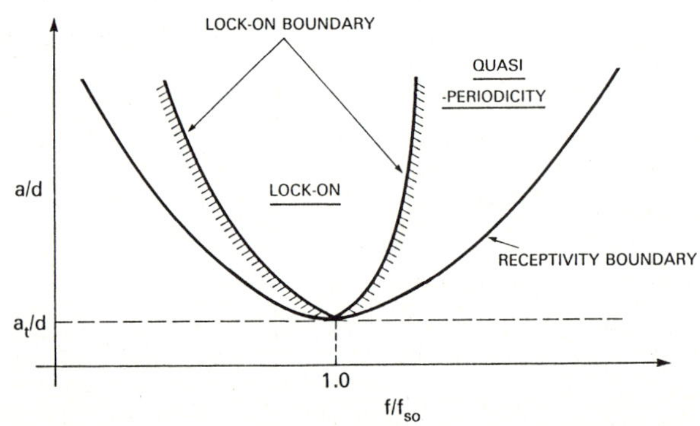

Figure 8 State selection (amplitude vs. frequency) diagram for laminar wakes. The plot should only be interpreted in a qualitative sense. Regions as the above are referred to in the literature as "resonant horns"; from Karniadakis and Triantafyllou (1989).

As the threshold amplitude a_t is approached, these finite regions shrink to a single frequency f_e. For cross flow oscillations of the cylinder $f_e = f_{so}$ while for in-line oscillations of the body and periodic perturbations of the mean flow $f_e = 2f_{so}$ as shown in Fig. 2. The small amplitude perturbations introduced by Armstrong et al. (1986, 1987) correspond qualitatively to those discussed by Karniadakis and Triantafyllou. The large amplitude cylinder oscillations and flow perturbations investigated by Griffin and Ramberg (1976) and Barbi et al. (1986) as shown in Fig. 1 introduce nonlinearities and complex changes in the near-wake flow field as shown by the flow visualization studies of Griffin and Ramberg (1974, 1976). At the largest amplitudes of oscillation a host of complex flow patterns were observed by the latter; and even more extensive kaleidoscope of complex vortex patterns over a wide range of frequencies and amplitudes was identified experimentally by Williamson and Roshko (1988).

The application of absolute-convective stability concepts has led to a promising new approach and a new theoretical framework for understanding the mechanics of vortex wake formation and development (Triantafyllou, et al. 1987; Monkewitz and Nguyen 1987; Chomaz et al. 1988; Rockwell 1987; Unal and Rockwell 1988; Karniadakis and Triantafyllou 1989). The most recent stability calculations based upon computed and measured mean velocities in the wakes of stationary circular cylinders show that the vortex formation region is absolutely unstable while the near-wake vortex street is convectively unstable.

Karniadakis and Triantafyllou (1989) conducted a linear stability analysis of the time-average flow in the near-wake which was derived from their direct numerical simulation of a stationary cylinder at $Re = 100$. The flow was assumed to be locally parallel and slowly varying in the downstream x-direction, both of which are reasonable assumptions. The average flow was found to be absolutely unstable for 2.5 diameters downstream from the cylinder. This corresponds directly to the vortex formation region. Further downstream the flow is convectively unstable. Thus the continuous formation of the vortex street is sustained by the near-wake absolute instability. Excellent agreement was found between the linear stability analysis and the direct numerical simulation in that the Strouhal number was $St = 0.179$ in both cases. This is slightly higher than most experiments (Roshko 1954; Williamson 1988), which may be due partly to finite computational grid size and domain extent, and to the three dimensional effects which invariably exist in cylinder wake experiments. Williamson (1988) clearly has shown the importance of the spanwise variation of the flow in the cylinder wake.

Further studies based upon these revolutionary approaches are likely to lead to new fundamental understanding of the near-wake vortex dynamics, which until now have been studied mostly by using traditional experimental techniques, (Gerrard, 1966; Bearman, 1967; Griffin and Ramberg, 1974, 1976).

SUMMARY AND CONCLUDING REMARKS

Some questions of bluff body near-wake similitude have been investigated for the case of vortex shedding from a circular cylinder, a flat plate, and a D-section cylinder in an perturbed flow consisting of a mean flow with an oscillatory component superimposed upon it. The cylinder base pressure measured in a perturbed flow is decreased markedly ($C_{pb} = -1.85$) from the measurement in the steady incident flow ($C_{pb} = -1.44$). There is a corresponding increase in the mean drag coefficient, from $C_D = 1.28$ to 1.52 when the flow perturbation amplitude is $2\Delta U/\omega d = 0.014$ for the circular cylinder. The predicted wake widths at the end of the vortex formation region for three cylinders (D-shaped, circular, flat plate) are in agreement with previous measurements when plotted against the separation velocity ratio $K = U_b/U$. The variation of the predicted wake width d' over the lock-on range in the perturbed or forced wake is similar to previous measurements in the wake of a cylinder which undergoes cross flow oscillations.

Computations of the downstream vortex spacing or wavelength λ are in excellent agreement with measurements in both forced and unforced, or natural, wakes for Reynolds numbers between 100 and 1000. The essentially constant vortex phase or convection speed demonstrates the basic non-dispersive character of forced and unforced vortex near-wakes.

The vortex lock-on in the perturbed flow exhibits a particularly strong form of resonance, with a relative perturbation amplitude of $2\Delta U/\omega d = 0.014$ producing a reduction in base pressure from C_{pb} = -1.44 to -1.85. A peak-to-peak amplitude of $2a/d = 0.20$ to 0.30 is required to effect the same reduction for a cylinder oscillating in cross flow. Conditionally-averaged wake measurements by Armstrong et al. (1987) have indicated that the vortex strength was increased by 29 percent for the above level of flow perturbation.

These results suggest that modification and control of the basic instability or formation mechanisms of the wake may provide a means for making substantial changes in the near-wake vortex pattern, and possibly the middle- and far-wake flow patterns as well which exist downstream from a bluff body. From a hydrodynamic stability standpoint, the vortex formation is absolutely unstable while the near-wake vortex street is convectively unstable.

ACKNOWLEDGEMENT

This work was conducted as part of a research program in fluid dynamics and bluff body flows supported by the Naval Research Laboratory.

REFERENCES

Armstrong, B.J., Barnes, F.H. and Grant, I., "The effect of a perturbation on the flow over a cylinder," Phys. Fluids, Vol. 29, 2095-2102, 1986.

Armstrong, B.J., Barnes, F.H. and Grant, I., "A comparison of the structure of the wake behind a circular cylinder in a steady flow with that in a perturbed flow," Phys. Fluids, Vol. 30, 19-26, 1987.

Barbi, C., Favier, D.P., Maresca, C.A. and Telionis, D.P., "Vortex shedding and lock-on of a circular cylinder in oscillatory flow," J. Fluid Mech., Vol. 170, 527-544, 1986.

Bearman, P.W., "Investigation of the flow behind a two dimensional model with a blunt trailing edge and fitted with splitter plates," J. Fluid Mech., Vol. 21, 241-255, 1965.

Bearman, P.W., "On vortex street wakes," J. Fluid Mech., Vol. 28, 625-641, 1967.

Bearman, P.W., "Vortex Shedding from Oscillating Bluff Bodies," Ann. Rev. Fluid Mech., Vol. 16, 195-222, 1984.

Blevins, R.D., "The effect of sound on vortex shedding from cylinders," J. Fluid Mech., Vol. 161, 217-237, 1985.

Chomaz, J.M., Huerre, P. and Redekopp, L., "Bifurcations to Local and Global Modes in Spatially Developing Flows," Phys. Rev. Let., Vol. 60, 25-28, 1988.

Cimbala, J.M., Nagib, H.M. and Roshko, A. "Large structure in the far wakes of two-dimensional bluff bodies," J. Fluid Mech., Vol. 190, 265-298, 1988.

Gerrard, J.H., "The mechanics of the formation region of vortices behind bluff bodies," J. Fluid Mech., Vol. 25, 401-413, 1966.

Griffin, O.M., "A universal Strouhal number for the'locking-on' of vortex shedding to the vibrations of bluff cylinders," J. Fluid Mech., Vol. 85, 591-606, 1978.

Griffin, O.M., "Universal Similarity in the Wakes of Stationary and Vibrating Bluff Structures", Trans. ASME, J. Fluids, Engrg. Vol. 103, 52-58, 1981.

Griffin, O.M., "Vortex Streets and Patterns", Mechanical Engineering, Vol. 104, No. 3, 56-61, 1982.

Griffin, O.M., "Flow Similitude and Vortex Lock-On in Bluff Body Near-Wakes," Phys. Fluids A, Vol. 32, 697-703, April 1989.

Griffin, O.M. and Ramberg, S.E., "The vortex street wakes of vibrating cylinders," J. Fluid Mech., Vol. 66, 553-576, 1974.

Griffin, O.M. and Ramberg, S.E., " Vortex shedding from a cylinder vibrating in line with an incident uniform flow," J. Fluid Mech., Vol 75, 257-271, 1976.

Hatfield, H.M. and Morkovin, M.V., "Effect of an Oscillating Free Stream on the Unsteady Pressure on a Circular Cylinder," Trans. ASME, J. Fluids Engrg., Vol. 95, 249-254, 1973.

Karniadakis, G.E. and Triantafyllou, G.S., "Frequency selection and asymptotic states in laminar wakes," J. Fluid Mech., Vol. 199, 441-469, 1989.

Koopmann, G.H., "The vortex wakes of vibrating cylinders at low Reynolds numbers," J. Fluid Mech., Vol. 28, 501-512, 1967.

Monkewitz, P.A. and Nguyen, L., "Absolute Instability in the Near-Wakes of Two-Dimensional Bluff Bodies," J. Fluids Struct., Vol. 1, 165-184, 1987.

Rockwell, D., Unpublished manuscript: A View of Hypotheses and Issues on Bluff Body Near-Wake Instabilities, 1987.

Roshko, A., "On the drag and shedding frequency of two-dimensional bluff bodies," Nat. Adv. Comm. for Aero., Washington, DC, Technical Note 3169, 1954.

Roshko, A., "On the Wake and Drag of Bluff Bodies," J. Aero. Sci, Vol. 22, 124-132, 1955.

Tanida, Y., Okajima, A. and Watanabe, Y., "Stability of a circular cylinder oscillating in uniform flow or in a wake", J. Fluid Mech., Vol. 61, 769-784, 1973.

Tatsuno, M., "Vortex Streets behind a Circular Cylinder Oscillating in the Direction of Flow", Bull. Res. Inst. Appl. Mech. Kyushu University, Vol. 36, 25-37, 1972 (in Japanese).

Triantafyllou, G., Triantafyllou, M. and Chrissostimidis, C., "Stability Analysis to Predict Vortex Street Characteristics and Forces on Circular Cylinders," Trans. ASME, J. Offsh. Mech. and Arctic Engrg., Vol. 109, 148-154, 1987.

Triantafyllou, G., Kupfer, K. and Bers, A., "Absolute Instabilities and Self-Sustained Oscillations in the Wakes of Circular Cylinders," Phys. Rev. Let., Vol. 59, 1914-1917, 1987.

Unal, M.F. and Rockwell, D., "On vortex formation from a cylinder. Part 1. The initial instability," J. Fluid Mech., Vol. 190, 491-512, 1988.

West, G.S. and Apelt, C.J., "The effects of tunnel blockage and aspect ratio on the mean flow past a circular cylinder with Reynolds numbers between 10^4 and 10^5", J. Fluid Mech., Vol. 114, 361-377, 1982.

Williamson, C.H.K. and Roshko, A., "Vortex Formation in the Wake of an Oscillating Cylinder," J. Fluids and Struct., Vol. 2, 355-381, 1988.

Williamson, C.H.K., "Defining a universal and continuous Strouhal-Reynolds number relationship for the laminar vortex shedding of a circular cylinder," Phys. Fluids, Vol. 31, 2742-2744, 1989.

WAKE STUDIES ON YAWED, STRANDED CABLES

Jose V. Nebres, Stephen M. Batill, Robert C. Nelson
Department of Aerospace and Mechanical Engineering
University of Notre Dame, Notre Dame, Indiana 46556 USA

ABSTRACT

The study of the near wake on a cable yawed with respect to a uniform flow was performed using five rigid cable models and a cylinder in a low speed wind tunnel. The models were tested at a Reynolds number of 6,000 based on diameter. Smoke-wire flow visualization was used to qualitatively and quantitatively evaluate the near wake flows. Hot wire measurements of the wake frequency spectrum were obtained. Results show that cable stranding affects shedding characteristics and near wake structure.

INTRODUCTION

Fluid dynamic forces on cables are observed in towed marine and airborne systems, mooring of offshore structures, and transmission lines subjected to strong winds, Griffin [1979]. The fluid forces acting on stranded cables can basically be characterized by the presence of drag and lift forces with both steady and unsteady components. The steady drag is due to fluid pressure and shear forces acting on the cable surface in the same direction as the flow. It is usually resolved into components which are tangent and normal to the cable. The steady lift force directed normal to the cable and the flow is primarily caused by the asymmetric pressure distribution along the cable surface. Due to the stranding, the cable acts as a "thick airfoil" and can produce significant lift forces. The unsteady components of the drag and lift forces which produce longitudinal and transverse vibrations, respectively, are induced by the shedding of vortices in the wake of the cable, Van Atta [1968].

A cable or wire rope is constructed by twisting wires together in a regular pattern. The basic design of a cable can be indicated by a two-number designation such as 7x7 or 3x19. The first number refers to the number of strands and the second number refers to the number of wires per strand. A "serrated" cable has a pair of wires within a given strand twisted together and results in a very rough external surface. The strands of a cable are twisted or laid in a direction about the cable axis which is specified by a characteristic "lay". A cable is called right lay if, as viewed along the cable axis, the strands pass from left to right as they move away from the viewer. A "left" lay cable has the opposite orientation of the strands. Another characteristic is the direction of the lay of the wires within each strand. A cable with the wires twisted opposite that of the lay of the strands is termed "regular" lay while that with the wires twisted in the same direction as the strands is called "lang" lay.

The orientation of a yawed cable with respect to the freestream can be defined by the cable angle, β, which is the angle between the longitudinal cable axis and the freestream velocity vector. Figure 1 shows a plan view and side view of a yawed cable. The effective cross-section of a stranded cable results from the intersection of a cutting plane with the cable. The cutting plane is normal to a plane containing the cable and the free-stream velocity vector and parallel to the free-stream. In the side view cross-section, the wires are not shown in detail and for simplicity, the strand is assumed to be a single flexible wire. The shape of the cross-section varies with position of the cutting plane along the length of the cable so that the effective cable geometry is complex, three-dimensional and periodic in the spanwise direction.

A cable yawed in a flow experiences both drag and lift. Horton, Ferrer, Watson and Charvoz [1987] conducted tests in a towing tank at the David Taylor Naval Ship Research and Development Center, where the lift and normal drag forces were measured for different cable angles. The normal drag and lift forces for five different cables were quantified as a function of cable angle given the fluid density, fluid velocity, cable diameter and cable length. A strong dependence on cable geometry and yaw angle was shown and lift coefficients as large as 0.3 (based on "planform" area) were achieved. The production of steady lift on either vibrating or rigid cables is not well understood. A number of theories have been proposed to explain the generation of steady lift, Simpson [1979], but none have been substantiated nor can the lift force be predicted.

The vortex shedding phenomenon has been studied with both yawed and unyawed cylinders. There are a number of distinct regions of wake formation which depend upon Reynolds number. A summary, describing the vortex wake formation at all Reynolds numbers can be found in Blevins [1977]. The work documented in this paper is in the range where the wake is characterized by a turbulent vortex street. The non-dimensional parameter used to quantify the periodic shedding is the Strouhal number, $S = f_V D / V$, where f_V is the vortex shedding frequency in Hz, D the cable diameter and V the fluid freestream velocity. The Strouhal number for circular cylinders normal to the flow in a Reynolds number range $300 < Re < 3x10^5$ is in the range from 0.18 to 0.22. There has been limited research on the shedding characteristics for the yawed rigid cable. However, the yawed rigid circular cylinder has been studied by Ramberg [1978] who

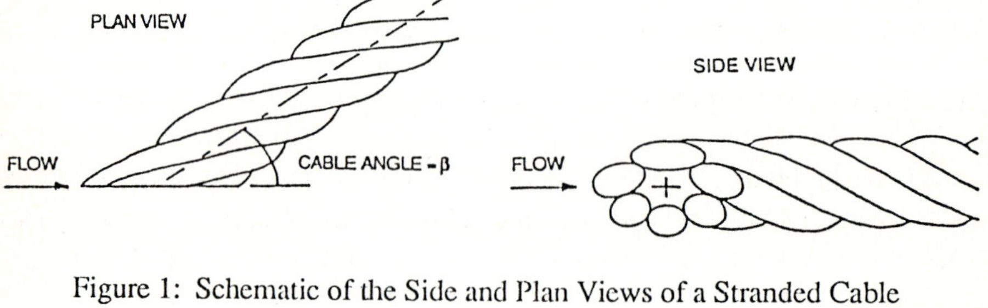

Figure 1: Schematic of the Side and Plan Views of a Stranded Cable

focused on the conditions under which the Strouhal number, when based on the normal component of the velocity, would be independent of yaw angle. It was found that it was necessary to place inclined end plates on the yawed cylinders to eliminate the dependence on the tangential flow component. The angle defining the orientation of the end plate was measured and details can be found in Ramberg [1978].

Cable stranding has been shown to have an effect on vortex shedding. Strouhal numbers measured by Horton, et. al. [1987] varied from a low S=0.18 for the "smooth" 1x19 cable to a high value of S=0.22 for the "rough" 4x7 cable. Yet the conclusion was that the effect did not seem very significant since the average Strouhal numbers were within 10 per cent of S=0.20 which is the circular cylinder value. Votaw and Griffin [1971] measured the Strouhal numbers of unyawed cables and observed slightly elevated values for cables with three and four strands as compared to values for cables with five and six strands. The differences also were not deemed very significant since the values were still within the range $0.18 < S < 0.22$.

Ramberg [1978] conducted flow visualization using yawed cylinders and has shown that the vortex filaments are oriented at a fixed, non-zero angle relative to the cylinder axis for certain cable angles. Flow visualization photographs showed striations in the wake caused by the shedding of vortex filaments. For a configuration approximating an infinite yawed cylinder in a flow, the angle which the filaments made with the flow which is called the shedding angle, α, was greater than the cable angle.

This brief review has summarized previous work on the flow around yawed stranded cables which involve the measurement and prediction of the drag force, lift force and vortex shedding characteristics. This current work was aimed at providing flow field information through flow visualization and near wake hot-wire measurement of the vortex shedding characteristics of five stranded cables yawed in air flow. The models used were the 1x19, 7x7, 3x19, 4x7 serrated, 6x25 lang lay cables, and a smooth circular cylinder. The cables were modeled to create rigid and straight replicas of actual cable geometries. The models were yawed at four different cable angles, 90°, 60°, 40°, and 20°. The wind tunnel tests were performed at velocities corresponding to a Reynolds number near 6,000 based on cable actual diameters. The flow Reynolds numbers were chosen to match those encountered in typical stranded cable applications.

Qualitative comparison of flow fields around the cables and the cylinder were accomplished through flow visualization. Still pictures were taken using the smoke wire technique. Quantitative analyses were also done on the flow visualization data. Plan view photographs were analyzed to obtain the shedding angles. The vortex shedding and spanwise wake characteristics were studied using hot-wire anemometry.

EXPERIMENTAL APPARATUS AND PROCEDURES

All tests were conducted at the University of Notre Dame Aerospace Laboratory wind tunnels. The cable models were assembled from hollow epoxy castings which were supported on a steel shaft to provide strength and stiffness. The fabrication technique has

MODEL	SURFACE CHARACTER	DIAMETER (INCHES)	CABLE ANGLE, β 90°	60°	40°	20°
			FIRST BENDING MODE (HZ)			
CYLINDER	Smooth	0.625	-	-	-	-
1X19 CABLE	Smooth	0.632	26.5	19.9	11.0	3.1
7X7 CABLE	Smooth	0.640	26.4	19.8	10.9	3.1
3X19 CABLE	Irregular	0.615	16.2	12.1	6.7	1.9
4X7 CABLE (SERRATED)	Irregular	0.700	28.7	21.5	11.9	3.4
6X25 CABLE (LANG LAY)	Smooth	0.625	25.6	19.2	10.6	3.0

Physical Characteristics of Cable Models.

proven to be efficient and the models quite accurately represented the complex, stranded cable surface. The wind tunnel cross section was 2 ft square and the models spanned the section, no end plates were used. The model lengths were therefore 2 ft/ sin(β). The properties of the cable models including estimates of the first bending natural frequencies for comparison with shedding frequencies are presented in the Table above. The Table also includes a qualitative assessment of the cable surface character or irregularity.

The tunnel was an indraft type which draws air from the laboratory area and exhausts to the atmosphere The inlet of the tunnel consists of 12 anti-turbulence screens and a contraction ratio of 24:1 based on area. This provided a near uniform freestream velocity profile in the test section with a turbulence intensity of less than 0.1%. Flow visualization was accomplished using the smoke wire technique, Batill and Mueller [1980]. The smoke wire technique produced very fine streaklines that afforded good flow field detail. The main objective of the flow visualization was to develop qualitative information. Nevertheless, some quantitative data was extracted using a digitizing table and a photographic enlarger.

Photographic data is presented using plan and side views of the flow field. For the plan view, a sheet of smoke from the smoke wire was introduced into the tunnel and the model was positioned in this plane. The photographs were taken with the camera directed normal to the smoke sheet. For the side views, the smoke was introduced in the "cutting plane" described earlier and the camera was directed normal to this plane.

A hot wire system was used to collect the spectrum and cross-correlation data. Two hot wire probes were used: one which was fixed and one which could be traversed with 3 degrees of freedom. Two DANTEC Type 55P11hot wire probes were used. The two probes were operated by a TSI Inc. Intelligent Flow Analyzer 100 anemometry system. The two signals from the IFA 100 were processed by a Scientific-Atlanta Spectral Dynamics 380 signal analyzer from which frequency spectrum information was determined.

DISCUSSION OF RESULTS

Flow visualization and hot wire experimental data are presented below. There was good agreement between the results from the two methods and the combined results provided a better foundation for the discussions. The following highlights the results of a more comprehensive study by Batill, Nelson and Nebres, [1988]. Additional flow visualization and hot-wire results are included in that reference.

Flow Visualization

The flow visualization experiments were conducted in order to qualitatively evaluate the difference in wake structure between each of the cable models. These results were acquired using the smoke-wire tracer technique at a Reynolds number of approximately 6000. Figures 2 and 3 illustrate the two basic types of visualization data. Figure 2 shows a plan view of the 6x25 cable at β = 40°. The smoke wire was positioned as closely as possible to the centerline of the model. As can be seen from the photographs, all of the smoke filaments appear to pass to the "front" of the model. The periodic shedding of vorticity into the wake is clearly visible. The location of the spanwise vortices is indicated by the regions of dense smoke in the plan view photograph. These regions are reasonably uniform across the wind tunnel.

Figure 3 presents a side view for the cylinder for a cable angle of 90°. The four white "triangles" in the photographs are used as alignment and spatial reference marks to identify the location where the flow impinges on the models. As the model is yawed, Figures 4 and 5, the near wake region is masked by the light reflected from the model.

The flow field associated with the circular cylinder was consistent with all previous studies on this configuration. The flow about the cylinder is laminar and transition takes place in the wake. Vorticity shed from the upper and lower separation points forms into discrete vortices of opposing sense in the wake. The transition in the

Figure 2: Plan View of a 6x25 Lang Lay Cable at $\beta = 40°$

wake is evident in Figures 3-5 and discrete vortices are present over the wide range of yaw angles. The vortex dominated wake is evident although the spacing appears to be modified for the yawed cylinder. For the circular cylinder there was no noticeable variation in the form or structure of the wake region for this range of cable angles.

The plan view photographs indicate a similar flow structure for each of the models but with some differences. The 1x19 and the 7x7 cables show a similar type of transition occurring in the wake shear layer with predominantly two dimensional behavior over the center section of the model. The transition appears to move closer to the cable for the 7x7 model. The 3x19 cable is very coarse and does demonstrate a

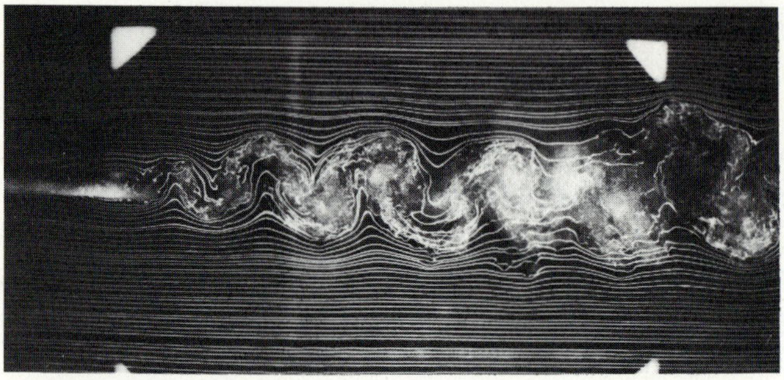

Figure 3: Side View of a Cylinder at $\beta = 90°$

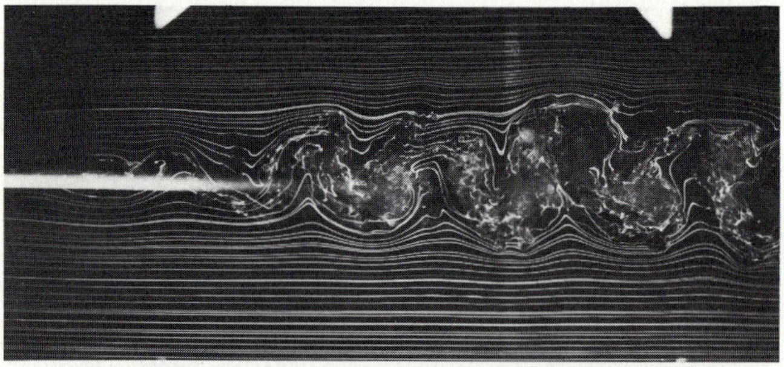

Figure 4: Side View of a Cylinder at $\beta = 60°$

Figure 5: Side View of a Cylinder at $\beta = 20°$

different wake behavior. Close inspection of the near wake region for the 60° cable angle, Figure 6, shows a very definite spanwise variation in the wake formation which correlates very well with the cable geometry. This correlation is not as apparent for the 40° and 20° cable angles. The 3x19 cable does not display the same regular vortex formation in the plan views and it would be very difficult to identify a spanwise filament from the photographs. The 4x7 serrated model also presents a very irregular profile to the oncoming flow. The wake structure is much less apparent for the 4x7 cable than for the other configurations. Close inspection of the near wake region at the 60° cable angle appears to indicate a spanwise periodicity. It is difficult to draw any significant conclusions from these two observations but there is an indication that the near wake structure is influenced by the cable configuration.

The side view photographs provide additional detail on the cable wake structure. The trends in the wake characteristics are similar for the cylinder, 1x19, 7x7 and 6x25 configurations for all yaw angles. Figures 7-12 are the sequence of side views of the 6x25 and 4x7 cables at cable angles of 90°, 60° and 20°. At the 90° cable angle, the formation of the periodic vortex wake is apparent and the wake width decreases and vortex spacing increases as the cable angle decreases. The diffusion and mixing of the smoke streaklines does seem to be more rapid for the cable designs than for the circular cylinder. The wake structures for the 3x19 and the 4x7 configurations differ from those of the other configurations. The vortex structure in the wake is not as distinct. The near field wake, within three to five cable diameters appears to be thinner than for the other configurations. At the 20° cable angle, these two configurations developed wakes which were narrower and again appeared to take longer to develop. One anomaly was the behavior of the 4x7 cable at the 60° cable angle. Unlike this cable at the other angles, the wake did grow quite rapidly and the vortices in the wake are almost aligned, not staggered as in the other cases.

Figure 6: Plan View of a 3x19 Cable at $\beta = 60°$

Quantitative information about the wake structure and in particular about the vortex shedding characteristics was extracted from the flow visualization photographs. This was done in order to provide a means of comparison with the hot wire results as well as to provide additional information on the flow field. The shedding angle, α, defined as the angle between the axis of the vortex filament convecting in the wake of the cable and the velocity vector in the plan view. Ramberg [1979] discussed the dependence of shedding angle on cable angle in his flow visualization data for yawed circular cylinders. However, the variation in shedding angles with cable angle, which is

Figure 7: Side View of a 6x25 Lang Lay Cable at $\beta = 90°$

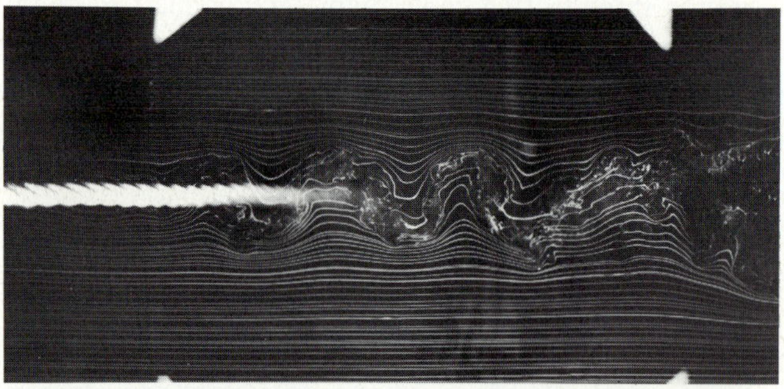

Figure 8: Side View of a 6x25 Lang Lay Cable at $\beta = 60°$

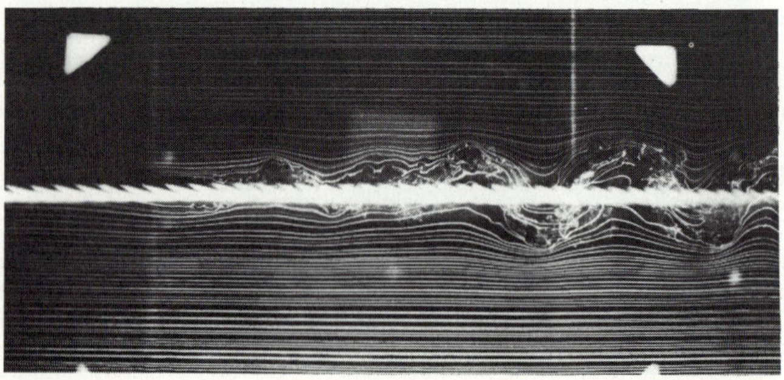

Figure 9: Side View of a 6x25 Lang Lay Cable at $\beta = 20°$

an additional indication of the three dimensional character of the wake flow, is not well understood. Plan view flow visualization photographs such as that in Figure 2 show filaments which are not straight lines. The flow is quite complex and the orientation of the vortex filaments appears to be a strong function of the upstream flow and end conditions on these finite length cables. For the purpose of this study, a straight line was fit through a number of points along a given filament. This information was collected for the circular cylinder and the 6x25 cable since the wake structure was distinct for these two configurations. The first three filaments downstream of the model were

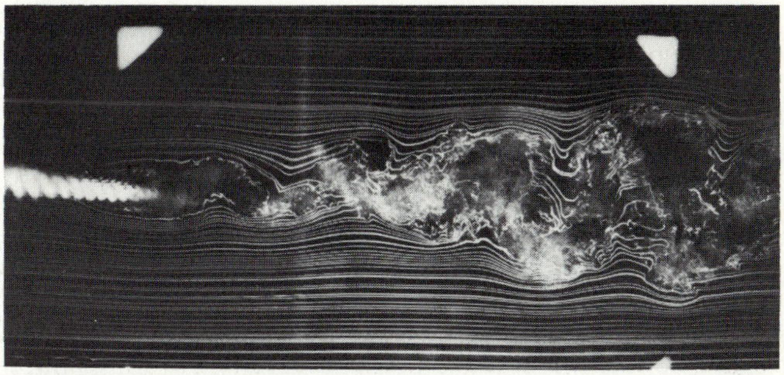

Figure 10: Side View of a 4x7 Serrated Cable at $\beta = 90°$

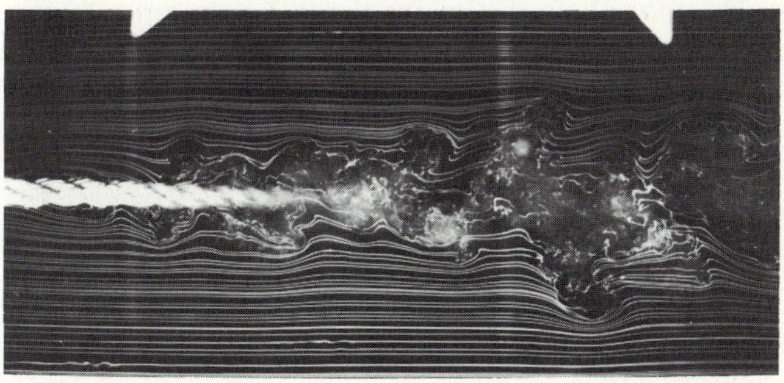

Figure 11: Side View of a 4x7 Serrated Cable at $\beta = 60°$

Figure 12: Side View of a 4x7 Serrated Cable at $\beta = 20°$

identified for each cable angle. The angle associated with a straight line fit to each filament is given a summary plot in Figure 13. For the 90°, 60° and 40° cable angles the shedding angle is equal to the cable angle, within the error bounds on this data. For the 20° case there is a marked difference between the shedding angle and the cable angle. The error bars shown are the standard deviation of the sample population of 12 filaments.

Hot Wire Measurements

All models were tested to obtain the vortex shedding characteristics. The wake frequency spectrum gives an indication of the character of the wake. At cable angles of 90°, the peak in the frequency spectrum is very well defined, as shown in Figure 14. As the cable angle decreases the peak is not as well defined and determination of the shedding frequency is more difficult. The wake frequency results are expressed in terms of the Strouhal number. The cables show a generally higher Strouhal number than the cylinder at 20°, 40°, and 60° cable angles. For the 90° case, the cables generally have lower Strouhal numbers than the cylinder. The Strouhal number is plotted against the cable angle in Figure 15. The curved line was fit to the data points for the circular cylinder. The 4x7 cable has the greatest Strouhal number of all of the models for a cable angle of 20° while it has the lowest Strouhal number at 90°. Horton, et. al. [1987] obtained almost the same results for the yawed cables in their experiments with cables free to vibrate. However their values for the 90° case indicate that the cables have higher values compared to the cylinder. This difference may be attributed to the flexibility of their models or to the difference in boundary conditions.

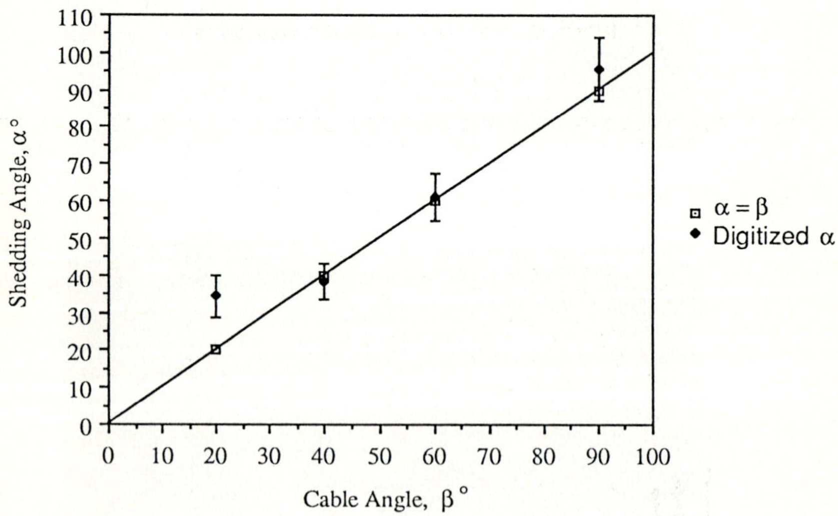

Figure 13: Shedding Angle vs. Cable Angle, From Digitized Plan View Photographs

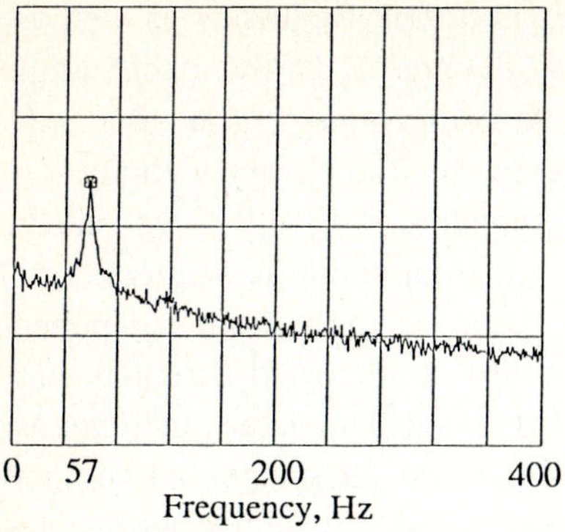

(a) 90° Cable Angle

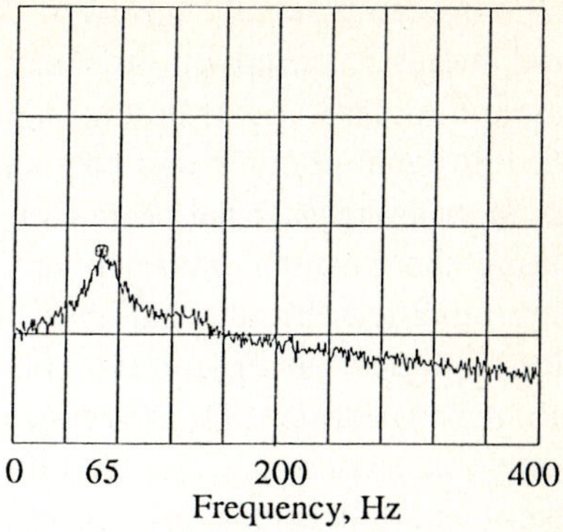

(b) 60° Cable Angle

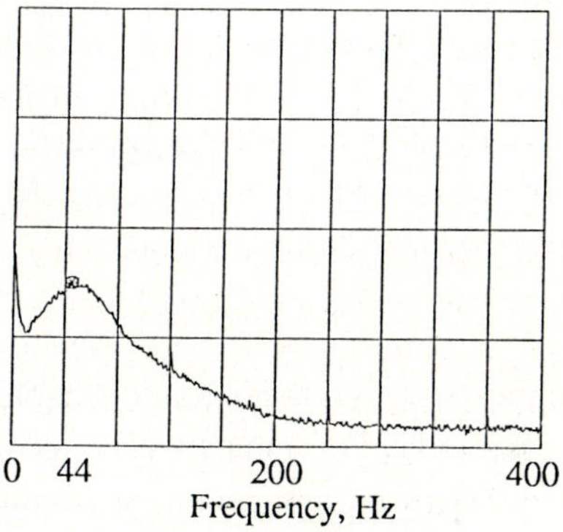

(c) 20° Cable Angle

Figure 14: Wake Spectra of a 4x7 Cable

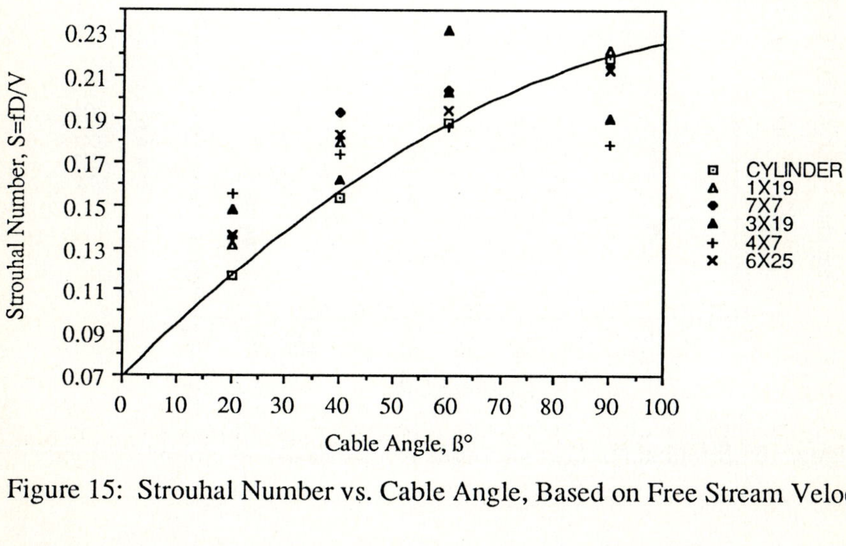

Figure 15: Strouhal Number vs. Cable Angle, Based on Free Stream Velocity

It has been generally observed that the vortex shedding frequency as well as the Strouhal number, based on free stream velocity, decreases as the cable angle is decreased. In order to evaluate this dependence of shedding frequency on cable angle, a modified Strouhal number was computed by replacing the free stream velocity V with Vsin β. In so doing only the normal component of the velocity, Vsin β, is considered in computing the Strouhal number and the tangential component is neglected. The resulting modified Strouhal number, S_N, which is based on the normal component of velocity is plotted in Figure 16. The plot shows that this modified Strouhal number does not remain constant for the range of cable angles tested. The values increase as the cable angle is decreased. Thus for the conditions in this test, the tangential component and other three dimensional effects influences the vortex shedding from yawed cables.

The vortex shedding frequencies were measured for a series of spanwise locations for three models: the cylinder, the 4x7 cable and the 6x25 cable. The results indicated that the shedding frequency was independent of the spanwise location. These measurements were conducted in a region near the center of the tunnel. The influence of the end-effects for the finite length cable models has been recognized, Ramberg [1978], and additional study is needed to evaluate that influence for the stranded cables.

CONCLUSIONS

The flow structure in the wake of the stranded cables was observed to be a function of the cable surface geometry. For cables with relatively smooth surfaces, i. e. the 1x19, 7x7 and 6x25, the wake resembled that of a smooth cylinder. However, for cables where the stranding created significant surface irregularities, i.e., the 3x19 and

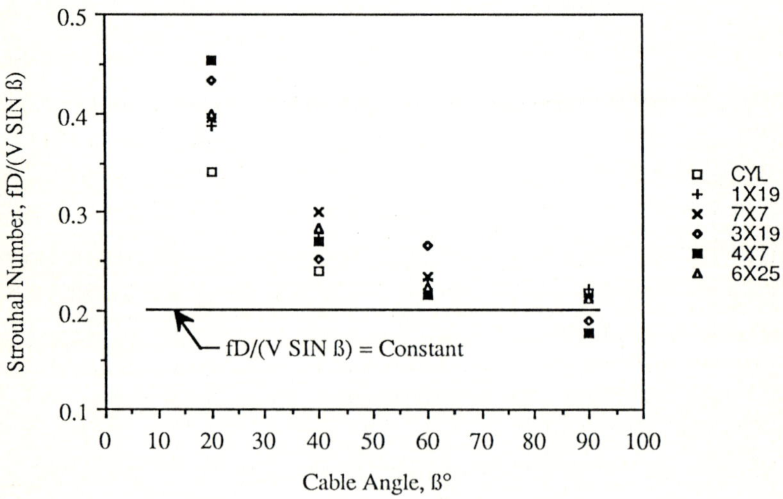

Figure 16: Strouhal Number vs. Cable Angle, Based on Normal Velocity

4x7 serrated cables, the wake structure was quite different from that of the smooth cylinder. Flow visualization photographs reveal a number of interesting features. A more diffused wake structure was observed for the 3x19 and 4x7 cables. Plan view photographs of these two cables showed vortex filaments which were less coherent. This qualitatively shows that the vortex shedding process for these two cables are more turbulent compared to that of the cylinder. The side view wake profiles of the 3x19 and 4x7 cables also differ from that of the cylinder. These cables have generally thinner wake widths and have longer formation regions compared to the cylinder. Plan view photographs showed, spanwise, cellular flow structures in the near wake of the 3x19 and 4x7 cables that follow the "peaks" and "valleys" in the cable geometry. This was not apparent in the other cable models. This has been observed in the 90° and 60° cable angle cases but not in the 40° and 20° cable angle cases. The flow visualization did not show any effect due to the direction of lay and there was no indication of the mechanism of lift development.

The hot-wire studies provided additional quantitative information. The characteristics of the vortex shedding frequency, the band width and the Strouhal number dependence upon model type, cable angle, Reynolds number and spanwise location were analyzed. The shedding frequency was independent of spanwise location. The band width of the shedding frequency increases with a decreased cable angle. The Strouhal number decreases as the cable angle decreases except for the 3x19 and 4x7 cables where the Strouhal number was higher at 60° than at 90° cable angle .

The hot wire results agreed well with the flow visualization observations. In most of the cases, the results have been affirmed by the two methods. Details from each method complemented results from the other. This resulted in a better understanding of the overall flow characteristics.

ACKNOWLEDGEMENTS

This work was supported in part by Department of Aerospace and Mechanical Engineering at the University of Notre Dame and by the Naval Coastal Systems Center (NCSC), Panama City, Florida on a grant from the Office of Naval Research. The authors wish to thank the Project Engineer at the NCSC, Mr. Kennard Watson, for his contributions to this work.

REFERENCES

Batill, S. M. and Mueller, T. J., "Visualization of the Laminar-Turbulent Transition in the Flow Over an Airfoil Using the Smoke-Wire Technique," AIAA 11th Aerodynamic Testing Conference, March 1980

Batill, S. M., Nelson, R. C., and Nebres, J. V., "An Experimental Investigation of the Flow Field Around Yawed, Stranded Cables", Naval Coastal Systems Center Technical Report, December 1988

Blevins, R.D., Flow Induced Vibration , Van Nostrand Reinhold Co. Publ., New York, 1977

Griffin, O. M., "Flow-Induced Oscillations of OTEC Mooring and Anchoring Cables: State of the Art," Naval Research Laboratory Memorandum Report 4766, May 1982

Horton, K. J., Ferrer, C. M., Watson, K. P., and Charvoz, D., "Measurement of the Hydrodynamic Force and Strum Characteristics of Stranded Cables," Naval Coastal Systems Center Technical Memorandum 471-87, December 1987

Nebres, J.V., "Flow Around Yawed Stranded Cables," M.S. Thesis, Department of Aerospace and Mechanical Engineering, University of Notre Dame, May 1989

Ramberg, S. E., "The Influence of Yaw Angle Upon the Vortex Wakes of Stationary and Vibrating Cylinders," Naval Research Laboratory Memorandum Report 3822, August 1978

Simpson, A., "Fluid Dynamic Stability Aspects of Cables," Proceedings of the Mechanics of Wave-Induced Forces on Cylinders, p. 90-132, T. L. Shaw, ed., Pitman Advanced Publishing Program, London, 1979,

Van Atta, C. W., "Experiments on Vortex Shedding From Yawed Circular Cylinders," AIAA Journal, Vol. 6, No. 5, p. 931-933, May 1968

Votaw, C. W., and Griffin, O. M., "Vortex Shedding From Smooth Cylinders and Stranded Cables," Transactions of the ASME, Journal of Basic Engineering, Vol. 93, p. 457-460, 1971

Lecture Notes in Engineering

Edited by C.A. Brebbia and S.A. Orszag

Vol. 1: J. C. F. Telles,
The Boundary Element Method
Applied to Inelastic Problems
IX, 243 pages. 1983.

Vol. 2: Bernard Amadei,
Rock Anisotropy and
the Theory of Stress Measurements
XVIII, 479 pages. 1983.

Vol. 3: Computational Aspects of
Penetration Mechanics
Proceedings of the Army Research
Office Workshop on Computational
Aspects of Penetration Mechanics
held at the Ballistic Research Laboratory
at Aberdeen Proving Ground, Maryland,
27–29 April, 1982
Edited by J. Chandra and J.E. Flaherty
VII, 221 pages. 1983.

Vol. 4: W.S. Venturini
Boundary Element Method in Geomechanics
VIII, 246 pages. 1983.

Vol. 5: Madassar Manzoor
Heat Flow Through Extended
Surface Heat Exchangers
VII, 286 pages. 1984.

Vol. 6: Myron B. Allen III
Collocation Techniques for Modeling
Compositional Flows in Oil Reservoirs
VI, 210 pages. 1984.

Vol. 7: Derek B. Ingham,
Mark A. Kelmanson
Boundary Integral Equation
Analyses of Singular, Potential,
and Biharmonic Problems
IV, 173 pages. 1984.

Vol. 8: Linda M. Abriola
Multiphase Migration of Organic
Compounds in a Porous Medium
A Mathematical Model
VIII, 232 pages. 1984.

Vol. 9: Theodore V. Hromadka II
The Complex Variable Boundary
Element Method
XI, 243 pages. 1984.

Vol. 10: C. A. Brebbia, H. Tottenham,
G. B. Warburton, J. M. Wilson, R. R. Wilson
Vibrations of Engineering Structures
VI, 300 pages. 1985.

Vol. 11: M. B. Beck
Water Quality Management:
A Review of the Development and
Application of Mathematical Models
VIII, 108 pages. 1985.

Vol. 12: G. Walker, J. R. Senft
Free Piston Stirling Engines
XIV, 286 pages. 1985.

Vol. 13: Nonlinear Dynamics
of Transcritical Flows
Proceedings of a DFVLR International
Colloquium, Bonn, Germany, March 26, 1984
VI, 203 pages. 1985.

Vol. 14: A. A. Bakr
The Boundary Integral
Equation Method in Axisymmetric
Stress Analysis Problems
XI, 213 pages. 1986.

Vol. 15: I. Kinnmark
The Shallow Water Wave
Equation: Formulation,
Analysis and Application
XXIII, 187 pages, 1986.

Vol. 16: G. J. Creus
Viscoelasticity – Basic
Theory and Applications
to Concrete Structures
VII, 161 pages. 1986.

Vol. 17: S. M. Baxter
C. L. Morfey
Angular Distribution
Analysis in Acoustics
VII, 202 pages. 1986.

Vol. 18: N. C. Markatos,
D. G. Tatchell, M. Cross, N. Rhodes
Numerical Simulation of Fluid Flow
and Heat/Mass Tranfer Processes
VIII, 482 pages. 1986.

Vol. 19: Finite Rotations
in Structural Mechanics
Proceedings of the Euromech
Colloquium 197, Jablonna 1985
VII, 385 pages. 1986.

Vol. 20: S. M. Niku
Finite Element Analysis
of Hyperbolic Cooling Towers
VIII , 216 pages. 1986.

Lecture Notes in Engineering

Edited by C.A. Brebbia and S.A. Orszag

Vol. 21: B. F. Spencer, Jr.
Reliability of Randomly
Excited Hysteretic Structures
XIII, 138 pages. 1986.

Vol. 22: A. Gupta, R. P. Singh
Fatigue Behaviour
of Offshore Structures
XXI, 299 pages. 1986.

Vol. 23: P. Hagedorn, K. Kelkel,
J. Wallaschek
Vibrations and Impedances
of Rectangular Plates
with Free Boundaries
V, 152 pages. 1986.

Vol. 24: Supercomputers
and Fluid Dynamics
Proceedings of the First
Nobeyama Workshop
September 3-6, 1985
VIII, 200 pages. 1986.

Vol. 25: B. Hederson-Sellers
Modeling of Plume Rise
and Dispersion –
The University of Salford
Model: U. S. P. R.
VIII, 113 pages. 1987.

Vol. 26: Shell and Spatial Structures:
Computational Aspects
Proceeding of the International Symposium
July 1986, Leuven, Belgium
Edited by G. De Roeck, A. Samartin Quiroga,
M. Van Laethem and E. Backx
VII, 486 pages. 1987.

Vol. 27: Th. V. Hromadka, Ch.-Ch. Yen
G. F. Pinder
The Best Approximation Method
An Introduction
XIII, 168 pages. 1987.

Vol. 28: Refined Dynamical Theories
of Beams, Plates and Shells and
Their Applications
Proceedings of the Euromech-Colloquim 219
Edited by I. Elishakoff and H. Irretier
IX, 436 pages. 1987.

Vol. 29: G. Menges, N. Hövelmanns,
E. Baur (Eds.)
Expert Systems in Production Engineering
Proceedings of the International Workshop
Spa, Belgium, August 18-22, 1986
IV, 245 pages. 1987.

Vol. 30: R. Doležal
Simulation of Large State Variations
in Steam Power Plants
Dynamics of Large Scale Systems
X, 110 pages. 1987.

Vol. 31: Y. K. Lin, G. I. Schueller (Eds.)
Stochastic Structural Mechanics
U.S.-Austria Joint Seminar, May 4-5, 1987
Boca Raton, Florida, USA
XI, 507 pages. 1987.

Vol. 32: Y. K. Lin, R. Minai (Eds.)
Stochastic Approaches
in Earthquake Engineering
U.S.-Japan Joint Seminar, May 6-7, 1987
Boca Raton, Florida, USA
XI, 457 pages. 1987.

Vol. 33: P. Thoft-Christensen (Editor)
Reliability and Optimization
of Structural Systems
Proceedings of the First IFIP WG 7.5
Working Conference
Aalborg, Denmark, May 6-8, 1987
VIII, 458 pages. 1987.

Vol. 34: M. B. Allen III, G. A. Behie,
J. A. Trangenstein
Multiphase Flow in Porous Media
Mechanics, Mathematics, and Numerics
IV, 312 pages. 1988.

Vol. 35: W. Tang
A New Transformation
Approach in BEM
A Generalized Approach for
Transforming Domain Integrals
VI, 210 pages. 1988.

Vol. 36: R. H. Mendez, S. A. Orszag
Japanese Supercomputing
Architecture, Algorithms, and Applications
IV, 160 pages. 1988.

Vol. 37: J. N. Reddy, C. S. Krishnamoorthy,
K. N. Seetharamu (Eds.)
Finite Element Analysis
for Engineering Design
XIV, 869 pages. 1988.

Vol. 38: S. J. Dunnett, D. B. Ingham
The Mathematics of Blunt Body Sampling
VIII, 213 pages. 1988

Vol. 39: S.L. Koh, C.G. Speziale (Eds.)
Recent Advances in Engineering Science
A Symposium dedicated to A. Cemal Eringen
June 20–22, 1988, Berkeley, California
XVIII, 268 pages. 1989